Bioenergetics

A Bridge across Life and Universe

Davor Juretić
Mediterranean Institute for Life Sciences
Split, Croatia

CRC Press is an imprint of the
Taylor & Francis Group, an **informa** business
A SCIENCE PUBLISHERS BOOK

Cover image by Davor Juretić

First edition published 2022
by CRC Press
6000 Broken Sound Parkway NW, Suite 300, Boca Raton, FL 33487-2742

and by CRC Press
2 Park Square, Milton Park, Abingdon, Oxon, OX14 4RN

Library of Congress Cataloging-in-Publication Data

Names: Juretić, Davor, 1944- author.
Title: Bioenergetics : a bridge across life and universe / Davor Juretić, Mediterranean Institute for Life Sciences, Split, Croatia.
Description: First edition. | Boca Raton : CRC Press, 2021. | Includes bibliographical references and index.
Identifiers: LCCN 2021017120 | ISBN 9780815388388 (hardcover)
Subjects: LCSH: Bioenergetics.
Classification: LCC QH510 .J87 2021 | DDC 572/.43--dc23
LC record available at https://lccn.loc.gov/2021017120

ISBN: 978-0-8153-8838-8 (hbk)
ISBN: 978-1-032-04529-0 (pbk)
ISBN: 978-1-351-17276-9 (ebk)

DOI: 10.1201/9781351172769

Typeset in Times New Roman
by Radiant Productions

Preface

During the 1990s, I wrote a book about bioenergetics in the Croatian language. It also contained some achievements from bioinformatics, and it was well-received, probably because, at that time, there were precious few professional articles and books about these topics in Croatia. When one young lady scientist approached me in the library of "Ruđer Bošković" Institute, introduced herself, and expressed her appreciation after reading my book, it was just as great a satisfaction to me as the "Matica Hrvatska (Matrix Croatica)" 1998 award for that book. The award was unusual because this old and venerable Croatian institution has previously awarded only fiction books and books dealing with humanistic studies. I am glad if my book helped to raise bioenergetics in Croatia from the realm of pseudoscience. It had no connection whatsoever with bioenergetic healing and bioenergetic psychotherapy. I would not go so far to deny any positive effect of these and other healing procedures, which are connected in the popular mind with bioenergetics, but bioenergetics as science is something else.

The present book is not an updated translation or new edition of that old book in the Croatian language. I also did not try to write a standard type of textbook or monograph about bioenergetics. It was a greater challenge for me and I hope it is more exciting for readers to understand bioenergetics in a broder context. We should all be interested in the question of what is life and why life originated and evolved in an often harsh external environment. Physics and chemistry should be able to see a more fundamental background for life development than just biology. However, we must admit that the mere existence of life is quite a difficult challenge for physics. I was convinced for a long time that bioenergetics is the best bridge between the mystery of life and the connection of life to its environment, hence comes the motivation for writing this book. This book's leitmotif is looking for applications in bioenergetics and enzyme kinetics of some universal physical principles. That is also the focus of most chapters. The work published over the past 20 years is brought together in this book. Many topics from standard texts on bioenergetics are omitted to explain better how bioenergetics fits the world order. In particular, I examined the relevance of different selection principles from irreversible thermodynamics to evolution, complexity, self-organization, and power-producing energy transformations in bioenergetics. In all of the studied examples of steady-state enzyme kinetics, I applied the same theoretical principle with the same goal of getting an approximate agreement with experimental results. The surprising insight stemming from such calculations is about elementary energy transformation steps crucial for increasing the amount of useful energy in the living system and wasted energy (entropy) in the universe. Thus, we found the best means for accelerating the thermodynamic evolution toward greater dissipation in the universe—it is the biological evolution of all living forms through microscopic dissipation-driven and dissipation-producing self-ordering. The fascination with bioenergetics of PhD student Davor Juretić in the 70s came to fruition in this book—biological energy transformations are indeed acting as a bridge between biology and the universe.

My dear wife, Ljiljana de Nutrizio, greatly aided in this book's writing by loving care and limitless support. I greatly appreciate collaborations and support from my colleagues at the Physics Department, Faculty of Science, University of Split, from the Mediterranean Institute for Life Sciences, also in Split, and from the Ruđer Bošković Institute in Zagreb, Croatia. Among them, professors Paško Županović, Željana Bonačić Lošić, Larisa Zoranić, Miroslav Radman, Dr. Fransoa-Xavier Pellay, Dr. Juraj Simunić, and Dr. Domagoj Kuić collaborated, contributed, supported, or took an active interest in research topics treated in this book. In turn, we inspired Andrej Dobovišek from Slovenia to explore with his Slovenian colleagues and us the possible connection between biological evolution and enzyme optimization based on the maximum entropy production principle. I appreciate the collaboration with Professor Roderick Dewar, who persisted in developing and promoting the principle throughout his research in Europe and Australia. I am also grateful for the comments about some topics in the book expressed by Professor Leonid Martyushev from Russia and Professor Ken Dill together with Jason Wagoner from the USA. I shall never forget the trust and unselfish help I obtained initially in my scientific career from professors Janko Herak in Croatia and Alec Keith in the USA, who were not repelled by my offbeat ideas from the early 70s on how entropy production connects life and universe. Anton Depope and my son Vitomir Juretić helped to arrange the web page where readers can find my publications discussed in this book (http://juretic.medils.hr/). The book would not be possible without greatly expanded open access to scientific publications promoted in this century by numerous individuals and organizations worldwide. Still, some topics were new to me when I was reading recent research papers, and I should be the only one to blame for possible mistakes in their interpretation.

Contents

Introduction

What is bioenergetics?

Any energy **transduction** (the bold and underlined terms are used throught this book to direct readers toward the "Definitions and explanations" section for the corresponding chapter) by living entities and biological macromolecules is a research subject for bioenergetics, which studies energy conversions performed by living cells and organisms. Bioenergetics is focused on the details of the initial energy transformation steps. Its traditional research topic is **ATP** synthesis by bioenergetic membranes with an essential role for membrane-located molecular motors responsible for protons' **active transport**. Bacterial cytoplasmic membrane, mitochondrial inner membrane, and chloroplast **thylakoid membrane** are examples of bioenergetic membranes, which perform an intensive conversion of freely available energy, or **free energy**, into ATP synthesis. In comparison to an equivalent volume of Sun-like star, the bioenergetic free energy transduction is millions of times more intensive (Metzner 1984). For living cells, energy transformations are an essential but inefficient process. Even when absorbed by cells, more than 80 or 90% of initial free energy is lost as the universe's **entropy** increases, regardless of free energy source.

Life sciences are all focused on what cells do with acquired free energy, often neglecting to consider the majority of lost or dissipated free energy. It is like dark energy or dark matter in physics. Nobody can see it or fully understand it, although it must have played an essential role in the universe's **evolution**. Similarly, when free-energy from photons or organic compounds is absorbed by cells and then mostly lost in the form of invisible infrared photons, we can only feel and measure a slight temperature increase in the environment, something that seems to be safe to ignore. In the long run, however, it does make a whole biosphere and a whole life-supporting atmosphere as a difference. Gradual changes, in the long run, are known as evolution. The evolution belongs to the most important scientific insights about nature, equally crucial in biology and physics. It is so pervasive, ubiquitous, and obligatory process for the emergence of any material structure that I would choose the sentence "Everything you see in nature emerged during evolution" if challenged to preserve just one short sentence out of all scientific knowledge. It is a rephrasing of the sentence: "Nothing in biology makes sense except in the light of evolution" (Dobzhansky 1973) when we take into account that all structures we study in geology, chemistry, physics, and astronomy also emerged during evolution.

At present, the definition of bioenergetics should be extended and modified, reflecting recent developments in this and related research fields. The essence of bioenergetics is its **holistic** ability to connect local and global changes. It uses interdisciplinary research to study how the structural organization of enzymes, biological circuits, and bioenergetic membranes leads to growth and **dissipation** processes within membrane-enclosed structures such as organelles and cells. From the point of view of living cells, the universe is everything around them. A drop of water or a snowflake is the universe for a single bacterium. From the universe's perspective, the biosphere exists in a thin membrane between the red-hot Earth interior and the deathly cold Earth exterior. The biosphere's ability to return excess energy received every day from the Sun in a degraded form spread in a wide space (mostly as infrared radiation) can be regarded as a thermodynamic justification for its existence.

Scientists appreciate bioenergetics as being essential for the origin and evolution of life (Russel 2007, Lane and Martin 2010, Lane 2010). Our atmosphere is the by-product of life and the best present-day evidence for the existence of life on Earth. Bioenergetics is essential for maintaining high oxygen concentration in air. A similar strong statement can be made about all large water bodies on the Earth's surface, from rivers and lakes to oceans. Water bodies have precariously balanced acidity, salinity, and oxygenation level suitable for life evolution.

When bioenergetics is considered a global activity of the biosphere, including humans and our civilization, there are then some global consequences in the current geological period aptly named Anthropocene. Unfortunately, we do not have a proper appreciation for free ecological services provided by millions of species. These global bioenergetics services maintain clean air with high oxygen and low carbon dioxide concentration, fertile soil, and abundant water supplies.

This book will present bioenergetics as analytic and holistic science connecting microscopic nanomotors with global scenery, thermodynamic evolution with biological evolution. The analogy with physics raises the question if lost or dissipated free energy by living entities is merely a waste that can be safely disregarded? Or is it a vital signature for the evolution of life? Is dissipation of the driving force enabling evolution and formation of living structures (Straškraba et al. 1999)? The answers offered in this book will not be found in the "orthodox" literature about bioenergetics.

Biological and thermodynamic evolution

Entropy is an important thermodynamic function used to quantify which part of a system's energy is not available for doing work. If you were ever baffled by entropy, do not be disheartened. Ilya Prigogine explicitly stated that "entropy is a very strange concept" (Prigogine 1989). Entropy increase is the constant feature of our universe intimately connected to its evolution. If left to itself, an **isolated system** will increase its entropy until it reaches the thermodynamic equilibrium when its entropy has the maximal value. This statement is the essence of the **Second Law** of Thermodynamics. The maximum entropy law describes the end of evolution: the state of thermodynamic equilibrium when a system meets its death. The evolution toward thermodynamic death seems to be opposite to the direction of biological evolution toward ever more complex structures and interactions among them. We can ask the question then, why the evolution of complex living structures is not in contradiction with the thermodynamic evolution towards ever higher entropy states? A large number of scientists discovered the answer to this question. However, it is still a controversial topic, connecting biology to physics. The evolution of beautiful macromolecules is one of many pathways nature found to speed up the thermodynamic evolution.

The evolution of life appeared as a by-product of the thermodynamic evolution. This insight will be elaborated in the book. It requires recognizing the tight connection between biological and thermodynamic evolution. I propose that a tight connection between biological and thermodynamic evolution persists far from equilibrium, a characteristic situation of an open living system capable of self-reproduction. An interdisciplinary approach is needed to unravel the nature of evolution coupling, which connects biology, chemistry, and physics.

Universal physical laws connect life and the universe. The Second Law of Thermodynamics is not limited to describing how some systems can reach thermodynamic death. It is relevant both for the evolution of the universe and life's evolution. A highly regarded expert for the irreversible evolution of the universe stated: "..the clearest characterization of life I know of is one given in thermodynamic terms" (Smolin 1997). Life is fragile and robust at the same time. Violent events in a nearby universe can extinguish life at any moment. Still, life on Earth persisted for almost four billion years: for about one-quarter of the universe age. After each near extinction, life was able to spread up again using the same old nanomotors connected through bioenergetic pathways.

My initial training was in theoretical physics. Although basic equations in classical and quantum physics are time-invariant, we all know that the past is beyond our reach. The time arrow points toward entropy increase. Entropy increase goes on while we are alive and proceeds within whatever is left from our bodies after death. Even if we can slow down, stop, or reverse aging in the future, this will not cancel the entropy we previously contributed nor stop the entropy export from our body to the universe. Even hypothetical immortals must continue contributing entropy increase to their environment to stay alive.

Entropy increase is not restricted to living cells or biological processes. Gravitational force may be an entropic force (Verlinde 2011), and black holes contribute to entropy increase in the universe more than anything else (Frautschi 1982, Egan and Lineweaver 2010). Since gravity and black holes direct the evolution of galaxies and stars with their planetary systems, without entropy increase and accompanied dissipation, we would not have any structure or life in the universe. Gravitational entropy offers the solution for the mysteriously **low initial entropy of the universe** (Patel and Lineweaver 2017). Verlinde recently proposed that entropy changes of the microscopic degrees of freedom are the solution to mysterious "dark matter" and "dark energy" (Verlinde 2017), which are thought to be responsible for 95% of matter and energy in the universe. In this picture, space time, masses, and gravity are no longer given features of the universe. All structures emerged through evolution and entropy changes starting from elementary particles and atoms. A seamlessly interwoven classical and quantum physics should be able to show that. It is fascinating how the entropy concept never stopped gaining in importance after being introduced by Clausius 150 years ago (Haddad 2017).

Most structures we can see in nature are unstable. They may have a long life like stars or a fleeting existence as some exotic elementary particles. In between are dust devils, tornados, whirlpools, swirling eddies of the fast-flowing river, and living beings. It would seem impossible to use the same name for so many different structures and phenomena. However, entropy increases, free-energy dissipation, the appearance of unstable structures, and a certain distance from thermodynamic equilibrium are common to all of them. It took the genius of the Nobel Prize winner Ilya Prigogine to name them all as dissipative structures (Prigogine and Nicolis 1967, Prigogine et al. 1969). Living entities are just one class of dissipative structures. In this book, we shall use several examples of simplified models for biological free energy converters to see how far one can go in understanding the bioenergetics of dissipative macromolecular systems when using our version of the maximum **entropy production** principle (Martyushev and Seleznev 2006) as a guiding physical principle.

For open systems, that can exchange energy and matter with their environment, we can postulate that every opportunity will be used by them to slide the system and its surroundings towards a more stable state with higher entropy. Every living system is such an open system constantly probing its ambience. It behaves as a catalytic agent that changes and grows at the expense of its environment. Charge separation by topologically closed membranes is the favorite way for a living cell to create the **protonmotive force** needed for maintenance and growth. Entropy production can be regarded as the required payment for the electrochemical proton gradient creation. Ultimately, mitochondria and chloroplasts, bacteria, and **archaea** cells use protonmotive force to generate far from equilibrium concentrations of adenosine triphosphate (ATP), adenosine diphosphate (ADP), and inorganic phosphate. These nonequilibrium concentrations are the driving force for all the work needed to stay alive and to multiply.

There must be some external free-energy source serving as the driving force to maintain a far-from-equilibrium state. Far from equilibrium state can be supported only by open systems free to exchange energy and matter with the environment. When all of this is fulfilled in an aquatic environment rich in minerals, it is thought that life can appear. It indeed appeared quickly on Earth between four and 3.5 billion years ago, possibly at alkaline hydrothermal vents well-protected at the bottom of oceans during Hadean or Archaean eon (Lane and Martin 2012). Acceleration of entropy increase, due to the appearance and continuous development of life, could have led only to an acceleration of thermodynamic evolution.

Bioenergetics as a challenge to physics

Terrel L. Hill (1917–2014) (Figure 1) recognized from the beginnings of his career as a physical chemist a need for interdisciplinary research across physics, chemistry, and biology (Chamberlin 2015). He was a shy, introspective, and sensitive man with a passionate creativity trait channeled through highly focused and disciplined research. I did not have the opportunity, but I would love to read his two books of poems, both devoted to his wife, Laura Eta Gano. He belongs to pioneers in several interdisciplinary research areas connecting physics with chemistry and biology. These research areas can best enrich one another when problem-solving in one of them requires and stimulates advances in another discipline as well. To study free energy transduction by biological systems required developing and using proper physical tools, such as small systems thermodynamics, far from equilibrium free energy transduction, and a creative combination of statistical mechanics and biochemical kinetics (Hill 1977). In a recent review paper about entropy production (Velasco et al. 2011), the authors assessed Hill's contribution in these words: "Over fifty years ago Terrel L. Hill formulated a very precise and conventional theory to undertake the study of the thermodynamics of small systems." We can add that the applications of his theory in biology (Hill 1977) are largely based on the conventional bilinear expression for entropy production (see Chapter 3). Interestingly, Velasco et al. (2011) considered the "entropy production" term as unfortunate and even meaningless, while for them, "the concept of entropy is, if not the most, certainly one of the most abused, misunderstood and polemic concepts in theoretical physics."

For most other scientists, entropy, entropy production, and dissipation are essential concepts in physics and biology likewise. An interesting question is whether evolution optimized biological free energy transduction in any way? Biochemists, who initiated bioenergetics (Lehninger 1945, Mitchell 1961), and physicists interested in life sciences (Schrödinger 1944) looked to physics to find some physical principle as a relevant candidate for optimization. Some thought that life optimizes everything to achieve minimal possible dissipation, while others embraced the seemingly opposite opinion that evolution optimized living structures for producing maximal possible dissipation. The

Figure 1: Terrel L. Hill (1917–2014) in Bethesda, Washington D.C., while relaxing among friends after his daily work at the National Institutes of Health in 1988. From a private collection of photographs with the kind permission of Hans and Anneke Westerhoff.

free-energy dissipation or entropy production took center stage in each case. In his well-known biochemistry textbook from 1975, Lehninger eloquently expressed preference, shared with his peers, for the minimal dissipation principle: ".... at least two general attributes of open systems in steady states have considerable significance in biology .... The most important implication is this: in the formalism of nonequilibrium thermodynamics, the steady-state, which is a characteristic of all smoothly running machinery, may be considered to be the orderly state of an open system in which the rate of entropy production is at a minimum and in which the system is operating with maximum efficiency under the prevailing conditions" (Hunt et al. 1987). Did life, in its attempts to escape entropic doom, choose to produce minimal entropy? Hunt et al. (1987) noticed the opposite tendency, namely that dissipation may increase in time in reacting chemical system.

The present-day life on Earth would not be possible in the absence of a huge temperature difference between the temperature at the Sun's surface and temperature at the Earth's surface. Each photon arriving from Sun can transfer quite a substantial amount of free energy to living cells capable of photosynthesis. This free energy is quickly dissipated, first to create the protonmotive force as an electrochemical proton gradient. Subsequently, ATP synthesis and all other thermodynamically uphill activities follow, which cell performs using the hydrolysis of ATP and other "high energy" compounds. The establishment of the protonmotive force requires the operation of proton pumps embedded in topologically closed bioenergetic membranes. It is possible that the emergence of topologically closed membranes, with a lipid core of membrane lipids, impermeable to water and protons, predated active proton transport and early bioenergetics (Segré et al. 2001). The bioenergetics required polypeptides and some early RNA-world catalysts as well (Higgs and Lehman 2015). According to the "Purple Earth" hypothesis, early photosynthesis was probably based on purple chromo-proteins like bacteriorhodopsins (DasSarma and Schwieterman 2018). Strictly speaking, it was not yet photosynthesis because it was not connected to carbon fixation, but ATP synthesis was possible if some early ancestor of ATP-synthase was present in the same membrane, which contained chromo-proteins with **retinal**-like pigment. It was less efficient in producing ATP molecules than present-day photosynthesis, and it did not produce oxygen. Modern photosynthesis is not very efficient, either. About 80 percent of photon free energy is dissipated in preparatory steps for ATP synthesis, while the synthesis of organic compounds and all other free-energy consuming work by a cell is responsible for at least 10% of additional entropy increase. Active transport of protons or other small cations against concentration gradients and electric field direction is the obligatory preparatory step establishing the membrane potential.

Active transport is certainly not an equilibrium situation or process. I would not be able to move, think, or write these sentences without proton, potassium, and sodium pumps continually re-establishing nerve cells' membrane potential after numerous depolarisation-repolarisation cycles caused by action potentials during my thinking process. After some time, I must eat something to renew the supplies of organic compounds (the glucose is a favorite food for the brain). We can conclude that there would be no biological complexity without dissipation, and there would be no biosphere. The major part of photon free energy Earth receives every day from Sun is returned to the universe as an entropy increase. Our biosphere's evolution gave it an important global role in producing entropy in the form of a large number of infrared photons (about 20) per a single photon received from the Sun. The cover page illustration from this book originated from my attempt at the symbolic presentation of how Sun's photons' dissipation drives particle fluxes used by life. Since dissipation is the major result of all processes leading to ephemeral living structures, one can certainly think about the question of why there is such a tight connection between dissipation and life.

Catalytic efficiency increases together with entropy production

At present, dynamic aspects of enzyme functions are getting increasing attention. Hundred years ago, the static "lock and key" image predominated in biochemistry with the enzyme as a lock and substrate as a key. Induced fit image (Koshland 1958) allowed for dynamic flexibility in enzyme action. Several enzyme cooperativity and allostery models were proposed in 60' to explain differences from standard Michaelis-Menten kinetics. Michaelis-Menten mass-action kinetics for enzymatic conversion of single substrate S to single product P leads to the equation for the reaction rate v: $v = V_{max}S/(K_M+S)$, where maximum reaction velocity V_{max} is the product $V_{max} = k_{cat}E_o$ of rate constant k_{cat} for irreversible product formation and total enzyme concentration E_o, and K_M is the **Michaelis constant** equal to substrate concentration when $v = V_{max}/2$. This curve has a hyperbolic shape. The k_{cat} is the turnover number, that is, the maximum number of substrate molecules converted to product per enzyme molecule per second. Lower K_M implies higher enzyme affinity for the substrate and higher k_{cat} implies higher enzyme activity. Enzyme catalytic efficiency (also named specificity constant) is defined as the ratio k_{cat}/K_M.

For catalysis, lowering **transition state** activation energy is essential, which is what all enzymes are doing. It leads to an acceleration of reaction rate, an increase in catalytic constant (the turnover number), and an increase in catalytic efficiency. It also leads to an increase in entropy production. Chapters 10–12 will present simulations of enzyme kinetics supporting these assertions. Some of these conclusions about entropy production changes during catalysis follow from the proof that maximal entropy production can be found in each transition between enzyme functional states (Dobovišek et al. 2011, Bonačić Lošić et al. 2017). The comparison between experimentally measured and predicted rate constants revealed a tight connection between enzyme performance parameters and entropy production, when important transitions were optimized using that proof.

During biological evolution, crucial steps can be connected to a steep rise in catalytic efficiency and entropy production. For instance, the triosephosphate isomerase from *Pyrococcus furiosis* with the highest known catalytic efficiency probably arose from its ancestor about three billion years ago. In the presence of an efficient enzyme, the reaction can be 10^{15} to 10^{19} times faster than in its absence (Wolfenden and Snider 2001). Entropy production associated with a substrate to product conversion must have also increased for a similarly huge factor due to enzymes' emergence.

The first model for the expansion of bioenergetic membranes in space and time lasted for about two billion years and was based on replication or doubling of cell numbers, volume, and membrane surface. After endosymbiosis of particular archaea and bacteria appeared about 2 to 2.5 billion years ago, there was an additional tremendous expansion of bioenergetic membranes in each of the eukaryotic cells compared to bacterial or archaea cells. This development allowed nuclear genes from eukaryotes to control about 10^4 higher amounts of free-energy transduction and dissipation than genes from archaea or bacteria (Lane and Martin 2010). It is a partial answer to the question of why such a lengthy time-period of two billion years was needed for the emergence of the first eukaryotic organisms and the subsequent rise of multicellular species. The revolution in biological evolution, leading to low probability symbiotic events, required a lot of preparation. The photosynthesis by seemingly insignificant and invisible photosynthetic cells had to overcome all brakes for the appearance of a highly nonequilibrium O_2 rich atmosphere. The living world also needed a lot of luck during billions of years for our Earth to escape catastrophic events (for instance, the collision with large asteroids or comets), which would wipe out the biosphere. Our large moon and giant outer planets probably protected us to some extent. Earth's inner composition enabled just the right level of dynamics for internal dissipative structures and for plate tectonics to provide proper mixing of minerals at Earth's surface and strong enough magnetic field for biosphere protection from harmful irradiation. Outer layers of Earth's surface were strongly influenced by proto-Earth collision with a Mars-sized planet, which probably created the present-day Earth-moon system. Active plate tectonics also had the potential to wipe out the life on Earth as it nearly happened about 252 million years ago as the **Permian-Triassic** extinction event during the formation of the supercontinent Pangaea (Bond and Grasby 2017). Whatever caused geological and biological changes at that time, it was so catastrophic and obvious in geological deposits that geologists use it to define the Permian-Triassic boundary. Because of all these fortunate circumstances during an extremely long time, it may be an over-optimistic scenario to expect to find multicellular life on other planets, let alone other civilizations, anywhere else in our corner of the Milky Way. This thought puts even more responsibility in our hands to refrain from destroying our planet's unique biodiversity despite all possibilities to make money in a present-day perverse economic, political, and legal climate, which describes destruction as progress.

Among many other examples connecting evolutionary innovations to rise in entropy production, birds and mammals' evolution provided a stable high-temperature environment for all cells in corresponding organisms. It, however, required much higher free-energy intake and concomitant entropy production jump regarding values characteristic for fishes and reptiles, for instance (Zotin and Zotin 1996). It is part of a broader story of how, during roughly 14 billion years, ever new features and structures emerged in the universe, each exhibiting a higher level of entropy production as the universe matured (Salthe 2015, Chaisson 2015). At the current Anthropocene geological period, the civilization has record-high entropy production as calculated per unit mass, unit volume, or single person, probably too high for our good. An analogy can be drawn here between our brain cells and our civilization controlling and influencing everything else on Earth for better or worse. When excited too much, our brain cells can endanger everything else in our body because neurons control all other cells and organs.

Brain bioenergetics has no competition in its intensity

Brains of large mammals contain nerve cells with long axons and intensive branching of dendrites connected to other nerve cells. All of the characteristic dimensions for a single large neuron are no longer measured in nanometers or micrometers, as for prokaryotic cells, but in meters. The length of neurons' life is comparable to animal lifespan, just as the length of a large motor neuron is comparable to animal body length. When extended, the length of DNA molecules from a single neuron (or any other cell with a nucleus) is comparable to our height.

Our brain consumes about 20% of available metabolic energy, which is ten times more than its share of the body weight (about 2%). The same secret behind all of these fascinating facts is that a single large neuron can contain more than ten million mitochondria (Rodriges et al. 2015). Thus, it controls metabolic flux and concomitant entropy production, which is several orders of magnitude higher than that of cells in seemingly much more active organs like the heart or muscle. The placenta is also a highly metabolic organ with an abundance of mitochondria. It is responsible for supporting the fast growth of the fetus's organs, including the developing brain. No wonder that creativity, sex, aging, and neurodegenerative diseases can all search for their origin in proper function or malfunction of bioenergetic membranes. In the next chapters, we shall see what is the nature of membrane-located protein pumps and channels that are mostly responsible for the generation and dissipation of ion gradients across bioenergetic membranes.

Life accelerates the thermodynamic evolution

To recapitulate this introduction in a more personal manner, let me stress again that I do not consider thermodynamic evolution as opposed to biological evolution. On the contrary, biological evolution leading to complex organic molecules and reaction networks increases entropy production much more than what would happen in the absence of living cells (Ulanowicz and Hannon 1987). Life also accelerates thermodynamic evolution. In 2005, I named this statement the evolution-coupling hypothesis (Juretić and Županović 2005) because it certainly needs much additional verification. It also needs to be better connected to bioenergetics. As a general concept, it is represented in Figure 2.

Assuming that life on Earth originated on Earth, it had initially only a minor effect on its environment. We can say that life was a side effect of far from equilibrium forces driving the evolution of hydrothermal alkaline vents in Hadean or Archaean oceans (Lane and Martin 2012). Eventually, life became the biosphere maintaining a favorable global environment for its own

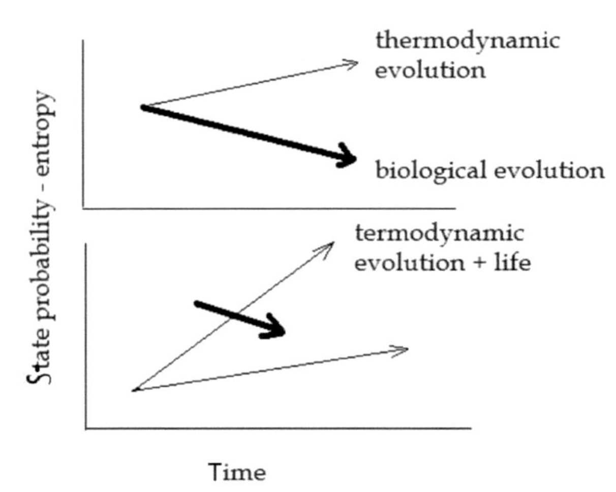

Figure 2: Thermodynamic and biological evolution are not separate processes, as presented in the upper part of this scheme. Instead, in the presence of life (lower part, upper arrow), the thermodynamic evolution is accelerated when life and its environment are considered as interconnected space-time structures.

needs according to the Gaia hypothesis introduced by James Lovelock in the 1970s (Lovelock and Margulis 1974, Lovelock 1988). It is still a minor side effect in our solar system driven by the Sun as its central engine. Is it destined to have an important role in accelerating the evolution of the solar system? Ideas about terraforming Mars seem to point towards a possibility of life spreading and transforming whatever it finds beyond Earth, but this cannot happen without the sustainable development of human science, technology, and economic strength needed for huge long-term investments into interplanetary travel.

Bioenergetics, ecology and global climate changes

Economic development has its bright and dark side. In developed nations, human needs and desires are satisfied better than in any of the past generations. We may have reached the peak of prosperity in 2019 when I first invested a serious effort in writing this book. This improvement has been achieved by two orders of magnitude higher entropy production per individual when compared to physiological entropy production (Stahl and Keller 1999). No other species accomplished the similar miracle of converting forests and fossil fuel into the multiplication of its individuals as a side effect of an artificial increase in dissipation through the development of agriculture and civilization. By burning fossil fuels, we have already prevented another ice age. Even if we do not survive the runaway greenhouse climate change we triggered, we may have avoided another snowball Earth episode. During snowball Earth episodes between 717 and 635 million years ago, the ice sheets advanced up to equatorial latitudes (Rooney et al. 2014). Runaway hothouse Earth (Venus-like) is an even more dangerous possibility, which could destroy Earth's whole life. We are told by climate scientists that even in the worst scenario of business-as-usual triggering numerous positive feedbacks for enhanced global warming, there is little chance of ever running into the apocalyptic Venus-like overheating. Let us hope they are right, despite frequent past underestimates of synergy, nonlinearity, and sources of greenhouse-driving events. Instead of a naïve, optimistic belief in ever-lasting economic growth fueled by limited natural resources, it would be wise for nations and world leaders to heed concerned scientists' advice (Ripple et al. 2020). It is good, however, to be aware, in the meantime, that we have released the fossil fuel as a dissipation monster from its underground safe prison. Not enough is known by scientists or anybody else about extreme far-from-equilibrium situations when dissipation is of such an intensity that most living structures will case to profit from strong driving forces and will start dying beyond recovery. The fact that there are tipping points when runaway cooling or heating can occur argues for a precarious balance and strong coupling between the biosphere and its inorganic environment, just as their evolutions were and still are tightly coupled. In the present Anthropocene age, human civilization evolution is fast becoming coupled with the evolution of the biosphere, atmosphere, oceans, and polar ice caps. Unwittingly, we took upon ourselves not only God's instruction from the Old Testament to steward all animals but also the need to steward Earth-biosphere coupled evolution towards a sustainable future.

The overview of presented topics

This Introduction has already touched upon most of the subjects discussed in the book. Nevertheless, let us mention here common themes and the organization of this monograph. In my view, the science of bioenergetics stands on several pillars. The first is understanding how living cells interact with their environment. A displacement from the thermodynamic equilibrium is the prerequisite for forming bridges between life and its inorganic ambient. The first three chapters contain helpful concepts from the irreversible thermodynamics and chemiosmotic theory for comprehending energy conversions essential to all life. The mystery of life question includes bioenergetics. Biochemistry alone is not enough to phantom what life is. Thus, Chapters 4 and 5 deal with Schrödinger's question of what is life, as the famous physicist stated it, and in terms of extremal principles from irreversible thermodynamics. Chapter 6 provides details about the protonmotive force, which drives "forbidden" biochemical reactions. The protonmotive force cannot exist without topologically closed membranes and membrane-embedded proteins performing active transport. Membrane proteins are the second pillar of bioenergetics. Chapter 7 offers numerous examples of membrane proteins' essential role in organelle formation, cell-signaling, and bioenergetics of diseased versus healthy states. The third pillar of bioenergetics is the evolutionary origin of bioenergetic membranes. Chapter 8 examines the conjecture that photosynthesis evolved by maximizing entropy production in the charge separation steps. The maximal entropy production theorem for enzymatic transitions is introduced in that chapter. It forms the theoretical backbone for all of my calculations presented in this book. Chapter 9 discusses the evolution-coupling hypothesis, which connects biological evolution to the acceleration of universal thermodynamic evolution. Chapter 10 presents how maximal partial entropy production requirement in rate-limiting steps can increase the catalytic efficiency. This chapter connects with the previous two chapters in the finding that more evolved enzymes exhibit higher entropy production. Chapters 11 and 12 builds on Chapters 7 and 8, respectively, by treating in-depth historical and molecular details of how researchers revealed the action of ATP synthase molecular nanomotor and the bacteriorhodopsin light-activated proton pump. Chapter 13 is all about the life origin question and how it could have been connected to the origin of bioenergetics. The integrative and holistic feature is the fourth pillar of bioenergetics. We used the example of hexokinases and alpha-synuclein in Chapters 14 and 15 to illustrate how former proteins integrate glycolysis with oxidative phosphorylation and how synucleins connect the bioenergetics of the brain, aging, and cancer cells. All of the bioenergetics' foundations are interconnected. Most of them (irreversible interactions, evolutionary history, and holistic features) are connected to our universe's evolution, as discussed in the last Chapter 16. The pattern-producing evolutionary pathways are self-selected as being the most efficient in increasing dissipation and speeding up universe's evolution. That is why we can expect that life-like phenomena are widespread in the universe, whether based on bioenergetic membranes, proteins, and nucleic acids, or some other structures.

Definitions and explanations from Introduction (when first mentioned in the text, all glossary items from this book are put into a bold and underlined font)**:**

Active transport is the vectorial transport across a bioenergetic membrane against a concentration and electric field gradient. Such transport does not obey Fick's law for the diffusion of small molecules, nor Ohm's law for electric current. Still, it leads to charge separation and additional accumulation of ions or molecules in the compartment, which already has higher concentrations.

Active transport needs free energy investment, and it is associated with higher dissipation than passive transport. It usually requires a free-energy input in the form of ATP hydrolysis or some other driving force. Membrane proteins known as pumps serve to perform active transport across the membrane. In contrast, membrane proteins known as channels can only mediate a downhill transport of molecules in the direction determined by the concentration gradient or electric field.

Archaea, Bacteria, and Eucarya domains of the living world replaced previous eukaryote-prokaryote dichotomy, mostly due to Carl R. Woese (Woese et al. 1990).

ATP or adenosine triphosphate is a small organic molecule. It is composed of a five-sided ring forming a sugar molecule ribose, adenine attached to one side, and a string of three phosphate groups attached to the other ribose side. Hydrolysis and release of the endmost phosphate group, producing ADP (adenosine diphosphate) during enzyme-catalyzed reactions in a cell, make available the free energy that organisms can use for protein synthesis or work.

Catalytic constant, k_{cat},is the turnover number or the maximal number of enzyme turnovers during one second. When Michaelis-Menten kinetics is sufficient to describe an enzyme's activity, the reaction rate for a given substrate concentration is proportional to the catalytic constant.

Dissipation is the product of entropy production and ambient temperature. It is the result of irreversible processes.

Entropy is a measure of energy dispersal at a specific temperature. In other words, it is a measure of a system's energy per unit temperature that is unavailable for doing useful work. Spontaneous changes are always irreversible and accompanied by energy dispersal and entropy increase.

Entropy production is the rate of entropy increase due to irreversible processes in a system.

Evolution is a process of gradual change. This concept refers mainly to biology or even more restrictively just to Darwin's theory of evolution. During the last 100 years, the developments in physics made it abundantly clear that everything in the universe, stars with their planets, galaxies with their black holes, emerged during the evolution of the universe. Even atoms in our body are the product of the evolution of stars and star explosions. Evolution is always associated with entropy increase and entropy production (see above).

Free energy is available energy, that is, the amount of work that a thermodynamic system can perform.

The **holistic** approach to research topics is comprehensive and integrative, in contrast to the opposite reductionistic method. The word was coined, and holism was presented as an idea, by Jan Smuts in 1926. Smuts described the holistic approach as a "process-oriented, hierarchical view of nature." The idea is very much alive today, for instance, from proponents of a more holistic approach in cancer research and therapeutics. Interestingly, some modern authors see in holism an underappreciated tendency of nature to form wholes that are greater than the sum of its parts, through creative evolution.

An **isolated thermodynamic system** cannot exchange mass or energy with its surroundings. Mass and energy cannot change after imposing the perfect isolation at the system, but entropy can until it reaches the maximal possible value when no further evolution is possible. It is the postulate of classical thermodynamics that ideally isolated systems occur in nature or can be experimentally prepared. However, gravity prevents the establishment of an ideal enclosure requirement. Isolated systems are useful but strictly hypothetical idealization in thermodynamics. All systems in the Universe are coupled to all other systems via gravity and are evolving together with concomitant fast entropy increases.

Low initial entropy of the universe is the mystery looking for ingenious solutions.

Michaelis constant K_m is numerically equal to substrate concentration at which the reaction rate of an enzyme is half of the maximal rate V_{max}. The Michaelis-Menten kinetics leads to the numerical estimates for the K_m. The main assumption is that enzyme concentration is much less than the substrate concentration. Other assumptions are a quasi-steady-state approximation and free diffusion.

The **Permian** geologic period lasted from 299 to 252 million years ago.

Protonmotive force is the phrase coined by Peter Mitchell to describe a proton electrochemical gradient across bioenergetic membranes of bacteria, mitochondria, and chloroplasts.

Retinal is an abbreviated name for retinaldehyde, which is an aldehyde of vitamin A.

In its most succinct form, **the Second Law of Thermodynamics** states that entropy production of an isolated system is always greater or equal to zero.

The **thylakoid membrane** is the name used for all membranes that are enveloping thylakoids. The thylakoids are well-protected organelles inside larger photosynthetic organelle in plants—the chloroplast, or inside cyanobacteria. Most chloroplasts have three membrane systems: the outer membrane, the inner membrane, and the thylakoid membrane system enveloped and protected by the inner membrane. Light-dependent reactions of photosynthesis take place in the thylakoid membrane.

Note that the **transduction** term in genetics has a different meaning of a DNA transfer from one cell to another via a viral vector.

The **transition state** of a chemical reaction has the highest potential energy along the reaction coordinate. The energy difference between the initial state and the short-lived transition state is called the activation energy. The transition state theory of chemical reactions has now a ripe old age of 85 years and still dominates the field of chemical kinetics.

The **Triassic** geologic period lasted from 252 to 201 million years ago.

References

Bonačić Lošić, Ž., Donđivić, T. and Juretić, D. Is the catalytic activity of triosephosphate isomerase fully optimized? An investigation based on maximization of entropy production. J. Biol. Phys. 43(2017): 69–86.

Bond, D.P.G. and Grasby, S.E. On the causes of mass extinctions. Palaeogeography, Palaeoclimatology, Palaeoecology 478(2017): 3–29.

Chaisson, E.J. Energy flows in low-entropy complex systems. Entropy 17(2015): 8007–8018.

Chamberlin, R.V. 2015. Terrel L. Hill 1917–2014. National Academy of Sciences Biographical Memoirs. http://www.nasonline.org/publications/biographical-memoirs/memoir-pdfs/hill-terrell.pdf.

DasSarma, S. and Schwieterman, E.W. Early evolution of purple retinal pigments on Earth and implications for exoplanet biosignatures. Int. J. Astrobiol. (2018): 1–10. doi.org/10.1017/S1473550418000423.

Dobovišek, A., Županović, P., Brumen, M., Bonačić Lošić, Ž., Kuić, D. and Juretić, D. Enzyme kinetics and the maximum entropy production principle. Biophys. Chem. 154(2011): 49–55.

Dobzhansky, T. Nothing in biology makes sense except in the light of evolution. Am. Biol. Teach. 35(1973): 125–129.

Egan, C.A. and Lineweaver, C.H. A larger estimate of the entropy of the universe. Astrophys. J. 710(2010): 1825–1834.

Frautschi, S. Entropy in an expanding universe. Science 2017(1982): 593–599.

Haddad, W.M. Thermodynamics: The unique universal science. Entropy 19(2017): 621. doi: 10.3390/e19110621.

Higgs, P.G. and Lehman, N. The RNA world: molecular cooperation at the origins of life. Nat. Rev. Genet. 16(2015): 7–17.

Hill, T.L. 1977. Free Energy Transduction in Biology. The Steady-State Kinetic and Thermodynamic Formalism. Academic Press, New York, NY, USA.

Hunt, K.L.C., Hunt, P.M. and Ross, J. Dissipation in steady states of chemical systems and deviations from minimal entropy production. Physica 147A(1987): 48–60.

Juretić, D. 1997. Bioenergetics: Work of Membrane Proteins. Informator, Zagreb, Croatia (in Croatian).

Juretić, D. and Županović, P. 2005. The free-energy transduction and entropy production in initial photosynthetic reactions. pp. 161–171. *In*: Kleidon, A. and Lorenz, R.D. (eds.). Non-equilibrium Thermodynamics and the Production of Entropy: Life, Earth and Beyond. Springer, Berlin, Germany.

Koshland, Jr. D.E. Application of a theory of enzyme specificity to protein synthesis. Proc. Natl. Acad. Sci. USA 44(1958): 98–104.

Lane, N. Chance or necessity? Bioenergetics and the probability of life. J. Cosmol. 10(2010): 3286–3304.

Lane, N. and Martin, W. The energetics of genome complexity. Nature 467(2010): 919–934.

Lane, N. and Martin, W. The origin of membrane bioenergetics. Cell 151(2012): 1406–1416.

Lehninger, A.L. The relationship of the adenosine polyphosphates to fatty acid oxidation in homogenized liver preparations. J. Biol. Chem. 157(1945): 363–382.

Lovelock, J.E. and Margulis, L. Atmospheric homeostasis by and for the biosphere—the Gaia hypothesis. Tellus 26(1974): 2–10.

Lovelock, J. 1988. The Ages of Gaia. Norton, New York, NY, USA.

Martyushev, L.M. and Seleznev, V.D. Maximum entropy production principle in physics, chemistry and biology. Phys. Rep. 426(2006): 1–45.

Metzner, H. Bioelectrochemistry of photosynthesis: a theoretical approach. Bioelectrochem. Bioenerg. 13(1984): 183–190.

Mitchell, P. Coupling of phosphorylation to electron and hydrogen transfer by a chemi-osmotic type of mechanism. Nature 191(1961): 144–148.

Patel, V.M. and Lineweaver, C.H. Solutions to the cosmic initial entropy problem without equilibrium initial conditions. Entropy 19(2017): 411. doi: 10.3390/e19080411.

Prigogine, I. and Nicolis, G. On symmetry-breaking instabilities in dissipative systems. J. Chem. Phys. 46(1967): 3542–3550.

Prigogine, I., Lefever, R., Goldbetter, A. and Herschkowitz-Kaufman, M. Symmetry breaking instabilities in biological systems. Nature 223(1969): 913–916.

Prigogine, I. What is entropy? Naturwissenschaften 76(1989): 1–8.

Ripple, W.J., Wolf, C., Newsome, T.M., Barnard, P., Moomaw, W.R. et al. World scientists' warning of a climate emergency. BioScience 79(2020): 8–12.

Rodriges, M., Rodriguez-Sabate, C., Morales, I., Sanchez, A. and Sabate, M. Parkinson's disease as a result of aging. Aging Cell 14(2015): 293–308.

Rooney, A.D., Macdonald, F.A., Strauss, J.V., Dudás, F.Ö., Hallmann, C. and Selbye, D. Re-Os geochronology and coupled Os-Sr isotope constraints on the Sturtian snowball Earth. Proc. Natl. Acad. Sci. USA 111(2014): 51–56.

Russel, M.J. The alkaline solution to the emergence of life: Energy, entropy and early evolution. Acta Biotheor. 55(2007): 133–179.

Salthe, S.N. 2015. Toward a natural philosophy of macroevolution. pp. 163–181. *In*: Serrelli, E. and Gontier, N. (eds.). Macroevolution, Interdisciplinary Evolution Research 2. Springer International Publishing, Switzerland.

Schrödinger, E. 1944. What is Life. The Physical Aspect of the Living Cell. Cambridge Univ. Press, Cambridge, UK.

Segré, D., Ben-Eli, D., Deamer, D.W. and Lancet, D. The lipid world. Origins Life Evol. B. 31(2001): 119–145.

Smolin, L. 1997. The Life of the Cosmos. Oxford University Press, Oxford, UK.

Smuts, J.C. 1926. Holism and Evolution. Macmillan, New York, NY, USA.

Stahl, A. and Keller, J.U. The entropic waste problem in energy engineering, economy, and ecology. J. Non-Equilib. Thermodyn. 24(1999): 260–279.

Straškraba, M., Jørgensen, S.E. and Patten, B.C. Ecosystems emerging: 2. Dissipation. Ecol. Model. 117(1999): 3–39.

Ulanowicz, R.E. and Hannon, B.M. Life and the production of entropy. Proc. R. Soc. Lond. 232(1987): 181–192.

Velasco, R.M., García-Colín, L.S. and Uribe, F.J. Entropy production: Its role in non-equilibrium thermodynamics. Entropy 13(2011): 82–116.

Verlinde, E.P. On the origin of gravity and the laws of Newton, J. High Energy Phys. (2011): 029. doi: 10.1007/JHEP04(2011)029.

Verlinde, E.P. Emergent gravity and the dark universe. Sci. Post Phys. 2(2017): 016. doi: 10.21468/SciPostPhys.2.3.016.

Woese, C.R., Kandler, O. and Wheelis, M.L. Towards a natural system of organisms: proposal for the domains Archaea, Bacteria, and Eucarya. Proc. Natl. Acad. Sci. USA 87(1990): 4576–4579.

Wolfenden, R. and Snider, M.J. The depth of chemical time and the power of enzymes as catalysts. Acc. Chem. Res. 34(2001): 938–945.

Zotin, A.A. and Zotin, A.I. Thermodynamic basis of developmental processes. J. Non-Equilib. Thermodyn. 21(1996): 307–320.

CHAPTER 1

Mitchell's Chemiosmotic Theory

The Background

1.1 The background and early developments

Present-day bioenergetics, systems biology, and cell biophysics are moving ever closer to the thermodynamics of irreversible processes, while initial expectations after the Second World War about the importance of the quantum biology approach did not come true yet. Quantum phenomena, such as tunneling and ultrafast nondissipative transitions, are relevant for the detailed discussion of photosynthesis and enzyme kinetics, for example. I do not doubt that life evolved ways and means for utilizing the quantum peculiarities like entanglement, superposition, wave-like properties, and tunneling, which even the best physicists don't know how to detect or employ at room temperatures. Some physicists argue for the need to use quantum physics in explaining consciousness. But on the whole, my viewpoint is closer to Delbrück's thoughts. Physicist Max Delbrück is today a legendary person among biologists due to the thoughtfulness of his contributions and developed methods in biology. However, in spite of being one of the pioneers in quantum physics applications, he did not use quantum physics in biology, and he had great respect for the wisdom of living species. In one 1949 lecture, he asserted: "You cannot expect to explain such a wise old bird as life in a few words." He had in mind the billions of years through which biological evolution created even the thinnest living particles, like bacterial viruses named phages, which he liked to use in his research. We must admit that on the mesoscopic and larger scale of macromolecules, cellular organelles, and cells, the classical physics predominates, and **irreversible thermodynamics** is the only physics subdiscipline allowing for irreversibility, evolution, and free-energy transduction, that are crucial attributes of biochemical processes in every living cell.

The first push in this direction was due to Aaron Katchalsky, Ilya Prigogine, Hermann Haken, Terrel Hill, and many other researchers from 1950 to 1980, who built upon the firm foundation for near-equilibrium irreversible thermodynamics by Lars Onsager, the winner of the 1968 Nobel Prize in Chemistry (Onsager 1931a,b). Although still of limited extent, these new physics developments were essential for a better understanding of complex biological structures and processes. General organizing, evolution, or selection principles from physics were not found during that period, which was not a disappointment to molecular biologists (they did not expect anything else in addition to Darwinian evolution and selection). Still, it was a disappointment to physicists, whose natural inclination is to derive as many predictions as possible from as simple and as general natural law as possible.

Another line of research focused on solving the problem of how oxidative or photosynthetic phosphorylation works. In that research field, Peter Mitchell (1920–1992) became my hero during the PhD thesis research I performed at the Penn. State Univ. from 1972 to 1976. Physicists love to see grand unification ideas either in their or in unrelated research topics. Peter was not a physicist, but he united membrane biochemistry with the biochemistry of soluble proteins, electrochemistry with biochemistry, mitochondrial with microbiological cytochrome research, respiration with photosynthesis, ion gradients with vectorial chemistry, electron transfer with active transport of protons, including conversions among electrical, osmotic, and chemical forms of energy (Mitchell 1977b). All these unifications led to the natural development of membrane bioenergetics as a mature research field. Remarkably, when his chemiosmotic hypothesis was published in 1961 (Mitchell 1961a,b) there was a little experimental evidence for it (Slater 1994). Mitchell's speculative and visionary intuition was vindicated later in experiments performed by him, his collaborator Jennifer Moyle, and other scientists. It turned out that the chemiosmotic theory by Peter Mitchell, the Nobel Prize winner in Chemistry 1978, was an excellent inspiration for numerous experiments probing the build-up of the electrochemical proton gradient (Mitchell 1979a,b). The chemiosmotic proposal initially contained some errors in mechanistic details. The chairman of the committee, which recommended Mitchell for the Nobel Prize in 1978, felt obligated to defend their decision with words: "Mitchell was right only on the phenomenological and not on the mechanistic level" (Malmström 2000).

One of my favorite books is: "Surely You're Joking, Mr. Feynman!" (Feynman 1985). The title could have been expressed as the question: "Are you serious, Mr. Feynman?" Except for the scientist's name, the question is identical to the one published by Leslie Orgel in the Nature journal: "Are you serious, Dr Mitchell?" (Orgel 1999). Mitchell's ideas seemed bizarre as Orgel commented, or chemically unattractive as put more kindly by Boyer (Prebble 2013), or weighted with perplexing physics as implicitly stated by Slater: "I did not understand why Mitchell believed that the translocation of protons contributed both to the pH difference and the potential across the membrane" (Slater 1994). Mitchell explained in the 1970s (Mitchell 1977b) why transmembrane nanomotors are sensing both the chemical and electric components of the electrochemical driving force across the membrane (see Chapter 6). Presently, the recapitulation of his arguments can be found in many textbooks or, for instance, in the review paper about the concept of the membrane proton well (also introduced by Peter Mitchell) (Mulkidjanian 2006) and the review paper about ATP synthase (Junge and Nelson 2015). However, in the 1960s, it was certainly disconcerting for research leaders in oxidative phosphorylation to give up their preconceived ideas because of the hypothesis that has come from an outsider to that field. Also, not enough experimental evidence has been accumulated to endanger their pet theories.

Contemporary research leaders considered as relevant the metabolic expertise in cell-free preparations of enzymes with as little membraneous contamination as possible. Mitchell did not agree. He was an expert in membranes, ion transport through membranes, and cytochromes embedded in the microbial cytoplasmic membrane (Mitchell and Moyle 1956). In his opinion, other **integral membrane proteins** also acted as enzymes or power-generating units: "(Membranes) contain the power-generating and power-consuming modules that are plugged through it and make contact with each other through the aqueous conductors on either side." (Mitchell 1991). He maintained this same opinion from the 1950s, and it gradually become widely accepted, but some of his other prescient proposals were contested before and after his Nobel Prize.

Mitchell's no-bull style in speaking and writing did not help. One example is the statement from his Nobel lecture: "By 1965, the field of oxidative phosphorylation was littered with the smouldering conceptual remains of numerous exploded energy-rich chemical intermediates" (Mitchell 1979b). Mitchell referred to the prevalent high-energy intermediate hypothesis for oxidative phosphorylation in the 1950s (Slater 1953), which was still the main framework for the unsuccessful experiments to find such intermediates in the 1960s. Characteristically, he stated: "The elusive character of the 'energy-rich' intermediates of the orthodox chemical coupling hypothesis would be explained by the fact that these intermediates do not exist" (Mitchell 1961a). A relatively small investment of his family money into Mitchell's private research institution, the Glynn House (see section 1.5), saved taxpayers from the USA and other countries of much greater additional useless spending of millions of dollars for funding research about the identity of the 'energy-rich' intermediates (Rich 2008).

Fundamental physical and chemical principles needed for the development of Mitchell's ideas were just around the corner or already present for some time in the scientific literature as free ingredients that only had to be recognized.

1.2 Living cells are equally brilliant chemists and physicists

The irreversible thermodynamics provided another main line of development for the concept of vectorial metabolism. It suggested some interesting answers about the question of why living cells are at the same time highly ordered and highly powerful entropy producers. It must be admitted that the intuition of Nobelist Ilya Prigogine was on the right track when he introduced the concept of dissipative structures (Glansdorff and Prigogine 1971, Prigogine and Lefever 1973), the structures spontaneously arising whenever there is a pressing need to increase the system's entropy production (due to increased driving force for instance). It appears that we do have something in common with such dissipative structures as whirlpools, eddies, tornadoes, and dust devils. However, for such a complex system as a living cell, the description in terms of dissipative structure is so abstract that it leaves out all details on how dissipation is performed, by what structures, and why it is functionally essential for a cell to have high dissipation.

There is no doubt that biological macromolecules are often capable of producing the increase in dissipation for many orders of magnitudes when stimulated with appropriate metabolites or free energy packages. The same nucleic acids or protein macromolecules are inert when outside the cellular environment. They can even form crystals in the hands of skilled biochemists interested in performing the **X-ray crystallography** to study their structure. We learn nothing new if we name all biomolecules as dissipative structures. But an excellent capability to regulate the dissipation level can be regarded as unique to living system bioenergetics.

"The living cell is a chemist," used to say my PhD thesis mentor, Prof. Alec Keith. One might add that each living cell, visible only under a microscope, is the best chemist in the whole world capable of speeding up some reactions for 10 to 20 orders of magnitude, and without the help of high temperatures. We have equally good reason to consider living cells as genius physicists. Simple bacterial cells are capable of creating and maintaining the strongest possible electric field across their cytoplasmic membrane. Just a little stronger field would ensure the cell's self-destruction. Cell's ability to separate charges is responsible for creating the electrochemical proton gradient or protonmotive force, which was mentioned before in the introductory chapter.

For a long time, biochemists did not appreciate that living cells are brilliant physicists, well versed not only in the synthesis of complex organic compounds but also in the art of charge separation and simultaneous generation of strong electric fields. Three domains of life, Archaea, Bacteria, and Eukaryota, generate similarly strong fields of 30 million volts per meter, corresponding to that discharged by a bolt of lightning. However, eukaryotic cells reserve their mitochondria organelles for creating such strong electric fields, while prokaryotic cells use their cytoplasmic membrane for the same purpose. It took the visionary disposition of Peter Mitchell to recognize the common energy-transduction principle of bacteria, mitochondria, and chloroplasts.

1.3 The chemiosmotic hypothesis

The powerhouses of our cells are organelles named mitochondria, while plant cells have chloroplasts too. Mitchell assumed that cells and organelles need an electrochemical proton gradient for ATP synthesis. He called it the protonmotive force (pmf). The ATP-synthase is a membrane-embedded protein, which couples the pmf and directional proton transport to ATP synthesis. Experiments in his and many other laboratories established the importance of charge separation ability in prokaryotes, mitochondria, and chloroplasts. Mitchell also recognized the importance of topologically closed membrane and vectorial active transport of protons by integral membrane proteins acting as proton pumps. Charge separation would be impossible if a bioenergetic membrane with proton pumps is not topologically closed and impermeable to protons. It would also be impossible if some input force did not serve to drive active transport of protons from the region of lower proton concentration and lower electric potential to the region of higher proton concentration and electric potential. Mitchell regarded the vectorial proton transport as the central tenet of his chemiosmotic hypothesis (Mitchell 1961a, 1966, 1979a, 1991).

Peter Mitchell proposed in 1958 that electron-transfer reactions are coupled to the translocation of protons across the membrane (Mitchell and Moyle 1958a,b). Such a general scheme of coupling electron to proton transfer is common to all fields of membrane bioenergetics, including respiration and photosynthesis. By invoking the physics of charge separation as the main driving force for metabolism, Mitchell created a profound paradigm shift in biology. A common opinion that biology is based on chemistry alone (bag-of-enzymes biochemistry) was no longer tenable. His Nobel Prize was well deserved, and it did not matter that he was wrong about some details. Ironically, his Nobel Prize was for work in chemistry, although the vital role of protonmotive force in all domains of life obviously transcended chemistry (Lane 2010). Leslie Orgel even claimed that no similar revolution in biology occurred after Darwin's (Orgel 1999). Somewhat similar to Darwin's creative isolation in the Down House, Mitchell wanted to see from 1962 onward if a small private research institute (the Glynn House, which he restored) could provide a place "of quiet haven for untrammeled scientific work and thought" (Slater 1994). As in Darwin's case, this move to relative isolation agreed splendidly both with Mitchell's health and with his unconventional thinking. Solitude, combined with voracious reading and willingness to exchange open-minded opinions with students and colleagues, can be a critical ingredient of creativity, whether forced on a person due to health problems (as in Mitchell's case) or not.

1.4 The impact after the Nobel Prize award to Peter Mitchell

Mitchell's influence increased after he became the Nobelist in 1978 for his fundamental contributions in formulating the chemiosmotic hypothesis and finding experimental evidence supporting it. This development tended to restrict the research subject of bioenergetics to his favorite topics of mitochondria and chloroplast organelles from eukaryotes, and equally intensive free-energy transduction by some prokaryotic cells. A high number of PhD theses during the last several decades examined structure-activity connection for membrane-embedded proton pumps. Several additional Nobel Prizes have recognized the fundamental and practical importance of getting a deeper understanding of how these vital nanomotors work. This research rediscovered structures already invented by blind biological evolution, as tasty, non-polluting solutions for sustainable growth at the time when closest human ancestors were single cells swimming in primordial oceans.

Four decades after P. Mitchell's Nobel Prize award, one can find many review papers about the chemiosmotic theory and its limitations or modifications (Ernster and Schatz 1981, Ädelroth and Brzezinski 2004, Kresge et al. 2006, Kocherginsky 2009, Junge 2013, Nath 2017, 2019, 2021; to mention just a few). Also, almost any textbook on biochemistry and bioenergetics contains its description too. Mitchell had high regard for the contributions of other research groups and visionary individuals. He also acknowledged the major influence the Popperian way of thinking in science (Popper 1934, 2002) had on his approach to forming hypotheses and in experiments carefully designed not only to test but to falsify them as well. In particular, Mitchell suggested that his chemiosmotic model proposal was the example of the way science progresses in theory-testing feedback cycles, according to Karl Popper (Mitchell 1977a). For instance, in his 14 February 1962 letter to Slater, he stated: "We have done a number of experiments here aimed to disproving the feasibility of the chemi-osmotic conception, but so far without success" (Weber 1991). One can only wish that the present-day scientific climate encourages young scientists to create and disprove their ideas in experiments more often than it is the case now. Even at Mitchell's time, some academic priorities such as publish-or-perish, or get funding, or ingratiate yourself to authorities were not agreeable with his personality. He often published his seminal discoveries in the form of abstracts for scientific conferences or in other than widely read scientific journals (Slater 1994).

1.5 Glynn Research Institute: An eccentric experiment

Mitchell's high productivity of ground-breaking proposals and discoveries took off after he became the fiercely independent but kind "sheriff" at the Glynn Research Institute. The Glynn House estate was established by Edmund John Glynn, who was the High Sheriff of Cornwall in 1799 (Weber 1991, Slater 1994). As the Glynn Research Institute, it became a highly attractive meeting place for more than 150 scientific visitors, including Mitchell's most vocal opponents. They disagreed among themselves but were all interested in relaxed discussions with Mitchell about mysteries of active transport, vectorial metabolism, and especially the oxidative phosphorylation. Mitchell was long attached to the emotional sentiment of how it would be nice to make peace between, in his words, transport people and metabolism people. In the 1979 interview, he disclosed his thoughts: "I think that the thing that influenced me most was noticing that the transport people, with whom I became associated because of Danielli, not only failed to understand the metabolism people, but they were positively scornful about their attitude. Conversely, the metabolism people not only did not understand the transport people, but they were scornful about them. So I think one of the strong feelings I can recognize or recollect is that of sadness that this should be the state of affairs. It was pretty clear to me that there was a transport aspect of metabolism" (Weber 1991). The eccentric experiment of establishing the Glynn Research Institute with his own money and seemingly equally eccentric experiments performed there under his guidance brought this utopian vision to fulfillment.

For the generalized analog of electromotive force (the chemicomotive force), Mitchell gave credit to Guggenheim (1933). In his Nobel lecture, he mentioned Overton (1902) and Gorter and Grendel (1925) for the painstakingly careful experiments and insight into composition, topology, permeability, and the bilayer organization of biological membranes. In his early career, Mitchell was fortunate to be the only PhD student of James F. Danielli, who was the expert for membranes (Danielli and Dawson 1935) and practiced a friendly and relaxed mentorship style (Slater 1994). In the 1950s, Mitchell loved to talk with David Keilin too (Mitchell 1979b), who was the expert for what was lacking in the Dawson-Danielli model of the plasma membrane–transmembrane proteins in general and the respiratory chain cytochromes in particular (Keilin and King 1958). Vectorial membrane-based metabolism coupled with transport was Mitchell's idea that profited too from the work of Henrik Lundegårdh (1945) on active transport of ions in plants (Larkum 2003).Thus, unconventional thinking and a highly developed intuition led Mitchell to have an original opinion on every subject considered by his teachers, something that unsurprisingly did not find favor with some of his examiners. They classified him as a poor examinee who got a satisfactory grade for his thesis in 1951 only after resubmitting it (Slater 1994).

1.6 Chemiosmotic hypothesis maturation into the chemiosmotic theory

For the transformation of the chemiosmotic hypothesis into the chemiosmotic theory, two publications dealing with the artificial bioenergetic membranes were the most important (beside Mitchell's observations and articles) (Morange 2007). Both came from the field of photosynthesis. <u>The first one</u> was the Jagendorf and Uribe paper from 1966 in which these authors described the paradoxical finding that <u>**vesicles**</u> prepared from chloroplasts could synthesize ATP in dark conditions if exposed to high enough pH gradient. They concluded in the Discussion section: "The experiments reported here show that ATP is formed by chloroplasts depending only on an artificial transition from acidic to basic medium. This ATP synthesis occurs certainly without the aid of light, and quite possibly without any electron transport.... These facts are, on the whole, consistent with the "chemiosmotic" hypothesis for the mechanism of phosphorylation, as elaborated by Peter Mitchell." The protonmotive force is mostly due to pH gradient in the case of chloroplasts, and Mitchell's hypothesis was general enough to encompass oxidative phosphorylation (when the membrane potential is more important) and photosynthetic phosphorylation as well. Both free-energy transduction mechanisms produced the electrochemical proton gradient as the first energy conversion step in the "currency", which can be easily used for all subsequent transduction hierarchy stages. The second one was discovered by Racker and Stoeckenius in 1973 and published in the 1974 issue of the Journal of Biological Chemistry. They created artificial vesicles with an audacious combination of ATP-synthase isolated from bovine heart mitochondria and **bacteriorhodopsin** isolated from the purple membrane of *Halobacterium salinarium*. When submitted to illumination, this model system produced ATP. It was the confirmation for the energy conversion according to chemiosmotic theory and the evidence that the bacteriorhodopsin functions as the light-driven proton pump.

One would expect that everybody agreed with this assessment about the importance of chemiosmotic hypothesis confirmation on simplified artificial membrane vesicles, but this was not the case. In the same year when the Nobel Prize award went to Peter Mitchell alone (1978), the inorganic chemist, professor Robert Joseph Paton (Bob) Williams, published a long paper in which he repeatedly discounted above mentioned experiments as being performed under artificial conditions that are not "the methods of biology" (Williams 1978). In his distinguished career, Williams published over 700 articles and books and established a new scientific discipline with a somewhat paradoxical title—Biological Inorganic Chemistry (Thompson 2016). Since he proposed in 1961 the energized proton-in-the-membrane scheme for energy transduction in oxidative and photophosphorylation (Williams 1961, 1989), the suggestion similar to the chemiosmotic hypothesis, one can wonder why he did not share the Nobel Prize with Mitchell. The similarity is only superficial. Localized proton theories of Williams certainly have some merit, but proved exceedingly difficult to confirm, while he rejected the chemiosmotic hypothesis as far as bulk protons are concerned. Likewise, after prolonged correspondence with Williams, Mitchell concluded that Williams' concept is a special case of substrate-level phosphorylation in which protons remained bound in the membrane (to some membrane proteins) instead of being transported across the membrane (Slater 1994).

Williams did not pursue experiments to prove or disprove his conjecture about high local concentrations of protons in the membrane. In his 1978 article, he acknowledged that his report omitted experimental details, although it was written as contrasting background to the influential review of oxidative phosphorylation and photophosphorylation (Boyer et al. 1977). Efraim Racker initiated the review due to the practical reason that funding agencies in the U.S. could lose interest in supporting bioenergetics because the field was in turmoil (Prebble 2013). Indeed, even some authors of this conciliatory review, which acknowledged the importance of the chemiosmotic hypothesis, assisted Williams in writing about numerous conceptual mistakes in chemiosmosis (Williams 1978).

As years passed, the research focus of Williams's supporters shifted from membrane-bound protons to proton pathways localized at the membrane surface. The tentative conclusion is that interfacial protons contribute to the protonmotive force (Kell 1979, Mulkidjanian et al. 2005, Xiong et al. 2010, Medvedev and Stuchebrukhov 2013). Also, Williams (1978), Westerhoff and van Dam (1987), and Kell (1979) pointed out that living cells and energy-transducing organelles are not in equilibrium with their surroundings. Thus, it would not be realistic to assume that the equilibrium distribution of ions across bioenergetic membranes is ever reached, although it is the underlying assumption of the chemiosmotic theory. Approximating the quasi-static with a steady-state or even equilibrium situation is a common practice in biochemistry. One must only be aware of these assumptions.

The chemiosmotic theory captured the imagination of Russian scientists too. Vladimir Skulachev became an ardent follower of Mitchell (Allchin 2002). His group focused on careful measurements of membrane gradients and put a lot of effort into calculating the protonmotive force from their data. They used **ionophores**, named Skulachev's ions, as convenient probes for the membrane potential (Skulachev 1985). These were cations enveloped with **hydrophobic** groups (tetraphenyl boron, tetraphenyl phosphonium, trinitrophenol) that facilitated their easy passage and equilibration across the cytoplasmic or inner mitochondrial membrane. When radioactively labeled or detected with an electrode sensitive to the specific ionophore, these probes reported how much they accumulated in respiring bacteria or mitochondria (Westerhoff et al. 1989).

One example of a natural Skulachev ion is the peptide antibiotic valinomycin, which lowers membrane resistance and specifically enhances the K^+ transmembrane diffusion for a large factor (Andreoli et al. 1967, Parsegian 1969). In retrospect, the paper by Adrian Parsegian, a gifted theoretical biophysicist, illustrates to what degree was structure and dynamics of biological membranes the *terra incognita*, well after Mitchell formulated his hypothesis about the functioning of membrane bioenergetics with firm foundation in what was to become the fluid-mosaic membrane model (Singer and Nicolson 1972). The concept of integral membrane proteins (see Chapter 7) was present in Mitchell's thinking already in the 1950s and 1960s, thus way ahead of the publication of the fluid-mosaic model. He was the first to propose the names uniport, symport, and antiport for corresponding transport roles performed by membrane proteins essential for membrane bioenergetics. Skulachev formulated the laws of cell energetics in 1992 and later published together with two other coauthors the "Principles of Bioenergetics" book in 2012. Dimroth group (Hilpert et al. 1984, Dimroth 1997) and Skulachev's group discovered that some bacteria could use the concentration gradient of sodium ions instead of protons to drive the ATP synthesis (Skulachev 1985, 1991).

Skulachev and Mitchell had a common interest in comparing bioenergetics' principles with economic laws in human society. Cells and bioenergetic organelles first transform available free energy into the convertible currency of ion electrochemical gradients and ATP molecules. Both currency types are needed as well as their facile interconversion and good regulation of corresponding fluxes. This is why the currents of protons, other ions, and ATP generation or hydrolysis are recognized as the dominant contribution to entropy production and also the signatures of the far from the equilibrium situation of metabolically active living systems. Research about entropy production of human society has more than academic interest and can undoubtedly profit from what we learned about bioenergetics (Junge 2013, Kümmel 2016).

Interestingly, Mitchell did not regard the chemiosmotic theory as the cornerstone of bioenergetics nor as the hypothesis invented to deal with the problem of the coupling mechanism in oxidative phosphorylation (Mitchel 1991). He explicitly stated in his 1991 paper that the chemiosmotic theory is more broadly based than bioenergetics. He regarded his early ideas on vectorial metabolism and chemiosmotic systems as qualitative solutions that inspired the search for appropriate biochemical and physiological problems!

1.7 Mitchell's last publication in 1991

Peter Rich reviewed Mitchell's last publication (Mitchell 1991) about vectorial metabolism, osmochemistry, and osmoenzymes (Rich 2008). In these fast-moving research fields, almost 30 years after Mitchell's demise, his ideas are still typically conducive to experiments. Karl Popper would be satisfied. However, Mitchell's concept of microscopic chemiosmotic coupling principle (1977c, 1979a,b, 1991) did not find traction with his colleagues during his life and afterward. Peter Rich did not mention it at all in his review, while Bob Williams declared already in 1978 that "Micro-chemiosmosis has no sense" (Williams 1978). I shall try to explain here why I consider it as prescient, inspirational, and unfortunately neglected.

Different terminology helped to obscure the connection to Mitchell's osmoenzymes and microscopic chemiosmosis. Osmoenzymes are presently described as biological motors (nanomachines) that are coupling chemical and mechanical processes. Mitchell distinguished molecular machines from osmoenzymes. He defined osmoenzymes as: ..."membrane-located enzymes that couple metabolism to chemical group, electron or solute translocation through part or all of an osmotic barrier (membrane)" (Mitchell 1991). His microscopic chemiosmosis terminology is more difficult to connect with current notation, but he helped by enlisting several examples. For instance, in his last paper, he stated: "A beautiful recent example of microscopic chemiosmotic coupling, supported by X-ray crystallography and kinetic studies, is provided by tryptophan synthase...." (Mitchell 1991). The example he mentioned is the paper by Dunn et al. in 1990. In this and two more recent review papers by the same first author (Dunn et al. 2008, Dunn 2012), the relevant terms are concerted proton transfer and substrate channeling.

In line with Mitchell's thinking, we should not allow that a preference for different names obscures the essential generality of dynamical and directional movements of electrons, protons, and groups of atoms during enzymatic catalysis. These microscopic nanocurrents can be called catalytic nanocurrents because transported particles within a single or several adjacent macromolecules involve substrates, enzyme domains, and products as reactants. Examples of triosephosphate isomerase, lactamase, hexokinase, ATP synthase, and bacteriorhodopsin kinetic cycles (Chapters 10, 11, 12) confirm that irrespective of their cytoplasmic or membrane location, the proton channeling mechanism is of crucial importance for the catalytic efficiency of different enzymes. Proton currents are always vectorial and always coupled to vectorial metabolism and molecular biodynamics that was forcefully promoted by Peter Mitchell (1991). From the thermodynamic perspective, it is impossible to divorce dissipation from currents, no matter what particles are mainly responsible for these currents.

Connecting these thoughts to section 1.2, we can conclude that enzymes create microscopic dissipative structures when stimulated by their substrates. Conformational changes and molecular dynamics occur as enzymes react with substrates, products, and water molecules to produce orchestrated directional microscopic movements. Individual enzymes or battery of connected enzymes can be considered dissipation gatekeepers that are ever ready for skyrocketing dissipation increase—the appearance of dissipative structures through microscopic nanocurrents. These catalytic nanocurrents are essential for the regulation of catalytic cycles in the nanoworld. The same thought, connecting microscopic transport events to metabolism, was expressed by Peter Mitchell in a different manner: "Most enzymes act as connectors for the conduction of a specific species of chemical group, such as phosphoryl or hydrogen, or of electrons, between different donor and acceptor species" (Mitchell 1977). In the presence of an **autocatalytic** growth, the number of enzymes multiplies until metabolism and physiology emerge as a manifestation of myriad regulated nanocurrents in cells and organisms.

Definitions and explanations from Chapter 1:

An **autocatalytic** reaction product is a catalyst for the same reaction. Autocatalysis produces its own catalyst.

Bacteriorhodopsin is an integral membrane protein similar to the rhodopsin from the human eye, but it does not function as a photon detector. Its main function is a light-activated proton expulsion from the cytoplasm of the bacterium *Halobacterium salinarum*. It is the simplest photosynthetic system nature developed because protons spontaneously return to bacterium through the ATP synthase so that their return induces the ATP synthesis.

Hydrophobicity is the tendency for the association of non-polar molecules in the presence of water. It arises from entropy increase associated with a release to the bulk solution of water molecules attached to a hydrophobic surface.

Integral membrane proteins are proteins so firmly embedded in cellular membranes to be considered as the integral parts of the membrane.

An **ionophore** is a chemical compound that can easily bind and release ions. Lipid-soluble ionophores are useful as ion-carriers and ion-transporters across the cell membrane.

Irreversible thermodynamics is the branch of thermodynamics dealing with irreversible processes, that is, with processes that produce a clear difference between past and future and are impossible to reverse back to the initial state. A common example is pouring a glass of wine in the ocean: nobody can reverse this process.

Vesicles in biology and biochemistry are either artificial or natural lipid structures. Their spherical shape forms spontaneously when phospholipids are mixed with the water solution. Phospholipids form a topologically closed membrane which separates internal from external water molecules. The membrane is composed of two lipid layers—external and internal, thus the name membrane bilayer for all phospholipid membranes. Each phospholipid molecule has a polar head and two lipid tails oriented so that only polar heads are in direct contact with water molecules. The membrane interior is highly hydrophobic with a low dielectric constant. A crude preparation of vesicles is called liposomes. Membrane biochemists are skilled in preparing unilamellar liposomes with only one membrane bilayer and the desired size from a small to medium and giant size. Small vesicles have a diameter smaller than the visible light wavelength and are invisible in solution. An artificial bioenergetic membrane has the advantage that the researcher can manipulate both its lipid and protein composition.

X-ray crystallography is the method to determine the atomic structure of molecules in a crystal. When extended to biological macromolecules, the method requires the use of sophisticated reconstruction techniques to produce the 3D picture from the 2D map of diffraction points obtained by incident X-rays beam.

References

Ädelroth, P. and Brzezinski, P. Surface-mediated proton-transfer reactions in membrane-bound proteins. Biochim. Biophys. Acta 1655(2004): 102–115.

Allchin, D. To err and win a Nobel Prize: Paul Boyer, ATP synthase and the emergence of bioenergetics. J. Hist. Biol. 35(2002): 149–172.

Andreoli, T.E., Tieffenberg, M. and Tosteson, D.C. The effect of valinomycin on the ionic permeability of thin lipid membranes. J. Gen. Physiol. 50(1967): 2527–2545.

Boyer, P.D., Chance, B., Ernster, L., Mitchell, P., Racker, E. and Slater, E.C. Oxidative phosphorylation and photophosphorylation. Annu. Rev. Biochem. 46(1977): 955–1026.

Danielli, J.F. and Davson, H. A contribution to the theory of permeability of thin films. Journal of Cellular and Comparative Physiology 5(1935): 495–508.

Dimroth, P. Primary sodium ion translocating enzymes. Biochim. Biophys. Acta 1318(1997): 11–51.

Dunn, M.F., Aguilar, V., Brzovic, P., Drewe, W.F., Houben, K.F., Leja, C.A. et al. The tryptophan synthase bienzyme complex transfers indole between the α- and β-sites via a 25–30 Å long tunnel. Biochemistry 29(1990): 8598–8607.

Dunn, M.F., Niks, D., Ngo, H., Barends, T.R.M. and Schlichting, I. Tryptophan synthase: the workings of a channeling nanomachine. Trends Biochem. Sci. 33(2008): 254–264.

Dunn, M.F. Allosteric regulation of substrate channeling and catalysis in the tryptophan synthase bienzyme complex. Arch. Biochem. Biophys. 519(2012): 154–166.

Ernster, L. and Schatz, G. Mitochondria: A historical review. J. Cell Biol. 91(1981): 227–255.

Feynman, R. and Leighton, R. (contributor). 1985. Surely You're Joking, Mr. Feynman!: Adventures of a Curious Character. Hutchings, E. (ed.). W.W. Norton, New York, NY, USA.

Glansdorff, P. and Prigogine, I. 1971. Thermodynamic Theory of Structure, Stability and Fluctations. Wiley-Interscience, New York, NY, USA.

Gorter, E. and Grendel, F. On bimolecular layer of lipoids on the chromocytes of the blood. J. Exp. Med. 41(1925): 439–443.

Guggenheim, E.A. 1933. Modern Thermodynamics by the Methods of Willard Gibbs. Methuen, London, GB.

Hilpert, W., Schink, B. and Dimroth, P. Life by a new decarboxylation-dependent energy conservation mechanism with Na^+ as coupling ion. EMBO J. 3(1984): 1665–1680.

Jagendorf, A. and Uribe, E. ATP formation caused by acid-base transition of spinach chloroplasts. Proc. Natl. Acad. Sci. USA 55(1966): 170–177.

Junge, W. Half a century of molecular bioenergetics. Bioenergetics in mitochondria, bacteria and chloroplasts. Biochem. Soc. Trans. 41(2013): 1207–1218.

Junge, W. and Nelson, N. ATP synthase. Annu. Rev. Biochem. 84(2015): 631–657.

Keilin, D. and King, T.E. Reconstitution of the succinic oxidase system from soluble succinic dehydrogenase and a particulate cytochrome system preparation. Nature, Lond. 181(1958): 1520–1522.

Kell, D.B. On the functional proton current pathway of electron transport phosphorylation. Biochem. Biophys. Acta 549(1979): 55–99.

Kocherginsky, M. Acidic lipids, H^+-ATPases, and mechanism of oxidative phosphorylation. Physico-chemical ideas 30 years after P. Mitchell's Nobel Prize award. Prog. Biophys. Mol. Biol. 99(2009): 20–41.

Kresge, N., Simoni, R.D. and Hill, R.L. Unraveling the enzymology of oxidative phosphorylation: the work of Efraim Racker. J. Biol. Chem. 281(4)(2006): e4–e6.

Kümmel, R. The impact of entropy production and emission mitigation on economic growth. Entropy 18(2016): 75. doi:10.3390/e18030075.

Lane, N. Why are cells powered by proton gradients? Nature Education 3(2010): 18. doi:10.1038/46903.

Larkum, A.W.D. Contributions of Henrik Lundegårdh. Photosyn. Res. 76(2003): 105–110.

Lundegårdh, H. Absorption, transport and exudation of inorganic ions by the roots. Archiv. Bot. 32A(1945): 1–139.

Malmström, B. Mitchell saw the new vista, if not the details. Nature 403(2000): 356.

Medvedev, E.S. and Stuchebrukhov, A.A. Mechanism of long-range proton translocation along biological membranes. FEBS Lett. 587(2013): 345–349.

Mitchell, P. and Moyle, J. The cytochrome system in the plasma membrane of *Staphylococcus aureus*. Biochem. J. 64(1)(1956): 19P.

Mitchell, P. and Moyle, J. Group-translocation: A consequence of enzyme-catalysed group-transfer. Nature 182(1958a): 372–373.

Mitchell, P. and Moyle, J. Enzyme catalysis and group-translocation. Proc. R. Phys. Soc. Edinb. 27(1958b): 61–72.

Mitchell, P. Coupling of phosphorylation to electron and hydrogen transfer by a chemi-osmotic type of mechanism. Nature 191(1961a): 144–148.

Mitchell, P. Chemiosmotic coupling in oxidative and photosynthetic phosphorylation. Biochem. J. 79(3)(1961b): 23P–24P.

Mitchell, P. Chemiosmotic coupling in oxidative and photosynthetic phosphorylation. Biol. Rev. 41(1966): 445–502.

Mitchell, P. Vectorial chemiosmotic processes. Annu. Rev. Biochem. 46(1977a): 996–1005.

Mitchell, P. Epilogue: from energetic abstraction to biochemical mechanism. Symp. Soc. Gen. Microbiol. 27(1977b): 383–423.

Mitchell, P. A commentary on alternative hypotheses on protonic coupling in the membrane systems catalysing oxidative and photosynthetic phosphorylation. FEBS Lett. 78(1977c): 1–20.

Mitchell, P. Compartmentation and communication in living systems. Ligand conduction: a general catalytic principle in chemical, osmotic and chemiosmotic reaction systems. Eur. J. Biochem. 95(1979a): 1–20.

Mitchell, P. Keilin's respiratory chain concept and its chemiosmotic consequences. Science 206(1979b): 1148–1159. Nobel Lecture, 8 December, 1978.

Mitchell, P. Foundations of vectorial metabolism and osmochemistry. Biosci. Rep. 11(1991): 297–344.

Morange, M. What history tells us XI. The complex history of the chemiosmotic theory. J. Biosci. 32(2007): 1245–1250.

Mulkidjanian, A.Y., Cherepanov, D.A., Heberle, J. and Junge W. Proton transfer dynamics at membrane/water interface and mechanism of biological energy conversion. Biochemistry (Moscow) 70(2005): 251–256.

Mulkidjanian, A.Y. Proton in the well and through the desolvation barrier. Biochim. Biophys. Acta 1757(2006): 415–427.

Nath, S. Two-ion theory of energy coupling in ATP synthesis rectifies a fundamental flaw in the governing equations of the chemiosmotic theory. Biophys. Chem. 230(2017): 45–52.

Nath, S. Integration of demand and supply sides in the ATP energy economics of cells. Biophys. Chem. 252(2019): 106208. doi: 10.1016/j.bpc.2019.106208.

Nath, S. Molecular-level understanding of biological energy coupling and transduction: Response to "Chemiosmotic misunderstandings." Biophys. Chem. 268(2021): 106496. doi:10.1016/j.bpc.2020.106496.

Orgel, L.E. Are you serious, Dr Mitchell? Nature 402(1999): 17.

Onsager, L. Reciprocal relations in irreversible processes. I. Phys. Rev. 37(1931a): 405–426.

Onsager, L. Reciprocal relations in irreversible processes. II. Phys. Rev. 38(1931b): 2265–2279.

Overton, E. Beiträge zur allgemeinen Muskel- und Nervenphysiologie. Pfluger's Arch. ges. Physiol. 92(1902): 115–280.

Parsegian, A. Energy of an ion crossing a low dielectric membrane: solutions to four relevant electrostatic problems. Nature 221(1969): 844–846.

Popper, K. 2002. The Logic of Scientific Discovery (from 1934 German original). Routledge, London, GB.

Prebble, J.N. Contrasting approaches to a biological problem: Paul Boyer, Peter Mitchell and the mechanism of the ATP synthase, 1961–1985. J. Hist. Biol. 46(2013): 699–737.

Prigogine, I. and Lefever, R. 1973. Theory of dissipative structures. pp. 124–135. *In*: Haken, H. (ed.). Synergetics. Vieweg+Teubner Verlag, Wiesbaden, Germany.

Racker, E. and Stoeckenius, W. Reconstitution of purple membrane vesicles catalysing light-driven proton uptake and adenosine triphosphate formation. J. Biol. Chem. 249(1974): 662–663.

Rich, P.R. A perspective on Peter Mitchell and the chemiosmotic theory. J. Bioenerg. Biomembr. 40(2008): 407–410.
Singer, S.J. and Nicolson, G.L. The fluid mosaic model of the structure of cell membranes. Science 175(1972): 720–731.
Skulachev, V.P. Membrane-linked energy transductions. Bioenergetic functions of sodium: H^+ is not unique as a coupling ion. Eur. J. Biochem. 151(1985): 199–208.
Skulachev, V.P. Chemiosmotic systems in bioenergetics. Biosci. Reps. 11(1991): 387–444.
Skulachev, V.P. The laws of cell energetics. Eur. J. Biochem. 208(1992): 203–209.
Skulachev, V.P., Bogachev, V. and Kasparinski, F.O. 2012. Principles of Bioenergetics. Springer, Berlin, Germany.
Slater, E.C. Mechanism of phosphorylation in the respiratory chain. Nature 172(1953): 975–978.
Slater, E.C. Peter Dennis Mitchell 1920–1992. Biogr. Mems Fell. R. Soc. 40(1994): 283–305.
Thompson, A.J. The science of RJP Williams. J. Biol. Inorg. Chem. 21(2016): 1–3.
Xiong, J.-W., Zhu, L., Jiao, X. and Liu, S.-S. Evidence for ΔpH surface component (ΔpH^S) of proton motive force in ATP synthesis of mitochondria. Biochim. Biophys. Acta 1800(2010): 213–222.
Weber, B.H. Glynn and the conceptual development: A retrospective and prospective view. Bioscience Rep. 11(1991): 578–617.
Westerhoff, H.V. and van Dam, K. 1987. Thermodynamics and Control of Biological Free Energy Transduction. Elsevier, Amsterdam, The Netherlands.
Westerhoff, H.V., Juretić, D., Hendler, R.W. and Zasloff, M. Magainins and the disruption of membrane-linked free-energy transduction. Proc. Natl. Acad. Sci. USA 86(1989): 6597–6601.
Williams, R.J.P. Possible functions of chains of catalysts. J. Theor. Biol. 1(1961): 1–17.
Williams, R.J.P. The multifarious couplings of energy transduction. Biochim. Biophys. Acta 505(1978): 1–44.
Williams, R.J.P. Energizing protons in membranes. Nature 338(1989): 709–711.

Chapter 2

Membrane Bioenergetics in a Nutshell

2.1 The importance of membranes and membrane proteins

The often-repeated dictum that nucleic acids (DNA, RNA) are the very foundation of life as we know it neglects to mention that even DNA or RNA in viruses need proteins to enter cells and to become alive inside a cell. The same is true for DNA found in cells and organelles, which can be considered as a memory cytoskeleton that becomes alive only when enveloped, nursed, manipulated, and stimulated with proteins, the cell's muscleman. Cellular membranes are not only protective envelopes and permeability barriers, but also the free-energy sources and almost intelligent signal transfer devices. Membranes are unique among other cellular biomacromolecules in having a noncovalent association of their constitutive lipids and proteins, making them extremely dynamic, flexible, and fast in responding to external signals or challenges.

Phospholipids are the main constituents of all biological membranes. All have two fatty acids of variable length and flexibility, and all have a polar head group. Phospholipids, like lecithin (phosphatidylcholine), spontaneously form bilayer structures when mixed with a water solution. Membrane bilayer has polar head groups of each lipid layer oriented toward the water solution, while fatty acid chains are oriented toward each other to form a water-impermeable lipid interior of the bilayer. Some proteins with enough hydrophobic amino acid segments easily enter within the bilayer, even to the extent that they are major membrane constituents per weight basis (phospholipids are major constituents in terms of their numbers). Membrane proteins are so large compared to phospholipids that many protrude across the whole lipid bilayer and have the majority of their amino acids outside one side, another side, or outside both sides of bilayer surfaces.

When we focus on function, the important function of membranes is the ability to compartmentalize cellular components, including small molecules, ions, and even water. This feat is achieved by the hydrophobic effect keeping together a wide variety of membrane lipids in noncovalent association forming bilayer structure, which is impermeable to protons, other ions, and other macromolecules. Only some hydrophobic peptides, lipids, and integral membrane proteins are welcome to enter the membrane and become membrane constituents.

The passage of protons, different ions, and other small molecules' cell needs to transport is cleverly organized by specialized membrane proteins and their association in a two-dimensional membrane envelope that forms the 3D topologically closed structure. This membrane function is so important that the membrane surface in eukaryotic cells is more than a thousand times higher (probably many million times higher for nerve cells) than in prokaryotic cells. As we mentioned before, it is needed not only for free-energy transduction but also for communication with other cells in multicellular organisms. For instance, extremely low calcium ion concentration is maintained inside the cytoplasm of most animal cells (around 0.1 micromolar). A small decrease in the absolute value of membrane diffusion potential activates voltage-dependent calcium channels. When calcium channels open during a short time interval, just a few calcium ions entering the cell are already enough to signal a cascade of far-reaching changes in cell behavior. All membrane channels are tightly regulated. External (toxins or antibiotics) or internal causes (**apoptosis**) can break the regulation of cation channels. The failure in membrane topological closure and cell death quickly follows (Voskuil et al. 2018).

2.2 Membrane proton pumps

For membrane bioenergetics, an active protons transfer through the membrane is even more important. An active transport process is performed by integral membrane proteins that are considered primary proton pumps. The cytochrome c oxidase is such a primary proton pump. It uses internal protons not only to synthesize water but also to eject them into the external environment against the direction of a transmembrane electric field (pointing to cytoplasm) and against the direction of a proton concentration gradient (Wikström and Sharma 2018). The energy source for this uphill proton transfer (thermodynamically speaking) is respiration because cytochrome c oxidase is the last protein in the respiratory chain enzymes located in the inner mitochondrial membrane and connected through multiple **oxidation-reduction** reactions and electron transport.

Ejected protons from mitochondria mostly stay close to the external membrane surface and are driven by an inside-oriented electrochemical gradient. They take the first opportunity to return to the matrix space. This opportunity is provided by the ATP synthase, which can be considered as the secondary proton pump. In the ATP synthesis mode, protons' re-entrance through ATP synthase starts its rotor rotation and brings previously attached ADP and inorganic phosphate in such proximity protected from water molecules that ATP synthesis occurs.

2.3 Energy transducing membranes

When one of the referees was reading my book on bioenergetics (1997, see Chapter 11) for the first time, his most important new insight was that by far the largest proportion of ATP synthesis is associated with membrane-bound enzyme complexes. Such complexes in the inner membrane of mitochondria and the thylakoid membrane of chloroplasts are responsible for respiration and photosynthesis, respectively. Such specialized membranes are "energy-transducing" membranes.

The bioenergetic function (free-energy transduction) is probably the most important function that membranes developed early in the evolution of life. The use of topologically closed membranes for free-energy transduction is just as universal and evolutionary old invention as the genetic code itself (Lane and Martin 2012, Lane 2017). Membrane bioenergetics enables creating and harnessing ion gradients through the chemiosmotic coupling. A single *Escherichia coli* cell turns over about 50 billion ATP molecules per division, which amounts to about 50 times its mass. While division (replication) is essential to life, it is only a side reaction when the cell's overall bioenergetics is considered.

Prokaryotic cells use their plasma membrane for free-energy transduction, but even in their case, specialized membrane patches exist with protein complexes devoted to particular bioenergetic function. One example is the thylakoid membrane performing photosynthesis in blue-green algae. Another example is the purple membrane from *Halobacterium salinarum* with tightly packed bacteriorhodopsin proton pumps activated by light. Present-day energy-transducing membranes in organelles of eukaryotic cells have evolved from a symbiotic relationship with primitive progenitors of mitochondria and chloroplast.

As mentioned before, the common goal of free-energy transducing membranes is ATP synthesis. All of the various membranes from different life domains are sufficiently related to form the core of classical membrane bioenergetics no matter what primary free-energy sources are used. Let us enumerate common features of these membranes even if we mentioned them before. Each energy-transducing membrane has integral membrane proteins that can be classified as primary or secondary proton pumps. The nature of primary proton pumps depends on the free-energy source. If small organic molecules are primary free-energy sources, then respiratory chain complexes catalyze the "downhill" transfer of electrons from substrates to final acceptors such as O_2. These **redox** reactions are used to create an electrochemical gradient of protons, which is called the protonmotive force. Photosynthetic bacteria and chloroplast use photon free-energy to drive "uphill" electron transport from water to acceptors such as nicotinamide adenine dinucleotide phosphate ($NADP^+$). Photosynthetic electron transport is also connected to "uphill" proton transport and the creation of a protonmotive force that can be used for ATP synthesis. The creation of protonmotive force is common to respiration and photosynthesis, but it can only happen if all proton pumps are embedded in topologically closed membrane impermeable to protons except through specialized transporters and secondary proton pumps such as the ATP synthase. For the ATP synthase, we say that it is the secondary proton pump because it can work both in the ATP synthesis mode when driven by a strong protonmotive force and in the proton active-transport ("uphill") mode when driven by an excess of ATP. Due to its importance, we should not be surprised with an exceedingly old evolutionary origin, conservation of its structure and function, and wide distribution of ATP synthases in practically all living cells.

2.4 Primary and secondary proton pumps

The essence of Peter Mitchell's chemiosmotic theory is that the primary proton pumps generate strong enough proton electrochemical gradient to force the secondary proton pump (ATP synthase) to work in the ATP synthesis mode. Let us clarify what an inside or negative side (N) is and what is outside or positive side (P). The negative side N is the side from which protons originate and are being pumped out of it, while the positive side P is the side to which protons are pumped. In the absence of protonmotive force, the ATP synthase would act solely as the primary proton pump in the ATP hydrolysis mode. In practice, for the ATP synthase to act continuously as the secondary proton pump (in the ATP synthesis mode), a proton circuit has to be established with protons linking primary and secondary proton pumps and with primary proton pumps being able to continuously use input free-energy packages. This is what occurs during photosynthesis or respiration when the gradient of protons is continuously replenished by the respiratory or photosynthetic electron-transfer chains, while cytoplasmic ATP is not accumulated to block further ATP synthesis, but is continuously being removed by cytoplasmic ATP-consuming reactions.

The basic postulate of Mitchell's theory, amply confirmed in numerous experiments, is that metabolism (respiratory or photophosphorylation) is tightly coupled to proton translocation: one cannot occur without the other. While electron circuits are basic ingredients of human technology, proton circuits connected to electron transport are basic ingredients of bioenergetics.

2.5 A bridge to the universe must not be blocked

From a more general viewpoint, a hierarchy of free-energy transduction steps starts with some initial microscopic and localized events, such as the photon absorption by the chlorophyll molecule. The excited surface and volume increase in ever slower and smaller steps until ultimately, many infrared photons are radiated away in the universe. A distant observer may guess that the biosphere exists after observing unusually high oxygen concentration and entropy production of the planet (Sagan et al. 1993). By the way, this insight into the main contribution of our home planet to the universe, the entropy increase, is another illustration of real meaning of mysterious quantity named entropy. Entropy, as a measure for a disorder, fails to explain why the universe is more disordered due to the Earth-Sun interaction. As a measure for energy dispersal, entropy increase easily explains how concentrated free-energy packages coming to the Earth from the Sun end up dispersed into a much larger number of low-energy photons, which Earth radiates everywhere as heat.

Connecting civilization progress with fossil fuel burning created an additional carbon dioxide and methane shield in the atmosphere, which destroyed the almost perfect balance between light absorption from Sun and heat release. Keeping for ourselves the useless energy turned out to be a bad idea leading to global warming with likely catastrophic consequences. "Energy production" or even "clean energy production" is common nonsensical terminology because energy cannot be produced or destroyed, and it is impossible to convert it into work with 100% efficiency without producing entropy. Unfortunately, increased entropy due to fossil fuel burning is not at all an immeasurable and unfathomable quantity. Its global effects, like warming, ocean acidification, sea rise, and irreversible melting of massive glaciers worldwide, are slow but irreversible. In later chapters, we shall see that all living cells and organisms persist by fine-tuning their energy uptake to internal entropy production and entropy release in the environment. A failure to control energy flows and entropy generation spell death. One would think that *homo sapiens*' civilization is intelligent enough to understand this simple law, which every living organism must obey.

Definitions and explanations from Chapter 2:

Apoptosis is an ordered process of programmed cell death in multicellular organisms. It is a highly regulated and irreversible chain of events triggered by cell stress or external signals.

Oxidation reactions involve the loss of electrons.

Redox term stands for the reduction-oxidation type of a chemical reaction.

Reduction reactions involve the gain of electrons.

References

Lane, N. and Martin, W.F. The origin of membrane bioenergetics. Cell 151(2012): 1406–1416.

Lane, N. Proton gradients at the origin of life. Bioessays 39(2017). doi: 10.1002/bies.201600217.

Sagan, C., Thompson, W.R., Carlson, R., Gurnett, D. and Hord, C. A search for life on Earth from the Galileo spacecraft. Nature 365(1993): 715–721.

Voskuil, M.I., Covey, C.R. and Walter, N.D. Antibiotic lethality and membrane bioenergetics. Adv. Microb. Physiol. 73(2018): 77–122.

Wikström, M. and Sharma, V. Proton pumping by cytochrome c oxidase—A 40 year anniversary. BBA—Bioenergetics 1859(2018): 652–698.

Chapter 3

Irreversible Thermodynamics and Coupled Biochemical Reactions

3.1 Entropy concepts and Onsager-Prigogine's description of driving and driven force-flux couple

In this chapter, we shall see that force-flux product has several important physical interpretations that are relevant for bioenergetic transformations. One of them is dissipative entropy change. The word entropy was coined by German physicists Rudolf Clausius (1822–1888) using the word *trope* (a turning point, transformation) and prefix *en-* (in, within) from the ancient Greek language. Dissipation results from an entropy-increasing irreversible process, such as transforming mechanical energy into heat due to friction. In an ideal case of a completely reversible process, entropy is the state function (depending only on the current state of the system) defined as

$$dS = đQ/T$$

where đQ is an infinitesimal exchanged heat, and T is the **absolute temperature**. For a given system going through a cyclic thermal reversible process, Clausius concluded that the integral over the exchanged heat must always vanish. In particular, entropy does not change during one reversible cycle. For all other (real) processes in a macroscopic world, entropy increases in an isolated system until its maximal value is reached in the thermodynamic equilibrium (The Second Law of Thermodynamics). There is no system ideally isolated from all long-range interactions in our universe, unable to exchange energy and matter with its environment. However, idealizations and simplifications are the trademarks of physics. Second Law violations do happen in the microworld when only a few particles are considered as a system, but in the world of classical macroscopic physics, it is the fundamental law.

Macroscopic systems are complex because they contain at least as many interacting particles as there are stars in our galaxy. The thermodynamics laws should be derivable from consideration of microscopic dynamics if we can reduce the complexity by selecting only a few parameters that are most useful in predicting macroscopic system properties. That is the job of statistical physics. For an isolated system, the total number of particles N, energy E, and volume V is still insufficient for its specification if the system is not in equilibrium. The missing state function is entropy. Equilibrium corresponds to the maximum entropy value. Fortunately, we can count the number of microstates comprising a macrostate with all four **extensive** properties by assuming equal *a priori* probabilities for all microstates. The N-particle distribution function, named the statistical weight Ω, is then obtained.

Austrian physicist Ludwig Boltzmann (1844–1906) proposed the statistical definition for entropy:

$$S_B = k_B ln\, \Omega$$

where $k_B \equiv 1.38 \cdot 10^{-23}$ joule/degree is Boltzmann's constant. For instance, S_B is the entropy of an isolated ideal fluid at thermodynamic equilibrium, with uniform density and temperature, but without interparticle forces (Jaynes 1965). Each particle within a fluid has the position and momentum coordinates in six-dimensional **phase space**. Thus, Boltzmann's entropy is essentially the logarithm of phase volume associated with a macroscopic state. The proportionality constant k_B provides a remarkable connection between the statistical entropy and the classical thermodynamic entropy of Clausius. Its value was first determined by German physicist Max Planck (1858–1947) when he derived the law of black body radiation (Planck 1901). Planck was also the first to publish the equation $S = k_B \cdot ln\Omega$ in 1906. However, he was inspired by Boltzmann, and graciously gave appreciation to him (Feistel 2019). The equation is inscribed on Boltzmann's tombstone as $S = k \cdot ln\Omega$. Planck's formula counts the discrete number of microstates. Thus, it is perfectly suitable for the quantization procedure introduced by Max Planck and for quantum statistics.

A more general entropy definition from statistical thermodynamics reads:

$$S_{GB} = -k \sum_{j,N} p_j(N) ln\, p_j(N)$$

N is the number of state components (atoms, molecules), p_j are probabilities of observing a given microstate, and GB index for entropy expression means that Willard Gibbs generalized the statistical interpretation of entropy attributed to Boltzmann. When fluctuations are possible, N is the mean value. The validity of this equation is not restricted to isolated systems and near equilibrium situations (De Groot and Mazur 1984). Also, the probability concept goes beyond classical mechanics and beyond reversible time-invariant processes. When the proportionality constant k is not the Boltzmann's constant k_B but is equal to +1, Shannon's information entropy can be defined, which has the same mathematical form:

$$S_I = -\sum_{i=1}^{n} p_i\, ln\, p_i$$

where n is the number of discrete set units with the probability p_i for each set element.

The S_{GB} entropy allows for irreversible relaxations through changes in microstates' occupation density. All energy forms contribute to S_{GB}, including the potential energy and interparticle interactions in the gas and condensed phase. S_{GB} becomes equal to S_B in the thermostatic limit. The macrostate of maximum Gibb's entropy is the thermal equilibrium characterized by a small number of physical constraints: energy, volume, and mole numbers. Modern chemical thermodynamics is based on that maximum principle. Williard Gibbs (1839–1903) was a mathematical physicist from the USA with a clear preference for applications-based theory. He declared once: "The pure mathematician can do what he pleases, but the applied mathematician must be at least partially sane" (Kline 1980, p 285). This witticism does not do justice to many mathematical discoveries that unexpectedly found interesting physics applications. Importantly, Gibbs's insights from the equilibrium theory enabled the generalization to nonequilibrium statistical mechanics.

Jaynes and other authors (Jaynes 2003) initiated applications of the principle of maximum entropy inside and outside physics. In Jaynes's extension of Gibbs' mathematical methods, the $S = k_B \cdot ln\Omega$ expression applies to arbitrary nonequilibrium states as a generalized entropy that can only increase in reproducible experiments (Jaynes 1965). Chapter 16, section 16.17, considers different entropy definitions in more detail. For now, we shall focus on entropy changes when a system is not isolated but can exchange energy and matter with its surroundings.

My favorite pathway to generalize entropy definition for bioenergetic applications is to consider open systems from the beginning and explicitly allow for entropy changes due to irreversible processes. In an open system, entropy change can be divided into two parts:

$$dS = d_i S + d_e S$$

where the first term on the right-hand side represents the entropy change due to irreversible processes in the system, while the second term represents entropy change due to entropy exchange with the system environment. Entropy production:

$$P = \frac{d_i S}{dt}$$

has a central role in the thermodynamics of irreversible processes. The inequality $P \geq 0$ is equivalent to the Second Law of Thermodynamics. In a steady nonequilibrium state, the system cannot avoid irreversible processes and corresponding entropy rise. Consequently, it can maintain its entropy ($dS = 0$) only by feeding itself with negative entropy change $d_e S$.

The central rule of the thermodynamics of irreversible processes is that entropy production of the system can always be represented as the sum of products of generalized forces X and corresponding fluxes J:

$$TP = \sum_k J_k X_k$$

where J_k are thermodynamic fluxes such as heat flux, electric current, chemical reaction rate, while corresponding forces are the temperature gradient, electromotive force, and chemical affinity. The TP product is the dissipation function: $\Phi = TP$, also called the dissipated power. For the example of transmembrane ion transport through a topologically closed membrane, the driving force has both electrical (membrane potential) and chemical component (ion concentration difference between two membrane sides).

The bilinear form $J_k X_k$ (sum implied) depends on thermodynamics (forces) and kinetics (flows). It connects causes and consequences, external forcing, and internal responses. As the cornerstone for the several chapters in this book, we should mention that the bilinear expression for dissipation is easily derived from fundamental Gibbs' equation connecting changes in entropy, internal energy, volume, and mass fractions (Prigogine 1967, De Groot and Mazur 1984, Essex 1984). For an open nonequilibrium system, there are interactions with the exterior. Prigogine took it into account by splitting a total entropy change into entropy exchange $d_e S$ through the system's boundary and a positive contribution $d_i S$ of irreversible internal processes. The external gradient of some **intensive** variable (temperature, chemical potential) is responsible for internal entropy production. In bioenergetics, words internal and external refer respectively to the system and its environment. The system can be composed of protein macromolecules, organelles, cells, organisms, or ecological systems, while corresponding environments are cytoplasm (for proteins and organelles) and extracellular biological or inanimate ambient (for remaining systems). The bilinear form for entropy production includes internal flows giving rise to irreversible entropy generation. At the same time, it includes all internal and external gradients of intensive variables for which we adopted a common name of generalized forces. For nonequilibrium steady-state situations of interest in bioenergetics, only the external forces can maintain such states outside equilibrium. External forces provide an obligatory power input for a persistent nonequilibrium state. Thus, an alternative (to useless waste) interpretation of entropy production is the power input from the external forces (Bruers et al. 2007).

In the presence of chemical reactions, the affinity or the chemical potential is a force causing chemical work delivery. An action integral also comes into play when entropy production is associated with a particular space-time pathway. Only an antisymmetric (irreversible) part of the action is then considered (see more about that topic in section 16.18). After Schnakeberg's (1976) and Hill's initiative (1977), the bilinear form for entropy production became essential in mapping between microscopic and macroscopic thermodynamics (Tomé and de Oliveira 2012, 2018). The violation of the bilinearity in thermodynamics, as a rare exception, was found by Essex (1984), when nonequilibrium radiation processes are taken into account. In any case, the canonical and almost universal identification of dissipation with the bilinear expression $J_k X_k$ is the starting point and a mandatory condition for the formulation of minimum (Prigogine 1977) and maximum (Martyushev 2010) entropy production situations.

There is one serious drawback of associating force-flux products with terms entropy generation, entropy production, or dissipation. In general, both forces and fluxes evolve with time, meaning that their $J_k X_k$ product must also change over time. Of course, stationary states with constant dissipation can be established and maintained by artificial means, and we do it daily by switching on the power supply. The current surge happens quickly then but not without any time delay, however short. Corresponding voltage-current product jumps from zero to some constant value and disappears again after the power switch is turned off. We use the interval for performing the desired function, such as washing dishes in a dishwasher. When output work is gained from "useless" dissipation, one can wonder how appropriate are dissipation and entropy production terms for the input force-flux products. Obviously, different work outputs can be obtained from a seemingly useless irreversible entropy increase. That confusion arises from a premature assignment of the entropy production terminology to the free energy influx multiplied by the internal flux response. Any living cell is no less intelligent than we are in channeling the input power into various self-serving functions, including energy storage, movement, biosynthesis, and replication. Cells also use the input energy flux to establish metastable steady states and nearly constant internal conditions in the face of potentially deleterious ambient changes.

It would be better to use the name potential entropy production, potential dissipation, or power influx for the force-flux product to take into account its full evolutionary potential. The "potential entropy" terminology has been used by Feistel (2019) to describe the amount of entropy that can be produced by future dissipation. In this book, we shall retain the traditional *entropy production* title for the $J_k X_k$ product while bearing in mind its inherent ability to evolve, transfer power, perform work, and store energy. In contrast, microscopic entropy definitions, such as Boltzmann's or others (see Chapter 16, section 16.17), do not imply any time evolution in the prescription that microstates number must be counted to find its value. The highest number of microstates in the thermodynamic equilibrium is in accord with the Second Law. Still, nothing can be declared about how fast the equilibrium is reached or how gradients of intensive variables can be dissipated to reach the equilibrium. Far outside thermodynamic equilibrium, dissipative flows can self-arrange into intricate patterns. The root cause for dissipative self-organization in the system interior, intertwined with the exported work and entropy generation output, is always the same: dissipation of external gradients. Hence, there is no alternative in the thermodynamic study of life-like phenomena but to focus on internal fluxes and conjugated forces (gradients of intensive variables) driving these fluxes.

Such an approach does not depend on the choice of entropy definition and in-built constraints for a chosen definition. Summing some or all flux-force products is essential for the study of persistent far-from-equilibrium states that evolved to maintain dissipative **homeostasis** in biological systems. The usual assumption is that flux-force products are additive **inner products**, each described as an additional entropy production. The assumption does not consider why the avalanche of response forces and fluxes originated, how it is structured, and the work performance potential for some of the driven fluxes. Not all of the considered flux-force products are wasted energy and lost work.

The bilinear expression $J_k X_k$ does not imply a linear connection between fluxes and forces. The opposite is true for enzymatic catalysis with highly nonlinear (exponential) relationships between affinity and substrate to product conversion rate. However, there are many well-known examples from physics for linear force-flux relationships. When noticed in experiments, such empirical relationships were named laws. For instance, these are Ohm's law for electrical currents, Fourier's law for heat flow, and Fick's law for diffusion. Corresponding forces are (in the same order) electromotive force, temperature gradient, and chemical affinity. Linearity is only approximate, but this approximation is pretty good near thermodynamic and **chemical equilibrium** for small gradients of some intensive parameters (the parameters like temperature that do not depend on the number of particles in the system). For strong forces, there is no longer a linear response. Then, life becomes really interesting and much more difficult to describe or predict. As a rule of thumb, the linear response appears when force X is considerably smaller than the thermal energy k_BT of a molecule (k_B is the Boltzmann's constant). That is not the case for chemical reactions. In chemistry and biochemistry, the reaction velocity is a nonlinear function of corresponding affinity (exponential function). That is one reason why intelligent-like behavior can be seen under a microscope, even for the simplest cells. Another reason is the mutual interaction of different processes. Let us examine how it happens first for the simple case of the linear relationship between forces X_1, X_2 and corresponding fluxes J_1, J_2:

$$J_1 = L_{11}X_1 + L_{12}X_2$$

$$J_2 = L_{21}X_1 + L_{22}X_2$$

where L_{ij} are phenomenological coefficients. An interesting conclusion from these relationships is that both forces can influence chosen flux, not only the force conjugated with that flux. For instance, the chemical potential gradient and corresponding diffusion flux can occur due to a temperature gradient (the Soret effect), and heat flux can occur due to a chemical potential gradient (the Dufour effect) (Mortimer and Eyring 1980). Another interesting and important finding by Lars Onsager in 1931 (see Chapter 1) is that the matrix of L_{ij} coefficients is symmetrical, namely that $L_{ij} = L_{ji}$ (the Onsager reciprocal relations).

The energy transduction is possible when two irreversible processes are coupled together. For the driving process 1, the input power or energy source is the product J_1X_1, while the output power J_2X_2 for the driven process 2 can be negative. The Second Law of Thermodynamics ensures that the output power (its absolute value) can never be greater than the input power. It is natural then to define the efficiency of free-energy transduction (or power transduction) as:

$$\eta = -\frac{J_2X_2}{J_1X_1}$$

Notice that η is the product of flux ratio J_2/J_1 with force ratio X_2/X_1. Since the output flux and output force are always smaller in the absolute value than the input flux and input force, these two ratios are also efficiencies. There is some confusion with respect to their names. We shall be using the name free-energy storage efficiency or thermodynamic efficiency for the X_2/X_1 ratio and flux transfer efficiency or flux yield for the J_2/J_1 ratio. From the definition of free-energy transduction efficiency, it is obvious that two **steady states** exist when output power vanishes together with the η. The first such state is established when $X_2 = 0$, that is, for vanishing free-energy storage efficiency. Energy is being constantly spent maintaining such nonequilibrium steady state. Still no useful work is ever obtained because it is the level flow (maximal secondary flux is achieved) short-circuited state. The other such state is established when $J_2 = 0$, that is, for vanishing flux transfer efficiency. It is the static head state with maximal (negative) secondary force $X_2 = \max < 0$, but zero efficiency of free-energy transduction. The driving force is constantly producing entropy in this state too, but also in vain, without creating any output power. Energy transduction has been blocked both in the level flow and in the static head state.

Ilya Prigogine (1967) noticed that static head state could be characterized with the minimum in entropy production, and this observation is today regarded as Prigogine's minimum entropy production theorem. It is easy to prove by performing the derivation of the dissipation function with respect to the secondary force X_2:

$$\frac{\partial \Phi}{\partial X_2} = L_{12}X_1 + L_{21}X_1 + 2L_{22}X_2 = 2J_2 = 0$$

High school mathematics is enough to verify that this is indeed the minimum for the entropy production (but different from zero as long as the primary force is different from zero). Notice that we have used Onsager relationships here ($L_{ij} = L_{ji}$).

Some well-known biochemists used this theorem to support their claim that each living cell, being unable to avoid entropy production, is at least choosing the lesser evil of ensuring minimal entropy production. Thus, erroneous claim entered into some biochemistry textbooks that static head state produces a maximum amount of useful work associated with maximal efficiency. Connecting maximal efficiency with minimal entropy production is suspect on several accounts. First, it is based on near-equilibrium approximation, but it breaks down at equilibrium when both entropy production and efficiency for any energy transformation disappear. Second, the static head state blocks any further free-energy transduction, bringing cellular metabolism and all movements to a standstill. Third, mixing different efficiency definitions, even by experts, introduced confusion and facilitated wrong interpretations. The free-energy storage efficiency X_2/X_1 is regularly called the thermodynamic efficiency. Readers are then left with the impression that authors are speaking about the efficiency of free-energy conversion or transduction $-(J_2X_2)/(J_1X_1)$, but that is not the case. The thermodynamic efficiency can become maximal for vanishing energy transduction efficiency and even negative for high enough membrane potential (Kim and Hummer 2012), which simply indicates that there is no net product flux or that some force-flux couple is not taken into account.

A few words are in order about units we used for entropy production in this book. As a physical state function, entropy is expressed in energy units (Joules) divided by the temperature (K for kelvins). An irreversible entropy change during some time (entropy production) requires dividing entropy units with time units (seconds). Entropy production density is taken into account when everything is additionally divided with moles, volume, or surface units (mol, m^3, or m^2). Figures presented in Chapters 8, 11, and 12 use J/(mol·K·s) units. In Chapter 10 figures, we used the dissipation density per molar RT or molecular thermal energy k_BT. Since dissipation $\Phi = TP$ and gas constant $R = N_Ak_B$ (N_A = Avogadro constant), only inverse seconds remained as relative dissipation density units. Note that constant temperature T is assumed in all of our optimizations presented in this book. The consequence is that estimated optimal parameters do not differ for extremums in entropy production and in dissipation function.

3.2 Far from equilibrium static head state analogy, slip coefficients and an effective degree of coupling

The static head state has been defined for near-equilibrium linear force-flux relationships when J_2 vanishes. The condition of zero net secondary flux J_2 can be used to define the nonlinear analogy of the static head state arbitrarily far from equilibrium (Juretić 1992) with a low but not necessarily minimum free-energy dissipation at that state (Juretić and Westerhoff 1987). Hill's diagram method is convenient to model simpler nonlinear transducers. Let us suppose that we have a kinetic model with two cycles, I and II joined in an overall cycle III with a slip transition "s". Figure 3.1 illustrates

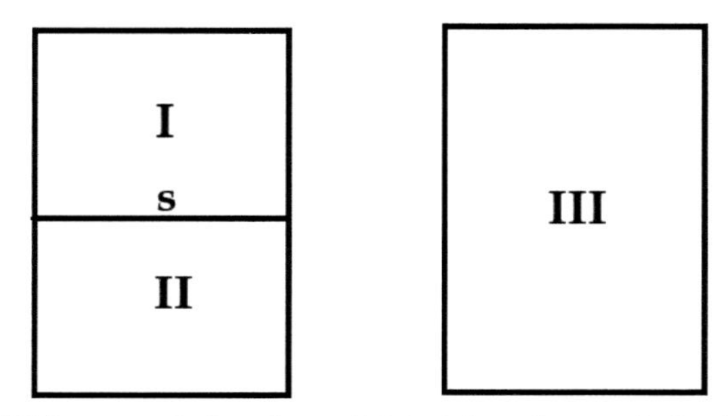

Figure 3.1: Two cycles I and II for, respectively, primary (driving) force-flux couple I and secondary (driven) force-flux couple II. The slip transition "s" connects them in the overall cycle III.

one possibility of how the catalytic cycle of the same enzyme can connect the driven cycle II with the driving cycle I. Both simpler and more complex diagrams (having more than six functional states) have been used by Terrel Hill (Hill 1977) to expose the basic principles of steady-state free-energy transduction far from equilibrium.

Fluxes have an exponential dependence on corresponding forces. For instance, the proton flux across the membrane has an exponential dependence on the electrochemical proton gradient. From $J_2 = 0$ condition, one can easily find the maximal secondary force as the free energy ΔG_2 stored in a forward static head state:

$$\Delta G_2(s_2) = RT \cdot \ln((s_2+1)/(s_2+\exp(\Delta G_1/RT)))$$

where s_2 is the forward slip coefficient. For all kinetic schemes containing the driving cycle I, driven cycle II, the slip connection between these cycles, and overall cycle III (as illustrated in Figure 3.1 for the simple case), the slip coefficient s_2 is:

$$s_2 = (\Sigma_{II}\, \Pi_{II})/(\Sigma_{III}\Pi_{III})$$

where Π_{II} and Π_{III} are the products of rate constants in the backward direction of pumping in a driven cycle II and overall cycle III, while Σ_{II} and Σ_{III} are, respectively, sum of appendages (products of rate constants) feeding into cycle II and cycle III. When it is possible to reverse the roles of driving and driven force, the condition $J_1 = 0$ can be used to find the maximal free energy stored in the reverse static head state:

$$\Delta G_1(s_1) = RT \cdot \ln\, ((s_1+1)/(s_1+\exp(\Delta G_2/RT)))$$

where s_1 is the reverse slip coefficient for the case when cycle I is the driven cycle:

$$s_1 = (\Sigma_I\, \Pi_I)/(\Sigma_{III}\Pi_{III})$$

To make it less abstract, a reader can have in mind how ATP-synthase can work either in the direction of ATP synthesis or in the direction of active proton transport (proton pumping mode) (see Chapter 11 for details about ATP-synthase structure and function). In the former case, the forward static head state is established when net ATP synthesis vanishes because the maximal possible nonequilibrium situation is reached for the ATP/ADP concentrations. In the latter case, the reverse static head state is established with vanishing net proton flux across the membrane and maximal possible electrochemical proton gradient.

With known rate constants and steady-state concentrations of substrates and products, an effective degree of coupling can be defined in the nonlinear range:

$$q = \{[1 + s_2 \cdot \exp(-\Delta G_1/RT)][1 + s_1 \cdot \exp(-\Delta G_2/RT)]\}^{-1/2}$$

All of the above nonlinear expressions are reduced to corresponding linear expressions in the limit of small forces. Forward and reverse slip coefficients are then:

$$s_2 = L_{22}/L_{12} - 1 \qquad s_1 = L_{11}/L_{21} - 1$$

while the degree of coupling depends only on slip coefficients (or L_{ij} coefficients):

$$q = [(s_2 + 1)(s_1 + 1)]^{-1/2} = L_{12}/(L_{11}L_{22})^{1/2}$$

In the linear range, knowing the degree of coupling is enough for finding the maximal possible efficiency:

$$\eta_{max} = q^2/[1+(1-q^2)^{1/2}]^2$$

An abrupt maximal efficiency decrease takes place when coupling diminishes. Thus, the increase in slip and a concomitant decrease in q may seem unsatisfactory for biological processes. Conversely, if slip is decreased to zero, a perfect coupling is achieved with maximal yield in free energy transfer.

For a nonlinear case, it is also generally true that the increase in slip causes a decrease in the degree of coupling and free-energy transduction efficiency. However, there are several advantages when slippage or a leak in parallel with a pump exists in coupled nonlinear flux-force relationships, as is the case for some membrane enzymes and enzymatic complexes in bioenergetics. First, maximal efficiency is always higher in a non-linear case when slip exists (Pietrobon and Caplan 1985). Second, the flux control coefficient (Westerhoff and van Dam 1987) can increase with increased slippage, and we know that tight control over ATP production is crucial for cell survival (Juretić 1992). Third, when free-energy dissipation is substantially increased, for instance, due to moderately increased slip, the proton pumps can operate at higher overall efficiency and are better regulated (Juretić and Westerhoff 1987). In the case of ATP-synthase, optimal metabolic control for product (ATP) synthesis is reached for maximal entropy production in the ATP synthesis/hydrolysis transition (Dewar et al. 2006, see Chapter 11 for details). Fourth, together with leaks, the slippage of proton pumps acts as a "safety valve," which ensures that membrane potential is never too high (Pietrobon and Caplan 1985). The membrane potential component of the electrochemical proton gradient is essential for all cellular free-energy transformations, but it must be tightly regulated in a certain range. When it is outside that range for plasma membrane or mitochondrial inner membrane, the malignant transformation of normal cells is stimulated (Yang and Brackenbury 2013), and there is an increase in the concentration of **reactive oxygen species** promoting the development of other complex diseases (Herst et al. 2017).

The coupling of two biochemical reactions is the most important way of how a cell can drive uphill reactions, which are thermodynamically unfavorable. In complex reactions, the uphill reaction is driven by a downhill reaction such as ATP hydrolysis into ADP and inorganic phosphate. That would not happen for a long time except in the presence of an enzyme which specifically accelerates that reaction (see Chapter 14). Even thermodynamically favorable reactions would not happen in a biologically relevant time frame if not accelerated by enzymes (Wolfenden 2003). For instance, the enzyme triosephosphate isomerase enhances the biochemical reaction it catalyzes for a billion times (see Chapter 10). In the absence of that enzyme, the half-time for the reaction is 1.9 days, an impossibly long time for functioning cellular glycolytic pathway (Radzicka and Wolfenden 1995). Another amazing enzyme performance is that of orotidine 5'-phosphate decarboxylase, which enables rate enhancement for the factor of 10^{17} (Callahan and Miller 2007). This enzyme catalyzes the last essential step in the biosynthesis of pyrimidines, the reaction with a half-life of 78 million years in the absence of the enzyme (Radzicka and Wolfenden 1995).

We should not neglect the other advantage enzymes provide when they connect the nonlinear flux-force reactions. A higher than 90% power transfer efficiency is achieved in a nonlinear mode (Juretić and Županović 2003; see Chapter 8). In contrast, the linear energy transfer regime restricts maximal power transfer efficiency to 50% according to the power transfer theorem (Boylestad 1999). That remark brings us to the question of how power production is related to entropy production.

3.3 Is power production equivalent to partial entropy production responsible for the emergence of the output work and force?

A few remarks about terminology will serve to connect this section with recent optimization results for biological ion pumps and nanomachines. The division of total entropy production into input and output force-flux couples and the realization that flux dependence on force is highly nonlinear for most biological systems survived the test of time. The efficiency term is predominantly reserved for the thermodynamic efficiency, that is, the negative ratio of output to input force: $-X_2/X_1$. Ion pumps convert one chemical potential into another. For instance, the F_0F_1-ATP-synthase converts electrochemical proton gradient into the output chemical potential $\Delta\mu_{ATP}$, which is calculated from [ATP], [ADP], and [P_i] concentrations. In this example, the generalized force terminology becomes specific. The protonmotive force (pmf) is the driving force, and the chemical potential of the ATP-ADP couple is the driven force. For the biological nanomachines producing mechanical work, the

$\Delta\mu_{ATP}$ is usually the input force, while work is considered the output force. A good example of such terminology can be found in Wagoner and Dill paper from 2019 (see Chapter 16, section 16.7). These authors retained the work designation for the $\Delta\mu_{ATP}$ when it is the output for the F_0F_1-ATP-synthase activity. It helps to clarify the difference between power production and obligatory entropy production contributions for producing the output work and power.

There are several ways to add partial entropy production contributions to get the total entropy production (Hill 1977). Sum of products of operational forces and fluxes is one way to do it—the way we used uptil now. The input flux and force are both positive, but the output force can be negative despite positive output flux. In its absolute value, the output force must be smaller than the input force to have positive total entropy production as required by the Second Law. Adding a negative contribution to the entropy production produces positive free-energy transduction efficiency in the range from zero to one, meaning that power can be extracted from the system. The product of output force and flux is that power. Strictly speaking, it is the proportionality relationship because each dissipation term must be divided with the temperature to get the corresponding entropy production. The output force-flux couple of biological nanomachines is the power when force produces work by acting over some distance during some time. All other ways of partitioning entropy production produce only positive contributions (Hill 1977). In all simulations presented in this book, we have calculated entropy productions associated with each transition between macromolecules' functional states. We shall see that productive rate-limiting transitions leading to charge separation and the establishment of the output force are the most fruitful for applying our maximal transitional entropy production theorem (Chapter 8).

Light-activated proton transport by bacteriorhodopsin pumps protons from the cytoplasm to the extracellular space working against the protonmotive force (pmf). In turn, the pmf is the power enabling ATP synthesis by F_0F_1-ATP-synthase. We have come now to the question from this section's title. Wagoner and Dill (2019) suggested maximizing the power output as a common evolutionary strategy of biological nanomotors and ion pumps. Optimal power output $X_2J_2 = X_{sec}J$ for the realistic model of bacteriorhodopsin photocycle is considerably lower (more than four times lower) and far away from the maximal power output (see Figure 12.3 from Chapter 12). Our scheme for the photosynthetic cycle's productive part did not contain an additional slip transition other than radiationless relaxation (see Figure 12.1 from Chapter 12). Still, the simulation results are essentially the same as when the additional slip was present (Juretić and Županović 2003). The maximal power output coincides with the maximal free-energy transduction efficiency at the optimal secondary force $X_{sec} = -123$ kJ/mol. Assuming that membrane potential is the main contributor to the proton electrochemical gradient X_{sec}, that pmf corresponds to –1.2 volts—much too high value to avoid the dielectric breakdown of the *Halobaterium salinarium* membrane. For almost all species, the biological evolution never ventured into the development of lipid membranes resistant to higher than 300 mV of membrane potential. The observed maximal output force of *Halobaterium salinarium* proton pumps can reach at most $X_{sec} = -27$ kJ/mol. That value corresponds to about 16% free-energy transduction efficiency, a far cry from the maximal possible efficiency of 72%. It also corresponds to the much higher total entropy production—an expected result because a larger difference between power input and output can only increase dissipation. Thus, the bacteriorhodopsin and similar ion pumps were not optimized for maximal power production during biological evolution. Therefore, the positive answer to this section's title question does not imply that we can replace physically founded variational principles with the selection principle imitating what is best for human needs (increased power production).

Definitions and explanations from Chapter 3:

Absolute temperatures are expressed in Kelvin degrees. All substances with T = 0 contain zero kinetic thermal energy, while T = 273.15 kelvins (K) corresponds to zero on the Celsius scale (0°C).

Chemical equilibrium of the chemical reaction is a dynamic equilibrium when the forward reaction proceeds at the same rate as the reverse reaction. It is the final state when concentrations of reactants and products do not change with time.

Extensive physical properties are additive when system size increases; **intensive** properties are size-independent. Mass, volume, and entropy belong to the former category of extensive, while temperature, density, and pressure belong to the latter type of intensive variables.

Homeostasis is the word describing the optimal and steady functioning of a living system. Despite external changes, our organism, for instance, maintains nearly steady-state values of internal parameters—pH, temperature, ion concentrations, blood sugar level, blood pressure, and so on, within some optimal range. Medical doctors can immediately recognize if some of the measured parameters are within or outside normal homeostatic values for your age and gender. Good regulation ability to maintain energy, bone, blood pressure, cholesterol, and blood sugar homeostasis beyond middle age is the key to maintain good health in older age.

The **inner product** concept is used here as the case of the scalar product or dot product of force and flux vectors.

Phase space of a mechanical system consists of all position and momentum values used to describe the system's evolution.

Power is defined in physics as the rate of doing work.

The **steady state system** does not change with time. Mathematically, it means that partial derivative with respect to time is equal to zero for all state variables.

The **reactive oxygen species** (ROS) are often free radicals derived from oxygen that have unpaired, highly reactive electrons. Excessive ROS production by mitochondria induces cell injury and death.

References

Boylestad, R. 1999. Introductory Circuit Analysis, Prentice-Hall, Upper Saddle River, NJ, USA.

Bruers, S., Maes, C. and Netočný, K. On the validity of entropy production principles for linear electrical circuits. J. Stat. Phys. 129(2007): 725–740.

Callahan, B.P. and Miller, B.G. OMP decarboxylase—An enigma persists. Bioorg. Chem. 35(2007): 465–469.

De Groot, S.R. and Mazur, P. 1984. Non-Equlibrium Thermodynamics. Dover, New York, NY, USA.

Dewar, R.C., Juretić, D. and Županović, P. The functional design of the rotary enzyme ATP synthase is consistent with maximum entropy production. Chem. Phys. Lett. 430(2006): 177–182.

Essex, C. Radiation and the violation of bilinearity in the thermodynamics. Planet. Space. Sci. 32(1984): 1035–1043.

Feistel, R. Distinguishing between Clausius, Boltzmann and Pauling entropies of frozen non-equilibrium states. Entropy 21(2019): 799. doi:10.3390/e21080799.

Herst, P.M., Rowe, M.R., Carson, G.M. and Berridge, M.V. Functional mitochondria in health and disease. Front. Endocrinol. 8(2017): 296. doi: 10.3389/fendo.2017.00296.

Hill, T.L. 1977. Free Energy Transduction in Biology: The Steady State Kinetic and Thermodynamic Formalism. Academic Press, New York, USA.

Jaynes, E.T. Gibbs vs. Boltzmann entropies. Am. J. Phys. 33(1965): 391–398.

Jaynes, E.T. 2003. Probability Theory: The Logic of Science. Cambridge University Press, Cambridge, UK.

Juretić, D. and Westerhoff, H.V. Variation of efficiency with free-energy dissipation in models of biological energy transduction. Biophys. Chem. 28(1987): 21–34.

Juretić, D. Membrane free-energy converters: The benefits of intrinsic uncoupling and nonlinearity. Acta Pharmaceutica 42(1992): 373–376.

Juretić, D. and Županović, P. Photosynthetic models with maximum entropy production in irreversible charge transfer steps. J. Comp. Biol. Chem. 27(2003): 541–553.

Kim, Y.C. and Hummer, G. Proton-pumping mechanism of cytochrome c oxidase: a kinetic master-equation approach. Biochim. Biophys. Acta 1817(2012): 526–536.

Kline, M. 1980. Mathematics: The Loss of Certainty. Oxford University Press, Oxford, UK.

Martyushev, L.M. The maximum entropy production principle: two basic questions. Phil. Trans. R. Soc. B 365(2010): 1333–1334.

Mortimer, R.G. and Eyring, H. Elementary transition state theory of the Soret and Dufour effects. Proc. Natl. Acad. Sci. USA 77(1980): 1728–1731.
Pietrobon, D. and Caplan, S.R. Flow-force relationships for a six-state proton pump model: intrinsic uncoupling, kinetic equivalence of input and output forces, and domain of approximate linearity. Biochemistry 24(1985): 5764–5776.
Planck, M. Ueber das Gesetz der Energieverteilung im Normalspectrum. Ann. Phys. 309(1901): 553–563.
Prigogine, I. 1967. Introduction to Thermodynamics of Irreversible Processes. Interscience, New York, NY, USA.
Radzicka, A. and Wolfenden, R. A proficient enzyme. Science 267(1995): 90–93.
Schnakenberg, J. Network theory of microscopic and macroscopic behavior of master equation systems. Rev. Mod. Phys. 48(1976): 571–585.
Tomé, T. and de Oliveira, M.J. Entropy production in nonequilibrium systems at stationary states. Phys. Rev. Lett. 108(2012): 020601. doi: 10.1103/PhysRevLett.108.020601.
Tomé, T. and de Oliveira, M.J. Stochastic thermodynamics and entropy production of chemical reaction systems. J. Chem. Phys. 148(2018): 224104. doi:10.1063/1.5037045.
Yang, M. and Brackenbury, W.J. Membrane potential and cancer progression. Front. Physiol. 4(2013): 185. doi: 10.3389/fphys.2013.00185.
Wagoner, J.A. and Dill, K.A. Opposing pressures of speed and efficiency guide the evolution of molecular machines. Mol. Biol. Evol. 36(2019): 2813–2822.
Westerhoff, H.V. and van Dam, K. 1987. Thermodynamics and Control of Biological Free Energy Transduction. Elsevier, Amsterdam, The Netherlands.
Wolfenden, R. Thermodynamic and extrathermodynamic requirements of enzyme catalysis. Biophys. Chem. 105(2003): 559–572.

CHAPTER 4

What is Life?

4.1 A physicist in love with life

Our tiny part of the universe became self-conscious when the scientific method of examining nature took off for good in the seventeenth-century largely due to Galileo Galilei's (1564–1642) efforts. Only then it became possible for scientists to communicate with a natural world by asking questions through experiments and constructing the theoretical edifices for explaining experimental results. Proper methods became gradually available for asking the fundamental questions: What am I? What is life? What is the universe? Our part of the universe had to wait for almost 14 billion years for this to happen, while life on Earth had to wait almost four billion years to develop self-consciousness. The living being who was the most influential in asking the question of what is life in the universe was also a scientist, a physicist in love with life, who hated war and discrimination.

During Second World War, when best physicists were harnessed into the feverish war effort work on the atomic bomb construction, great Austrian physicist Erwin Schrödinger published in 1944 an inspired booklet with the title: "What is life?" (Schrödinger 1944). Schrödinger was the theoretical physicist with some early experience in conducting experiments. It convinced him that his forte is the theory, not the experiments. But the assistantship in experimental physics proved an invaluable asset to his theoretical work. He never did any biological experiments (Dronamraju 1999, O'Connor and Robertson 2017). After 76 of mine and "What is life?" years, we can say with confidence that a hundred pages of Schrödinger's little book inspired a number of pioneers of molecular biology to perform the experiments, which identified the molecular hardwire for the hereditary code-script (Dronamraju 1999, Sigmund 2019).

The book arose from his lectures in February 1943 at Trinity College, Dublin, Ireland. Schrödinger proposed that life maintains itself by negative entropy influx and that genes are real physical objects that should be associated with the structure of an aperiodic crystal. At that time, it was still not known that DNA is indeed a biopolymer macromolecule, which can be crystallized and which is responsible for the preservation of biological identity through genes and their interconnections. One fruitful idea in his book is that genes are stable biomacromolecules of the greatest importance for life and that the chemical nature of genes must be established. It does not matter that he wrongly proposed a large protein molecule as a likely candidate for a gene (most prominent contemporary biologists thought so too).

Schrödinger remembered his 16 years in Dublin, Ireland, as very good years. His unorthodox lifestyle with two wives and his daughters' shared upbringing was not found acceptable by several other distinguished institutions in the UK and the USA. Until the end of his life in 1961, he never addressed the remarkable progress his book initiated in molecular biology. It was a hobby to him, and afterward, he moved on to other preoccupations (Symonds 1986). Schrödinger did not return to biology and did not answer a kind letter in which Francis Crick acknowledged the impact of *What is Life?* on him and Jim Watson (O'Connor and Robertson 2017, Sigmund 2019). In my opinion, he returned to his priority research interests in physics, such as the unified field theory, the research topic that stymied the best efforts of him and Einstein (O'Connor and Robertson 2017). The less likely explanation is that he was stung by the unsparing criticism of his booklet offered by some of his contemporary colleagues, who had a better knowledge of chemistry and biology. After exposing some weak points of Schrödinger's booklet, Linus Pauling concluded that, nevertheless, Schrödinger is basically responsible for the development of molecular biology by introducing new ideas from physics (Pauling 1987).

The calamity of the Second World War prevented proper communication among scientists. Schrödinger was not aware that his highly-regarded colleague, physicist Max Delbrück, had founded the phage group at Cold Spring Harbor in 1940 (Dronamraju 1999). The experiments there already answered some questions that Schrödinger raised in his booklet well before he published it. Another group in the USA published the proof that protein-free material consisting of DNA is responsible for transforming the bacteria heredity (Avery et al. 1944), just several months before Schrödinger's booklet was published. Schrödinger himself confessed in his book that he was influenced by "beautiful work" of geneticist Nikolai Timoféeff-Ressovsky (1900–1981) and his coauthors (Timoféeff-Ressovsky et al. 1935). Strongly inspired by that 1935 paper: "The nature of genetic mutations and the structure of the gene" is probably a better assessment (Moberg 2020).

Timoféeff-Ressovsky called himself the grandfather of the scientific approach to find the molecular nature of genes (Timoféeff-Ressovsky 2000, Ratner 2001). He implanted the ideas about the molecular nature of genes in Max Delbrück's and Erwin Schrödinger's mind. The blooming of interdisciplinary sciences in the 1920s and early 1930s had seemingly no limits in Germany. Beauty manifested itself not only in equations but also in the microevolution theory of female beauty jokingly named "The Isolines of The Female Beauty" (Timoféeff-Ressovsky 2016). Schrödinger was in charge of grading beauties in Austria, Hungary, Czechoslovakia, and Switzerland. Another polymath physicist, Franko Rasetti from Italy, distributed marks "with passion" to locations in Italy, Yugoslavia, and Greece, because everything he did during 100 years of his life had hallmarks of elegance, simplicity, beauty, and passion (Kerwin 2002). The Copenhagen mathematical method, used by that playful theory, suggested to him that my native town Split is a very high pick for female beauty. The pilots from allied forces did not know about that theory. On June 3, 1944, they bombed again and destroyed the church of Saint Peter, the largest church from Split, located at the city center, next to the Diocletian's Palace. It happened about a half year after the wedding ceremony of my beautiful mother and my father in that church. Thus, they were not among 277 civilians who died, and I got the opportunity to become living proof of their love when I was born on October 23, 1944.

Schrödinger likely learned during his lifetime about the inhuman treatment that Timoféeff-Ressovsky received from the totalitarian Nazi and communist regimes but could not do anything about it. Gestapo executed Dmitrij, Nikolai's eldest son and a member of the underground resistance, in the Mauthausen concentration camp on May 1, 1945, just before the war ended. After the war, the starvation regime introduced by Joseph Stalin in one of the most terrible Gulag concentration camps was so severe that Timoféeff-Ressovsky permanently lost most of his eyesight (Paul and Krimbas 1992). Mass killings, tortures, and genocides perpetrated in the East by communists, during the first several years after peace was proclaimed, did not bother Western humanists and peacekeepers.

Still, communists did not succeed in transforming Timoféeff-Ressovsky into a monster like all his jailors were. He believed that the Russian people's dignity would bring about desperately needed change for the time of slavery to pass away (Berg 1990). He was still teaching science to his fellow inmates when he and all other prisoners were at the brink of dying from starvation, physical and mental abuses (Solzhenitsyn 1973, 1974)! With the words: "When will this madness finally cease," he resisted both Hitler's and Stalin's tyranny. All Timoféeff-Ressovsky papers were published afterwards with the help of his faithful wife Helena Alexandrovna Timoféeff-Ressovsky, also the talented biologist, who discussed everything with her half-blind husband and typed what he dictated in the paper. These two brave souls were dragged through the harshest possible discrimination, just because they loved each other, their country, and their science. Their "crime" was the lack of praise from them for the slave masters of totalitarian regimes, who already tortured and killed millions of innocent beings. But their love for science won before they died. Almost blind Timoféeff-Ressovsky played a vital role in saving genetics from near destruction of that science in the Soviet Union brought by Joseph Stalin and Trofim Lysenko. All this may look distant and unrelated to the peak of prosperity and democracy we had in 2019, but it is not. As long as the human spirit is alive, we shall always have with us great scientists persecuted by the authorities (Berg 1990), but some will be legally rehabilitated only many years after their death (Ratner 2001).

4.2 Schrödinger's passion for understanding the secrets of life initiated molecular biology

As Salvador Luria's first graduate student, James Watson was one of the readers enchanted with Schrödinger's booklet, which he later called "that little gem." Salvador Luria (medical doctor, microbiologist, and Nobel laureate in 1969 together with Max Delbrück) had good reasons to believe that DNA is a better candidate as a heredity carrier than proteins, and this thought was impressed on young Watson during his PhD study under Luria's mentorship. All scientists know that James Watson and Francis Crick (a physicist) got the 1962 Nobel Prize precisely for their contribution in discovering the chemical nature of genes. Francis Crick was also inspired by Schrödinger's booklet that convinced him and many other physicists just how interesting are questions offered by biology.

DNA is indeed a much more stable molecule than most proteins. The present-day decoding success of 38 thousand years old Neanderthal genome from Croatia's Vindija cave, by Svante Pääbo, and of 41 thousand years old Denisovan genome from the Siberian Denisova cave, also by Svante Pääbo group in Germany, testifies to the stability of macromolecules responsible for the inheritance, which Schrödinger anticipated. However, he did not anticipate that the stability of biomacromolecules can be due to the stability of functional processes leading to hydrolysis/degradation and re-synthesis of the same complex structures. It is the reason why our kind of human species is quite similar to the first Homo sapiens individuals who met Neanderthals in Europe about 50 thousand years ago. On average, we have today lesser physical strength and slightly smaller brain volume than 40 thousand years ago. At that time, both our physical strength and brain volume were significantly lesser than corresponding Neanderthals' qualities. Thus, it is not clear why we survived and Neanderthals did not (that question inspired science fiction writer Robert J. Sayer to create the trilogy: "The Neanderthal Parallax").

Schrödinger did not consider the stability of functional processes maintained during eons, most likely due to earlier mentioned aversion of physicists toward the idea that function may dominate structure; in other words, that structure is enslaved by the function it performs. This idea is routinely rejected in the physics of nonliving matter. However, when we examine what function means for biologists, the answer is straightforward. Whatever is the function of interest, it is always performed by ephemeral organic structures as a set of interconnected well-regulated processes. Accordingly, life is basically a process (Woodger 1962). A functional account of life is that the living system is a self-regulated functional network of processes regardless of the nature of processed molecules. Astrobiologists, scientists dealing with the origin of life question, and scientists attempting to create synthetic life in the laboratory (synthetic cell) are all guided by some working hypothesis about what minimal set of chemical reactions life needs to live (Bedau 2010).

The importance of function in the living world is a deeper problem for a present-day scientific mind, despite a quite common occurrence of terms "function" and "purpose" in biological scientific publications. Aristotle's final causes cannot be easily distinguished from function or purpose as the ultimate reasons for something to exist. However, as the ultimate reason for something, the final cause invocation has historically been the question of why or how God did things in a particular way. This term (final cause) has been purged from natural sciences. Rosen (1991) revived the final cause idea in biological sciences to explain how an organism continuously recreates itself, thus invoking circular processes of creation and destruction driven by externally available free-energy sources without recourse to any external creator.

Physicists are well aware that Schrödinger (1887–1961) is the second father of quantum mechanics (after Heisenberg), whose method prevailed in explaining the properties of atoms and small molecules. Schrödinger was awarded the Nobel Prize in Physics on December 10, 1933. He was then a guest professor at Oxford University. Schrödinger pointed out in his booklet how quantum physics and genetics development started in the same year. The year was the turn of the 20th century. Max Planck's discoveries in quantum physics in 1900 used the entropy concept. Mendel's laws in genetics were rediscovered, also in 1900, by Hugo de Vries in Holland, Carl Correns in Germany, and Eric Tschermak in Austria. Gregor Mendel was the scientist and monk from what is today the Moravia region of the Czech Republic, who published his paper "Experiments on plant hybrids" in 1866.

Despite his Nobel Prize, Schrödinger's excursion in biology was considered inappropriate by some scientists, like Max Perutz. Perutz could not restrain himself from sarcastic witticism to the effect that what is correct in Schrödinger's booklet is not original, while what is original is not correct (Perutz 1987). That may be the reason why Schrödinger's booklet is forgotten and omitted as a reference today in most popular books dealing with molecular biology. I shall try to show in the next chapters that Schrödinger's ideas are undeservedly forgotten and neglected. In science, just as in life, what fruits are stimulated to grow through our efforts is more important than whether our ideas are mostly correct or incorrect.

Nobody will deny that molecular biology is a valuable fruit initiated by the question "What is life"? The question belongs among the most difficult and most fruitful scientific questions ever asked. In his book about the same question, Addy Pross (2016) stated that the system chemistry is the new research field, which shall bring us closer to the answer. In his words, life is defined as: "a self-sustaining kinetically stable dynamic reaction network derived from the replication reaction". Thus, the focus changed during the seven decades from looking for extraordinary stable macromolecules to the analysis of how dynamic **kinetic stability** can ensure the long-term survival of structural complexity (Moberg 2020). The analysis of how chemistry can beget biology still lacks physical foundations. That might be the reason why more than 100 of better-known definitions of life can be found in the scientific literature (see the next chapter and Chapter 13).

Schrödinger freely admitted that he is an amateur in biological research. However, his excursion in biology, from the viewpoint of physics, helped him to realize that life is different from all other systems that physicists examined before. About 76 years later, most scientists are still unable to take the bold step Schrödinger took, of looking at life from outside and asking the question of what is the role of life in the universe, and why it exists in the universe. This question brought me to the realm of biophysics when I started my career as a young scientist at the Ruđer Bošković Institute in Zagreb, Croatia.

How is it possible that Schrödinger's hobby initiated so many benefits in scientific fields where he freely admitted being only a naïve physicist dabbling in biology? Freeman Dyson (1923–2020) took the trouble to find a total of five references to the technical literature and less than ten equations in Schrödinger's booklet, and praised it for being "clearly and simply written" as a "fine piece of English prose" (Dyson 2009). He pointed out that Schrödinger was woefully ignorant of chemistry, isolated in Ireland, and that his thoughts about biology were based on brilliant but already old experimental discoveries of Nikolai Timoféeff-Ressovsky and Max Delbrück. However, Schrödinger knew how to ask the right questions even when he did not know how to answer them. Thus, Schrödinger's creation in biology outlived the creator and started a long life independent of its creator, making friends and enemies in the process of self-perpetuated reproduction. The bravery to embark on an interdisciplinary synthesis of facts and theories with incomplete knowledge was the saving grace and noble drive for the *What is Life?* creator. The other saving grace was that Schrödinger was a passionate man, a poet, and philosopher who believed that his scientific work was an approach to godhead, a fire spirit never to be labeled and classified in the same box with other "boring" scientists (Moore 1989). His 1926 ***annus mirabilis***, when he published four groundbreaking papers on wave mechanics, reverberated through all his later works and shaped the opinions about them.

Definitions and explanations from Chapter 4:

Annus mirabilis means miraculous or amazing year. In science, that phrase is usually applied to years 1666 and 1905 to describe, respectively, the revolutionary discoveries of Isaac Newton and Albert Einstein.

Kinetic stability is metastability—the out-of-the equilibrium state in which a driven system persists if not disturbed too much. Both classical and quantum systems can be stabilized by dissipation (Doaré and Michelin 2011, Spagnolo et al. 2018).

References

Avery, O.T., Macleod, C.M. and McCarty, C.M. Studies on the chemical nature of the substance inducing transformation of pneumococcal types. Induction of transformation by a deoxyribonucleic acid fraction isolated from pneumococcus Type III. J. Exp. Med. 79(1944): 137–158.

Bedau, M.A. An Aristotelian account of minimal chemical life. Astrobiology 10(2010): 1011–1020.

Berg, R.L. In defense of Timoféeff-Ressovsky. The Quarterly Review of Biology 65(1990): 457–479.

Doaré, O. and Michelin, S. Piezoelectric coupling in energy-harvesting fluttering flexible plates: linear stability analysis and conversion efficiency. J. Fluids Struct. 27(2011): 1357–1375.

Dronamraju, K.R. Erwin Schrödinger and the origins of molecular biology. Genetics 153(1999): 1071–1076.

Dyson, F. 2009. Origins of Life. Cambridge Univ. Press (Second Edition), Cambridge, UK.

Kerwin, L. Franco Rasetti (1901–2001). Nature 415(2002): 597–597.

Moberg, C. Schrödinger's What is life?—The 75th anniversary of a book that inspired biology. Angew. Chem. Int. Ed. Engl. 59(2020): 2550–2553.

Moore, W. 1989. Schrödinger. Life and Thoughts. Cambridge Univ. Press, Cambridge, UK.

O'Connor, J.J. and Robertson, E.F. Erwin Schrödinger and quantum wave mechanics. Quanta 6(2017): 48–52.

Paul, D.B. and Krimbas, C.B. Nikolay V. Timoféeff-Ressovky. Sci. Am. 266(1992): 86–92.

Pauling, L. 1987. Schrodinger's contribution to chemistry and biology. pp. 225–233. *In*: Kilmister, C.W. (ed.). Schrödinger: Centenary Celebration of a Polymath. Cambridge University Press, Cambridge, UK.

Perutz, M.F. Physics and the riddle of life. Nature 326(1987): 555–558.

Pross, A. 2016. What is Life. How Chemistry Becomes Biology, 2nd Ed. Oxford Univ. Press, Oxford, UK.

Ratner, V.A. Nikolay Vladimirovich Timofeeff-Ressovsky (1900–1981): Twin of the century of genetics. Genetics 158(2001): 933–939.

Rosen, R. 1991. Life Itself: A Comprehensive Inquiry into the Nature, Origin, and Fabrication of Life. Columbia University Press, New York, NY, USA.

Schrödinger, E. 1944.What is life. Cambridge Univ. Press, London, UK.

Sigmund, K. The physicist and the dawn of the double helix. Science 366(2019): 43–43.

Solzhenitsyn, I.A. 1973, 1974. The Gulag Archipelago I and II. Harper & Row Publishers, New York, NY, USA.

Spagnolo, B., Carollo, A. and Valenti, D. Stabilization by dissipation and stochastic resonant activation in quantum metastable systems. Noise induced phenomena in quantum metastable systems. Eur. Phys. J. Special Topics 227(2018): 379–420.

Symonds, N. What is Life? Schrodinger's influence on biology. Q. Rev. Biol. 61(1986): 221–226.

Timoféeff-Ressovky, N.W., Zimmer, K.G. and Delbrück, M. Über die Natur der Genmutation und der Genstruktur, Nachrichten von der Gesellschaft der Wissenschaftenzu Göttingen: Mathematische-Physikalische Klasse, Fachgruppe VI, Biologie Bd. 1, Nr. 13(1935): 189–245.

Timofeeff-Ressovsky, N.V. 2000. The Stories Told by Himself with Letters, Photos and Documents. Soglasie (in Russian), Moscow, Russia.

Timofeeff-Ressovsky, N.W. 2016. Some Stories Told by N.W. Timofeeff-Ressovsky. pp. 3–12. *In*: Korogodina, V., Mothersill, C., Inge-Vechtomov, S. and Seymour, C. (eds.). Genetics, Evolution and Radiation. Crossing Borders: The Interdisciplinary Legacy of Nikolay, W. Timofeeff-Ressovsky, Springer, Cham, Switzerland.

Woodger, J.H. Biology and the axiomatic method. Ann. N. Y. Acad. Sci. 96(1962): 1093–1116.

CHAPTER 5

Some Answers to Schrödinger's Questions

5.1 The stability paradox

The first question Schrödinger asked, "What are the chemical constituents of life?", was largely solved during the past 60 years, but of course, there is a lot of additional work for molecular biologists, biophysicists, and bioinformaticians to fill the remaining gaps in our knowledge. Attempts to unify known facts about life constituents in the turn of century book of the same title "What is life?" (Margulis and Sagan 2000) pointed toward new insight about the common origin and present-day genetic similarity of all living beings. It arose during a very long evolution and selection process. However, Schrödinger's second question, how and why life is different from all other phenomena in the universe, deserves additional attention and should not be forgotten.

Schrödinger asked that question in the form of a paradox: how complex nonequilibrium structure of living beings may be maintained in the more or less same fashion for many years, even eons. Based on the Second Law of Thermodynamics, physicists expected that simplification would occur spontaneously until the equilibrium state is reached. That is the process leading to entropy rise until the maximal entropy value is reached in the equilibrium state. The equilibrium state is incompatible with life. When the system is removed from the equilibrium state, it is impossible to avoid entropy production, dissipation, and degradation of the system when one can neglect system interaction with its environment (the rest of the universe). Entropy increase is often associated with a simple interpretation that disorder has increased in the system, and its complexity is decreased.

The obvious answer to this apparent paradox is that we cannot neglect system interaction with its environment when the system is a metabolically active cell or organism. We have mentioned in the Introduction that the intensity of free-energy transduction in living cells is very high even when compared with the intensity of energy transformations taking place in the Sun-like stars at incomparably higher temperatures. Therefore, an essential property of a living cell is that it is an open system that exchanges energy and matter with its environment at a very high rate. Thus, it can counter the Second Law tendency toward disorganization and maintain its far-from-equilibrium structure in a quasi-steady state. Steady-state is a stationary state by definition, which means that thermodynamic functions such as entropy or free-energy remain the same or very similar during shorter periods. Internal entropy production is efficiently exported into the environment when a steady state is not the equilibrium state. The corresponding free-energy decrease is just replaced with what Schrödinger called the parcels of negative entropy taken from the environment. In our everyday language, this is nothing else but food taken from the outside and processed by the system in such a way that the system receives the free energy needed to drive "uphill" metabolic processes. Concomitantly, its entropy is decreased to balance natural entropy increase and free-energy decrease due to intensive life-supporting cellular processes. Every living system is capable of this balancing act leading to an entropy increase in the rest of the universe and corresponding delicate dynamic compensation between internal entropy increase and decrease. Just a small overflow of free energy in the cell is enough to prepare it for growth, reproduction, and development. This process of regulated free-energy transduction is essential for being alive and should not be neglected in endeavors to define what life is (Cornish-Bowden and Cárdenas 2019).

What is the "food" performing the miracle of cells maintaining their internal structure and also growing in some cases into whole complex organisms, such as 100 meters high trees? Photons free-energy is "food" for plants and all other organisms or single cells performing photosynthesis. Free energy in the form of photons cannot be used directly by cells or cellular organelles such as chloroplasts. It must be first converted into electron current, charge separation, and then into electrochemical proton gradient, responsible for creating a strong electric field and concentration gradient into which energy is temporarily stored. For chloroplasts, the greater part of free-energy conversion and storage is due to the creation of proton concentration imbalance or pH difference between the inside (thylakoid space) and outside space (cytoplasm in plant cell) (see Chapter 1, section 1.6). The smaller part is the build-up of electrical potential difference across the thylakoid membrane. Such protonmotive force, consisting of the chemical end electrical part, enables the second step of free-energy conversion responsible for the ATP synthesis. The subsequent ATP hydrolysis in the cytoplasm (as the third step in the hierarchy of free-energy conversions) delivers large free energy packages necessary for the synthesis of all cellular macromolecules, operation of all molecular motors in the cells, cell movements, development, and division. Animals and all their cells have found a shortcut by eating plants or other animals and using their stored free energy in the respiration process, leading to the formation of electrochemical proton gradient (created by mitochondria in our cells).

Humans cannot synthesize nine out of 20 proteinogenic amino acids. Evolution has found many other shortcuts appearing to us as errors (Lents 2018). Most of them are survival-promoting adaptations to the preferred environment. Some vitamins and essential unsaturated fatty acids stimulate our brain's development and should be present in our diet from an early age. Our diet must have a sufficient variety for us to remain healthy. Some scientists connected the taste developed for freshwater or seawater food sources (containing essential omega-3 fatty acids) to the initial separation of hominid from pongid lines and gradual transformation of tree dwelling apes into creatures with better aquatic capabilities and bigger brains (Tobias 2011).

Following in the footsteps in Schrödinger's booklet, we shall focus our attention on the question of what is different about living systems, such as metabolically active cells, from all other systems in the nature that were exposed to the scrutiny of physicists?

1) Complex, low probability structures (macromolecules) are characteristic features found in every living cell. Biomacromolecules are born, die, and are synthesized again during their life cycle that can last from minutes to years and are faithfully repeated producing more of less unchanged structure-function relationships.
2) Through even longer periods (hundred millions or billions of years), the evolution toward greater complexity is a characteristic feature for life, in contrast to evolution toward thermodynamic equilibrium, which is the best-known example of evolution in physics.
3) For active living cells, free-energy transduction and entropy production are extraordinarily intensive—millions of times more intensive than energy transduction occurring in the equivalent volume of Sun-like star operating at higher temperatures for many orders of magnitude.
4) We can associate numerous costly (in terms of free-energy) positive and negative feedbacks with living systems. Such control circuits, developed during biological evolution, enable life to spread in a supportive environment as a well-controlled explosion. The strict control of biochemical reaction networks is responsible for growth and the conservation of self-similar complex functions through millions of years.

5.2 Photons as "food" for life

To be more specific, let us concentrate on the thermodynamics of photosynthesis because we know that photosynthesis is a key for the survival of life forms that predominate today in our planet's biosphere. By studying the process of photosynthesis, biophysicists realized that it is a surprisingly low-efficiency process—only a small part of Sun's radiation energy that reaches Earth's surface is captured by photosynthetic cells or organelles, and only a small part of captured photon free-energy is used to produce new organic compounds. On the other hand, it is quite clear that nature had millions, if not billions, of years to perfect the photosynthesis process by aiming for maximal efficiency. The question suggested by these observations is that possibly some other physical quantity, instead of efficiency, needed to be extremal during the biological evolution of photosynthetic structure-function relationships. In other words, may be nature did not even try to produce photosynthesis apparatus of maximal efficiency but was forced by physical laws to choose some optimal efficiency under given constraints. Extremum principles are well known in physics, but maximum efficiency is usually mentioned as an upper-efficiency limit, rather than the maximum efficiency principle.

In the field of thermodynamics, it is known that increased efficiency can be realized by decreasing free-energy dissipation. One can naturally be tempted to ask if nature aimed to achieve minimal dissipation associated with photosynthesis during evolution. Very fast and localized first steps, associated with photons captured by photosynthetic cells, are almost certainly of quantum nature and, therefore, dissipationless. Latter conversion of photon free energy into electron current, electrochemical proton gradient, ATP synthesis, and synthesis of all other organic molecules in a cell produces free-energy dissipation. Still, may be that dissipation or entropy production is minimal.

5.3 Prigogine and Ziman dispute about minimum or maximum entropy production

The best-known theorem about minimum entropy production has been formulated by Ilya Prigogine (Prigogine 1945). When speaking how firmly the Prigogine's theorem is established for a linear force-flux relationship, I do remember the elegant proof when the opposite has been found. It is Ziman's proof about the equivalence of Boltzmann's equation to the maximum entropy production principle (Ziman 1956, see Chapter 9 too). **Boltzmann transport equation** evaluated by Ziman describes the linear response of charged particles under the influence of the not too strong electric field. Ziman noticed that entropy production is the basic integral invariant of the Boltzmann equation. Prigogine was present at the conference where Ziman discussed his proof while mentioning the failure of Prigogine's theorem. Prigogine did not explain his critical remark that he doubted the validity of part of Ziman's thermodynamic interpretation. Whenever he discussed the Boltzmann transport equation in his later papers, Prigogine never mentioned the principle Ziman constructed: "Consider all distribution of currents such that the intrinsic entropy production equals the extrinsic entropy production for a given set of forces. Then, of all current distributions satisfying this condition, the steady state distribution makes the intrinsic entropy production a maximum." The effect of collisions in always increasing the entropy is in agreement with the Second Law of Thermodynamics. The irreversible microscopic processes that cause intrinsic or internal entropy production are the scatterings inside the material (phonon-electron scattering, for instance). Extrinsic entropy production is calculated from measurements of forces and currents made outside the system. In essence, it is the electric field's product with the electric current. The product is equal to Joule heat. In the steady-state of constant entropy, an external electric field tends to produce more order in the electron distribution (directional drift of electrons). It removes the amount of entropy inside the material equal to the amount generated by the irreversible transport process. Removed internal entropy appears externally as heat, which is swept into the reservoirs to maintain isothermal conditions.

John Michael Ziman (1925–2005) was a highly regarded expert for a wide range of problems in the theory of metals and semiconductors. He established the close analogy between the formalism of the Onsager theory of irreversible processes and the theory of the Boltzmann equation (Ziman 1960). After generalizing the above mentioned variational principle, it reads in his 1960 book: "In the steady state, currents are such that the production of entropy has the largest value consistent with its subsequent conservation." That masterpiece book is still alive in numerous subsequent editions and has been cited almost ten thousand times, while Ziman's 1956 paper was cited over 200 times. The variational method he developed for finding maximum entropy production has the power of the general variational principle of the transport theory. It has been repeatedly used for easier solving of problems in that field, in a sharp distinction to the inability of the Prigogine theorem to be helpful in practice (Martyushev and Seleznev 2006). Strangely enough, these facts did not raise the skepticism about the validity of Prigogine's theorem for a long time. Bertola and Cafaro mentioned both Ziman's and Prigogine's contributions in their 2008 paper, concluding that the minimum entropy production of the system in a stationary state cannot be different from zero. Admittedly, Rolf Landauer concluded already in 1975 that "Even in the case of strictly linear circuits, where we might expect the validity of the minimum entropy production theorem to be unlimited by the restrictions to systems 'not too far from equilibrium,' we find that minimal entropy production does not apply."

5.4 Andriesse and Juretić dispute about minimal entropy production in photosynthesis

The assumption that minimal entropy production does apply to photosynthesis was made by the Dutch astrophysicist and scientific historian Cornelis Dirk Andriesse (Andriesse and Hollestelle 2001). After publishing calculations with the imposed requirement that entropy production must be minimal in the case of stellar mass loss (Andriesse 2000), these authors claimed that photosynthesis is in quantitative agreement with the principle of minimum entropy production. The minimum entropy production principle usually refers to the corresponding theorem by Ilya Prigogine. For its derivation, see Chapter 3 and Prigogine's book (Prigogine 1967). Prigogine's theorem defines a nonequilibrium steady state for linear relationships between fluxes and forces. It is the unique close-to-equilibrium nonequilibrium state, which is the closest analogy of the equilibrium state in the nonequilibrium region. For the example of two input and output fluxes and forces, Prigogine's steady state is defined by zero output flux and maximal output force. Such steady-state is well known in mechanics and electrodynamics as well, and it is called the static head steady state when the motor has the highest potential to do some work, but zero work is produced due to the open circuit condition. The stall force is established that does not give any power output (see Chapters 1 and 3). In terms of efficiency, the static head state is associated with maximal efficiency for free-energy storage and zero efficiency for free-energy transduction. Entropy production is maintained at the minimal possible value greater than zero value as the consequence of zero output flux condition. Energy input is being constantly used and spent only to maintain low but unproductive dissipation. In effect, when static head steady state is created and maintained, it prevents any further free-energy transduction. For instance, in the static head steady-state, ATP synthase cannot use the protonmotive force to produce any net ATP synthesis. All subsequent free-energy transduction steps leading to the synthesis of biomacromolecules and activation of cellular molecular motors cannot happen when ATP synthase is stuck in the static head steady state. The hierarchy of free energy transductions, which is the hallmark of the metabolically active cell, is blocked if we impose the condition of minimum entropy production for any of the initial stages in free-energy transduction.

All of this analysis, well known to scientists using irreversible thermodynamics in chemical and biological applications, was neglected by Andriesse. We published our critique of Andriesse's calculations and sent it to the first number of BioComplexity journal. It was accepted, but the publication of the journal was canceled. I made the paper available on the internet (Juretić 2002), and (as an abbreviated version) in our Juretić and Županović (2003) paper. The critique claimed that the minimum entropy production state, described in two papers (Andriesse 2000, Andriesse and Hollestelle 2001), is, in reality, an equilibrium state with zero entropy production. For the assumed linear relationship between fluxes and forces, entropy production becomes a quadratic function of both forces and can be minimized with respect to each force. After such minimization in Andriesse's papers, two extremal conditions are obtained, which must be fulfilled simultaneously. If Onsager reciprocal relations (see Chapters 1 and 3) hold, one can easily recognize these conditions as a requirement for both fluxes to vanish. The entropy production then vanishes too. Authors are specific in their claim that Onsager reciprocal relations may be (Andriesse and Hollestelle 2001) or should be (Andriesse 2000) violated and that both fluxes are different from zero when entropy production has reached its minimum. However, their flux ratio is then equal to the negative force ratio. If the primary force-flux couple is labeled with X_1, J_1, and secondary force-flux couple with X_2, J_2, the result $J_2/J_1 = -X_1/X_2$ follows from their equations (9) and (10) (Andriesse 2000). Since the entropy production is proportional to $X_1J_1 + X_2J_2$, it must be equal to zero in their minimum entropy production state. Therefore, it does not matter whether Onsager relations are broken in the linear range near equilibrium or not. In both cases, the minimal entropy production state, described by the authors, is the zero entropy production state of chemical or thermodynamic equilibrium. This is only to be expected. When all forces are varied in the minimization process, then none of them is subject to a constraint, and thermodynamic equilibrium is obtained as the only possible steady-state. The equilibrium state cannot, by any feat of imagination, be associated with a metabolically active living cell. If we accept the authors' claim that fluxes (and forces) are different from zero in the state of minimum entropy production, as defined by them, then the photosynthetic efficiency must be 100%. This is seen when their $J_2/J_1 = -X_1/X_2$ result is inserted in the ratio $-X_2J_2/X_1J_1$, which is the efficiency of free-energy transduction (Hill 1977).

Andriesse and Hollestelle (2001) avoided this strange result (100% efficiency and zero entropy production) by assuming that the force ratio is equal to 2.72 in favor of the secondary force, although it is fixed from the outset by their minimum entropy production requirement. With their best estimate for the flux ratio of 0.022, their secondary (driven) force would have to be 45 times higher than the primary (driving) force to produce 100% efficiency, or about three times higher, to produce 6% efficiency. In reality, the situation is just the opposite. The flux ratio is close to one, and the force ratio is considerably smaller than one. These authors achieved less than 2% difference between measured and predicted efficiency values for C4-plants due to almost perfect compensation of huge overestimate in force ratio multiplied with a huge underestimate in the flux ratio. In conclusion, the author's minimum entropy production requirement has nothing to do with the minimum entropy production theorem (Prigogine 1967). It merely describes the equilibrium state. Since all fluxes and forces vanish in equilibrium, free-energy transduction is impossible. It is the thermodynamic death of the system, as much opposed as possible to bioenergetic processes maintaining far-from-equilibrium energy conversions essential for life, such as photosynthesis. We can conclude that for photosynthesis, the approaches promoted by the Nobel Prize winner Ilya Prigogine and astrophysicist C.D. Andriesse are irrelevant and completely wrong, respectively.

In 2008, Dr. Andriesse made available on the internet the reply to my critique of his calculations (Andriesse 2008). He admitted that there is no entropy production when both forces are allowed to vary in the case of an assumed linear relationship between fluxes and forces. The mathematical conditions he used are so restrictive that only the equilibrium state can be established within such constraints. His reply focused on differences in the significance of symbols used in physics and biology. According to Andriesse, when Terrel Hill used the equation for entropy production in enzyme-catalyzed reaction (Hill 1977):

$$TP = \sum_k J_k X_k \qquad \text{(see Chapter 3, section 3.1)}$$

it was the biological example of inappropriate units used for flows, namely J_k (s^{-1}) instead of particle flux per unit surface and unit time. He cited the well-known monograph about nonequilibrium thermodynamics (De Groot and Mazur 1984) to support this claim. Thus, Andriesse made the distinction between real (physical) fluxes and "biological fluxes", by omitting to mention the consistent usage of numbers N (molecules or moles) for all definitions of fluxes in Hill's book. He also did not mention that De Groot and Mazur (1984), Prigogine (1967), and other experts for nonequilibrium thermodynamics consistently spoke about generalized fluxes and forces whenever they presented the entropy production as exactly the same sum of flux-force products (the equation above has the same meaning for one, two or more flux-force pairs corresponding to $k = 1, 2,..$, respectively). All fluxes do depend inversely on the time variable, but flux identity determines the remaining units. For heat flow, the remaining unit is J/m^2, and for chemical reactions, molar units are used. Different units for particular flows do not imply any discordance with the energy conservation law, as Andriesse suggested in his reply. The basic bilinear expression for the entropy production per unit time can be derived using the balance equations for matter, momentum, and energy when local equilibrium can be assumed (Prigogine 1969a; see section 3.1). Dr. Andriesse's arguments presented in his reply do not carry enough weight to call into question Terrel Hill's expertise in founding the field of **nanothermodynamics** (Hill 1964), in the domain of physical chemistry and statistical mechanics (Hill 1987), and in entropy production calculations for biological free-energy transductions (Hill 1977).

Charge separation, electrochemical proton gradient, chemical processes, and bioenergetic membranes do not play any role in Dr. Andriesse's calculations, who considered only photon flux and glucose diffusion flux through cell walls. In comparison to the cytoplasmic membrane, the cell wall is not the permeability barrier to glucose efflux. Glucose efflux is just one among myriad other processes initiated by photon absorption and plant's photosynthetic machinery. Consequently, it is no wonder that the ratio of output flux (calculated from glucose diffusion value across cell walls, which is not much different from glucose diffusion in water) to input photon flux is extremely small (0.022, as we mentioned above). Andriesse correctly pointed out in his reply that he did not multiply the flux ratio with the output to the input force ratio. Andriesse and Hollestelle (2001) used their own efficiency definition by multiplying the flux ratio (the flux of glucose molecules divided by the flux of photons) with the ratio of output Gibbs energy per molecule to photon energy hν. They named it "the ratio of energies in these fluxes", and did not explain why the ratio of forces should not enter into the definition of transduction efficiency, or why the term *forces* is always put in parenthesis ("forces"), as if forces are less well defined than fluxes. Anyway, the ratio of output to input energies used by them is considerably greater than 1 (2.72, as we mentioned above) despite the overestimate for available energy from absorbed photons by chlorophyll molecules. Hence, they entered into the contradictory requirement that output and input energies or "forces" are freely variable and constant at the same time. Either requirement prevents relaxation to the nonequilibrium steady state, such as the static head state, whether it can be characterized with the Prigogine's minimum entropy production value different from zero or not—a good reason for these authors to omit explicit mentioning of the Prigogine's theorem! Another omission, which also escaped the attention of reviewers for Andriesse and Hollestelle manuscript, is the fact that the paper neglected entropy increase in each of many intermediate energy conversion steps between photons absorption and glucose efflux from plant cells containing chloroplasts.

5.5 Jennings' dispute with Lavergne, Kox, and Parson about negative entropy production in photosynthesis

Outside equilibrium, the claim that entropy production is always a positive definite quantity is equivalent to the Second Law of Thermodynamics, while the statement that photosynthesis is associated with negative entropy production is equivalent to the claim that Second Law is not obeyed by photosynthesis. Unfortunately, such a claim was published by Jennings et al. (2005) in the high-quality journal (see Chapter 8 too for the discussion of Jennings et al. papers). It did not help that many authors presented strong arguments against abrogating Second Law in photosynthesis (Juretić and Županović 2003, Lavergne 2006, Knox and Parson 2007a,b). Jennings and his collaborators published additional papers (Jennings et al. 2007, 2014) in which these authors stressed that the very initial photosynthetic steps must have extremely high efficiency. It is something to be expected from the partially quantum nature of these steps and the quantum efficiency calculations they used instead of free-energy transduction efficiency.

5.6 Prigogine's theorem is not the physical principle, but it is mixed up with his concept of dissipative structures

What are the physical principles relevant to life emergence and development?

The Prigogine's theorem is not a physical principle at all, but it is only a theorem specifying a unique steady state close to equilibrium. Even worse, when stated as the principle of minimum entropy production, it was shown to be based on a mathematical error (Hunt and Hunt 1987, Ross and Vlad 2005). Kondepudi (1988) opined that the principle does not break down when the system relaxes to the state of higher entropy production. Experts' opinion about the relevance of Prigogine's excursions into self-organizing scenarios akin to the emergence of life is nicely summarized in the recent paper by Jeremy England (2015). He described Prigogine's minimum entropy production principle as "elegant" and "the most abidingly fascinating," but one of the least practical applications of the Onsager's reciprocal relations. I support that opinion, but do not agree with England's assessment attributed to Prigogine: "many of the properties of living things might be explainable as 'dissipative structures' that arise from a general thermodynamic tendency to reduce the rate of entropy production" (England 2015). On the contrary, after exhaustive analysis, Prigogine and his collaborators repeatedly expressed the same conclusion that the formation of dissipative structures far from equilibrium may be coupled to the increased rate of dissipation (Prigogine 1967, Prigogine and Lefever 1975). Their model calculations showed the possibility of a sharp increase of dissipation by several orders of magnitude when new structure-forming irreversible processes appear in the system (Prigogine et al. 1972). When Prigogine mentions that the occurrence of chemical instabilities depends on a minimum level of dissipation (Prigogine 1969a), it is not the requirement for minimal dissipation. Instead, it is the observation that some minimal threshold for a dissipation value must be reached as the prerequisite for the fast growth of fluctuations and chemical instabilities connected to the formation of new structures capable of producing higher dissipation (Prigogine et al. 1972).

When Prigogine introduced the concept of dissipative structures in time (oscillatory reactions) or space (pattern formation), it was a radical departure from the minimum dissipation concept (Prigogine and Nicolis 1967, Prigogine and Lefever 1968, Prigogine 1969a,b). He was inspired to drive a reorientation of the research topic pursued by his group toward pattern formation dissipative structures and self-organization phenomena (Lefever 2018) when he became aware in the spring of 1966 of Alen Turing article about the possible chemical basis of morphogenesis (Turing 1952). Prigogine realized that the *dissipative structure* concept's importance goes much beyond the morphogenetic problem originally studied by Turing. The concept is still popular in recently published research about biological rhythms, but there are also many examples of biological self-organization that do not represent dissipative structures. The self-organization of biological membranes is one such example when maintenance of lipid bilayer state does not require increased dissipation (Goldbeter 2018). The claim about all biological macromolecules and their aggregates being dissipative structures is also present in recent publications, although it goes beyond the careful classification of structures whose origin involves the passage by an instability into some **attractor** state exhibiting spatio-temporal pattern. Such claims are so broadly careless that the concept of dissipative structures loses any meaning.

Admittedly, Prigogine himself helped to introduce confusion. He was naturally attached to the idea of how to generalize thermodynamics to the non-linear region. In his words: "In order to relate the concept of *evolution* to the concept of *structure* .. it is necessary to generalize thermodynamics to the non-linear region.....The main idea of the approach is to extend in some sense the minimum entropy production theorem" (Prigogine and Nicolis 1971). It is the reference cited by Jeremy England (2015). Unfortunately, Ilya Prigogine published self-contradictory opinions and opened avenues for invoking his Nobel Prize authority to support diametrically different interpretations. He nicely and honestly illustrated his meandering approach in the Noble Prize lecture (Prigogine 1978): "It should be emphasized that the theorem of minimum entropy production requires even more restrictive conditions than the linear relations. For many years, great efforts were made to generalize this theorem to situations further away from equilibrium. It came as a great surprise when it was finally shown that far from equilibrium, the thermodynamic behavior could be quite different, in fact, even opposite to that indicated by the theorem of minimum entropy production." The contradiction is best expressed in two sentences from his 1972 paper: "As evolution proceeds to a new stable regime, the system tends to increase its dissipation. Once in the new regime, the system will again adjust to the constraints and will tend to decrease its dissipation." (Prigogine et al. 1972). If *decreasing the dissipation* phrase meant going back to the situation when minimum entropy production theorem is valid (as some experts interpreted it), the contradiction is laid bare as the "unscrambling an egg" proposition. In thermodynamics, it means that the adjustment of the macroscopic system to the imposed external far-from-equilibrium constraints can happen only through decreasing the entropy production. It is sheer speculation.

Perhaps the best expression of Prigogine's reorientation has been published in his 1975 contribution (Prigogine and Lefever 1975). I collected relevant thoughts from that article and put them together while underlining his novel concept about increased dissipation during the formation and after the emergence of dissipative structures that may have been relevant for prebiotic evolution (Figure 5.1): "The essential property of the instability.... corresponds to an increase in the interactions with the environment and in the specific dissipation (i.e., dissipation per unit mass). In other words, this instability, which is triggered by nonequilibrium environmental conditions... maintaining a continuous energy dissipation, increases the level of dissipation further, and thereby

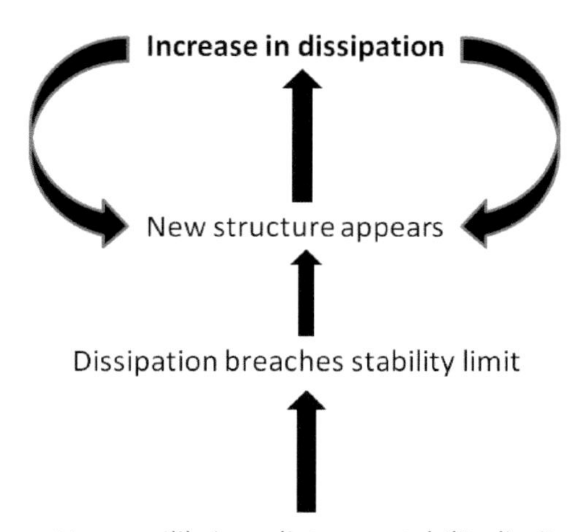

Figure 5.1: Prigogine's proposal of how evolution (prebiotic) may have been coupled to symmetry-breaking instabilities through evolutionary feedback loops creating new (more complex) structures and increasing dissipation at the same time. The stability limit is the threshold for the appearance of instabilities. Its exact value depends on the level of dissipation. When a new irreversible process does appear in the system, the threshold may be breached, while dissipation is increased above the minimally required level to maintain stability. A hierarchy of dissipative structures is bound to appear due to coupled structure formation and dissipation increasing processes. I draw this picture after reading Prigogine's papers and books published in the period from 1967 to 1975. See text for details.

creates conditions that are favorable to the appearance of other instabilities.....the interactions of living systems with the external world.... suggests a relation between structure and energy dissipation, which can be summarized by the following *evolutionary feedback*. Each time that an instability is followed by a higher level of energy dissipation, one may consider that the driving force for the appearance of further instabilities has been increased. Some irreversible processes taking place inside the system are functioning more intensely..."

Coming back to the minimum entropy production theorem, among restrictive conditions needed for its proof (linear flux-flow relationships, vicinity to equilibrium, the constancy of phenomenological coefficients, Onsager reciprocity relations) we should not forget (in the simplest case of driving and driven force-flux pair) the applied restriction of one firmly fixed external force (the driving force) and other completely free to change (the driven force) until driven flux vanishes. It is not easy to find in nature the examples of dissipative systems when all of these conditions are faithfully obeyed. Instead of giving up, most authors dealing with this topic appealed to some of the ill-defined transformations of the minimum entropy production theorem into the minimum entropy production principle. Since Prigogine himself mentioned many different examples in which far-from-equilibrium systems do not behave according to the minimum entropy production principle, it is unjustified to present his modeling results or natural examples for dissipative systems as the support for that principle. Nevertheless, the statements of the type "Prigogine proved that chemical species dynamically form an oscillatory pattern of minimum of entropy production, *MinEP*" can be easily found in recent publications (Veveakis and Regenauer-Lieb 2015).

5.7 Prigogine's authority promoted skepticism toward the study of increased entropy production response after external forcing

Ilya Prigogine was well aware that photosynthesis corresponds to quite a complex set of non-linear chemical reactions taking place far-from-equilibrium. Additionally, the initial interaction between radiation and matter produce excited nonequilibrium states hard to describe with existing thermodynamic tools (Prigogine 1969a). He never attempted to apply his minimum entropy production theorem (or any of the attempted generalizations labeled as principles) to photosynthetic systems. He encouraged others to find evidence about the formation of dissipative structures due to photosynthesis. It is not clear if he was aware of Ziegler's formulation of MaxEP principle (Ziegler 1958, 1961), or of Jaynes conjecture (Jaynes 1980) that perhaps MaxEP principle for prescribed conservation laws is much closer to the spirit of Gibbs' work than the MinEP principle. It took quite some time for researchers in the field to realize how Prigogine's authority promoted the skepticism toward well-founded contributions that seemingly contradicted Prigogine's principle of minimum entropy production (Dewar 2003, Martyushev and Seleznev 2006, Bordel 2010). It also took some time until maximum entropy production was used to characterize a body's behavior in a radiation field (Würfel and Ruppel 1985). The analogy with chemical reactions was found to be helpful by these authors, as we also found for photosynthesis (Juretić and Županović 2003). While comparing Ziegler's and Prigogine's search for a postulate governing chemical reactions, Würfel and Ruppel (1985) concluded that the application of Ziegler's MaxEP principle is a better choice (Ziegler 1983a,b). Bordel (2010) concluded that the maximum entropy production principle should be accepted by the scientific community as the fundamental principle of nonequilibrium thermodynamics.

More recent formulation of the maximal entropy production (MaxEP) physical principle (Dewar 2003, 2005, Dewar and Maritan 2014) is intimately connected with well known maximal information entropy principle (Edwin Jaynes references from 1950 to 2003, which can be found at http://bayes.wustl.edu/etj/etj.html). Despite serious critique by some physicists (Gemmer et al. 2009), the MaxEnt principle is widely accepted. While it is restricted to the steady-state in its present formulation, it is not restricted to linear flux-force relationships. The nonlinear flux-force relationships are not exception but a rule in enzymatic kinetics and most of biology. Nonlinearity is also the feature that makes climate prediction so difficult, and it is of interest that one of the first papers about the maximum entropy production principle appeared in "Nature" concerning climate prediction (Paltridge 1979).

5.8 Where we stand with defining life?

Historically speaking, defining life was anything but a trivial task. More than 100 non-redundant definitions were published from the year 1802 onward (Trifonov 2011, Koh and Ling 2013). Even NASA went into this field by adopting the Joyce definition from 1994: "Life is a self-sustained chemical system capable of undergoing Darwinian Evolution." NASA was not in the least troubled by the obvious consequence of this definition that mules and distinguished scientists that are past reproductive age should not be considered alive. More importantly, none of these definitions mentioned electrochemical processes, nor entropy production, as if these essential physical manifestations of active living systems on Earth are not always associated with a frenzy of cellular metabolic processes. Cornish-Bowden and Cárdenas (2019) observed that definitions of life do not mention metabolic regulations either, but there are some exceptions. Macklem and Seely did consider the dissipation and self-regulation of genes and metabolic networks (Macklem and Seely 2010). They repeated the older conclusion that "nature abhors the gradient" (Schneider and Sagan 2005) and takes the most efficient way to abolish it. The pathways of least resistance are chosen for the flow of energy and matter. The next chapters will expand on the connection among metabolic regulation, entropy production, and origin of life theories after we first learn in more detail how the electrochemical process of charge separation establishes the protonmotive force and membrane potential.

5.9 Dissipative adaptation concept

Let us pick up all thematic threads from this chapter to see if we can create a coherent common picture. We first interpreted the stability paradox noted by Schrödinger as the necessity for living systems to maintain dynamic stability in the face of internal entropy rising tendency and external instability-causing disturbances. The steady-state nonequilibrium condition is never fully fulfilled by living systems, but it is approximately maintained during short periods. It helps, of course, if we can imagine that only one external force exists to keep the system from relaxing toward thermodynamic equilibrium and that it does not change during the same time period. Suppose it suddenly disappears because the system became isolated. We can expect the system to relax with a maximal possible speed toward equilibrium, thus choosing the maximum entropy production pathway (Kuić et al. 2012). If the force is set up to persist, the microscopic irreversible events in the system (nanocurrents, collisions) will adapt a distribution of microscopic fluxes, which can most efficiently increase its internal entropy production until it becomes equal to externally observed entropy production in the established stable steady-state, just as Ziman described (Ziman 1956, 1960). Further increase of external force can cause additional instabilities to appear, leading to a fast transformation of microscopic into macroscopic fluctuations, a structure-forming transition toward new steady-state coupled to considerably increased entropy production of emerged dissipative structure, as Prigogine observed (Prigogine et al. 1972). "Far-from-equilibrium chemistry leads to possible 'adaptation' of chemical processes to outside conditions", wrote Prigogine and Stengers (1984). Far-from-equilibrium chemistry from a thermodynamic perspective cannot mean anything else but high dissipation. It appears that Jeremy England's phrase "dissipative adaptation" (England 2015, Perunov et al. 2016, Kachman et al. 2017) is not a truly novel concept. It also covers much broader ground then he intended, starting from early insights of Ziman and Prigogine to recently found examples of chemical or light-stimulated dissipative self-organization (Ragazzon et al. 2015, Te Brinke et al. 2018, Ropp et al. 2018, Ragazzon and Prins 2018, Fornalski 2019, Bochicchio et al. 2019). Dean Astumian was rightly disappointed (Astumian 2018a) that the *dissipative adaptation* phrase was so eagerly adopted in an explosion of publications after Jeremy England presented it in his 2015 paper, without mentioning Dean's much earlier contributions to stochastic pump theory (Astumian et al. 1987). Professor Astumian went a step further with the claim that "the basic claim of the dissipative adaptation model is wrong" (Astumian 2018b).

I can agree with the spirit of Astumian's assessment in the case of enzymatically-induced structure formation, namely that dissipation alone is not sufficient if kinetic asymmetry constraints are not taken into account too. Paradoxically, far-from-equilibrium steady-state condition of life-like systems can persist if maintained through dynamic kinetic stability, the stability through constant change (Pross 2012, Lehman 2013). Jeremy England's idea about "dissipated work" energy alone driving the nonequilibrium scenario of self-assembly is rooted in Crooks' finding (Crooks 1999) that the entropy production is simply related to the work performed by the applied field and free energy change. Chemistry and enzyme-catalyzed biochemical reactions, in particular, introduce the need to take into account kinetics as well. Biological systems are adept at proliferating chemical affinities as new derived forces whenever new means of accelerating chemical reactions are discovered during biological evolutions. It can lead both to additional internal entropy production and additional structure creation, while catalytic efficiency is increased. We shall discuss these topics in the following chapters.

The dissipative adaptation concept can be applied to microscopic dissipation events as well, just as Ziman described them in his insightful contributions (Ziman 1956, Ziman 1960–2003). When any condensed matter system is subjected to constant external field and firmly kept at steady-state isothermal condition, it must adjust its internal dissipation caused by particles' collisions. The internal dissipation must be exactly balanced with exported heat and imported entropy-lowering drift due to field effects. Ziman's formulation of the general variational principle for transport theory is equivalent to the statement that microscopic dissipative adaptations take place in the shortest possible time, thus producing the maximal amount of entropy subject to imposed constraints. It remains to be seen to what extent Ziman's principle can be extended to living systems, which also have the capability to maintain the steady-state by balancing internal entropy production with negative entropy import (as Schrödinger observed) and positive entropy export. Eventually, the steady-state becomes incompatible with the instability of biological macromolecules and free-energy storage processes preceding self-replication. As Jeremy England observed, replication does not require less but more of the work done on the system to be dissipated (England 2015). The microscopic dissipative adaptation is transformed into macroscopic dissipative adaptation. Living cells are not passive receivers for work done on them but are actively seeking richer free-energy sources. Through mutations expressed as novel structures and enhanced catalytic capabilities of enzymes, cells can gain the ability to extract more power even from an unchanged environment. Higher dissipated power has the advantage of making it impossible for a living system to go back the way it came. Hence, the evolutionary history of the living world, be it a single cell, single organism, single species, or ecological system of many interacting species, plays a crucial role in the magnificent symphony of life.

5.10 Self-emergence models

We shall devote several of the following chapters to explore how the MaxEP principle can be applied to biological macromolecules and bioenergetic systems. Close to the end of this chapter, let us turn back to the question of what is life and how physics can help to illuminate that question. The succus of Schrödinger's questions comes down to two life mysteries he clearly exposed in his 1944 booklet: first, which biological macromolecules are responsible for hereditary characteristics and how inherited information is maintained through eons, and second, how the spontaneous emergence of self-organized order is possible at all (Macklem 2008). The first question initiated revolutions in genetics, molecular biology, and biophysics. The reductive approach sufficed to tackle that challenge with complete success. The second question led to the realization that reductionism cannot solve the mystery of self-emergence. To many distinguished physicists, who have chosen to enter biology by attempts to solve the problem of consciousness, it was not clear how important is the profound understanding of bioenergetics, biochemistry, system biology, evolutionary systems theory, neurophysiology, and far-from-equilibrium thermodynamics in the nonlinear regime. It is far better to first study simpler emergence models such as locomotion and self-aggregation of amoebae.

Amazingly enough, the pattern formation by slime molds *Physarum polycephalum* inspired the building of the Physarum-machine to test how well the 3D version of that algorithm can predict the observed Cosmic Web network for 37662 galaxies (Burchett et al. 2020). The results supported the conjecture that gravitationally relaxed matter from the intergalactic medium has the tendency to form optimal transport networks, just as unicellular *P. polycephalum* does, while it forages to connect food sources. The gravitational attraction of galaxies effectively serves as a "food" source for mostly invisible circumgalactic medium and completely invisible dark matter that molds the Cosmic Web—a fascinating example of how bioenergetics bridges life and the universe!

Physarum polycephalum is a protist, not a fungus. Its cell can form an extensive network of interconnecting veins through which cytoplasm streams in pulsatile contractions covering an area of about one thousand cm^2 (Beekman and Latty 2015). We can be disgusted with the yellowish plasmodium of *P. polycephalum* forming over tree chunk, which is so radically different from the beautifully sculptured human body with just the right amount of smooth-looking muscles. Or is it so different? Actin-myosin fibers from muscles move our arms or legs as we wish. Respiring mitochondria satisfy the voracious energy needs of an athlete's muscles. The surface area of the inner mitochondrial membrane with respiratory complexes is around 14 thousand square meters in an average adult male, who consumes around 380 liters of oxygen per day. Top athletes can sustain ten times greater oxygen consumption rates during the speed burst (Rich 2003). Locomotion bioenergetics of *P. polycephalum* also uses numerous mitochondria and the contraction relaxation cycles of actin-myosin fibers in food foraging.

The expressed opinions in the current literature are sharply divided regarding the thermodynamics of unnervingly high brainless intelligence exhibited by *P. polycephalum* in finding the best solution to complex networking problems (Beekman and Latty 2015). Most authors claim that optimized flow reduces power consumption (Oettmeier and Döbereiner 2019). Others (Satterwhite-Warden et al. 2015, Kondepudi et al. 2015) recall that entropy production is increased due to increased particle motion. Collisions cannot be avoided during co-operative behavior and collective foraging for free energy. A time-evolution takes place toward states of higher dissipation of complex self-organizing nonequilibrium systems, just as Ziman described for a considerably simpler case of the transport theory (Ziman 1956).

5.11 Life as self-organized dynamical order

While attempting to answer Schrödinger's question about the emergence of self-organized order, Ramstead and collaborators (2018) felt compelled to cite the influential concept of self-organized criticality (Bak et al. 1988). Per Bak (1947–2002) and his collaborators reported the discovery of a general organizing principle governing a class of dissipative coupled systems. Self-organized criticality (SOC) is a specific bridge among life, bioenergetics, and the universe, the bridge which points toward the MaxEP principle, according to Roderick Dewar (Dewar 2003). Per Bak's personality combined the best traits of universality seeking physicists who refused to recognize boundaries between soft and hard sciences, reductionism and holism, theory, and observations. His SOC theory excelled in interdisciplinarity to such an extent that it could be applied to every branch of science, and it became one of the most cited works in theoretical physics. As an aside, he watched with growing concern how the fertile ground for original thoughts, the Niels Bohr Institute in Copenhagen, was gradually strangled by heavy-handed management of science by politicians and bureaucrats (he was collaborating there with Niels Bohr's grandson Thomas Bohr in the physics of simple systems with chaotic behavior). It was an almost identical situation with budget cuts every year to his group at the Brookhaven National Laboratory in New York, where he accepted the permanent position in 1983. However, it suited him just fine to work below the radar of gargantuan scientific programs shaped by scientific administrators with some specific applications in mind. Bak's no-nonsense pugnacious personality was never agreeable to pompous fools who crossed his path.

But let us see how bioenergetics can fit within the SOC theory. Examples he mentioned for SOC application are sand piles, rice piles, connected pendulums, current flow through resistors, the flow of rivers, earthquakes, wildfires, landslides, flicker noise in the intensity of sunspots and the light from quasars, solar flares, pulsar glitches, and starquakes. From life sciences and humanistic sciences, Bak and other authors studied the punctuations in biological evolution, extinction events, epidemics, economics, stock market crashes, and traffic jams within the SOC theory framework when the system is in a subcritical state and is ripe for collapse. Bak himself stayed away or did not have enough time to apply the SOC theory in the case of the metabolically active cells that are always poised at the brink of catastrophic autodestruction (see Chapter 14).

Initial studies of toy brain models by Per Bak and collaborators were considerably extended in recent years and compared with observations, but are still controversial (Hesse and Gros 2014, Valverde et al. 2015, Bettinger 2017). Less controversial is that the supplied energy must exceed a certain threshold for a remarkable sudden onset of coherence to occur (Pietruszka and Olszewska 2020). Why should biological systems operate close to the edge of dynamic instability? Put in different words, why should self-organization of growth benefit from an orchestrated instability or be endangered by stress? Various answers have been offered recently in this active research field. An effective response to sudden environmental change must be met by adapting power production (and associated dissipation) by coherent state transition to new spatiotemporal patterns of self-organization. The functional advantages of operating near criticality are the ability to react to highly diverse stimuli, long-range communication, associated optimal computational capabilities, and seemingly intelligent behavior in the absence of neuronal cells (Bettinger 2017, Calvo et al. 2020).

Per Bak pointed out that phase transition theory in the equilibrium statistical mechanics is fundamentally different from self-organized criticality transitions in the dynamical systems far from equilibrium. The 1982 Nobel Prize in Physics awarded to Kenneth Wilson testifies the importance of the phase transition concept when it is restricted to the equilibrium statistical mechanics. Biologists who observe and model transitions in bioenergetics often name them phase transitions without being aware of the difference. For instance, Aon and collaborators (2004) considered the isolated cardiac cell's mitochondrial network operating near the critical state. They claimed that even mild stress could induce a phase transition in the entire cell (Aon et al. 2004). When a threshold for reactive oxygen species is locally reached, the membrane potential becomes unstable, and the instability can spread throughout the mitochondrial and cellular network of the heart leading to dangerous or often fatal cardiac arrhythmia (Aon et al. 2009). Another confusion about SOC theory instabilities was expressed by theoretical physicist Smith and biophysicist Harold Joseph Morowitz (who passed away in 2016) in their recent book (2016). They decided to describe the SOC theory with the name granularity in order to distinguish granular flows from fluid flows (pages 21 and 576 from their book). The authors who studied fragile granular matter (Cates et al. 1998) carefully described generic fragility as reminiscent of SOC and never mentioned any resemblance to different types of dynamics in an evolving cell. That was probably the reason why Smith and Morowitz did not consider the SOC theory important enough for their book to deserve a serious discussion.

Schrödinger's two questions, in fact, both deal with the stability paradox. Freeman Dyson opined that the second question about metabolism had a lesser priority in Schrödinger's mind (Dyson 2009). Metabolism and bioenergetics are connected to stable processes and unstable structures. The stationary state of an active cell is not so stable as Prigogine imagined. It is the self-organized and self-regulated state with a certain degree of marginal stability maintained by dissipative processes that enable the communication throughout the entire system. Occasional catastrophic avalanches can span the whole system, just as it happens in the sand pile balanced in the critical (SOC) state. According to Dewar, the SOC emerges as the special case of the MaxEP principle in the flux driven systems (Dewar 2003). Living cells are incomparably more complex than sand piles. Still, we know that seemingly insignificant changes in intercellular or extracellular conditions can cause the inhibition of bioenergetics and cell destruction or autodestruction. Human society is also poised at the edge of autodestruction, as the COVID-19 pandemic just proved.

The SOC theory is certainly not the last word about the stability-instability dilemma for living systems. The implication of the Global Consensus Theorem (Grossberg 1980) is that the growth of any complex system will be inhibited by dwindling free-energy resources until it ends up with self-structuring to maintain homeostasis (Rosenfeld 2013). In his perspective article from 2013, Rosenfeld concluded that SOC theory and the GCT concept are both needed to study dissipative self-organization, swarm intelligence, neuronal networks, and mechanisms of carcinogenesis. Denmark and the world lost prematurely one of their most prominent scientists, "the most American of Danes," as Croatian-American scientist Predrag Cvitanović quipped. Per Bak had smoldering leukemia. Leukemia may well be the SOC theory phenomenon, together with some other types of cancer cell proliferation (Tsuchiya et al. 2016). Finally, this chapter's considerations can be summed up in the form of an answer to the question "What is Life?" *Life is a set of self-organized and self-regulated electrochemical processes capable of controlling the entropy production rate through growth and evolution of marginally stable non-equilibrium structures that enable the communication throughout the entire system* (see Chapter 13 for additional definitions of life).

Definitions and explanations from Chapter 5:

An **Attractor** is a set of numerical values toward which all the neighboring trajectories converge.

Boltzmann transport equation for charged particles is quite complex, but it can be cast into the linear relationship between forces and currents when the electric field is not overly strong.

Nanothermodynamics is the term Terrel Hill used as a shortened and more fashionable title of the research field "thermodynamics of small systems" (Hill 2001a,b). However, it is a crucial generalization of the thermodynamics of Gibbs to include small systems relevant for nanophysics and biology when fluctuations and surface effects are significant (Chamberlin 2015, Lucia 2015).

References

Andriesse, C.D. On the relation between stellar mass loss and luminosity. Astroph. J. 539(2000): 364–365.

Andriesse, C.D. and Hollestelle, M.J. Minimum entropy production in photosynthesis. Biophys. Chem. 90(2001): 249–253.

Andriesse, C.D. 2008. http://cdandriesse.byethost6.com/nabla.html?i=1.

Aon, M.A., Cortassa, S. and O'Rourke, B. Percolation and criticality in a mitochondrial network. Proc. Natl. Acad. Sci. USA 101(2004): 4447–4452.

Aon, M.A., Cortassa, S., Akar, F.G., Brown, D.A., Zhou, L. and O'Rourke, B. From mitochondrial dynamics to arrhythmias. Int. J. Biochem. Cell Biol. 41(2009): 1940–1948.

Astumian, R.D., Chock, P.B., Tsong, T.Y., Chen, Y.D. and Westerhoff, H.V. Can free energy be transduced from electric noise? Proc. Natl. Acad. Sci. USA 84(1987): 434–438.

Astumian, R.D. Stochastic pumping of non-equilibrium steady-states: How molecules adapt to a fluctuating environment. Chem. Commun. (Camb.) 54(2018a): 427–444.

Astumian, R.D. Stochastically pumped adaptation and directional motion of molecular machines. Proc. Natl. Acad. Sci. USA 115(2018b): 9405–9413.

Bak, P., Tang, C. and Wiesenfeld, K. Self-organized criticality. Phys. Rev. A 38(1988): 364–374.

Beekman, M. and Latty, T. Brainless but multi-headed: Decision making by the acellular slime mould *Physarum polycephalum*. J. Mol. Biol. 427(2015): 3734–3743.

Bertola, V. and Cafaro, E. A critical analysis of the minimum entropy production theorem and its application to heat and fluid flow. Int. J. Heat Mass Transf. 51(2008): 1907–1912.

Bettinger, J.S. Comparative approximations of criticality in a neural and quantum regime. Prog. Biophys. Mol. Biol. 131(2017): 445–462.

Bochicchio, D., Kwangmettatam, S., Kudernac, T. and Pavan, G.M. How defects control the out-of-equilibrium dissipative evolution of a supramolecular tubule. ACS Nano 13(2019): 4322–4334.

Bordel, S. Steepest entropy increase is justified by information theory. The relations between Ziegler's principle, Onsager's formalism and Prigogine's principle. Physica A 389(2010): 4564–4570.

Burchett, J.N., Elek, O., Tejos, N., Prochaska, J.X., Tripp, T.M., Bordoloi, R. et al. Revealing the dark threads of the Cosmic Web. Astrophys. J. Lett. 891(2020): L35, March 10: doi.org/10.3847/2041-8213/ab700c.

Calvo, P., Gagliano, M., Souza, G.M. and Trewavas, A. Plants are intelligent, here's how. Ann. Bot. 125(2020): 11–28.

Cates, M.E., Wittmer, J.P., Bouchaud, J.-P. and Claudin, P. Force chains, and fragile matter. Phys. Rev. Lett. 81(1998): 1841–1844.

Chamberlin, R.V. The big world of nanothermodynamics. Entropy 17(2015): 52–73.

Cornish-Bowden, A. and Cárdenas, M.L. Contrasting theories of life: Historical context, current theories. In search of an ideal theory. Biosystems (2019): 104063. doi:10.1016/j.biosystems.2019.104063.

Crooks, G.E. Entropy production fluctuation theorem and the nonequilibrium work relation for free energy differences. Phys. Rev. E 60(1999): 2721–2726.

De Groot, S.R. and Mazur, P. 1984. Non-Equlibrium Thermodynamics. Dover, New York, NY, USA.

Dewar, R.C. Information theory explanation of the fluctuation theorem, maximum entropy production and self-organized criticality in non-equilibrium stationary states. J. Phys. A: Math. Gen. 36(2003): 631–641.

Dewar, R.C. Maximum entropy production and the fluctuation theorem. J. Phys. A: Math. Gen. 38(2005): L371–L381.

Dewar, R.C. and Maritan, A. 2014. A theoretical basis for maximum entropy production. pp. 49–71. *In*: Dewar, R.C., Lineweaver, C.H., Niven, R.K. and Regenauer-Lieb, K. (eds.). Beyond the Second Law, Springer-Verlag, Berlin-Heidelberg, Germany.

Dyson, F. 2009. Origins of Life. Cambridge Univ. Press (Second Edition), Cambridge, UK.

England, J. Dissipative adaptation in driven self-assembly. Nat. Nanotechnol. 10(2015): 919–923.

Fornalski, K.W. Radiation adaptive response and cancer: From the statistical physics point of view. Phys. Rev. E 99(2019): 022139. doi:10.1103/physreve.99.022139.

Gemmer, J., Michel, M. and Mahler, G. 2009. Quantum Thermodynamics. Emergence of Thermodynamic Behavior Within Composite Quantum Systems (Second Edition). Springer, Berlin, Germany.

Goldbeter, A. Dissipative structures in biological systems: bistability, oscillations, spatial patterns and waves. Phil. Trans. R. Soc. A 376(2018): 20170376. http://dx.doi.org/10.1098/rsta.2017.0376.

Grossberg, S. Biological competition: Decision rules, pattern formation, and oscillations. Proc. Natl. Acad. Sci. USA 77(1980): 2338–2342.

Hesse, J. and Gross, T. Self-organized criticality as a fundamental property of neural systems. Front. Syst. Neurosci. 8(2014): 166. doi: 10.3389/fnsys.2014.00166.

Hill, T.L. 1964. Thermodynamics of Small Systems—Two Volumes Bound as One. Dover, New York, NY, USA.

Hill, T.L. 1977. Free Energy Transduction in Biology. The Steady-State Kinetic and Thermodynamic Formalism. Academic Press, New York. NY, USA.

Hill, T.L. 1987. Statistical Mechanics: Principles and Selected Applications, 2nd ed., Dover. New York, NY, USA.

Hill, T.L. Perspective: Nanothermodynamics. Nano Letters 1(2001a): 111–112.

Hill, T.L. A different approach to nanothermodynamics. Nano Letters 1(2001b): 273–275.

Hunt, K.L.C. and Hunt, P.M. Dissipation in steady states of chemical systems and deviations from minimum entropy production. Physica1 47A(1987): 48–60.

Jaynes, E.T. The minimum entropy production principle. Ann. Rev. Phys. Chem. 31(1980): 579–601.

Jennings, R.C., Engelmann, E., Garlaschi, F., Casazza, A.P. and Zucchelli, G. Photosynthesis and negative entropy production. Biochim. Biophys. Acta Bioenerg. 1709(2005): 251–255.

Jennings, R.C., Belgio, E., Casazza, A.P., Garlaschi, F.M. and Zucchelli, G. Entropy consumption in primary photosynthesis. Biochim. Biophys. Acta 1767(2007): 1194–1197.

Jennings, R.C., Santabarbara, S., Belgio, E. and Zucchelli, G. The Carnot efficiency and plant photosystems. Biophysics 59(2014): 230–235.

Joyce, G.F. 1994. Foreword. pp. xi–xii. *In*: Deamer, D.W. and Fleischacker, G.R. (eds.). Origins of Life, Jones and Bartlett, Boston, MA, USA.

Juretić, D. Comment on "Minimum entropy production in photosynthesis." BioComplexity 1 (2002): http://mapmf.pmfst.unist.hr/~juretic/Juretic-revised-comment.pdf.

Juretić, D. and Županović, P. Photosynthetic models with maximum entropy production in irreversible charge transfer steps. J. Comp. Biol. Chem. 27(2003): 541–553.

Kachman, T., Owen, J.A. and England, J.L. Self-organized resonance during search of a diverse chemical space. Phys. Rev. Lett. 119(2017): 038001. doi: 10.1103/PhysRevLett.119.038001.

Knox, R.S. and Parson, W.W. Entropy production and the Second Law in photosynthesis. Biochim. Biophys. Acta 1767(2007a): 1189–1193.

Knox, R.S. and Parson, W.W. On "Entropy consumption in primary photosynthesis" by Jennings et al. Biochim. Biophys. Acta 1767(2007b): 1198–1199.

Koh, Y.Z. and Ling, M.H.T. On the liveliness of artificial life. Human-Level Intelligence 3(2013): Article 1.
Kondepudi, D.K. Remarks on the validity of the theorem of minimum entropy production. Physica A 154(1988): 204–206.
Kondepudi, D., Kay, B. and Dixon, J. End-directed evolution and the emergence of energy-seeking behavior in a complex system. Phys. Rev. E 91(2015): 050902(R). doi:10.1103/physreve.91.050902.
Kuić, D., Županović, P. and Juretić, D. Macroscopic time evolution and MaxEnt inference for closed systems with Hamiltonian dynamics. Found. Phys. 42(2012): 319–339.
Landauer, R. Inadequacy of entropy and entropy derivatives in characterizing the steady state. Phys. Rev. A 12(1975): 636–638.
Lavergne, J. Commentary on photosynthesis and negative entropy production by Jennings and coworkers. Biochim. Biophys. Acta 1757(2006): 1453–1459.
Lefever, R. The rehabilitation of irreversible processes and dissipative structures' 50th anniversary. Phil. Trans. R. Soc. A 376(2018): 20170365. doi:10.1098/rsta.2017.0365.
Lehman, N. Kinetics to the rescue. Trends Evol. Biol. 5(2013), February. doi: 10.4081/eb.2013.br1.
Lents, N.H. 2018. Human Errors: A Panorama of Our Glitches, from Pointless Bones to Broken Genes. Houghton Mifflin Harcourt, New York, NY, USA.
Lucia, U. A link between nano- and classical thermodynamics: dissipation analysis (the entropy generation approach in nano-thermodynamics). Entropy 17(2015): 1309–1328.
Macklem, P.T. Emergent phenomena and the secrets of life. J. Appl. Physiol. 104(2008): 1844–1846. See also Letters to the Editor from J. Appl. Physiol. 104(2008): 1848–1850.
Macklem, P.T. and Seely, A. Towards a definition of life. Perspect. Biol. Med. 53(2010): 330–340.
Margulis, L. and Sagan, D. 2000. What is Life. University of California Press, Berkeley, CA, USA.
Martyushev, L.M. and Seleznev, V.D. Maximum entropy production principle in physics, chemistry and biology. Phys. Rep. 426(2006): 1–45.
Oettmeier, C. and Döbereiner, H.G. A lumped parameter model of endoplasm flow in *Physarum polycephalum* explains migration and polarization-induced asymmetry during the onset of locomotion. PLoS ONE 14(2019): e0215622. doi.org/10.1371/journal.pone.0215622.
Paltridge, G.W. Climate and thermodynamic systems of maximum dissipation. Nature 279(1979): 630–631.
Perunov, N., Marsland, R.A. and England, J.L. Statistical physics of adaptation. Phys. Rev. X 6(2016): 021036. doi: 10.1103/PhysRevX.6.021036.
Pietruszka, M. and Olszewska, M. Extracellular ionic fluxes suggest the basis for cellular life at the *1/f* ridge of extended criticality. Eur. Biophys. J. (2020): Mar 24. doi: 10.1007/s00249-020-01430-3.
Prigogine, I. Modération et transformation irréversible des systèmes ouverts. Acad. R. Belg. Bull. Cl. Sci. 31(1945): 600–606.
Prigogine, I. 1967. Introduction to Thermodynamics of Irreversible Processes. Interscience, New York, NY, USA.
Prigogine, I. and Nicolis, G. On symmetry-breaking instabilities in dissipative systems. J. Chem. Phys. 46(1967): 3542–3550.
Prigogine, I. and Lefever, R. On symmetry-breaking instabilities in dissipative systems II. J. Chem. Phys. 48(1968): 1695–1700.
Prigogine, I. 1969a. Dissipative structures in biological systems. pp. 162–185. *In*: Marois, M. (ed.). Proceedings of The Second International Conference on Theoretical Physics and Biology, Versailles, France.
Prigogine, I. 1969b. Structure, dissipation and life. pp. 23–52. *In*: Marois, M. (ed.). Theoretical Physics and Biology. Wiley Interscience, New York, NY, USA.
Prigogine, I. and Nicolis, G. Biological order, structure and instabilities. Q. Rev. Biophys. 4(1971): 107–148.
Prigogine, I., Nicolis, G. and Babloyantz, A. Thermodynamics of evolution. Physics Today, December (1972): 38–44.
Prigogine, I. and Lefever, R. 1975. Stability and self-organization in open systems. pp. 1–28. *In*: Nicolis, G. and Lefever, R. (eds.). Membranes. Dissipative structures and Evolution. Wiley, New York, NY, USA.
Prigogine, I. Time, structure and fluctuations. Science 201(1978): 777–785.
Prigogine, I. and Stengers, I. 1984. Order out of Chaos. Bantam Books, Toronto, Canada.
Pross, A. 2012. What is Life. How Chemistry Becomes Biology. Oxford Univ. Press, Oxford, UK and USA.
Ragazzon, G., Baroncini, M., Silvi, S., Venturi, M. and Credi, A. Light-powered autonomous and directional motion of a dissipative self-assembling system. Nat. Nanotech. 10(2015): 70–75.
Ragazzon, G. and Prins, L.J. Energy consumption in chemical fuel-driven self-assembly. Nat. Nanotech. 13(2018): 882–889.
Ramstead, M.J.D., Badcock, P.B. and Friston, K.J. Answering Schrödinger's question: A free-energy formulation. Phys. Life Rev. 24(2018): 1–16.
Rich, P. Chemiosmotic coupling: The cost of living. Nature 421(2003): 583.

Ropp, C., Bachelard, N., Barth, D., Wang, Y. and Zhang, X. Dissipative self-organization in optical space. Nat. Photonics 12(2018): 739–743.

Rosenfeld, S. Global consensus theorem and self-organized criticality: Unifying principles for understanding self-organization, swarm intelligence and mechanisms of carcinogenesis. Gene Regul. Syst. Biol. 7(2013): 23–39.

Ross, J. and Vlad, M.O. Exact solutions for the entropy production rate of several irreversible processes. J. Phys. Chem. A 109(2005): 10607–10612.

Satterwhite-Warden, J.E., Kondepudi, D.K., Dixon, J.A. and Rusling, J.F. Co-operative motion of multiple benzoquinone disks at the air–water interface. Phys. Chem. Chem. Phys. 17(2015): 29891–29898.

Schneider, E.D. and Sagan, D. 2005. Into the Cool. Univ. of Chicago Press, Chicago, USA.

Smith, E. and Morowitz, H.J. 2016. The Origin and Nature of Life on Earth. The Emergence of the Fourth Geosphere. Cambridge Univ. Press, Cambridge, UK.

Te Brinke, E., Groen, J., Herrmann, A., Heus, H.A., Rivas, G., Spruijt, E. et al. Dissipative adaptation in driven self-assembly leading to self-dividing fibrils. Nat. Nanotechnol. 13(2018): 849–855.

Tobias, P.V. 2011. Revisiting water and hominin evolution. pp. 190–198. *In*: Vaneechoutte, M. (ed.). Was Man more Aquatic in the Past? Bentham Science Publishers, Dubai, U.A.E.

Trifonov, E.N. Vocabulary of definitions of life suggests a definition. J. Biomol. Struct. Dyn. 29(2011): 259–266.

Tsuchiya, M., Giuliani, A., Hashimoto, M., Erenpreisa, J. and Yoshikawa, K. Self-organizing global gene expression regulated through criticality: Mechanism of the cell-fate change. PLoS One 11(2016): e0167912. doi: 10.1371/journal.pone.0167912.

Turing, A.M. The chemical basis of morphogenesis. Phil. Trans. R. Soc. Lond. B 237(1952): 37–72.

Valverde, S., Ohse, S., Turalska, M., West, B.J. and Garcia-Ojalvo, J. Structural determinants of criticality in biological networks. Front. Physiol. 6(2015): 127. doi: 10.3389/fphys.2015.00127.

Veveakis, E. and Regenauer-Lieb, K. Review of extremum postulates. Curr. Opin. Chem. Eng. 7(2015): 40–46.

Würfel, P. and Ruppel, W. The flow equilibrium of a body in a radiation field. J. Phys. C: Solid State Phys. 18(1985): 2987–3000.

Ziegler, H. An attempt to generalize Onsager's principle, and its significance for rheological problems. Z. Angew. Math. Phys. 9b(1958): 748–763.

Ziegler, H. Zweiextremalprinzipien der irreversiblenthermodynamik. Ing. Arch. 30(1961): 410–416.

Ziegler. H. 1983a. An Introduction to Thermomechanics, North-Holland, Amsterdam, The Netherlands.

Ziegler, H. Chemical reactions and the principle of maximal rate of entropy production. J. Applied Math. Phys. (ZAMP) 34(1983b): 832–844.

Ziman, J.M. The general variational principle of transport theory. Can. J. Phys. 34(1956): 1256–1273.

Ziman, J.M. 2003. Electrons and Phonons: The Theory of Transport Phenomena in Solids (First published 1960, Oxford University Press). Selwood Printing, West Sussex, UK.

Chapter 6

Protonmotive Force

Maintaining a distance from the thermodynamic equilibrium is obviously essential for metabolically active living organisms. They synthesize and contain biomolecules whose concentrations and structures are often far from equilibrium one. Some biological macromolecules can be very long-lasting if left alone in a suitable cold and dry environment. With some rare exceptions, none are left intact during cell or organism lifetime. Proper regulation, repair, and adaptation to environmental changes require intensive metabolic and anabolic processes to maintain persistent recycling procedures (degradation and re-synthesis). It requires a constant free-energy input, necessarily connected with related dissipation and far-from-equilibrium condition. It is not clear how to measure or calculate the displacement from equilibrium. The entropy production vanishes in equilibrium, and we expect it to be higher with higher displacement from equilibrium. We can claim that entropy production is a measure of the system's distance from equilibrium. There are other candidate functions too. The Gibbs energy change due to, for instance, electrochemical ion gradient across the membrane can also be considered as a function of displacement from equilibrium.

In the presence of a membrane permeability barrier and some charge separation device, there are two forces acting on an ion. One is due to ion's concentration gradient across a membrane, and another is due to the electrical potential difference between the aqueous phases separated by the membrane. Consider the Gibbs energy change ΔG for the transfer of 1 mol of solute across a membrane from the outer phase concentration c_A to the inner phase concentration c_B, where the volumes of the two compartments are sufficiently large that the concentrations do not change significantly. In the absence of membrane potential, ΔG is given by:

$$\Delta G(kJ/mol) = 2.3RT\log_{10}\left(\frac{c_B}{c_A}\right)$$

where $\Delta G = G_B$ (for inside ion concentration) – G_A (for outside ion concentration) and R is the gas constant.

The other force is due to the transfer of an ion driven by membrane potential in the absence of a concentration gradient. In this case, the Gibbs energy change when 1 mol of a cation with a charge m^+ is transported down an electrical potential of $\Delta\Psi$ (mV) is given by:

$$\Delta G\ (kJ/mol) = mF\,\Delta\Psi$$

where F is the Faraday constant and $\Delta\Psi = \Psi_B$ (inner solution) – Ψ_A (outer solution).

In the general case, the ion will be affected by *both* concentrative *and* electrical gradients, and the net ΔG when one mol of cations with a charge m^+ is transported down an electrical potential of ΔΨ (mV) from a concentration of c_A^{m+} to c_B^{m+} is given by the general electrochemical equation:

$$\Delta G(kJ/mol) = mF\Delta\Psi + 2.3RT\log_{10}\left(\frac{c_B^{m+}}{c_A^{m+}}\right)$$

where 2.3 factor is the approximate value for natural logarithm of 10 (ln 10 = 2.30258509…).

In this equation, ΔG is often expressed as the electrochemical potential or ion electrochemical gradient: $\Delta\mu_H$

In the case of protons, pH is the logarithmic function of proton concentration [H⁺]: pH = –log[H]), and m = +1, so that this equation becomes:

$$\Delta G = \Delta\mu_H = F\Delta\Psi - 2.3RT\cdot\log_{10}(H_{in}/H_{out})$$

or

$$\Delta\mu_H = F\Delta\Psi - 2.3RT\cdot\Delta pH$$

In the research literature, the protonmotive force is most often expressed in volts as:

$$\Delta p = \Delta\mu_H/F = \Delta\Psi - 2.3(RT/F)\cdot\Delta pH$$

The ΔpH is defined as the pH in the N-phase (negative inner matrix space in the case of mitochondria) minus pH in the P–phase (positive outer cytoplasmic space in the case of mitochondria). The Δp is the protonmotive force (pmf) or proticity, first introduced by Peter Mitchell in 1966 when he published his review on chemiosmotic coupling known as the first "Grey Book". According to this definition, it has a negative value of about –220 mV in actively respiring bacteria and mitochondria.

The pmf, as defined above, can also be considered as the work per unit charge required to move a proton from the inside to the outside of the cell. To find the relationship between work and free-energy change, we can use the equation: $\Delta G = -W_{max}$ for a reaction carried out reversibly at constant temperature and pressure. It will give us the maximum positive work that the process can perform. In that case, Mitchell's definition of the pmf (and Nicholls and Ferguson's (2002) definition):

$$\Delta p = -\Delta\mu_H/F$$

is in accord with the definition of thermodynamic force or affinity for a chemical reaction: $X = \Delta p = \mu_{out} - \mu_{in} = -\Delta\mu$, which is positive when $\mu_{out} > \mu_{in}$. In respiring mitochondria, it is indeed $\mu_{out} > \mu_{in}$, because then $\Delta\Psi < 0$ according to our ΔΨ definition (outside Ψ is usually set up at the zero value, while inside Ψ is negative) and ΔpH is positive in agreement with our ΔpH definition because matrix space becomes depleted in protons during active respiration. For active mitochondria, the pH outside is about 1.4 units lower than inside, and the membrane potential is about 0.14 V, the outside being positive. Positive work per unit charge is obtained when protons are spontaneously driven to enter the internal space by inside directed electric field and higher outside proton concentration. The Δp of 200 mV corresponds to a $\Delta\mu_H$ of 19 kJ/mol.

At room temperature, usually taken to be 25°C, and when everything is expressed in millivolts, the pmf reads:

$$\Delta p = \Delta\Psi - 59\cdot\Delta pH$$

The pmf in this expression is the work per unit charge required to move a proton from the inside to the outside of the cell, the ΔΨ is the difference in electrical potential between the inside and outside of the cell, and ΔpH is the difference in pH between the inside and outside of the cell. In the case, if we are interested in the pmf created by chloroplast, protons would go spontaneously in the opposite direction from the inside (lumen space) to the outside (cytoplasmic space).

Unfortunately, there is no uniform convention for the Δp sign. Mitchell originally defined Δp as positive when protons tend to flow from the external medium into bacteria or mitochondria. However, in the field of electrophysiology, the ΔpH is defined as $pH_{in} - pH_{out}$, while measuring electrode potential in the cytoplasm is stated relative to a reference electrode in the external medium. It would make Δp a negative quantity in mitochondria and bacteria.

When only absolute Δp value is considered or reported in experiments, it becomes clear that the transmembrane circulation of more than one proton (three or four) is needed for the synthesis of one ATP molecule by the ATP synthase. At equilibrium, the free energy change of the driving reaction (for ATP synthesis) would just balance the electrochemical potential of protons: $\Delta G = nF\Delta p$, where n is the H^+/ATP stoichiometry. Real (nonequilibrium) concentrations of ATP, ADP, and inorganic phosphate are such that ΔG for ATP hydrolysis provides 57 kJ/mols of free-energy, which means that ATP synthesis due to ATP synthase rotor rotation induced by the protonmotive force requires at least that amount of free-energy.

References

Mitchell, P. Chemiosmotic coupling in oxidative and photosynthetic phosphorylation. Biol. Rev. 41(1966): 445–502. Reprinted in Biochim. Biophys. Acta 1807(2011): 1507–1538.

Nicholls, D.G. and Ferguson, S.J. 2002. Bioenergetics 3. Academic Press, London, UK.

Chapter 7
Membrane Proteins

7.1 Introduction to membrane proteins

Membrane proteins are usually defined in biochemistry as proteins that are difficult to separate from biological membranes. Such proteins consist up to 25–30% of all cellular proteins, which are coded by 25% of known genes (Stansfeld et al. 2015), but appear as a relatively small number of identical units, which is an additional reason why their study is more demanding than the study of soluble proteins. Their structure determination by the use of X-ray crystallography is much more challenging than the crystallography of soluble proteins. Low expression of membrane proteins enhances the difficulty in isolating these proteins with hydrophobic segments from their native hydrophobic membrane environment. An additional complication in working with membrane proteins is that after isolation, it is not possible to test their transport or enzymatic activity (possibly destroyed during the isolation process) without inserting them into some artificial membrane.

Membrane proteins that are named integral membrane proteins pass their polypeptide chain one or more times through the membrane bilayer. They can perform their energy-transduction function and their transport or enzymatic function only when they are an integral part of the membrane phospholipid bilayer. Their length (total number of amino acids in their sequence) is often longer than the soluble proteins' length. It can easily reach several thousand amino acids.

After this short introduction to membrane proteins, it is only natural to ask why they are important. In short, the answer is that specialized membranes and their membrane proteins have constantly been gaining in importance during the natural evolution and selection process from archaic and prokaryotic to eukaryotic cells and multicellular organisms. So it is not an accident that today about 50 to 60% of all drugs used in medical practice have integral membrane proteins as their molecular target. The biological reason for medical importance is often their role in bioenergetics. Membrane proteins channel the protonmotive force created by respiration or photosynthesis into ATP synthesis to such an extent that about 95% of cellular ATP molecules are produced by integral membrane proteins such as the ATP synthase.

The participation of membrane proteins in a membrane bilayer organization is anything but insignificant. It differs widely among different membranes, from 18% in the case of myelin membrane to 76% weight fraction in the case of tightly packed proteins of the inner mitochondrial membrane (Guidotti 1972). At a lipid/protein ratio from 50:1 to 100:1, membrane proteins cover membrane area as large as 30% (Dupuy and Engelman 2008). Recent molecular dynamics simulations (Corradi et al. 2018) confirmed the experimental evidence of specific lipid-protein interactions. Each protein associates with a preferred lipid shell when presented with a realistic lipid environment of asymmetrically distributed lipid species between two bilayer leaflets ("fingerprint"). Corradi et al. (2018) concluded that the fingerprint concept for lipid-protein interaction is valid for membrane channels, transporters, receptors, growth factors, and even monotopic proteins such as the prostaglandin H2 synthase (see Table 7.1 for the classification of membrane proteins).

How many membrane proteins are known in the year 2020 (April)? Among 163 thousand solved Protein Data Bank (PDB) protein structures, the percentage of membrane proteins is considerably smaller from the above-mentioned estimate between 25 and 30%. There are 473 monotopic, 11958 transmembrane alpha-helical, and 1011 transmembrane beta-barrel proteins (see Table 7.1 for the classification of membrane proteins). The redundancy is quite extensive. When only the unique proteins are selected among these three classes, there are 1033 integral membrane proteins (see: https://blanco.biomol.uci.edu/mpstruc/). Thus, less than 10% of all proteins with the known tridimensional (3D) structure are membrane-embedded proteins, and less than 1% are unique proteins in the PDB database. It is still a huge advance compared to the situation I encountered 30 years ago while attempting to predict which protein segments will enter into and become an integral part of the membrane. At the Department of Biochemistry, University of Oxford, England, researchers created a database containing a total of over four thousand membrane proteins of known 3D structure (updated by using references: Stansfeld et al. 2015, Newport et al. 2019). It is still a small percentage out of 562 thousand manually annotated Swiss-Prot sequences and 178 million computationally annotated TrEMBL primary structures from the UniProt Knowledge database (UniProtKB), which are likely to contain at least 25% of proteins with the subcellular membrane location (Fagerberg et al. 2010).

Additional analysis is needed to determine the membrane association topology even for the fully resolved membrane proteins' structures. For drug applications, one must know the location of frequent drug targets: extracellular and intracellular loops. The topology determination is accomplished when the sequence location of all transmembrane segments is found and the protein N-terminal subcellular location. Early attempts to solve this problem used only the hydrophobicity of protein segments and an educated guess how the most hydrophobic membrane-associated segments are cut by virtual external and internal membrane surface (Zucić and Juretić 2004). Researchers in Sansom's Oxford laboratory realized that a much more sophisticated computational method is needed to orient a membrane protein relative to a lipid bilayer of fixed dimensions (Stansfeld et al. 2015). They used fast coarse-grained molecular dynamics simulations for automatic insertion of a membrane protein into the lipid membrane.

The Swiss-Prot is often considered to be the best-annotated protein data bank for protein sequences. Still, mistakes in the annotation of membrane proteins have been frequent. Antiquated, low accuracy tools were sometimes used to annotate membrane protein topology (sequence location and orientation of transmembrane spanning segments). Some mix-ups happened due to different trivial errors that had a tendency to be maintained and carried over to new Swiss-Prot versions. It is probably one of Murphy's laws: "The authority of errors tends to be greater than the authority of corrections" for those who believe in humans' innate fallibility, whether scientists or not. My experience, too, is that errors have truly incredible capacity to survive and to be believed. By the way, Matthews' paper from 1995 got the 1996 Ig Nobel Prize for Physics for his study if a toast ending with the butter-side down, when tumbling from a table to the floor, is the manifestation of Murphy's Law: "If anything can go wrong, it will." Errors are frequently found when one goes deep enough into the sequence annotation problem. If and when a paper is published correcting these errors, it is still far from certain that old mistakes will not be reproduced in other scientific papers or in the new versions of the same database. Students will have no doubt whatsoever that all the database data are 100% accurate.

With all these caveats taken into account, an accurate prediction of protein 3D structure based on its sequence is still a precious goal in biophysics. A excellent and highly touted AlphaFold tool for *de novo* predicting 3D protein structures (Senior et al. 2020) is not devoid of weaknesses. Two examples of weak AlphaFold performance in the **CASP**13 competition (Lepore et al. 2019) are two target proteins important in biotechnological applications. They have in common the presence of unstructured loops and either transmembrane helix (*Arabidopsis thaliana* xylan O-acetyltransferase 1) or dual topology possibility: membrane-anchored and extracellular

(*Aspergillus niger* α-xylosidase). It is no wonder. No matter how sophisticated, neural network-based tools like AlphaFold can only learn to predict similar features presented to them for training. The training set consisted predominantly of soluble proteins that are not overly difficult to crystallize. Membrane proteins and partially disordered proteins were mostly excluded from the training set. This chapter shall show that cell signaling, interactions, and bioenergetics have its microscopic origin in ordering dynamics and promiscuous interactions among highly flexible membrane-loving proteins—the most challenging structures for examining and predicting. Dual or flexible topology (section 7.3), intrinsically disordered segments (section 7.8.2, first paragraph), and self-organization of homo and heteropolymers (sections 7.7 and 7.8) are such difficult to study but crucial features.

7.2 Classification of membrane proteins

The classification of membrane proteins in Table 7.1 offers examples of peripheral proteins, monotopic, bitopic, and polytopic proteins, that are either already heavily used as drug targets (G-protein coupled receptors, prostaglandin H_2 synthase) or are potential drug design targets. I omitted less frequent proteins with transmembrane beta sheets and included membrane-active antimicrobial peptides, among other self-inserting polypeptides, as the lead compounds for drug

Table 7.1: Classification of membrane proteins that form membrane-associated helices (bold examples = important pharmaceutical targets or lead compounds for drug design).

Membrane protein type (reference)	Alternative name or common classification	Binding character	Examples
Peripheral proteins	Extrinsic protein Amphitropic protein	Weak and reversible	**Myelin basic protein,** **Annexins,** **Phospholipase C,** **Ewing sarcoma protein**
Lipid-anchored proteins		Protein has covalently linked lipid groups which insert into membrane	GPI-linked protein (GPI = glycosylphosphatidylinositol), **Penicillinase,** **cAMP protein kinase,** **Src family kinases,** G proteins
Monotopic proteins (Allen et al. 2019)		Insert in but do not span the membrane	**Caveolins,** **Monoamine oxidase,** **Prostaglandin H_2 synthase,** **Fatty acid amide hydrolase,** **Squalene-hopene cyclase,** **Microsomal cytochrome,** **P450**
Bitopic proteins (Pogozheva and Lomize 2018)	Integral membrane proteins	Have only one transmembrane (TM) span	**Glycophorin,** **Syndecans,** **Cadherin,** M13 coat protein, Phospholamban, FtsQ cell division protein, MotB chemotaxis protein
Polytopic proteins	Integral membrane proteins	Have more than one transmembrane (TM) span	**G-protein coupled "serpentine" receptors,** **Ion channels,** **ABC transporters** **Presenilins**
Toxins and self-inserting polypeptides	Soluble amphitropic	Perturb bilayer	Diphteria toxin, Colicins, α-hemolysin, **Antimicrobial peptides,** Ionophores

design (Juretić and Simunić 2019). Before considering integral membrane proteins, let us mention the class of monotopic membrane proteins (Bracey et al. 2004, Allen et al. 2019), which is very important in regulating different functions in complex organisms such as ourselves. The widespread and long-term aspirin usage (acting on the prostaglandin H_2 synthase) with minimal side effects (such as increased bleeding after some surgical procedure) testifies to that effect. There are likely many more essential monotopic membrane proteins. At present, their monotopic membrane structure cannot be easily predicted due to difficulties in distinguishing them from soluble and integral membrane proteins. Researchers interested in discovering additional monotopic membrane proteins by using some rational method would have to develop dedicated bioinformatics tools. One can take the challenge to modify the algorithm SPLIT (Juretić et al. 1999, 2002) toward this aim of discovering additional potential candidates for monotopic membrane proteins because SPLIT is very fast in finding potential membrane buried helices that are not transmembrane, but its accuracy for predicting such short membrane-associated helices has not been tested. An additional complication is that membrane-associated helices often exhibit different inclination with respect to membrane surface from horizontal to almost perpendicular. The inclination depends very much on helix **amphipathicity** and net charge.

7.3 The reentrant topology of monotopic membrane protein

One can find only 25 nonredundant monotopic membrane proteins in the PDB database (Allen et al. 2019). They are underrepresented among all nonredundant (less than 70% sequence identity) PDB proteins perused in 2019: ~ 35000 soluble, 1700 integral, ~ 1400 polytopic, ~ 1350 peripheral, ~ 260 bitopic, and ~ 25 monotopic (0.06%). The membrane association frequently occurs through electrostatic and hydrophobic interaction of amphipathic helices oriented parallel to the membrane plane. Deeper entry into a single leaflet of the lipid bilayer is usually through the reentrant membrane helix or long hydrophobic loop. Due to the paucity of solved structures, the prediction of surface helices and reentrant segments is an important but difficult and rarely attempted undertaking. The additional complications are experimentally detected dual or flexible topology cases—the same sequence can code for more than one topology (Granseth 2010). Dual and multiple topologies are often associated with one or more helices with marginal ability to become membrane buried due to insufficient hydrophobicity or an excess of asymmetrically positioned charged residues.

The "positive-inside" rule helps to predict the transmembrane topology of integral membrane proteins (von Heijne 1989, Juretić et al. 2002, Granseth et al. 2005). Positively charged amino acids, lysines (K), and arginines (R), and their motifs are predominantly located in the cytoplasmic loops according to that rule, thus ensuring the topological stability when the KR bias is either < -2 or > 2. The corollary of the "positive-inside rule" with a tentative title "weak positive-inside rule" applies to dual-topology proteins that have lesser positive charges in their extra membrane loops and the KR bias close to zero. Dual topology is convenient for some integral membrane proteins from viruses because it enables the dual function as well: receptor binding and virus assembly (without counting additional regulatory functions) (Lambert et al. 2004). Multifunctionality evolved in viruses in order to have a maximal amount of information packed in the small size of their genome.

A low KR bias does not help to predict the sequence location of reentrant domains, but the recognition of topological determinants does. Small bacterial phosphoglycosyl transferases, exemplified by the *Campylobacter* PglC, have been predicted with N-terminal transmembrane helix extending from residue 12 to 35 (von Heijne 1986). This annotation survived in the UniProt (code: A7ZET4_CAMC1) up to the present day, as a disconcerting example of the above mentioned Murphy's law. Paradoxically, the X-ray structure from PDB (code: 5W7L), and corresponding reference by Ray et al. (2018), can be found in the same (unreviewed) UniProt entry, although these authors clearly presented the 7 to 36 segment as a reentrant helix-break-helix penetrating 14 Å into the cytoplasmic face of the membrane and reemerging on the same face. The helix 7-36 is broken into two helices A and B by a Ser-Pro motif. The Ser-Pro motif has the deepest membrane penetration. It produces the "V" shape with an inter-helix angle of 118° attuned to return the B helix C-terminal to the membrane surface.

Both SPLIT tool versions (3.5 and 4.0) make the same mistake of misrepresenting the membrane-associated N-terminal segment as a transmembrane helix. Only the zero charge bias and OUT topology predicted by SPLIT 4.0 indicates that something may be wrong with that prediction. Indeed, predicting monotopic PglC as bitopic membrane protein causes the wrong prediction for the subcellular location assignment (extracytoplasmic) for more than 80% of residues, including the catalytic domain. The N-terminal domain's topological determinants are the break between N-terminal helices containing the serine-proline kink, a characteristic hydrogen bond network, and conserved basic residues flanking the helices. The cytoplasmic location of those and all other basic residues is seemingly in accord with the "positive-inside rule" (von Hejne 1989). Common tools for the topology prediction of membrane proteins can recognize some of these topological determinants but not all of them, making it likely that some other monotopic membrane proteins are also wrongly assigned as bitopic proteins.

Lomize et al. (2012) server https://opm.phar.umich.edu/ comes handy for quick examinations of which sequence segments form reentrant membrane loops and how deep they enter into the membrane bilayer. Excreted small phospholipases are multifunctional proteins with a potent bactericidal activity (Weinrauch et al. 1998). Only shallow entry in the outer membrane leaflet is seen from modeling experiments (Lomize et al. 2012). Break-inducing Pro residue or Ser-Pro residues are found just after the calcium-binding site and before amphipathic helices of excreted mature phospholipase A2 with assigned monotopic structure. For sequence comparison, 3D visualization, and sequence location of helices "kissing" the membrane, see, for instance, the human enzyme (codes: 1N28, PAG2A_HUMAN), and porcine enzyme (codes: 1P2P, PA21B_PIG). SPLIT algorithm errs for these monotopic enzymes in predicting the transmembrane helix for immature protein N-terminals known to be **signal peptides** (see sections 7.8.2 and 7.8.6 for other examples of signal peptides). Mature phospholipases can be attached or released from the membrane. Without molecular dynamic simulations, it is understandably difficult to predict what will happen even with solved X-ray or NMR structure (Balali-Mood et al. 2009). Considerably larger calcium-independent phospholipase A2 (iPLA$_2$) has three membrane docking regions: the segments 552–555, 643–646, and 710–724 (Bucher et al. 2013), each containing at least one proline. The longest segment has the motif Pro-Ser-Asn-Pro that directs the entrance of short amphipathic helix to shallow depth into the membrane.

7.4 Helix-break-helix structure of membrane-associated antimicrobial peptides

Nature found the simpler usage for the Ser-Pro kink, or just the Pro kink, in numerous antimicrobial peptides (AMPs) that form the helix-break-helix or helix-kink-helix structure after coming in contact with the membrane (Kozić et al. 2018, Juretić et al. 2018). The "positive-outside rule" can be formulated for those AMPs that have both the central kink-causing proline and basic residues close to sequence ends. Why "positive-out" and not "positive-in" rule? The empirical "positive-in" rule emerged because, firstly, the proteins destined for cytoplasmic membrane association are synthesized inside in the cytoplasm, and secondly, membrane bilayer is asymmetric in such a way that the inner leaflet has a net negative charge for eukaryotic cells (von Heijne 1992). Basic proteins or peptides acting from the outside naturally find that the anionic external membrane surface from prokaryotic cells is much more attractive for them than the neutral surface of eukaryotic membranes. Hence, the "positive-outside rule" can apply for their reorganization upon attachment to the membrane surface.

We would expect that helical antimicrobial peptides with a net positive charge and central kink, break, or V-turn are selective peptide antibiotics that can distinguish bacterial from the eukaryotic membrane. To see how active and selective they are, let us calculate their relative overall performance with respect to pexiganan—the best-known peptide antibiotic. Overall performance is defined as the product of mean antibacterial activity and selectivity index (Juretić and Simunić 2019). A quick perusal of natural AMPs with membrane-associated helix-(Pro-break)-helix structure rank ChMAP-28 (Panteleev et al. 2018), papiliocin (Lee et al. 2015), cecropin-B (Wang et al. 2018), buforin II (Kobayashi et al. 2000), and malectin (Čeřovský et al. 2008), respectively, as 1600, 190, 10, 8, and 1.5 times better than pexiganan in the overall antibacterial performance. Also, designed helical AMPs with inserted central proline maintain potent antibacterial activity and gain the ability to differentiate between prokaryotic and eukaryotic cells. Examples are P18 (Shin et al. 2001), HPA3P2 (Lee et al. 2014), and RL with rigid D-Pro-Gly turn (Shao et al. 2018). Cationic "boomerang" peptides often have anticancer and antiviral activity, which is not surprising when corresponding external membranes have a net negative charge (Juretić et al. 2020). The break motif between helices can be devoid of proline residues and is more flexible (even able to form helical hairpin structure) when it does not contain proline (Juretić et al. 2018, Kozić et al. 2018). Antimicrobial and cytotoxic activity of AMPs (Juretić et al. 2020) is mostly attributed to the catastrophic collapse of cellular bioenergetics after these peptides produce large-enough membrane permeability increase to prevent the ATP synthesis by mitochondria (Westerhoff et al. 1989). Using the design methods I recently described (Juretić et al. 2020), some membrane proteins' reentrant regions can be transformed into potential peptide antibiotics. The native function of membrane-docking domains can be co-opted and re-directed toward desired medical applications. However, a note of caution is needed here. Anything that may mimic the important part of some hub protein from animals can be potentially dangerous for the host organism.

7.5 Membrane-associated amphipathic helices

There is no sharp boundary between peripheral and monotopic membrane proteins. Amphipathic helices can serve either for weaker membrane association in peripheral proteins or for stronger association with a single bilayer leaflet when part of the helix-break-helix motif. For instance, the longest alpha-amphipathic segment 586–617, from the C-terminal of peripheral Ewing sarcoma (EWS) protein, can serve for weak attachment to the membrane (possibly after arginine methylation). This protein functions as the RNA-binding transcriptional repressor. It shuttles among the nucleus, cytoplasm, and cell surface (Zakaryan and Gehring 2006), acting as an important hub that interacts with 94 disease-associated proteins (Belyanskaya et al. 2001). This protein offers the challenge of recognizing glycine-rich motifs and arginine-rich motifs that may be important in promoting shuttling through and reversible surface attachment to different membranes.

We used the SPLIT bioinformatic tool applied to UniProt entry Q01844 (EWS_HUMAN) to find the sequence location of the EWS membrane anchor segment: GMFRGG**R**GGD**R**GGF**R**GG**R**GMD**R**GGFGGG**R**RGG. Four arginine-rich motifs of the type RXXXR and RXXXXXXR (where X means any amino acid residue) are highlighted using the bold font for corresponding arginines. These motifs may promote the interaction of cationic helices with negative charges of membrane phospholipids. There are also 10 "small" motifs of the type GXXXG (where X can be any amino acid) that promote interaction of helices in a membrane environment and the oligomerization of amphipathic alpha-helices after peptide-membrane interaction occurs (Walters and DeGrado 2006). These motifs are highlighted by using the gray background when they appear in the 586–617 amphipathic segment. The membrane anchor segment is located next to the PGGP motif, which is known to bind to the Src-homology-3 (SH3) domain frequent in some lipid anchored proteins such as the Src signaling kinase myristoylated on its N-terminal glycine.

7.6 Pore loop domains from ion channels

The reentrant loop may have a more complex secondary structure, such as the loop-turn-helix or helix-turn-loop conformation. The helix-turn-loop reentrant domain contains a vitally important selectivity filter close to the external face of voltage-gated channels. The TVGYG selectivity filter motif of the KcsA potassium channel assumes an extended conformation. It is located in the sequence after the short helix-turn region that enters less than halfway into the membrane (Zhou and MacKinnon 2003). A tetrameric potassium channel is formed out of four monomers. Four TVGYG motifs construct a tight passage for ions. Potassium ions are specifically selected to enable fast facilitated diffusion when the channel is open. The proline residue appears shortly after that motif to ensure the turn toward the second transmembrane helix. Thus, reentrant loops are present in polytopic membrane proteins too, where they play the fundamental role in controlling the bioenergetics of our muscles and brain. The KcsA channel does not contain the voltage sensor, but it was very useful for solving the structure of voltage-gated channels.

Roderick MacKinnon got his Nobel Prize in Chemistry 2003 "for structural and mechanistic studies of ion channels." During the Nobel Prize lecture, Roderick admitted that he became "fascinated with understanding the atomic basis of life's electrical system" (MacKinnon 2004). He knew that polytopic membrane proteins with voltage-gated channel function must have at least four essential segments: the voltage sensor, pore, selectivity filter, and gate. It was his great pleasure to deduce from experiments the tetrameric architecture of K^+ channel in which four subunits form the central pore for ion pathway. The existence of a "pore loop" in each of four subunits was the next conclusion in the deduction process. It brought together the meaning of pore-inhibition experiments with toxins, a high conduction rate of 10^8 potassium ions per second, and amino acid substitution experiments within the "loop", which affected the channel's ability to discriminate between K^+ and Na^+ ions in better than 1000:1 ratio. It became clear that four pore loops formed the selectivity filter as an integral part of the pore. His group started thinking about voltage-gated channels as remarkable molecular machines able to combine high selectivity and high conductivity at the same time.

Exceptional conservation of the signature sequence for the pore loop selectivity filter (such as the TVGYG motif mentioned in the previous paragraph) was found from bacteria to higher eukaryotic cells. When biological evolution preserves something essentially unchanged for billions of years, notwithstanding all extinction events, it speaks to you in a laud language whose meaning can be decoded. It turned out that humans have many different ion channels, some of them associated with medically important channelopathies (Ashcroft 1999, Spillane et al. 2016). The activity of billions of ion channels must be coordinated and regulated to move your pinkie as you wish! Among potassium channels, some have six transmembrane helices in one domain and identical selectivity filter TVGYG to the one present in the bacterial potassium channel domain that has only two transmembrane helices.

MacKinnon and his group made an educated guess that the bacterial KcsA channel has many features with good enough similarity to eukaryotic potassium channels. Thus, it warranted the long-time commitment of producing its high-resolution structure—the expensive and difficult effort with an uncertain outcome in the 1990s. Books and reviews dealing with ion channels' biophysics always mention the KcsA K^+ channel as the first selective ion channel for which the high-resolution structure has been determined by MacKinnon and his group (Doyle et al. 1998, Jackson 2006). In the recent review about the Handbook of Ion Channels (Zheng and Trudeau 2015), McClintock and Kaufman (2018) mention that more than 300 thousand scientific papers have been published about ion channels! In my experience, almost all voltage-gated ion channels form selective membrane pores by using the reentrant loop with pore helix and selectivity filter. After the application of the SPLIT online tool, the pore helix sequence location can be recognized in seconds for most of them.

7.7 Monotopic lipid-anchored protein caveolin

7.7.1 How dynamic beauty of eukaryotic plasma membrane decorations emerged during evolution?

Small plasma membrane invaginations named caveolae were discovered about 70 years ago. These Ω-shaped invaginations, with flask-like or bulb-like appearance, decorate the internal leaflet of a cytoplasmic membrane with "little-caves" in a dynamic fashion some authors named "kiss-

and-run" type of behavior (Mohan et al. 2015). Some cell types have 30% to 50% of their cytoplasmic membrane face covered with caveolae (Drab et al. 2001, Parton 2018), about 200 thousand of them per single cell (Sampson et al. 2007). At the beginning of the 21st century, caveolae were upgraded to the *bona fide* status of organelles when scientists finally recognized their extraordinary wide range of cellular functions (Head et al. 2014). These membrane structures are veritable signaling hubs at the plasma membrane surface composed of many different proteins with specific lipid and protein preferences (Lamaze et al. 2017). The constant recycling of caveolae requires the uninterrupted free-energy input accompanied by a high dissipation rate.

Apart from the physical connection to continuously recycled far-from-equilibrium dissipative structures, inverted flower-like-bulbs of multiple caveolae have a fascinating artistic beauty and complex symmetry (Golani et al. 2019). The plasma membrane of epithelial cells is richly adorned with inside oriented caveolae and deep tubules, and outside projections named cilium and microvilli (Head et al. 2014). The outside surface is ornamented too with carbohydrate branches of glycoproteins, while the inside surface is connected to the energy-dependent and dynamic structure of the **cytoskeleton** network. I would say that all these cellular structures in animals have the unique beauty inseparably connected with their dynamic free-energy requiring functions of providing the interface to the outside environment (surface of epithelial cells) or the blood flow (surface of endothelial cells).

Regarding molecular details of how such an intricate structure-function connection developed during biological evolution, one can point the finger toward the introduction of cholesterol, sphingolipids, and phosphatidylcholine that allowed for the complexity increase of eukaryotic membranes about 2.5 billion years ago when environmental oxygen concentration increased. Caveolins are major caveolae proteins known as cholesterol-binding proteins. Caveolins appeared together with invertebrate and vertebrate animals as a relatively late invention in eukaryotic evolution (Kirkham et al. 2008). They are of fundamental importance in mammalian cells for mammary gland development and lactation. Their small size is at the borderline between longer peptides and small proteins. But what a punch these small proteins have within themselves when they form caveolae at the cell surface! We can postulate the emergence of lipid rafts as precursors of caveolae and other specialized surface structures. For that to happen, lipid and protein evolutionary innovations' synergy had to translate into their physical associations. Indeed, all three known caveolins have the cholesterol-biding motif and associate with cholesterol, sphingomyelin, phosphatidylserine, and phosphatidylinositol rich lipid rafts at the cytoplasmic membrane face (Hirama et al. 2017).

7.7.2 Mechanoprotection by flattening the caveolae

Caveolae bulbs enhance the asymmetry between the external and cytoplasmic face of the plasma membrane. For caveolae rich cells, the cytoplasmic leaflet area exceeds that of the outer membrane leaflet for about 15 to 20% (Parton et al. 2020). Whatever the reason for imposed flattening, this area difference cannot persist after caveolae become transformed into protein-lipid rafts. The elimination of area difference can be achieved by rapid and directional flip-flop transbilayer movement of membrane lipids from internal to the extracellular leaflet. Cholesterol is the most likely candidate for fast flipping, depletion of the cytoplasmic leaflet, and cholesterol enrichment of the extracellular monolayer. The cholesterol transport in the opposite direction, toward intracellular lipid storage droplets (Le Lay et al. 2009), or other internal organelles (Mundy et al. 2012), is also mediated by caveolae dynamics. Caveolae then detach from the membrane to form cavicles (caveolin-coated vesicles). Cavicle trafficking to and from the plasma membrane by using cellular cytoskeleton (tubulins and actins) maintains cholesterol homeostasis (Mundy et al. 2012). Newly synthesized cholesterol molecules are transported from the endoplasmic reticulum to the plasma membrane. On the other hand, caveolin-1 is the major player in endocytosis and transcytosis (Mergia 2017, Krishna and Sengupta 2019)—vitally important processes that are also linked to the pathogenesis and viral infections.

Caveolin oligomers are instrumental in forming higher-order caveolae structures with about 144 caveolin-1 molecules neatly arranged as dimmers. When stretched by mechanical stress, the programmed caveolae collapse occurs to save the cell from membrane rapture. Remaining caveolae lipids and integral membrane proteins cannot be distinguished then from almost planar lipid rafts. For instance, giant vacuolated cells of the notochord have a high density of caveloae to buffer mechanical tension and safeguard spine development (Garcia et al. 2017). It is a common protective mechanism in fish, humans, and all other chordates having the notochord. The notochord regulates organogenesis and the maintenance of structural spine integrity despite mechanical stress.

This nice protective mechanism depends on caveolins and associated cavin proteins' ability to induce the negative curvature in the neck region close to the membrane inner surface and positive curvature in the internally extended bulb region (Parton et al. 2020). Cells' mechanoprotection acts by flattening caveolae in response to increased membrane tension. A remarkable device to protect the topology of a fragile membrane extended for thousand of nanometers in any direction but no more than about four nanometers thick in its hydrophobic interior! Those who worked with black lipid membranes, as human-made imitations of biological membranes, know very well how quickly and easily they rapture even without some obvious mechanical cause.

7.7.3 Caveolin-1 multitasking roles are implicated in about 300 biological processes

The mechanoprotection is far from being the only benefit provided by membrane-lipid rafts and caveolae as plasma membrane signaling platforms and organelle hubs integrating multiple biological processes. According to UniProt annotation, the caveolin-1 (Q03135, CAV1_HUMAN) is an impressive protein hub. Caveolin-1 is involved in at least 30 binary interactions with other proteins and more than 300 biological processes (GO annotations). It is no wonder that virtually ubiquitous tissue distribution is found for caveolins and caveolae with many interesting examples of high and low presence for particular caveolins in certain cell types (Cohen et al. 2004).

One would think that mammals should not be able to survive without caveolin-1. Surprisingly so, the disruption in the caveolin-1 gene was not lethal for the knockout mice despite the complete absence of caveolae organelles in endothelial and epithelial cells (Drab et al. 2001). However, the abnormal functioning of the lungs and the cardiovascular system was evident in the laboratory setting, leading to reduced ability to perform any physical work, a good enough reason for the preservation of caveolin-1 during natural evolution. All mysteries and paradoxes associated with the structure and function of caveolins in health and disease are far from being resolved, but hard work in many laboratories identified the most important amino acid residues and sequence motifs in the primary structure.

The connection between caveolin-1 and regulation of organs and tissues in the body (for instance, the organogenesis during embryonic development) gained additional support in recent discoveries by the Lotte Pederson group from the University of Copenhagen (Schou et al. 2017). Another organelle at the membrane surface, the cilia, contains caveolin-1 at its base. Common to caveolae organelles' malleability is the dynamics of cilia formation and disassembly (Jefries et al. 2019). The primary cilia are antenna-like structures projecting from the surface of most non-dividing cells. The process of cilia formation and disassembly is coordinated with the progression of the cell cycle. The caveolin-1 deficiency induces irreversible cell cycle arrest, inhibition of cilia disassembly, cellular senescence, and the increased possibility for cancer cell growth (Jefries et al. 2019). Conversely, caveolin-1 has the central role in regulating cilium's ability to activate the Sonic hedgehog signaling pathway, the crucial pathway for the development of major organs (Schou et al. 2017). It follows that caveolin-1 must have a previously unrecognized role in "ciliopathies," diseases causing severe defects in the brain and kidney.

Our brain is a strange organ with ravenous free-energy needs (Attwell and Laughlin 2001) and curious limitations in the energy storage ability. I would say that self-organized criticality (Chapter 5) is the brain's hallmark. Normal brain function requires a sufficient blood supply from other organs at precisely the right time in the right place, so much so that functional magnetic resonance imaging technique purports to detect mendacity by tracking blood flow to brain areas activated by lies (Langleben et al. 2002). The underlying mechanism is actually a process named neurovascular coupling that couples fast recognition of heightened neural activity in some locations to equally fast blood flow increases to these brain areas. Critical for this process is the tight regulation of the protective blood-brain barrier.

Until quite recently, the conventional wisdom was that caveolae formation was actively suppressed in all of the specialized endothelial cells from the blood-brain barrier. This year (2020), Gu and colleagues (Chow et al. 2020) published their results about opposite characteristics of capillary and arterial endothelial cells with the conclusion that the suppression of caveolae is not uniform. They observed negligible numbers of caveolae in capillary and abundant caveolae organelles in arteriolar endothelial cells. Arteries account for only five percent of the blood vessels in the brain but carry oxygen-rich blood. Their caveolae are now implicated in precisely targeted blood flow increases, probably by their ability to cluster the specific ion channels and receptors. As it is usually the case, this piece of fundamental research opened new avenues of study how caveolae and caveolins affect health, aging, and chronic disorders, such as hypertension, diabetes, muscular degeneration, cancer, lipodystrophy, and Alzheimer's disease. It should be possible to build further upon finding that caveolin-1 overexpression improved hyppocampal functions that are greatly compromised with age (Mandayam et al. 2017). These are essential cognitive abilities such as short-term recollection, working memory, and executive function.

7.7.4 Surprisingly sparse knowledge and abundance of wrong predictions about the caveolin-1 structure

Let us briefly examine the features of interest from caveolin-1 sequence:

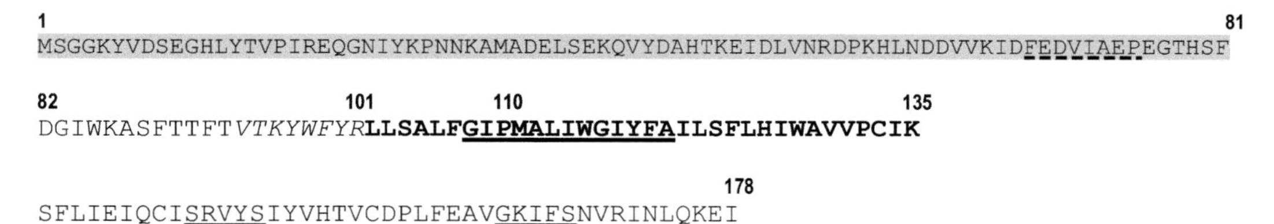

Figure 7.1: Caveolin-1 primary structure with illustrated four major domains: N-terminal (gray shade) up to residue 81, scaffolding 82–101, intermembrane 102–135 (bold letters), and C-terminal domain residues 136–178. Oligomerization-promoting segments are underlined.

Common wisdom about domain organization of caveolins changed through the years. Razani et al. (2002) used the name "membrane-spanning segment" and "transmembrane domain" for the L-102 to I-134 hydrophobic stretch of 32 residues, but recognized that its inferred hairpin loop structure is not accessible to the extracellular milieu. These authors also named the "N-terminal membrane attachment domain," "C-terminal membrane attachment domain," "oligomerization" (61–99), and "scaffolding" (90–101) domain with evident overlap between these functionally important segments. Hence, caveolin-1 offered the challenge how to recognize from the primary structure that the segment with a high propensity for membrane buried α-helix is not forming the transmembrane helix, but instead forms the hairpin broken-helix structure that returns to the same membrane side. Molecular dynamics simulations and NMR experiments confirmed that the break GIP-110 motif, together with the C-terminal of the first helix and the N-terminal of the second helix, can enter more or less deeply into the membrane interior, but never reach the opposite membrane face (Krishna and Sengupta 2019, Root et al. 2019). According to Root et al. (2015), the break region GIP-110 regulates the helix-break-helix intermembrane domain's (residues 85–136) orientation and inclination. The current thinking considers the intramembrane domain of caveolin-1 to extend from L-102 to K-135 (Figure 7.1 bold letters).

Root et al. (2019) offered two classifications for not-overlapping caveolin domains. The first one is the more traditional separation into four major regions: the N-terminal domain, the scaffolding domain, the intramembrane domain, and the C-terminal domain. In caveolin-1, these domains extend from M-1 to F-81, from D-82 to R-101, from L-102 to K-135, and from S-136 to I-178. Out of these four major domains, only the N-terminal domain is significantly different in length (shorter) for caveolin-2 and caveolin-3. The second definition is more based on experimental data about secondary structure than the function of protein segments. It separates the N-terminal disordered domain M-1 to F-81, the tripartite intramembrane domain with membrane helix-1 from D-82 to F-107, G-108 to P-110 break, and membrane helix-2 from M-111 to A-129, and bipartite C-terminal domain with V-130 to P-132 break and amphipathic helix-3 extending from C-133 all the way to the C-terminal residue I-178. It is the proposal based on the collation of data for that "woefully under-characterized" protein as authors admitted. The amphipathic helix of shorter length can be approximately located between D-50 and N-60 within the N-terminal disordered domain, and of longer length between R-146 and I-166 within the membrane-associated domain of the C-terminal (Figure 7.2). For the other two caveolins, the homology cannot help much for structure determination. The pairwise sequence identity is 52% for caveolin-1 and -3, 32% for caveolin-2 and -3, and only 29% for caveolin-1 and -2 (Root et al. 2019). Similarity and identity calculations depend on what bioinformatic tools have been used for sequence alignment. Older calculations produced somewhat higher similarity and identity values (Razani et al. 2002, Cohen et al. 2004).

The SPLIT 4.0 algorithm prediction (Figure 7.2) finds the sequence location of the "transmembrane helical segment" from R-101 to L-125. Thus, this tool extends the N-terminal for the Q03135 UniProt annotation of "intramembrane segment" A-105 to L-125 and predicts all of 101–125 residues to be in the TMH conformation, while flanking residues 88–100 and 126–142 are respectively assigned to the beta-strand and helical conformation. The consensus sequence location L-102 to K-135 for the intramembrane domain is presented in bold letters for the corresponding caveolin-1 segment (Figure 7.1; Root et al. 2019). Tusnády and Simon tool HMMTOP (2001)

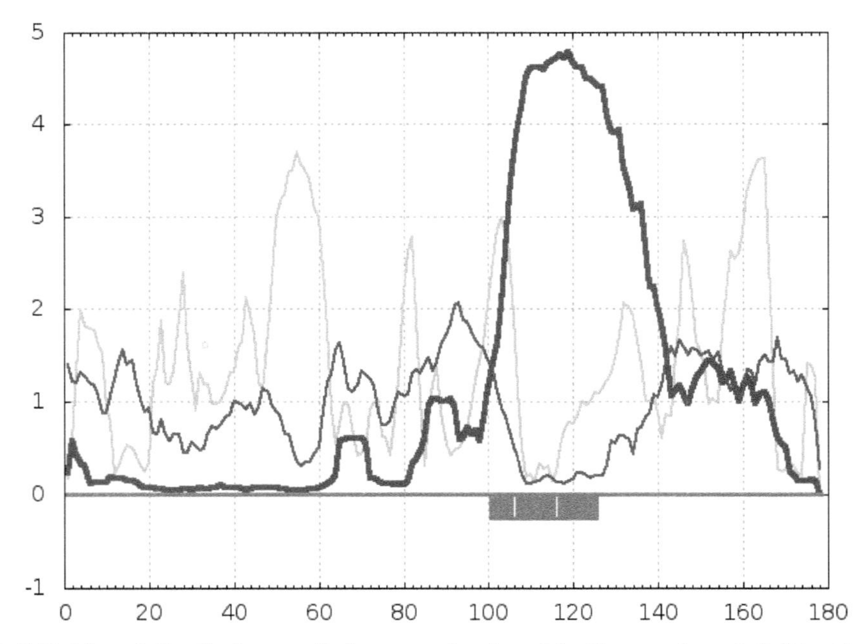

Figure 7.2: Split 4.0 prediction for the caveolin-1 sequence location of the "transmembrane helix" (residues R-101 to L-125) produces a high peak in the profile of TMH preferences (red) and the digital TMH prediction (pink rectangle just below the x-axis). The predictions of beta structure (blue profile) and amphipathic alpha structure (gray profile) are of unknown accuracy.

predicts the TMH-break-TMH pattern at respective sequence locations 120–139, 140–145, 146–169. The improved CCTOP tool by Dobson et al. (2015) predicts a single TMH extending from L-102 at the internal membrane side to the A-129 at the external side. The advantage of using the consensus CCTOP web server is the ability it provides to compare the predictions of 10 "state-of-the-art" topology prediction methods for integral membrane proteins. All of them, including SPLIT, HMMTOP, and CCTOP predict wrong transmembrane topology for caveolin-1, that is, one to three transmembrane helices. The PSIPRED workbench (http://bioinf.cs.ucl.ac.uk/psipred/, Buchan and Jones 2019) also predicts three transmembrane helices. They do not exist! The correct topology confirmed in numerous experiments is the reentrant helix-break-helix pattern with the N-terminal of the first helix and the C-terminal of the second helix, both located close to the cytoplasmic membrane surface (Root et al. 2019). The first helix of the helix-break-helix pattern is likely to extend uninterrupted from the scaffolding domain into the intramembrane domain, thus weakening the idea about the structural and functional separation of 20 amino acid residues' long scaffolding domain. The scaffolding domain and equally well preserved intramembrane domain have in common deeper membrane entrance than remaining domains. The presence of the small motifs [G,A,S]-XXX-[G,A,S] (underlined in the Figure 7.1) within these two domains probably contribute to caveolin propensity for oligomerization (Walters and DeGrado 2006).

7.7.5 Caveolin signatures

The dimerization and oligomerization ability of caveolins is likely to depend on conserved structural motifs. Experimental proofs exist that the motif F(68)EDVIAEP(75) allows for oligomerization within a caveolar coat (Head et al. 2014). This motif is located within the oligomerization domain (Razani et al. 2002) mentioned above (a low dash underscore in Figure 7.1). It is perfectly conserved in all caveolins (Root et al. 2019) and used as the consensus pattern F-E-D-[LV]-I-A-[DE]-[PA] by the PROSITE tool to collect almost all known caveolin sequences. Thus, research workers in this field adopted the name "caveolin signature" for the PROSITE pattern PS01210 with this motif. Since the eight amino acid residues are not long enough to exclude accidental hits from unrelated proteins, I recently (26 April 2020) repeated the PROSITE scan using the entire UniProt database and found 145 bacterial species with the PS01210 motif. These are all overpredictions, too many of them because I found them in unrelated bacterial proteins. A total of 869 hits were correct positive predictions for caveolins from Bilateralia (861 for jawed vertebrates and the remaining eight hits for mites, ticks, and mollusks).

I asked myself whether the scaffolding domain would be a better caveolin signature? The consensus scaffolding domain can be constructed by using only human caveolins (Root et al. 2019). It is the PROSITE pattern: D-[GK]-[IV]-W-[KI]-[ACV]-S-[FHY]-[TA]-[TL]-F-[TE]-[VI]-[TS]-KY-[WV]-[FMC]-Y-[RK]. This pattern is indeed more specific for caveolins despite seemingly lesser conservation. It finds a total of 868 hits on 866 sequences after scanning the entire UniProtKB database. There was only one hit in the non-caveolin protein (Uniprot code A0A498NX10), the hepatocyte growth factor receptor-like isoform X1 from rohu fish species (carp) found in rivers in South Asia. Intriguingly, the N-terminal of that protein with 1578 amino acid residues contains the region 46–177 recognized by InterPro, PFAM, and PROSITE resources as the caveolin. The SPLIT profiles for the first 200 amino acids are almost identical to those shown in Figure 7.2 for human caveolin-1. The incorrect assembly of caveolin and growth factor receptor genes from the rohu fish cannot be excluded because all open reading frames are computationally extracted as the preliminary (unreviewed) data from the shotgun entry. Anyway, the scaffolding domain merits additional discussion, which will show that it is truly a unique peptide.

7.7.6 The minimal scaffolding domain named surrogate peptide

Pleasant-Jenkins et al. (2017) used a new name for scaffolding 82–101 caveolin-1 domain—the surrogate peptide. They argued for the desirability to develop this peptide as a novel treatment to reduce fibrosis and improve the ventricular function in patients suffering from congestive heart failure. Caveolin-1 is underexpressed in heart fibrosis associated with pathological cardiac hypertrophy, and the surrogate peptide can replace the beneficial effect achieved by using the full-length virus-encoded caveolin-1. In an animal model (mice), the surrogate peptide was able to reverse the abnormal thickening of the heart wall leading to heart failure. It is estimated that half of the human patients with pathological cardiac hypertrophy die within five years after the diagnosis. Unfortunately, presently approved therapeutic options are limited in halting disease progression and nonexistent in reversing it. Anti-fibrotic properties of caveolin-1 and its surrogate peptide have been seen in the skin, lung, kidneys, heart, and possibly exist in other organs as well.

The minimal scaffolding domain 82–101 is crucial for inhibition of clathrin-independent endocytosis (Chaudhary et al. 2014). The proposed mechanism is decreased membrane protein lateral mobility due to perturbed diffusion within membrane lipid microdomains. One of the many functions of complete caveolin-1 is to negatively regulate the endocytosis in a selective manner, thus acting as a gatekeeper for toxins and pathogens. Under the hypoxia condition of a hostile environment inside the tumor, caveolin-1 inhibits protein internalization for most proteins with some rare but important exceptions (Bourseau-Guilmain et al. 2016). The exceptions can be used for specific targeting of cancer cells to deliver cytotoxic drugs without harming healthy cells. Hence, the knowledge about what caveolin-1 is doing in healthy and diseased tissues can make a difference between life and death in treating severe medical conditions.

7.7.7 The importance of posttranslational modifications

The posttranslational modifications of caveolin-1 include the palmitoylation sites C-133, C-143, and C-156 (Parton et al. 2006). Structurally, the palmitoylations make caveolins both monotopic and lipid anchored membrane proteins. When palmitoylated, caveolin-1 is better able to sequester cholesterol and associate with the membrane (Krishna and Sengupta 2019). The cholesterol binding motif from caveolin formed by residues V(94)TKYWFYR(101) from the scaffolding domain (italic letters in Figure 7.1) (Head et al. 2014) is, however, distant in sequence from C-terminal cysteines, which can be palmitoylated. Thus, an indirect promotion of a larger cholesterol occupancy at the palmitoyl tails is also possible. The functional role of palmitoylation is less well understood. Caveolae can be formed by caveolin-1 devoid of palmitic acids. Still, reversible palmitoylation is likely to be a powerful regulatory mechanism for the caveolin-1 role in organizing the signaling molecules and controlling the membrane dynamics (Eisinger et al. 2018).

Another recently discovered role for caveolin-1 is anti-metastatic surveillance (Celus et al. 2017). Metastasis causes at least 90% of all cancer-related deaths, so that each new untangled puzzle piece about natural inhibitors of metastatic growth can open promising therapeutic perspectives. The puzzle was why metastasis-associated macrophages express a high level of caveolin-1 despite the well-known association of cancer cell lines with low levels of expressed caveolin-1. In humans, the consequences of cav-1 gene mutations can lead to tumor progression of human breast cancers (Fiucci et al. 2002) or even to premature aging symptoms (Schrauwen et al. 2015). Massimiliano Mazzone and his collaborators discovered that the upregulation of this protein at the surface of metastasis-associated macrophages hinders metastatic growth in the lung environment (Celus et al. 2017). It is an additional confirmation for the caveolin-1 role as a gatekeeper multi-tasking protein with the ability to protect the body against diseases, toxins, and pathogens. The protection can be executed through the activation of apoptosis by phosphorylated caveolin-1 Tyr-14 residue (Shajahan et al. 2007).

Sex, death, and programmed cell death (apoptosis) are all evolutionary inventions of multicellular organisms, indispensable for the growth, development, and propagation of eukaryotic species. We do not feel any loss when millions of our cells die each day of our lives to prevent cancer, viral infections, inflammatory and autoimmune diseases. Intrinsically initiated apoptosis is regulated by the Bcl-2 family of small proteins that control mitochondrial membrane integrity.

They are even more versatile and multi-tasking than all caveolins and cavins together because they form the watch-dog system for preventing the survival of cancer cells (Goodsell 2002), stressed cells (Chong et al. 2014), and cells infected by viruses (Kvansakul and Hinds 2013). Their structure is much better known than caveolins' structure, but not the dynamics of conformational changes determining their multiple cellular locations and weaker or stronger membrane association with different organelles. Interaction among 20 members of the Bcl-2 family, their oligomerization in a membrane environment, interaction with other proteins, and phosphorylation of serine and threonine residues are all crucial for a remarkable number of molecular functions and biological processes (345 GO annotations for the Bcl-2 protein P10415 alone). It would be indeed surprising if the crucial hub protein with a protective role against self-inflicted capital punishment never met with the hub protein caveolin-1 for multiple signaling pathways. Researchers who examined this question concluded that phosphorylated caveolin-1 triggers apoptosis by inactivating the Bcl-2 inhibition of apoptotic cell death (Shajahan et al. 2007). The tyrosine-14 phosphorylation of CAV1 enhances paclitaxel-mediated cytotoxicity of breast cancer cells. Taxanes, including **taxol** and paclitaxel, inhibit **microtubules**' bioenergetics by stabilizing them and arresting cells from dividing. When cell cycle progression is halted, the apoptotic signaling molecules of Bcl-2 family and tumor suppressor proteins p21 and p53 can be activated. It is undoubtedly desirable to learn more about how caveolins enter into that complex story. Finding the best personal treatment for breast cancer patients requires a deep understanding of molecular and system biology background.

7.7.8 Disease-causing mutations in COV1 gene reveal molecular details about caveolin-1 structure-function connection

Earlier mentioned disruption in the caveolin-1 gene (Drab et al. 2001) is the excision of exon 3, which encodes for the scaffolding domain, intramembrane domain, and the palmitoylation sites. Although not lethal, the disruptions and mutations in the cav-1 gene cause severe deleterious effects, such as impaired nitric oxide and calcium signaling in the cardiovascular system. A loss of caveolae and low caveolin-1 levels was observed in some tumor cells. One example is the loss of caveolae in Ewing's sarcoma tumor cells (Huertas-Martinez et al. 2016). This disease mainly affects children. A promising therapeutic option in the future is to cause cancer cell death by triggering caveolae formation with genomic editing tools.

Another example is a very low level of caveolin-1 expressed in highly metastatic mammary carcinoma cell line Met-1 (Williams et al. 2004). The scaffolding domain (residues 82–101 of caveolin-1 presenting the surrogate COV1 peptide) was delivered into the cytoplasm of these cells by fusing it with the cell-penetrating peptide. The addition of the scaffolding domain inhibited the invasive ability of Met-1 cancer cells. Presently, the role of caveolins in health and disease remains a challenging research subject after the tumor-promoting role of caveolin-1 was found for other cancers, such as the prostate cancer and hepatocellular carcinoma (Sohn et al. 2016, Parton et al. 2018, Chai et al. 2019). The exit pathway from this confusion seems to point toward the need to address the complex signaling-platform role of caveolae, as distinct from caveola components.

Caveolin-1 does not act alone in everything it does. Dimerization and oligomerization of caveolins in coordinations with cavins lead to membrane invagination and creation of caveola (Mohan et al. 2015). The sensitivity of that process is best illustrated after conserved prolines 110 or 132 are substituted with some other amino acid residues. The natural variant 132 P→ L is found in up to 16% of all breast cancers. Lee et al. (2002) observed that Pro-132-Leu mutant protein forms misfolded oligomers. These are retained within the Golgi complex and unable to reach plasma membrane caveolae. NMR experiments suggested that P132L mutation extended the second membrane helix for an additional four residues (Rieth et al. 2012). More dramatic conformational change happens when proline 110 is mutated to alanine (Aoki et al. 2010, Aoki and Epand 2012). The re-entrant helix-break-helix loop is converted into the transmembrane helix forcing the conversion of caveolin-1 membrane topology from monotopic to bitopic (Root et al.

2015). Admittedly, caveolin-1 is a truly tough target for structure determination (Root et al. 2019) due to its high intrinsic dynamic (especially residues 62–80) and high hydrophobicity (residues 98–141) connected to the facile adaptation to different intracellular environments. For almost all research about caveolins structure-function connection, the Paul Boyer words from 2002 still ring true: "Occasionally in biochemical research one encounters a property of a system that seems designed to confuse and thwart the researcher."

Grave disease conditions connected to caveolin-1 inspire the determination to persist in decoding molecular details despite all difficulties. Lipodystrophy familial partial 7 (FPLD7; Cao et al. 2008), congenital generalized lipodystrophy 3 (CGL3; Kim et al. 2008), and pulmonary hypertension primary 3 (PPH3; Austin et al. 2012) are diseases caused by mutations in the CAV1 gene. All three conditions are associated with nonsense or frameshift mutations leading either to premature stop of protein synthesis or to the scrambled sequence after the mutation site. Such mutations have been found in severe genetic diseases. For caveolin-1 expression, the omission of all amino acids after Glu-38 or replacement with different residues after Cys-133 resulted in severe metabolic derangements, impaired growth, and wheel-chair bound condition since age 20 in one case (Cao et al. 2008). Reduced caveolin-1 expression led to pulmonary hypertension with an increased danger of heart failure in mice and humans (Zhao et al. 2009).

7.7.9 Gain-of-function mutations in caveolin-3 can cause various skeletal muscle diseases and congenital heart syndrome

We can compare loss-of-function disease-causing mutations in caveolin-1 to gain-of-function mutations in caveolin-3. Mutations T78M, A85T, and F97C in the intramembrane caveolin-3 region cause the channelopathy named long-QT syndrome (Vatta et al. 2006). Corresponding residues in caveolin-1 are A105, A112, and F124. Long-QT is a potentially fatal heritable disease. The diagnosis is based on an electrocardiogram. There are about ten subtypes of a congenital long-QT syndrome. Most of them are caused by genetic defects in potassium or sodium voltage-gated channels. An electrocardiogram of a healthy person reflects voltage-gated regulation and synchronization of billions of cardiac ion channels. A natural question then is why the caveolin-3 is involved at all (LQT9 subtype)? None of the caveolins have known channel or pore function in the plasma membrane, but caveolin-3 influences the heart's rhythm through the regulation of cardiac potassium and sodium channels, which is disrupted, for instance, in the rare T78M mutant (Campostrini et al. 2017). Direct physical association is proposed for specific channels and caveolin-3 in cardiac caveolae organelles (Vaidyanathan et al. 2018).

Our heart is a sophisticated muscle. If the heart benefits from the caveolin-3 presence, this can be the case for all other muscles too. Indeed, CAV3 has a critical role in muscle cell physiology. Minetti et al. in 1998 and Betz et al. in 2001 connected Caveolin-3 defects to skeletal muscle disease phenotypes. Betz et al. (2001) identified the CAV3 mutants responsible for rippling muscle disease (Ricker et al. 1989). As usual in science, a new word had to be invented to describe different muscle diseases connected to caveolin-3 mutations (limb-girdle muscular dystrophy, distal myopathy, and hyperCKemia in addition to rippling muscle disease). Ozawa et al. (2001) and other research groups proposed the term caveolinopathies about 20 years ago. It was adopted in a review paper by Gazzerro et al. (2010). Thus, disease-causing protein defects are tools in molecular medicine for the design of knowledge-based treatments. The prerequisite is to uncover the essential roles of amino acid residues or motifs from corresponding intact proteins.

Let us pick up just one caveolin-3 mutant for closer examination. The same Pro→Leu mutation is sometimes labeled as P104L and sometimes as P105L. The difference is caused by an uncertain initiation site because caveolin-3 has two methionines at its N-terminal. To be consistent with previously used alignment and numbering of amino acid residues from caveolins in this chapter (according to Root et al. 2019), we shall assume the first methionine as the initiation site for the start codon during the translation process. The assumption identifies P105L as the cause for the rippling muscle disease (RMD2) previously connected to P104L mutation (Minetti et al. 1998, Betz et al. 2001). Where is it located in the protein, and what is it doing to organelles, cells, and tissues needing intact caveolin-3? We have seen previously that corresponding caveolin-1 mutation P132L slightly extends the second membrane helix because Pro-132 is located in the C-terminal of the intermembrane region. A seemingly small conformational change led to misfolded caveolin-1 oligomers incapable of trafficking from Golgi complex to plasma membrane caveolae, but the avalanche of negative consequences did not stop there. The P132L is a mutant found in human breast cancers. The caveolin-3 protein with identical Pro→Leu mutation in the homologous intermembrane region causes very different disease-phenotype. And it is not only a relatively benign RMD myopathy but also the limb-girdle muscular dystrophy-1C expressed as weakness and wasting of limb musculature (Deng et al. 2017). About 40 different pathogenic mutations of CAV3 have been observed over the years. Muscle bioenergetics is inhibited by most of these mutations, including P105L, because the caveolin-3 is muscle-specific protein. It is a veritable protein hub like caveolin-1 endowed with multiple regulatory functions, all requiring the presence of caveolins in the sarcolemma membrane.

7.8 Syndecans: Bitopic membrane proteins

7.8.1 Communication mediated by exosomes and syndecans

In this section, we shall continue using selected membrane protein to trace how scientific discoveries connected its structure to organelle biogenesis, metabolism, and signaling in health and disease. While studying biophysics at Penn State University (1972–1976), I never heard about syndecans and exosomes. Exosomes were first identified in the 80s as membrane-bound vesicular structures, organelles capable of fusing with the plasma membrane, and generating small extracellular vesicles (Pan and Johnstone 1983, Johnstone et al. 1987, Saunders et al. 1989). Bitopic membrane proteins-the syndecans- are essential for exosomes' biogenesis (Baietti et al. 2012). It is a highly regulated process requiring free-energy input in the form of ATP hydrolysis. Exosome vesicles are composed of specific proteins and carry a diverse cargo of molecules, including nucleic acids. We can consider them as cell-created virus-like structures or cell-created drug delivery systems (Familtseva et al. 2019).

Exosomes are promising models for drug delivery that may be amenable for therapeutic targeting. A total of 6514 exosome-located proteins and many different RNA species have been detected in human exosomes (http://www.exocarta.org/). Caveolins and syndecans are among these proteins. That such a high percentage of the human **proteome** is connected to exosomes testifies to their functional importance. For comparison, the crucial organelle in human cell bioenergetics—the mitochondria, is formed by approximately 1500 proteins coded by less than 5% of the protein-coding human genome (Guda et al. 2004, see cytochrome c oxidase section too). More than six thousand proteins connected to exosomes are well in the range of bacterial proteomes. For example, *Escherichia coli* strain K12 has less than five thousand protein-coding genes (https://www.uniprot.org/proteomes/UP000000625).

Human bitopic membrane protein syndecan-1 (SDC1_HUMAN, P18827) belongs to the single-pass type I integral membrane proteins, which have only one TM helix and extracellular (outside) location of their N-terminal. It is cell surface proteoglycan that bears both heparan sulfate and chondroitin sulfate and links the cytoskeleton protein network to the extracellular matrix molecules and growth factors (Bernfield et al. 1992). Syndecans are aptly named from the Greek *syndein* to bind together (Saunders et al. 1989). Mammals have four syndecans, all of them sharing the same membrane topology with the syndecan-1. Syndecans form an important hub in the cellular network named syndecan interactome, which encompasses an astonishing number of 351 partners (Gondelaud and Ricard-Blum 2019). In healthy cells, exosomes with their network of glycosylated syndecans serve for intercellular communication. The other means for information-transfer in animal cells are signaling via the transfer of soluble molecules or through the shedding of microvesicles distinct from exosomes (Cocucci et al. 2009), and contact-dependent signaling by certain types of intercellular junctions (Franke 2009) or tunneling nanotubes (see Chapter 15). Cell-to-cell adhesion is mediated by intact syndecan-4. Syndecan ectodomains (extracellular domains) often have an independent life when generated by alternative splicing or released in the extracellular milieu by limited proteolysis called ectodomain shedding (Manon-Jensen et al. 2010). Ectodomains lack the transmembrane and cytoplasmic domain, but also regulate numerous biological processes.

7.8.2 Simple bioinformatic tools can locate islands of order in syndecan sequences

The paucity of experimental structural data for syndecans is probably related to their high intrinsic disorder. Up to 86% of residues are predicted to be in a disordered state for syndecan-1 (Peysselon et al. 2011). These authors proposed the role of intrinsic disorder in promoting the extracellular matrix organization at the supramolecular and tissue level. The shape shifting ability of intrinsically disordered proteins (IDP) with their quick-on-quick-off binding to different partners is also essential for their role as cellular messaging hubs. IDP encompasses at least 30% of the human proteome (https://darkproteome.wordpress.com/). One can infer for syndecans that the inherent disorder enables the dynamic coherence of signaling pathways regulated by the syndecan protein family—a beautiful example of order-promoting disorder far from thermodynamic equilibrium!

When 3D data from NMR or X-ray crystallography are not available, simpler methods can be used to look for structure-function relationships. The SPLIT server prediction clearly delineates the N-terminal signal peptide and the transmembrane domain of syndecan-1. It predicts the 251–274 segment as the only transmembrane α-helix. This protein offers the challenge of how to distinguish the signal domain from transmembrane helix domains. In this case, the SPLIT algorithm prediction is the correct one. However, the percentage of false-positive TMH prediction at known sequence N-terminal location of the signal domain is higher for SPLIT compared to the other top-ranking topology-prediction algorithms. This weakness diminishes SPLIT **prediction accuracy** (Juretić et al. 1999, 2002). Methods of solving the problem to distinguish with high accuracy signal sequences from transmembrane helices have been published (Käll et al. 2004, Reynolds et al. 2008) and can be used to improve the SPLIT 4.0 algorithm. The present SPLIT version is useful for predicting the correct membrane topology of a mature protein sequence, the sequence without the signal peptide. The latest version of the signal peptide prediction software (SignalP-5.0, see Nielsen et al. 2019) can always be used to remove this N-terminal peptide (if it is predicted) before the SPLIT application. Then, the unique advantage of our software, the ability to recognize excess lysines and arginines and excess motifs containing these cationic amino acid residues at the cytoplasmic membrane side, comes to the forefront among other sequence features known to increase the prediction accuracy for correct transmembrane topology. It is the case for all four syndecans. They are all correctly predicted with the extracytoplasmic N-terminal topology. All syndecans have a high charge bias in favor of cationic residues in their highly conserved cytoplasmic domain (see below about the functional importance of the cytoplasmic domain). Syndecan-4 has one of the highest charge biases (-15).

7.8.3 Syndecans are both hubs and molecular switches in the protein-interaction network responsible for normal development or cancer progression

The deletion of a gene for expressing such hubs as SDC1 is expected to be lethal because something so centrally connected probably affects many crucial cellular processes. Indeed, it is difficult to imagine how a protein can interact with hundreds of partners unless it has a global regulatory and housekeeping function. To the surprise of researchers, SDC1-null mice turned out to be ostensibly normal and fertile (Alexander et al. 2000). Even more surprising, loss of SDC1 is a gain-of-function mutation (Aquino et al. 2018). Mice deficient in the syndecan-1 expression are resistant to tumorigenesis of mammary glands (Alexander et al. 2000), to upregulation of macropinocytosis by pancreatic cancer cells (Yao et al. 2019), and to several bacterial infections (Aquino et al. 2018).

The context-specific operation of syndecans, which transiently appear and disappear on the cell surface for some cell types, was presciently described by Guido David (1993). It seems like the magic of evanescence. Classical philosophers argued that "nothing could be more offensive to the intellect" than evanescence (Earle 1984). Still, the context-dependent changes of on-off states and dynamic "personality" of proteins are steadily gaining observational support in modern biophysics (Henzler-Wildman and Kern 2007). Order-to-disorder transition can be an ultrasensitive regulatory mechanism that is paradoxically responsible for forming higher-order dynamic structures (Csizmok et al. 2016). One can guess that syndecan-deficient mice have lost some complex functions that can be subverted by cancer cells and are not essential for animals' survival in the artificial laboratory environment.

When information-transfer is exploited by cancer cells, syndecans and exosomes can promote metastasis (Syn et al. 2016). Elevated serum levels of SDC1 predict shortened survival of patients with lung cancer, multiple myeloma, colorectal cancer, bladder cancer, and prostate cancer (Gharbaran 2015). Human prostate cancer types differ in their ability to cause harm. Most prostate cancers (80%) grow so slowly that they will not harm their hosts. The syndecan-1 concentration level is then similar to that of healthy prostate cells. But it is typically 16 times higher in the late stage of prostate tumor cells that are quickly invading surrounding tissues and spreading to distant organs (metastasizing), eventually leading to patient death (Heath et al. 2009). This example suggests again the system biology reason why membrane proteins are so important. Some of them may be responsible for a shift from health or benign tumor to lethal disease. When directing the activity of many other proteins in local protein-interaction networks, then single-point mutation or change in the level of expression in them or interacting partners can trigger a disease.

In the case of syndecans, these multifunctional proteins act as molecular switches, receptors, and co-receptors for channeling signaling, metabolism, and concomitant dissipation for building either healthy tissues or for disease progression when co-opted by microorganisms or cancer cells. There are some devastating cancers with low survival rates, such as glioma cancer of the nervous system and pancreatic ductal adenocarcinoma. The 8-percent survival rate at five years is observed for that type of pancreatic cancer (Yao et al. 2019). Cancers cells often upregulate macropinocytosis to scavenge extracellular nutrients and satisfy their insatiable need for consuming molecular "food" for fast growth. Macropinocytosis is an old evolutionary pinnacle for some eukaryotic cells' gain-of-function ability to ingest extracellular liquid, dissolved molecules, and particulate material (Bloomfield and Kay 2016). It is connected with the formation and recycling of specialized organelles: macropinosomes and macropino-lysosomes (Canton 2018). The metabolism reprogramming through enhanced macropinocytosis is critical for the survival and growth of cancer cells. Yao et al. (2019) identified syndecan-1 as an essential mediator of macropinocytosis in invasive prostate cancer. In common with metastatic cancer cells, certain bacteria and viruses developed means for the subversion of normal macropinocytotic function to invade host cells. Unfortunately, pharmacological inhibition of macropinocytosis has never been accomplished, but the discovery of the molecular switch role for syndecan-1 opens the feasibility of making that process vulnerable to therapeutic intervention. This perspective alone is enough to focus our attention on finding what different domains of syndecan-1 are doing. Trafficking of syndecan-1 to and from the cell surface enables exocytosis (release of exosomes), endocytosis, and macropinocytosis. When activated, these highly dynamic processes require continuous recycling of relevant proteins and high free-energy input. Each function-process can be regarded as a chain reaction and a new dissipation pathway triggered by molecular switches residing inside the SCD1 structure.

7.8.4 Structure-function connection for syndecan interactome

Let us go briefly through what is known about structure-function connection for syndecan interactome while bearing in mind how fast-moving is this research field and how vast are likely to be yet uncovered research avenues. One illustration for this statement is recently published observation (Gondelaud and Ricard-Blum 2019) that all four syndecans share 18 common interacting partners, yet the binding sites on syndecans are known only for a few of them. These are T-lymphoma invasion and metastasis-inducing protein (TIAM1), and syntenin-1 (SDCBP). Corresponding binding motifs are the C-terminal sequence T(303)KQEEFYA(310) in the case of syndecan-1 (P18827) and A(191)PTNEFYA(198) in the case of syndecan-4 (P31431), both located in the conserved C2 part of the short cytoplasmic domain. TIAM1 connects extracellular signals to integrin α6β4, which is another common interacting partner of all syndecans. TIAM1 regulates bidirectional signaling between the extracellular space and the cellular cytoskeleton. Syntenin-1 influences the recycling of syndecans.

The cytoplasmic domain of syndecan-4 contains only 28 amino acid residues with the sequence that can be separated into conserved parts C1 and C2 flanking the variable (V) region in the middle:

The variable domain still contains highly conserved Tyr, Pro, Gly, and Lys residues found in the same locations after aligning human with a whale (and many other) syndecan sequences. The conserved C1 domain is used as a consensus syndecan signature (the PROSITE tool pattern PS00964) when the last residue from the transmembrane domain is added at the N-terminal: [FY]-R-[IM]-[KR]-K(2)-D-E-G-S-Y. Conserved residues and residue motifs always implicate biologically important function. Indeed, tyrosines (Y) from C1 and V domains are connected to posttranslational phosphorylation, wound healing, and cell migration in health and disease (Morgan et al. 2013). When syndecan signature pattern PS00964 was constructed, it worked with high accuracy by predicting 24 known syndecans out of all UniprotKB/Swiss-Prot sequences without any false positive or negative hits. In May of 2020, the same pattern found more than 1460 syndecans in mammals, birds, frogs, insects, and worms, even including the tardigrades (water bear)—the most resilient animals. It is the testimony of fast recent advances of genomics and proteomics and the confirmation that syndecans have a long evolutionary history (about 600 million years) in multicellular organisms belonging to the Bilateria clade of the Animalia kingdom. Overpredictions include several uncharacterized proteins from the fin whale, magnesium transporter from Indian major carp, and a protein from the Proteobacteria *Kangiella spongicola*. Despite the similarity to syndecans, the bacterial protein almost certainly has an entirely different function.

Syndecan-4 acts as a molecular antenna and mechanosensor trigger. It senses the environment outside the cell, detects outside force-caused movements, and transforms them into the conformational change of its small but essential cytoplasmic domain (Chronopoulos et al. 2020). In this recent work, the lead researcher Armando del Río Hernández discovered with his coworkers the mechanotransduction pathway for long-distance signaling across the plasma membrane. It consists of syndecan-integrin crosstalk, which is synergistically triggered by a localized external force to elicit a remarkable cell-wide cascade of responses. In the eukaryotic cell's microworld, it is the striking example of long-distance (about 1000 nm) direct physical contacts and force transfer among extracellular matrix, syndecans, integrins, α-actinin, and F-actin molecular scaffold of the cytoskeleton. Lipid second messenger phosphatidylinositol 3,4,5 triphosphate (PIP_3) and α-actinin bind to the syndecan-4 variable (V) cytoplasmic domain. With such detailed knowledge of the force-transfer mechanism and relevant regulatory molecules, it is possible to force the cells to move in fast forward motion or stop them altogether (Morgan et al. 2013). Fine control of cell life and movement in wound healing and diseased conditions should also be possible in the foreseeable future.

7.8.5 The application of syndecan-4 in neovascularization

A promising medical application is the regrowth of damaged blood vessels. Aaron Baker, with his coworkers, improved the neovascularization by delivering syndecan-4 proteoliposomes with fibroblast growth factor to rats (Jang et al. 2012). Proteoliposomes are artificial vesicles containing lipids and proteins in their membrane. The proteoliposomes are used in vaccine development and experimental cancer treatment to deliver encapsulated and concentrated therapeutic molecules to specific body locations. Jang et al. (2012) used the standard protocols from membrane biochemistry to incorporate syndecan-4 into the vesicles containing the combination of phosphatidylcholine and phosphatidylethanolamine phospholipids with unsaturated fatty acid chains, cholesterol, and sphingomyelin.

Syndecan-4 served as the coreceptor for growth factors. It enabled their efficient delivery to blood vessels, stimulation of neovascularization, and restoration of the blood supply. This treatment has the potential to bypass surgery by injecting lipid nanoparticles into a patient. A significant percentage of the older population suffers from chronic myocardial ischemia (reduced blood flow to the heart), and presently used surgical methods have some long-term limitations.

7.8.6 Syndecans role in brain maturation and neurogenesis

Syndecans have a decisive role in the brain maturation of embryos. Abortion promoting activists, who are presenting it as a human right, disregard the fact that we have all been embryos once, and that the wiring of the brain starts early during embryonic development. We can never be grateful enough to our mothers for everything they provided to us both before giving birth and afterward. One of the mother's gifts to her unborn infant is the peptide with the sequence YDPEAASAPGSGNPCHEASAAQKENAGEDP named Y-P30 survival-promoting peptide. This 30 amino acids long peptide is located next to the signal peptide in the N-terminal segment of primate-specific innate peptide antibiotic dermcidin secreted by our sweet glands and in breast milk (Neumann et al. 2019). After the signal peptide is cleaved off, one of the mother's immune cell classes performs the proteolytic cleavage of mature dermcidin and produces bioactive Y-P30 during pregnancy. Except for the pregnancy period, and possibly suckling period, the Y-P30 is virtually absent in adults. From the mother's blood, the peptide passes the blood-placenta barrier. Together with other neurite growth-promoting peptide, the pleiotrophin, Y-P30 accumulates in neurons of the infant's brain and stimulates axonal growth so that both peptides bind to the syndecan-3 extracytoplasmic glycosylated domain (Landgraf et al. 2008). Downstream signaling through syndecan increases the amount of F-actin in axonal growth cones. In this way, the mother's pregnancy hormones ensure her help in developing the infant's brain before birth. After birth, the proteolytic processing of mature dermcidin produces the C-terminal peptide DCD-1L, which is an unusually long anionic antimicrobial peptide with the sequence: S(63)SLLEKGLDGAKKAVGGLGKLGKDAVEDLESVGKGAVHDVKDVLDSVL(110) (Paulmann et al. 2012). It has antimicrobial activity against a wide spectrum of pathogenic microorganisms, even in high salt concentrations.

By the way, a long string of letters in mature dermcidin sequence of 110 amino acid residues (the P81605 code from Uniprot) looks boring and meaningless until scientists translate its message to us: "I am a precious gift to you. I nurse your soul and defend you from infections. I bind all of you in my embrace" (my free interpretation). Neuman et al. (2019) speculated that after birth, mothers can still provide the Y-P30 peptide to their infants through breast milk and that this peptide can pass through the blood-brain barrier in newborns. If confirmed, it will help to explain the biochemical mechanism, which ensures strong social interaction in primates between mother and her suckling.

Several decades ago, it was believed that adult brains in mammals ceased with new neurons' production. Common knowledge was that about 85 billion nerve cells and as many glial cells originated from brain stem cells during embryogenesis. It was deemed to be more than enough to guarantee the proper function of our central nervous system. Occasional loss of several million neurons during binge drinking or some disease was not a cause for concern. A change in paradigm occurred after recognizing that adult mammals continue to produce neurons in some brain regions. Neural stem cells persist in the brain's neurogenic niches (mainly the hippocampus). They generate there a large number of immature neurons throughout adulthood but with a decrease in the rate of neurogenesis with increasing age (Aimone et al. 2014). Neurogenesis may have a key role in memory formation. With a confirmed role in brain maturation, syndecans should not be neglected as possible candidate-proteins involved in the regulation of neurogenesis that takes place in an adult brain. Kazanis et al. (2010), and Morante-Redolat and Porlan (2019) observed the upregulation of syndecan-1 in neural stem cells and astrocytes that are actively proliferating. It is a complex regulation enabled by syndecan-1 recruiting soluble growth factors, chemokines, and cytokines at the extracellular side. The intracellular amplification of these signals is mediated by the flexible cytoplasmic domain of that protein. The proteoglycans' ability to shield protein growth factors from circulating proteases is also instrumental in maintaining their stores and establishing protein gradients that guide cell migration (Elfenbein and Simons 2010).

7.8.7 Are syndecans involved in establishing protein gradients during organogenesis?

Establishing protein gradients is a metabolically costly but indispensable process for morphogenesis and organogenesis. Crucial morphogens are sonic hedgehog proteins. The heparan sulfate proteoglycans connect syndecans to sonic hedgehog proteins, thus implicating them in organogenesis (Chang et al. 2011). The connection can be with glycosylated ectodomains of intact mature syndecans, or with their proteolytically released heparan sulfate by sulfatase or heparanase enzymes (Gondelaud and Ricard-Blum 2019). In each case, the electrostatic interaction is very strong because heparin and heparan sulfate belong to bioactive molecules with the highest negative surface charge. The ubiquitous binding of heparin and its high molecular weight cousin heparan sulfate to many different proteins with cationic domains is still under investigation concerning their normal role in the body, notwithstanding well-known anticoagulant medical applications of heparin. The sonic hedgehog protein has a highly cationic Cardin-Weintraub amino acid motif of the XBBBXXBX type, where B is a basic residue, and X is any residue (K(32)RRHPKK(38) in humans). It is a consensus sequence for heparin and syndecan-4 binding. Of course, protein binding sites are not restricted to continuous sequence regions. With the 3D conformation available for human sonic hedgehog proteins, Chang et al. (2011) answered in experiments their question whether other residues are involved in binding heparan sulfate. They concluded that several additional cationic residues are involved. Of these residues, only the first one (Arg-28) is spatially and sequentially close to the Cardin-Weintraub motif, while the other one (Lys-178) is spatially close but distant in sequence.

7.8.8 Homodimeric and heterodimeric interactions among syndecans

Turning back to strikingly numerous interaction partners and ligands of syndecans, Li et al. (2016) conjectured that the vast majority of them interact with highly diverse and tissue-specific heparan sulfate decorations of syndecan ectodomains. An additional complexity level is due to homodimeric and heterodimeric interactions among four syndecans (Kwon et al. 2016). The transmembrane domains (TD) of syndecans are the first to engage in dimerization with cytoplasmic domains often following. Here, we present the transmembrane domains of human syndecans with dimerization promoting motifs and amino acids in gray and bold font, respectively:

```
SDC1: V(252)LGGVIAGGLVGLIFAVCLVGFMLY(276)
SDC2: V(145)LAAVIAGGVIGFLFAIFLILLLVY(169)
SDC3: V(385)LVAVIVGGVVGALFAAFLVTLLIY(409)
SDC4: V(146)LAALIVGGIVGILFAVFLILLLMY(170)
```

The best conserved is the GGXXG small motif (Walters and DeGrado 2006; X is the hydrophobic residue L, V, or I) and the central phenylalanine residue (F) with the same TD location but different sequence position which is conserved in all mammalian syndecans. As always, high conservation indicates some critical function. A total of four small motifs (Ala also projects a small residue) may have a role in enhancing the heterodimer formation between syndecan-2 and syndecan-4. The conserved Phe residue strengthens homodimer formation (Kwon et al. 2016). It remains to be seen how unique is the regulatory mechanism in syndecan signaling when mediated by the oligomerization of transmembrane domains (Jang et al. 2018). The last several years of research efforts illuminated some details of how the dimerization induced by transmembrane domains regulates the oligomeric status of the cytoplasmic domain *in vivo*. A large scale molecular dynamics computational study with almost six million atoms was performed to answer the question about principal dynamic modes of motion for the mechanotransduction (Chronopoulos et al. 2020) conducted by the TM syndecan-4 dimmer (Jiang et al. 2020). The result was the insight about scissor-like motion and bending of SDC-4 transmembrane helices with the pivot area kept together with the GGXXG adjusting screw.

7.9 Presenilin-1

7.9.1 Presenilin and amyloid cascade hypothesis

The Latin phrase "*nomen est omen*" (the name is a sign) seems appropriate for the serpentine presenilin-1 protein (PSN1_HUMAN, **P49768**), which is connected to the Alzheimer's disease. It waves in and out of the membrane almost ten times, according to older cartoons illustrating the topology of that polytopic integral membrane protein. The presenilin forms the membrane protease named γ-secretase together with several other membrane proteins. One of its multiple functions is to cleave more than a hundred type I membrane proteins. Included among them is the amyloid precursor protein (APP). After cleavage, amyloid-β peptides can be secreted by neuronal cells (Tomita 2017). One of the cleaved peptides, the 42-amino acid residue long Aβ, regulates bioenergetic pathways (Wilkins and Swerlow 2017), neuronal electrophysiology, neurogenesis, neuronal survival, learning and memory, and antimicrobial host defense function, but it is also toxic to neuronal cells when overproduced (see Chapter 15). The amyloid cascade hypothesis proposed that accumulated Aβ plaques in the brain, after gain-of-function mutation in presenilin, drive the cascade of synapse loss, neuronal death, cognitive dysfunctions, and ultimately death (Hardy and Higgins 1992). This hypothesis led to intensive efforts during the last several decades to slow down or find a cure for Alzheimer's disease by reducing the amount of Aβ peptide in the brain. Unfortunately, such efforts have failed to show a clear clinical benefit (Banerjee et al. 2018, Panza et al. 2019). Researchers realized that all previous insights into the complexity of how the brain regulates its bioenergetics are still not good enough. Other γ-secretase and presenilin functions should be considered to understand why numerous mutations in presenilin sequence are associated with neurodegenerative diseases.

Caveolin and syndecan deficient mice are alive, but the knockout of presenilin in mice and flies is lethal, indicating some essential presenilin function during the development. It can be calcium homeostasis, cell-cell adhesion, or membrane trafficking. Axonal trafficking is the essential bioenergetic function of nerve cells mediated by molecular motors. The internal highway system of neuronal axons must properly function to ensure the long-distance transport of proteins and organelles from one part of a nerve cell to another—an indispensable bioenergetic process for survival and signaling function. Presenilin regulates how quickly molecular motors travel along microtubule tracks, possibly influencing their binding or release (Banerjee et al. 2018). This proposed presenilin role is independent of its catalytic protease activity within the γ-secretase complex. The mutations and deletions influencing presenilin scaffolding and trafficking function are located at its hydrophilic loop region between transmembrane domains 6 and 7.

Interestingly, the presenilin hypothesis postulates a loss of presenilin function as the leading cause of neurodegeneration and familial Alzheimer's disease (FAD) (Shen and Kelleher 2007, Xia et al. 2015). It is just the opposite of the amyloid cascade hypothesis, which implies increased presenilin and γ-secretase activity. Instead of using knockout mice, Jie Shen and Raymond Kelleher used "knocked in" mice. The FAD-associated mutation was introduced in both presenilin genes of these animals: presenilin-1 and presenilin-2. Mutations in these genes account for 90% of FAD cases in humans. Unfortunately, animals with both genes mutated did not survive after birth. However, mice survived with a single presenilin gene mutated. In this way, Shen and coauthors created the first animal model for the FAD disease. The "knocked in" animals with L435F mutation in the presenilin-1 (PS1) showed the deficiencies in learning and memory compared to normal mice. This mutation abolished the ability of PS1 to support neuronal survival during aging. Impaired synaptic plasticity and age-dependent neurodegeneration paralleled the same pathology found in human FAD patients. I would say that lady-scientist Jie Shen with her coauthors showed how a decades-long detour from revealing normal presenilin functions led to useless spending of billions of dollars without clear benefits to FAD patients. It is partly the fault of funding agencies with their shortsighted vision almost entirely focused on finding cures for pathological conditions, instead of understanding the complexity of beneficial presenilin interactions in the brain.

7.9.2 Transmembrane topology and 3D structure

Presenilin is the catalytic subunit of the γ-secretase complex. It is found in the plasma membrane, endosome, Golgi apparatus, and endoplasmic reticulum. When located in organelles, all presenilin-1 segments from the polypeptide chain can be divided into three classes concerning their spatial location: cytoplasmic, transmembrane, and lumenal. Plasma membrane location means that internal (luminal) and external (cytoplasmic) segments have reversed their role: lumenal segments are then in contact with small molecules from extracellular space. Thinakaran et al. discovered in 1996 that presenilin mostly exists as cleaved N-terminal and C-terminal fragments, which remain associated in the secretase complex. The γ-secretase complex assembly and maturation to the catalytically active form proceeds through obligatory auto-endoproteolysis steps (auto-cleavages) of presenilin-1 or presenilin-2 (Fukumori et al. 2010).

Even after finding its detailed 3D structure utilizing advanced electron microscopy techniques (Bai et al. 2015), presenilin-1 still presents the challenge to determine the exact sequence location, total number, and orientation of each transmembrane segment. This task is important because a large number out of more than 100 known disease-causing mutations (Alzheimer's disease 3) are located inside or close to predicted transmembrane helices. When mutated residues reside inside transmembrane segments, additional challenges appear in the design of drugs targeting lipid-binding sites. For instance, a novel allosteric modulating drug, AZ3451, binds to the PAR2 receptor within the membrane (Cheng et al. 2017).

The interplay between theoretical predictions supported with experimental data and structure determination supported by theoretical predictions has been quite exciting during attempts to find membrane topology of this and other integral membrane proteins. For a long time, the Swiss-Prot annotation of transmembrane segments was done using the Kyte-Doolittle sliding algorithm (1982). That algorithm used mean hydrophobicity of 21 residues and assigned sequence location of 8 TM segments for presenilin-1, each with 21 residues. More sophisticated topology predictors did not agree with each other, nor with Swiss-Prot annotations. For instance, the Split 4.0 algorithm predicted 10 TM segments with a variable number of amino acids present in each putative transmembrane α-helix. Unfortunately, both predictions agreed that this protein has the cytoplasmic location for its N and C-terminal, which turned out to be an erroneous prediction. As annotated by the current Swiss-Prot version, the presenilin-1 has 9 TMH, and its C-terminal is located in the lumenal (extracellular) space. The annotation was greatly helped with structure determination (Bai et al. 2015). The PDB structure 5A63 for the human gamma-secretase complex containing presenilin-1 was used to obtain the transmembrane topology of the presenilin-1 by research groups in England, Hungary, Germany, and the USA. All helices are inclined, some for about 45° regarding the perpendicular axis to bilayer surface. The transmembrane topology is not easy to see from the PDB structure. Molecular dynamics simulations were needed to define transmembrane topology and the procedure of stepping back and forth from the atomistic model to the coarse-grained model, as we mentioned in the introduction to this chapter. The final result is similar but not identical to Swiss-Prot annotation. See, for instance, the topology obtained with the CCTOP server http://cctop.enzim.ttk.mta.hu/ (Dobson et al. 2015).

The difference in predicting ends of transmembrane helices is essential in predicting interactions of presenilin-1 with other proteins. For instance, the nonconservative V82K or V82E substitution abolished binding to the cytoskeletal Glial Fibrillary Acidic Protein (GFAPε) (Nielsen et al. 2002). It appears that the Val 82 in a helical conformation is critical for the interaction with GFAP. SPLIT algorithm predicts a high TMH conformational preference for Val 82, which is considerably decreased after V82K or V82E substitution. Simulations based on 3D structure confirmed that Val 82 is well inside the N-terminal of the first transmembrane helix (Stansfeld et al. 2015) but the Swiss-Prot annotation located Val 82 in the cytoplasmic space just outside the helical transmembrane domain. We can conclude that sophisticated modeling of integral membrane proteins in a simulated membrane environment can help in understanding interactions among membrane proteins of great importance in neurobiochemistry.

7.9.3 Mutations associated with Alzheimer's disease

Let us analyze mutations that can be presently found in the Swiss-Prot database (P49768, June 2020) to see where these Alzheimer's disease-causing mutations (for the familial type 3 disease) are located in the protein sequence (Figure 7.3). For the presenilin plasma membrane location, the luminal segments became extracellular regions, as mentioned above. Out of 121 mutations associated with the familial early-onset form of Alzheimer's disease (AD3), the majority (71) is located inside TMH segments. A complete set of over 300 mutations and their sequence locations can be seen at the link: https://www.alzforum.org/mutations/psen-1. The cartoon of presenilin waving in and out of the membrane confirms the conclusion that most of the red-colored residues (substituted with pathogenic mutations in AD3) are located in the protein membrane-dwelling segments. It is a nice proof that TMH segments in at least some polytopic membrane proteins are not just structural elements but also play a fundamental functional role. The link provided above confirms that one of the most terrible diseases in older persons (Alzheimer's disease, type 3) is mostly associated with changes for transmembrane residues of presenilin-1.

How these changes influence the presenilin mechanism of action as the intermembrane protease? First of all, FAD-associated mutations from presenilin transmembrane segments can be grouped into several hotspots by looking at the presenilin sequence (Figure 7.3) or its 2D cartoon. Two hotspots are identified by looking at the much more realistic 3D-presenilin structure with highlighted residues mutated in Alzheimer's disease (Bai et al. 2015). The first one is from the neighborhood of the catalytic residues D257 and D385 that encompass THMs 6–9. These mutations can cripple the protease activity of γ-secretase complex despite an intact pair of catalytic aspartates for which no AD3-associated mutations have been found. The second one affects inward-facing residues from TMHs 2–5 with possible transport function. Both hotspots are deeply embedded in the membrane but require access to polar molecules, such as water or lipid-shy cations, for the functions impaired by these mutations. The 3D structure based on cryo-electron microscopy results published by Li et al. (2013) and Bai et al. (2015) did not reveal any details about the water-access pathway nor the substrate site for the proteolytic activity. Even bigger guns had to be used than the state of the art cryo-electron microscopy.

7.9.4 Elusive operation of the hatchet enzyme

Raquel Lieberman and Volker Urban collaborated in using neutron beams from the high flux isotope reactor to decipher the 3D structure for the bacterial analog of presenilin (Naing et al. 2018). They were well aware of MacKinnon's breakthrough with voltage-gated channel structure, which resulted after the decision to study the bacterial KcsA channel (see Section 7.6). Thus, they decided to investigate the archaeal ortholog of the intermembrane aspartyl protease. Once again, scientists used structural and functional similarities that bridge the several billion years of huge gulf of biological evolution across life domains. They fully expected to learn more about the Alzheimer's disease after revealing the 3D structure of the non-soluble aspartyl protease from *Methanoculleus marisnigri.*

That ancient prokaryote loves to produce methane and may belong to the first free-living ancestors of archaebacteria (Souza et al. 2013; see Chapter 13 for more details about life-origin speculations). Low homology of archeal presenilin from *Methanoculleus marisnigri* (only 18.3%, Li et al. 2013) did not discourage attempts to get an insight into the human presenilin structure through the study of archeal presenilin (Li et al. 2013, Naing et al. 2018).

After scholarly research, Naing et al. (2018) confirmed that water enters the active site and honestly admitted the lack of knowledge about how it happens. Other unanswered questions are how the substrate moves to the active site to interact with an activated water molecule and how

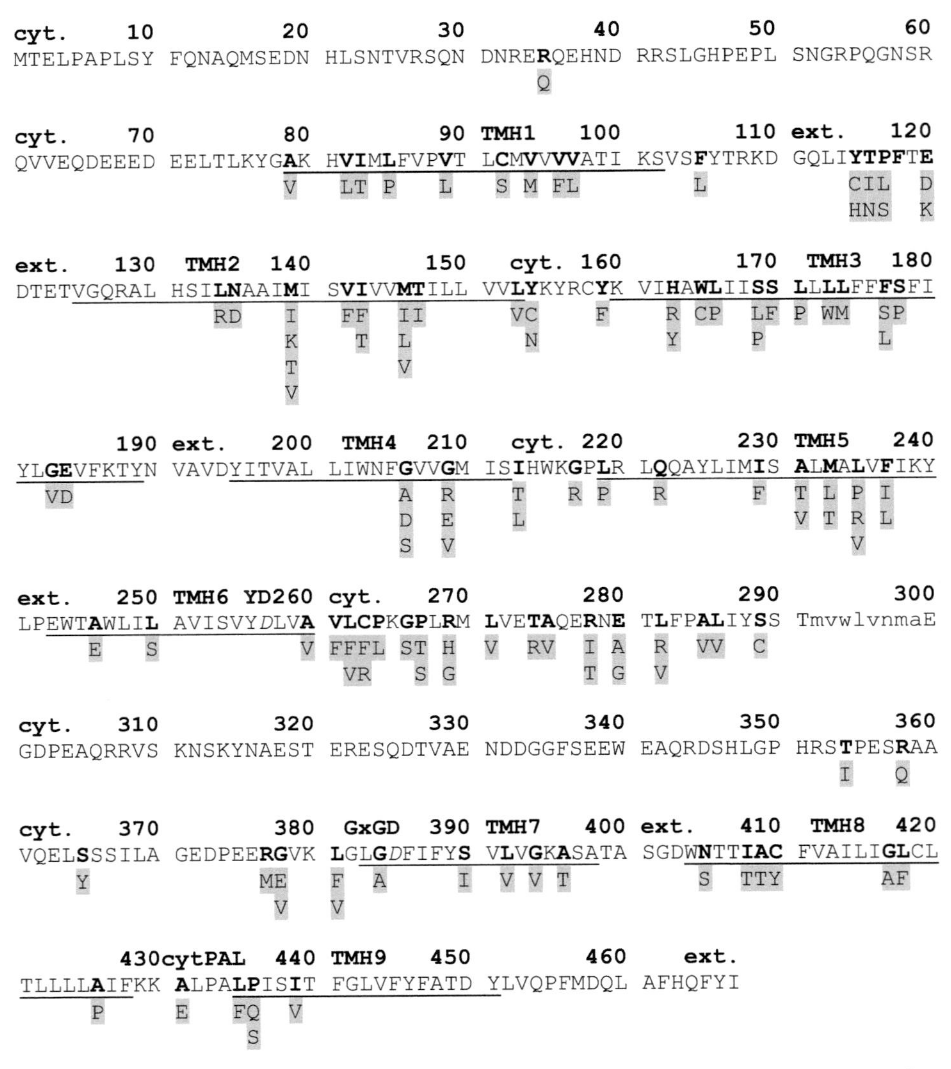

Figure 7.3: Presenilin-1 mutations associated with familial Alzheimer's disease AD-3 (FAD). Bold letters are for residues represented with the single amino acid codes substituted with some other residues in FAD mutations (below them with the gray background). Underlined segments are transmembrane helices determined in Sansom's laboratory and labeled above as TMH1 to TMH9 (Stansfeld et al. 2015). The cyt. and ext. abbreviations stand, respectively, for cytoplasmic and extracytoplasmic space location. The codes for aspartate residues in italic font (D257 and D385) are introduced to highlight two crucial catalytic residues for intermembrane cleavage of amide bond (protease activity). Conserved YD, GxGD, and PAL motifs (bold codes above sequence) are essential for catalytic and substrate binding function in related intermembrane aspartyl proteases. Lowercase letters from M292 to A299 define the presenilin region subjected to multiple auto-cleavages during γ-secretase maturation.

the substrate's processive trimming occurs after binding to γ-secretase. The successive trimming leads to the also unknown mechanism of product release (Liu et al. 2020). However, molecular dynamics simulations are getting ever better as an appropriate tool for tracing dynamic conformational changes that must be associated with the initial binding of substrates and the role of substrate flexibility (Li et al. 2017). The presenilin flexibility is likely to be increased near the active site after the auto-cleavages in the M292-A299 region (Figure 7.3), allowing for a better sampling of substrate conformations prone to similar cleavages (Fukumori et al. 2010). The active site conformation of presenilin is dependent on the P(433)AL(435) motif. It is not so distant from the cleavage region in the 3D γ-secretase complex, as it looks like in the presenilin sequence (Figure 7.3). The substrate helical conformation must be locally unwinded to offer the scissile peptide bond in an extended structure to presenilin's molecular aspartate scissors (Selkoe and Wolfe 2007, Liu et al. 2020).

In this chapter, we frequently mentioned signal peptides and the importance of recognizing when they are chopped off to create mature membrane proteins or exported bioactive peptides. Proteolytic enzymes that are performing this job - signal peptide peptidases (SPP)—are related to the presenilin family of aspartyl proteases (Golde et al. 2009). Despite minimal homology, both families contain the conserved active site motifs YD and GxGD located in adjacent transmembrane helices that form the aspartate scissors. Once again, vitally important proteins such as SPP can be misused. SPP promotes the final processing of the core protein from the hepatitis C virus and the hepatitis G virus (GB virus). Other examples are the proprotein convertase furin and the transmembrane serine protease TMPRSS2. Furin protease collaborates with the γ-secretase in forming the functional insulin receptor (Kasuga et al. 2007). TMPRSS2 is relevant for the prostate's normal physiological function but less critical for other aspects of development and homeostasis (Kim et al. 2006). Both furin and TMPRSS2 have been implicated in the SARS-CoV-2 virus cell entry—the virus responsible for the current pandemic (Hoffmann et al. 2020, Wrobel et al. 2020). In contrast to other CoV viruses infecting mammals, SARS-CoV-2 has the unique four amino acid insertion motif PRRA within the spike protein. This motif with the subsequent arginine is the potential cleavage site for mammalian furin. A protease cleavage triggers the membrane fusion between the viral membrane and our susceptible cells' plasma membrane.

7.10 Rhodopsin

7.10.1 The remarkable sensitivity of rhodopsin's quantum detector function

In the mammalian retina, the main **photoreceptor cells** are rods and cones. Rods have extreme sensitivity to light and are responsible for vision in dimly lit conditions. As we all know, the light is composed of photons. The quantum nature of photons makes them waves and particles (quanta) at the same time. How nature solved the problem of detecting photons? Rhodopsin (gene Rho) is the major rod protein acting as the photoreceptor, which evolved for dim light vision in vertebrates. Rhodopsin is another example of polytopic membrane proteins. Evolutionary biologists found out that the last common ancestor of all jawed vertebrates, which probably lived about 400 million years ago, already developed a similar complex eye structure with the rhodopsin as the photodetector (Kevany and Palczewski 2010). Similar visual pigments are found in rod outer segments of all vertebrates and invertebrates with a functional role to act as quantum detectors and initiate the transformation of light into vision (Yau and Hardie 2009). Related microbial rhodopsins developed billions of years ago to mediate the energy conversion needed for cell survival and cell signaling. Rho has similar transmembrane topology to bacteriorhodopsin (see Chapter 12 for details about bacteriorhodopsin light-activated kinetics and bioenergetics) with seven transmembrane helices (TMH) and the retinal pigment attached to 7th TMH (Palczewski et al. 2000). The pigment changes its conformation immediately after photon absorption. The conformational changes of the whole Rho apoprotein opsin follow in a slower time frame.

Essential to Rho function is the interaction with the G-protein G_T in the visual phototransduction pathway. Activated rhodopsin triggers a well-known signaling cascade with several amplification steps, ultimately leading to electric signal transmission from the rod cell to the synaptic terminal in the form of an action potential. The brain cortical centers for visual perception integrate and interpret generated electric potential in the retina extracting signal from noise in all transmission steps. Evolution has perfected vision sensitivity to the ultimate threshold so that after dark adaptation, our brain can detect a single photon (Hecht et al. 1942, Field et al. 2005). Upon light activation of Rho, the Rho interaction with transducin protein G_T becomes quite involved encompassing a total of 1042 $Å^2$ interface area (Gao et al. 2019). As a result, Rho elicits a 10^7-fold increase in the nucleotide GDP-GTP exchange rate. The chain reaction avalanches triggered by photon absorption represent a huge bioenergetics' amplification of the initial free-energy delivered from a photon.

7.10.2 Rhodopsin's structure

Cytoplasmic Rho loops connecting TMH 3 with 4, and TMH 5 with TMH 6 are named intracellular loops 2 and 3, respectively (ICL2 and ICL3). The conserved E(D)RY motif of ICL2 interacts via R135 with C347 from transducin G_T protein, stabilizing the active-state Rho conformation. The ICL3 residue T243 forms hydrogen bonds with K341 and D337 from G_T $\alpha5$ helix, while ICL3 residue Q237 forms a hydrogen bond also with K341 from that helix. These interactions stabilize the shift to active-state G_T conformation, which, with its $\alpha5$ helix, enters deeper into Rho active-state cavity.

The ICL2 loop has surprising conformational plasticity. In the Rho-G_T activated complex, it adopts an extended-loop conformation, but it is also capable of folding into an α-helix when the Rho-arrestin complex is formed (Gao et al. 2019). Arrestin protein prevents damage to the photoreceptor cell when the activation is not well balanced with deactivation and regeneration processes (Chatterjee et al. 2015). The Rho crystal structure alone (Palczewski et al. 2000, Okada et al. 2004) does not provide clues about the functional importance of its second and third cytoplasmic loop. A curious indication of their functional importance is that these loops are structurally poorly defined, probably due to their inherently stronger dynamics, flexibility, and conformational plasticity. Is it possible to see any such indication in the absence of heroic efforts to find all-atom locations for the whole activated Rho-transducin complex (Gao et al. 2019)? The SPLIT tool we developed at the University of Split, Croatia, can predict TMH sequence location. Besides, it can predict short membrane-associated, mostly amphipathic, helices, which are either extended alongside the membrane surface or bend back from the membrane surface toward the membrane interior (Juretić et al. 1999). For the bovine Rho sequence below, the SPLIT 4.0 version works better than the SPLIT 3.5 version. It predicts correct transmembrane topology with only slightly different sequence location of seven TMH from those annotated by Uniprot OPSD_BOVIN entry and PDB 1U19 entry (underlined segments), as can be seen from Table 7.2.

The SPLIT output of preference profiles from Figure 7.4 provides positional information for THM preferences (red profile), amphipathic helix preferences (gray profile), and beta-strand preferences (blue profile). One can see that the N-terminal of ICL2 and C-terminal of ICL3 are predicted as extramembrane helical extensions of TMH3 and TMH5 into cytoplasmic space. The N-terminal segment of ICL2 contains R135 in a predicted helical conformation. Interestingly, R135 has by far the highest predicted turn preference (2.57) and the highest predicted product of turn and helix preference (5.89) among all other rhodopsin residues. We have introduced the product of turn and membrane-associated helix preferences as the numerical output for SPLIT 4 (Table 7.2; Juretić et al. 2002). It is a convenient parameter for finding highly flexible short segments, which can assume different conformations near the THM entrance or exit position into the membrane. These predictions are easy to obtain and in accord with recent confirmations that the second and third cytoplasmic loops have functional importance for rhodopsin signal transduction dynamics.

Table 7.2: Bovine rhodopsin sequence with the found and predicted TMHs locations. OPSD_BOVIN sequence has underlined TMH segments according to PDB entry 1U19. The numbers below each single letter amino acid code are products of preferences for turn and membrane-associated helix conformation predicted by SPLIT 4.0. The two highest values for that product are for E134 and R135, the residues at the N-terminal of the second internal loop ICL2 (gray background).

```
>sp|P02699|OPSD_BOVIN Rhodopsin OS=Bos taurus OX=9913 GN=RHO

MNGTEGPNFYVPFSNKTGVVRSPFEAPQYYLAEPWQFSMLAAYMFLLIMLGFPINFLTLY
000000000000000000000000000000000111122222211111111223332222
VTVQHKKLRTPLNYILLNLAVADLFMVFGGFTTTLYTSLHGYFVFGPTGCNLEGFFATLG
211000000011222133311333122332222111111111111111111111112222
GEIALWSLVVLAIERYVVVCKPMSNFRFGENHAIMGVAFTWVMALACAAPPLVGWSRYIP
333222111222366411111111000000111123212221122222222221111000
EGMQCSCGIDYYTPHEETNNESFVIYMFVVHFIIPLIVIFFCYGQLVFTVKEAAAQQQES
000000000000000000000122211102220000000111123322111110000000
ATTQKAEKEVTRMVIIMVIAFLICWLPYAGVAFYIFTHQGSDFGPIFMTIPAFFAKTSAV
000000000123310000000001122233322322111100112222212233222111
YNPVIYIMMNKQFRNCMVTTLCCGKNPLGDDEASTTVSKTETSQVAPA
232122211110001111111111100000000000000000011121
```

Predicted topology for bovine rhodopsin

TMH	SPLIT prediction	Uniprot P02699	PDB 1U19	IN loops
1	OUT **38–61**	**37–61**	**35–60**	
2	**73–93**	**74–96**	**72–99**	
3	**115–139**	**111–133**	**109–133**	**134–150** ICL2
4	**153–174**	**153–173**	**151–173**	
5	**203–228**	**203–224**	**202–224**	**225–252** ICL3
6	**251–276**	**253–274**	**253–277**	
7	**285–304**	**287–308**	**286–308**	

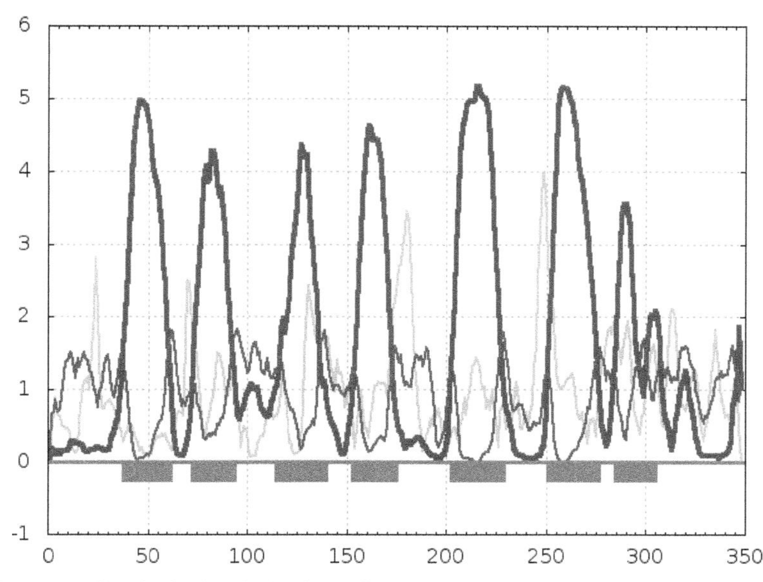

Figure 7.4: Preference profiles for bovine rhodopsin predicted by SPLIT 4.0. Pink rectangles below the y = 0 axis are assigned sequence locations of transmembrane helices. See the main text or Figure 7.2 legend for the meaning of red, blue, and gray preference profiles.

The preferences for the last of predicted membrane-spanning segments are split into two peaks, reflecting the partially hydrophilic nature of the active site where retinal is bound.

7.10.3 Why phototransduction cascade is tightly coupled to high entropy production?

Building upon recently published rhodopsin-transducin structure (PDB 6OY9), the signal transduction dynamics can be inferred (Gao et al. 2019). It provides plentiful additional insights on how protein-protein interaction channels amplifies initial signals through a cascade of bioenergetic conversions. It is one of the first examples of how G-protein-coupled receptors (GPCR) function with hard to overestimate influence in discovering similar interactions for over 800 members of GPCR superfamily in humans (Fredriksson et al. 2003). The phototransduction cascade triggered by photon absorption is not restricted to humans or mammals. Other species can also detect just a few photons in a completely dark room. Let us briefly describe how high sensitivity is achieved and what it has to do with bioenergetics.

Tightly-packed rhodopsins are located in photoreceptor cells. It was recently realized that rhodopsin organization in the disc membrane is embedded through several hierarchical levels starting with dimers (Gunkel et al. 2015). Rhodopsin dimers are arranged in the rod outer segments as ordered supermolecular tracks at a density of about 20 to 25 thousand molecules per square micrometer (Koch and Dell'Orco 2015). Photoreceptor cells are the first-order neurons of the visual pathway. Like most other neurons, photoreceptor cells cannot divide. Unlike other neurons, photoreceptor cells daily regenerate 10% of their rhodopsin containing segments by using selective autophagy (Kevany and Palczewski 2010, Boya et al. 2016, Léveillard and Sahel 2017). Ten days are enough for complete renewal of outer rod segments in our eyes, almost as if we are getting new eyes three times in each month (Koch and Dell'Orco 2015). Intensive autophagy is metabolically expensive but critical to minimize oxidative stress and protein aggregation because the retina is exposed to a variety of environmental insults, unlike brain parts with a deeper location. Alterations in autophagy are involved in age-associated decline and many eye diseases such as glaucoma, retinal dystrophies, autosomal dominant optic atrophy, and age-related macular degeneration (Boya et al. 2016).

The importance of vision for collecting the information regarding the outside world is reflected in an immense free-energy cost, and high entropy production required not only for activation cascade triggered by photon absorption, but also for aerobic glucose metabolism, regeneration of antenna discs with concomitant synthesis, recycling, protein transport costs, and active transport of ions against their concentration and electrical gradients. A single rod cell consumes at least 100 and 20 million ATP molecules per second, respectively, in darkness and light (Okawa et al. 2008). Interestingly, light-activated photoreceptor cells consume less free-energy than while resting in the dark. Neurons in the brain's gray matter are metabolically the most active while signaling, but both neuron classes are comparable in free-energy use to that of a human leg muscle running the marathon (Hochachka 1994, Attwell and Laughlin 2001). In all neurons, the active transport of ions is the most substantial free-energy consuming function. For instance, in a single neuron, the integral membrane protein named Na^+/K^+ pump hydrolyzes more than 300 hundred million ATP molecules per second to restore ion gradients to nonequilibrium resting levels. Only the continuous ion pumping and associated high entropy production can prepare the neuron to receive another signal to initiate the action potential (Attwell and Laughlin 2001).

The first amplification step in the photoreceptor cell is the activation of about 10 to 20 transducin molecules by single activated rhodopsin (Okawa et al. 2008, Koch and Dell'Orco 2015, Gunkel et al. 2015). Activated G protein transducin, in turn, activates the effector phosphodiesterase PDE6, which performs hydrolysis of the second messenger cyclic nucleotide guanosine 3',5'-cyclic monophosphate (cGMP) with high turnover rates. The relative concentration of Rho, transducin, and PDE in rod cells is approximately 100:10:1 (Yau and Hardie 2009). More than 100 thousand cGMP molecules are hydrolyzed during 100 ms per one rhodopsin activated by a one-photon absorption event (Zhang and Cote 2005). Decreased cGMP concentration leads to the closing of cGMP-gated cation channels. A fast drop occurs in the concentration of free Ca^{2+} and Na^{+} in the cytoplasm leading to the cell hyperpolarization and the inhibition of glutamate neurotransmitter release at the synaptic terminal of the photoreceptor cell (Stephen et al. 2008). There are about 10 thousand excited rhodopsins per second at a high light intensity, and almost all cGMP-gated channels are closed. The concentration of Ca^{2+} decreases about ten times from about 400–600 nM to about 10–50 nM after channel closure (Zhang and Cote 2005). When free-energy consumption is compared for rod cells in dark and bright light, the total ATP hydrolysis rate in the photoreceptor cell declines from 10^8 ATP molecules per second to about a quarter of that value (Okawa et al. 2008). In the darkness, cGMP concentration is high enough to maintain non-selective cation channels in the open state and steady inward "dark current" sufficient to depolarize the plasma membrane at the membrane potential of about –30 mV and to maintain steady glutamate release. Hyperpolarization initiated by photon absorption increases the membrane potential to about –70 mV. The high density of Na^{+}/K^{+}-ATPase pumps in the inner segment of rod cells enables these specialized neurons to restore and maintain nonequilibrium internal concentrations of low Na^{+} and high K^{+} in the face of large inward Na^{+} flux (through cGTP-gated channels) and outward K^{+} flux (through nongated selective K^{+} channels). A high ATP hydrolysis rate by Na^{+}/K^{+}-ATPase active transport pump is obligatory in all neuronal cells. It is the major contributor to high metabolic costs for all types of neuronal signaling activity. The amplification of the light signal does not stop at the retina. Several brain structures are almost entirely devoted to processing and interpreting signals received by our eyes with concomitant high free-energy costs and entropy production (Attwell and Laughlin 2001).

Activation steps are tightly regulated with efficient shut-off mechanisms to avoid the accumulation of harmful by-products and to control sensitivity (Koch and Dell'Orco 2015). These steps include a 10-fold increase in cGMP synthesis, phosphorylation of activated rhodopsin to turn it off, and binding of arrestin to phospho-rhodopsin, which prevents rhodopsin-transducin interaction. The resulting increase in cytoplasmic Ca^{2+} is sensed by Ca^{2+}-binding recoverin protein and by calmodulin. Recoverin controls the phosphorylation of rhodopsin by rhodopsin kinase, while Ca^{2+}-calmodulin decreases the cGMP sensitivity of cationic channels. This intelligent-like behavior uses a large number of fast-acting feedback loops (in milliseconds to seconds time scale), each connected to some metabolic cost. For instance, the aerobic glycolysis in mammalian retina is used for the daily renewal of 10% of the outer segments of photoreceptors—a huge bioenergetic challenge to maintain optimal membrane fluidity for optimal phototransduction (Léveillard and Sahel 2017). In conclusion, phototransduction's highly nonequilibrium resting condition has its price—as high entropy production as an athlete's leg muscle while running the marathon. This mental image is not entirely appropriate because an athlete would have to breathe four times faster while waiting for the starter's pistol sound than during running to resemble resting (dark) and active (light period) phototransduction bioenergetics.

7.11 Cytochrome c oxidase

7.11.1 Oxygen-mediated water synthesis coupled to proton pumping

Cytochrome c oxidase is also the polytopic integral membrane protein. Since it is composed of several subunits, which are important for the respiratory function, it belongs to membrane-embedded respiratory complexes. Out of four such complexes located in the inner mitochondrial membrane, the cytochrome c oxidase is the terminal complex IV. The respiratory complex IV is responsible for 95–98% of total oxygen consumption in mammalian cells. This enzyme is found in mitochondria of all eukaryotic cells and also in the plasma membrane of aerobic bacteria. The reduction of O_2 molecules to water molecule H_2O is just one of its essential planetary functions. A total of ~ 90% of the oxygen taken up from the atmosphere by living beings is used by cytochrome c oxidase to catalyze water synthesis (Chan and Li 1990, Pereira et al. 2001). Why different animals, fish species, invertebrate species, and all other aerobic species, including plants during their active respiration, need to perform the synthesis of water molecules? Water synthesis looks like the most useless process, which one can imagine in a biosphere rich with water in the form of rain clouds, rivers, lakes, oceans, and glaciers.

The overall reaction catalyzed by this enzyme found in biochemistry textbooks involves electron transfer from cytochrome c:

$$4\text{Cyt c}^{2+} + 4H^+ + O_2 \rightarrow 4\text{Cyt c}^{3+} + 2\,H_2O$$

However, the reaction written in this form does not provide any clue that cytochrome oxidase is doing anything else besides oxygen uptake from the atmosphere, cytochrome c oxidation, and water synthesis as a non-polluting output. This equation describes well enough what cytochrome oxidase is doing in solution. The distinction of crucial importance is what this enzyme is additionally doing when incorporated in a membrane. Mårten Wikström discovered about 45 years ago (Wikström 1977) that pumping protons together with water synthesis by cytochrome oxidase creates the electrochemical proton gradient across mitochondrial or bacterial membranes. Proton pumping was confirmed with preparations of artificial membranes when purified enzymes became available. To specify the difference between protons taken up from inside space for water synthesis and pumped-out protons, Wikström named them "chemical" and "pumped" protons. Together with the received electron, the chemical proton participates in the catalytic chemistry.

It was not a new concept. The influential Peter Mitchell's paper about chemiosmotic coupling was entirely based on the concept of vectorial transmembrane transport for chemical protons (Mitchell 1961). Both cytochrome oxidase actions contribute to creating the protonmotive force across topologically closed membrane impermeable to protons except through proton pumps. Decreased proton concentration in the internal space makes that space and internal membrane surface negative with respect to the external space and the external membrane surface. When all this is taken into account, the attention can be focused at what happens to electrons, protons, oxygen, and water molecules in different compartments as illustrated in the modified equation for cytochrome oxidase catalytic action:

$$4e^-_P + 4H^+_{c,N} + 4H^+_{p,N} + O_2 \rightarrow 4H^+_{p,P} + 2H_2O$$

where subscripts c and p denote "chemical" and "pumped-out" protons, while subscripts N and P denote negatively and positively charged side of the membrane, respectively (Wikström et al. 2018). Molecular oxygen has excellent solubility in water and even five times better solubility in lipid membranes. Thus, it can easily reach the cytochrome oxidase's reaction center, which is deeply buried in the central enzyme location with Fe-Cu_B binuclear couple where oxygen chemistry occurs. The binuclear center location corresponds to the most hydrophobic middle position between two phospholipid layers forming the membrane bilayer.

7.11.2 Cytochrome c oxidase structure

The 3D structure of cytochrome oxidase is well known today thanks to scholarly contributions of dedicated research teams during the last several decades (Iwata et al. 1995, Tsukihara et al. 1995, 1996, Yoshikawa et al. 1998, Koepke et al. 2009, Zong et al. 2018). Before going into the origin and functional arrangement of all 13 to 14 polypeptide chains forming this enzymatic complex with its metal cofactors, it is helpful to present a simplified scheme of how it works in the membrane according to the second equation shown above (Figure 7.5).

Hard work on isolation, purification, and crystallization of cytochrome oxidase from bovine hearts enabled present-day insights into how its structure and function are connected. This work started in the laboratory of American scientist Winslow Caughey in 1962 (Caughey 1971). Its first phase ended up in 1995 in Japan and Germany when the structure of cytochrome oxidase from the bovine heart and *Paracoccus denitrificans* was solved by using

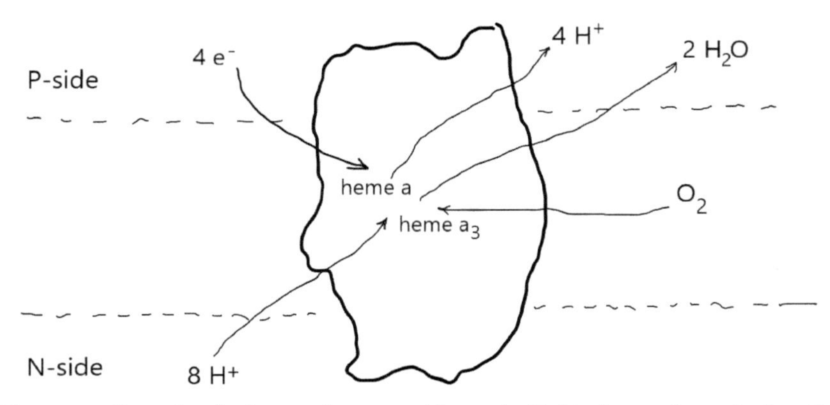

Figure 7.5: The cartoon illustration for the cytochrome c oxidase embedded in the membrane (horizontal dotted lines). Arrows illustrate transfers of electrons, protons, oxygen, and water molecules to or from low-**spin** heme a and a high-spin heme a_3 with a net result of proton extrusion and water synthesis. The oxygen molecule is soluble in the membrane and can easily pass through lipid molecules to reach centrally located active site. The exit pathway for protons is from the mitochondrial matrix (the N-side) toward the external cytoplasmic side (the P-side). Synthesized water molecules can also exit toward N-side, although some authors put it toward P-side as we did.

inventive co-crystallization, inspired choice of detergent, and the most advanced techniques of X-ray crystallography which existed at that time. It was no small feat because, in 1995, cytochrome oxidase complex was the largest integral membrane protein whose structure was determined with an excellent resolution for membrane proteins (Iwata et al. 1995). Ten years before, in 1985, the same author, Hartmut Michel, solved with his student Johann Deisenhofer the 3D structure of photosynthetic reaction center (Deisenhofer et al. 1985), a milestone in understanding structure-function connection for membrane proteins, which was quickly recognized with the Nobel Prize in 1988 to him, Johann Deisenhofer and Robert Huber. The location of more than ten thousand and more than six thousand protein atoms was determined for cytochrome oxidase and photosynthetic reaction center, respectively. Interestingly, first advances in resolving structures of membrane proteins were achieved by cryo-electron microscopy (for bacteriorhodopsin, see Henderson et al. 1990 and refs therein). Recently, cryo-electron microscopy technique has become competitive with X-ray crystallography thanks to the development of new electron detectors, allowing high-resolution studies. Researchers from Austria and UK used cryo-electron microscopy to solve a giant structure of **respiratory complex I** (the first of respiratory proton pumps from the inner mitochondrial membrane) with a total mass of 970 kilodaltons (Fiedorczuk et al. 2016).

7.11.3 The illogical manner in which nature puts together the cytochrome c oxidase polypeptides

There are thousands of examples of how scientists obtained a detailed insight into the evolution of complex structures performing complex functions. Biological evolution often led to a much more sophisticated design and mechanism of action than any known human design, although it did not require the intelligent design with some final goal as a design target. We mentioned in a previous section (7.10) how similar visual transduction pathways developed in different species (Yau and Hardie 2009). All the same, the absence of a design target often led to overly complicated and strange processes and structures (Lents 2018). In this section, we shall shortly mention the illogical manner of how cytochrome c oxidase polypeptides are synthesized and put together in a mitochondrial membrane. Three catalytic subunits are synthesized by mitochondrial ribosomes using mitochondrial circular DNA for instructions. All other subunits are synthesized in the cell nucleus and need to go through a labyrinthine pathway to reach the mitochondrial matrix and to find catalytic subunits in the inner mitochondrial membrane. We already mentioned the evolutionary rationale of why Occam's razor reasoning did not apply in this case—a whole mitochondrion is mostly synthesized by the cell's nuclear machinery. During the evolution of 1.5 to 2 billion years, eukaryotic cells advanced from learning how to tolerate the symbiosis with alpha-proteobacterium (Lane and Martin 2010) to learning how to synthesize it.

Three catalytic subunits from cytochrome oxidase, I, II, and III, and 10 additional polypeptides synthesized by mitochondria are vestiges of several thousand genes and proteins, which the bacterial precursor of mitochondria possessed billions of years ago. It is a living-fossil proof of how blind biological evolution combined old with new structures to create something better, bigger, and more complex without ever having such a goal. Just imagine what waste it is to synthesize 78 mitochondrial ribosome subunits in the cell cytoplasm and transport them into mitochondrial matrix space to construct the mitoribosome (Lightowlers et al. 2014). The only function of mitochondrial ribosomes is to translate 13 out of 1500 mitochondrial proteins (about 1% of all mitochondrial proteins, all serving as subunits of the respiratory chain) from the mitochondrial genome. That illogical feature survived through eons despite frequent mutations in the mitochondrial translation system that cause various pathologies such as neurodegenerative disorders, cardiomyopathies, and metabolic syndrome (Sylvester et al. 2004). However, all free-energy input and vastly increased dissipation to perform this feat (fast mitochondrial creation-destruction recycling must also be taken into account) is amply paid off in terms of cell signaling regulated by mitochondria, regulation of metabolic functions, **respiratory control**, control of cell survival, and many other new functions these free-energy producing organelles gained in the same time period (Lim et al. 2016). The most important advantage of having a large number of mitochondria inside the eukaryotic cell is the vast increase in surface area of bioenergetic membranes, which enabled a roughly 200000-fold increase in host genome size compared to bacteria and about 1000-fold greater power per genome (Lane and Martin 2010). Greater power is mostly released as free-energy dissipation and entropy production, which may indeed be considered as a waste, but in eukaryotic cells, this waste is inseparably connected to the compartmentalization of energy coupling and a critical redistribution of DNA in relation to bioenergetic membranes, a prerequisite to eukaryote complexity.

7.11.4 The proton pumping cycle requires the participation of metal atoms

The function of 10 to 11 "accessory" subunits (Zong et al. 2018) is presumably regulatory, and we shall shortly examine only how the structure of three catalytic subunits and their cofactors is connected to the proton-pumping function of cytochrome oxidase. Four electrons flow from soluble cytochrome c to Cu_A atom pair bound to polypeptide II. Subsequently, they reach heme a and heme a_3 where the binuclear heme-copper Cu_B center is attached at the polypeptide I. Subunit III stabilizes Cu_B by making many hydrophobic contacts with subunit I, which enhances enzyme turnover rate. It contains three tightly bound polar lipids proposed to facilitate O_2 channeling. Oxygen binds to heme a_3. The oxygen molecule is converted into two water molecules after accepting electrons and four protons. Protons are taken up from the mitochondrial matrix. Additional four protons are exported from the matrix into the cytoplasm through this enzyme complex. Besides two iron atoms as central metal atoms of hemes, the cytochrome oxidase also contains three copper atoms, one manganese atom, one zinc atom, and several polar lipid ligands.

Amino acid residues from polypeptide I participate in the proton pumping cycle. An important active site residue is tyrosine 244 from transmembrane helix VI, which interacts with histidine 240 also from helix VI. His 240 is ligated to the Cu_B center, and this is also the case with histidines 290 and 291 from transmembrane helix VII. Helix VI sequence motif -G**H**PEV**Y**- with functionally critical Tyr 244 is highly conserved in cytochrome oxidases from all eukaryotic species (residue numbering according to a bovine enzyme). Blomberg (2016) suggested that this active site residue delivers an electron and a proton toward catalytic cleavage of the O-O bond, thus leaving the tyrosine unprotonated during several reduction steps of the catalytic cycle.

7.11.5 Is the active site "breathing" in the cytochrome c oxidase?

What is the pathway for proton uptake and extrusion? Three channels have been proposed: D, K, and H channel. Twelve transmembrane helices of subunit I are arranged in three clusters with each cluster containing four TM helices. Each helical cluster encircles a hydrophilic pore—the candidate for proton channel (Wikström et al. 2018). Proton pathway is a better name than a proton channel because the word "channel" brings to mind a drain or tube with clearly delineated borders. First of all, protons are quantum particle-wave objects that can easily tunnel across distances shorter than one Å on biologically relevant time scales. With their almost two thousand smaller mass, electrons can perform quantum tunneling much more quickly than protons. There are no clearly delineated borders for electrons and protons. Secondly, protons hopping between water molecules inside and between hydrated cavities found in cytochrome oxidase structure enable long-range proton transfer (tens of angstroms) (Xu and Voth 2005). It is something that larger cations cannot do. Larger cations have to "drag" a solvent shell with them. They do require either selective or nonselective channel inside membrane protein if they are to transfer across the membrane. Thirdly, protons moving through a network of hydrogen bonds do not have labels telling us that a specific proton accomplished the N to the P-side passage. Instead, the "proton-wire" inside membrane protein facilitates a cooperative transfer of protons between successive molecules (Ball 2008), the proton flow mechanism similar to electron flow in electric current.

The so-called Grotthuss mechanism for proton hopping between water molecules (Nagle and Morowitz 1978) has a long history (Marx 2006, de Grotthuss 2006), and it experienced recent revival through improved molecular dynamics simulations for proton transfer across membrane proteins (Xu and Voth 2005, Goyal et al. 2015). Despite many efforts, recent and current doctoral dissertations, it is fair to say that cytochrome oxidase, and its high efficiency in oxygen reduction coupled to proton pumping still belongs to the most poorly understood among different types of membrane proton pumps. Fortunately, fine details of molecular dynamics for peptides and proteins can be now examined with advanced molecular dynamics (MD) simulations, which offer a new microscopy technique for visualization and prediction of a functionally critical nanosecond to microsecond events (Goyal et al. 2015, Ulmschneider and Ulmschneider 2018).

MD predictions are still tentative, but I find them highly exciting, much more exciting than a small statistically-insignificant high energy peak recently found at CERN (it inspired more than 100 distinguished physicists to publish scientific papers). For proton pumping of cytochrome oxidase, such calculations suggested the intermittent and transient nature of proton pathways that are converging toward the proton loading site, probably the propionate D of heme a_3 (Goyal et al. 2015) or propionate A of heme a_3 (Blomberg and Siegbahn 2012). A combination of site-directed mutagenesis and simulations determined that glutamate 286 (E-286) from *Rhodobacter sphaeroides* cytochrome oxidase subunit I (E-242 in the bovine sequence) is the potential proton valve. This critical glutamate residue receives an excess proton from the proton uptake site (D-91) after it delivers the proton to the proton loading site. The residue D-91 is highly conserved. It is a starting point of a "proton wire" named the D-channel, which extends from the N side of the membrane to E-242. The flexibility of the active site is essential to this "kinetic gating" mechanism, also named the "water-gated" mechanism (Wikström et al. 2003, Son et al. 2017, Wikström and Sharma 2018).

The central cavity squeezed among E-242, heme a, and heme a_3 with the binuclear center can "breathe," and that is the essential reason in the nanoworld why we are able to breathe in the macroscopic world. In each catalytic cycle, the central hydrophobic cavity undergoes a reversible transition between dry and wet configuration. In the dry configuration, very few water molecules are present there (up to three), while in wet configuration, two synthesized and possibly some additional water molecules are added, leading to regular cavity expansion and contraction. Calculations indicated that the active-site cavity undergoes expansion and compression for up to150 $Å^3$ during one turn of the proton-pumping cycle (Goyal et al. 2015). It is a very significant and very fast oscillatory conformational dynamics in a nanoworld of bioenergetics (up to one thousand turnovers per second). From crystal structures alone, it is not clear how many water molecules are present in the central cavity (if any) and how protons are transferred to the active site. The dynamics suggested above is the result of MD simulations. Such calculations also implicate equally fast oscillations in electric field strength and possibly its direction in the space between two hemes and glutamate 242.

7.11.6 How enzyme controls the strength and direction of the colossal electric field in critical proton-gating situations?

When oriented from heme a_3 toward heme a, roughly perpendicular to heme surfaces and to the direction of the electric field created by respiring mitochondria, the electric field strength in the central cavity can reach the colossal strength of billion volts per meter (Kaila et al. 2010, Wikström and Sharma 2018). Such electric fields are created at the focus of a laser beam, which can change the medium's properties so that nonlinear effects can occur. At its peak, the internal electric field surge created by cytochrome oxidase probably facilitates electron transfer to the binuclear center after the proton has been placed at the proton-loading site. According to the watergate proposal of Wikström et al. (2003), electron transfer to heme a_3 would cause the switch in the direction of the electric field. Chemical-proton transfer follows from re-protonated E-242 to the binuclear center in order to complete oxygen reduction and water synthesis.

The fast decrease of the cavity's electric field happens due to the increase in the dielectric constant of wetted-cavity. The electrostatics of the proton-loading site changes enough for the pump-proton to be released toward the P-side during the millisecond time range. Since cytochrome c donates four electrons, this whole sequence of events repeats itself four times. Repetitions are not identical. In two steps, chemical protons are accepted from the D-channel and in two steps from another proton pathway named the K-channel.

Five charge transfer steps occur in each of the four electron-proton coupling and proton extrusion steps. Firstly, an electron is transferred from cytochrome c to heme a. Secondly, a pump-proton is transferred from glutamate 242 to the proton-loading site. Thirdly, an electron is transferred from heme a to heme a_3 and to the binuclear reaction center. The fourth charge transfer is chemical proton transfer from E-242 to the binuclear reaction center. The last charge transfer is the extrusion to the P-side of the pump-proton from the proton loading site. Mutation of glutamate-242 to any other residue disables cytochrome oxidase from performing chemical and proton-pumping work.

The proton transport pathway D connecting the N-side with glutamate 242 in the A-class cytochrome oxidases (human-like) contains approximately ten water molecules and at least nine functionally important residues for that pathway. These residues are the proton uptake site D-91, the asparagine gate N-98 next to N-90, N-163, Y-19, M-71, and serines 101, 156, and 157 (Wikström et al. 2018). The transport pathway K starts at the N-side with proton uptake site E-62. The pathway contains a number of functionally important water molecules and amino acid residues that facilitate proton transport up to tyrosine-244 and the binuclear reaction center. The exit pathway for protons is less well known. We can even say that the details of charge transfer pathways and the mechanism of action for cytochrome oxidase remain one of the most formidable unsolved problem in the interdisciplinary fields encompassing biochemistry, electrochemistry, enzyme kinetics, and bioenergetics. It is also the problem that all fields of classical physics and modern biology cannot solve alone without asking for help from density functional calculations from quantum mechanics (Blomberg and Siegbahn 2012).

The main disputed points are three critical gating situations and how natural evolution solved them. It is assumed that electron transfer, via Cu_A to heme a triggers the proton transfer from N-side to proton loading site and that proton transferred to proton loading site triggers electron transfer from heme a to the binuclear center. Measured lifetimes of electron and proton transfers are used together with the transition state theory to estimate heights of corresponding transition state barriers. The specific location where protons are temporarily stored (pump loading site) is not known. It is generally assumed to be close but separate from the site where oxygen chemistry and water production occur (the binuclear center) (Goyal et al. 2015). The first gate prevents protons from the P-side to reach the proton loading site. It only requires a fairly constant and high barrier between the proton loading site and the P-side. The barrier is high enough to block protons returning from the P-side and low enough to allow proton transfer from the proton loading site to the P-side in the last rate-limiting step of about millisecond duration. It is unknown where is the specific location of the transition state corresponding to this barrier, except that it should be above heme a_3 propionates and below the P-side. The second gate or proton valve guides the transfer of the first proton (pump-proton) from the N-side toward the proton loading site and not to the water production site. In contrast, the chemical proton from the N-side is guided toward the binuclear center. The branching point (valve) must exist near glutamate-242. At that site, which can be E-242, pump-proton takes one "proton-wire" and the chemical proton another one. The third gate prevents the pumped proton from the loading site returning to the N-side when chemical protons cause destabilization. It also must prevent pump-proton hopping to the binuclear center.

Obviously, everything is precisely coordinated in time and space so that fast oscillations produced in central cavity volume, dielectric constant, and enormous electric field cause only those changes of transition state barrier heights for proton valves which would produce proton pumping and water synthesis. During one catalytic cycle in about one millisecond, four lightning strikes illuminate the central cavity, when an electron from heme a is accelerated in an electric field of about billion volts per meter to reach heme a_3 and binuclear reaction center (Kaila et al. 2010, Wikström and Sharma 2018). No wonder that these high energy electrons can occasionally wreak havoc and produce oxygen radicals instead of fusion of hydrogens with an oxygen atom in the water molecule. Reactive oxygen species are the byproduct of mitochondrial electron transport, partially useful as the mitostress signals and partially injurious in the long run (Herst et al. 2017). Anyway, a lot of additional research seems to be needed for a seamless connection of classical and quantum calculations to gain even better insight into how this classical-quantum machine works. The weakness of quantum chemistry calculations on large cluster models with 390 atoms are boundary problems (protons are not allowed close to artificial outer boundary) and the absence of the electrochemical gradient (Blomberg and Siegbahn 2012). These are important considerations when we study thermodynamics, kinetics, and efficiency of this proton pump.

7.11.7 The thermodynamic efficiency of cytochrome c oxidase

Molecular oxygen has a high affinity for electrons. As a terminal electron acceptor, it is an excellent choice. The primary driving force for cytochrome oxidase work is the difference in redox potentials between donor cytochrome c redox couple (ferrocytochrome-ferricytochrome) and acceptor oxygen-water redox couple. The difference of about 550 mV in physiological conditions corresponds to almost 13 kcal/mol per one electron. This driving force is quite enough to drive two protons against a maximal protonmotive force of about 220 mV that mitochondria can develop in its state 4 when lack of ADP limits the respiration rate (see Chapter 16 explanations). Driving two protons across the membrane using the transfer of one electron corresponds to what cytochrome oxidase is doing during one of its four proton-pumping catalytic steps. However, only the pump-proton is transferred from N to the P-side. The chemical proton used for water synthesis also increases the electrochemical proton gradient because it causes a decrease in internal proton concentration. With thermodynamic stoichiometry set up at the fixed ratio of two extracted protons from the N-side per one-electron transfer toward oxygen for establishing the secondary protonmotive force, the free-energy transfer efficiency appears to be very high, 70 to 80% at maximum load. Yet, the experimentally measured ratio for net uptake of protons per one-electron transfer is about 1.6 (Kaila et al. 2010)—the value which would indicate lower free-energy transfer efficiency of close to 50%.

From considerations of cytochrome oxidase thermodynamic efficiency, it is clear that powerful tools must be used to study this molecular machine far from equilibrium. For realistic simulations, the slippage possibility must also be included. We performed such calculations already in 1988, and the results were quite interesting (Juretić et al. 1988). We concluded that goals of maximal efficiency and minimal dissipation tend to exclude each other and that intrinsic uncoupling (slippage) may play an important role in fine-tuning the dissipation level and biologically optimal efficiency. For moderate slip values and physiological values of increasingly strong protonmotive force, we found the regime in which overall efficiency increases with increased entropy production until maximal entropy production is reached. Admittedly, this was calculated for a simple 6-state kinetic scheme with a redox-driven cycle and proton pumping cycle connected together in the presence of a nonproductive slip transition.

By using more complex kinetic schemes and kinetic master-equations, Kim and Hummer (2012) have seen that increasing the complexity of the kinetic models improved efficiency. A complicating feature in these simulations is a high number of functional states and their interconnections. The model with the two electron-acceptor sites (heme a and binuclear center) and three proton sites (Glu 242, pump-site, and the binuclear center) already has $2^5 = 32$ functional states and about 100 transition rate coefficients. Corresponding steady-state equations can still be easily solved, but simulations have too many possibilities due to a lack of experimental data. In a simpler model with the two proton-acceptor and one electron-acceptor site, and with a total of eight microscopic conformational states, Kim and Hummer (2012) also obtained interesting results. Branched kinetic gating of the proton transfer to the pump site and subsequently to the binuclear center improved pumping efficiency of cytochrome c oxidase in accord with the proposed water-gated mechanism of pumping (Wikström et al. 2003). The full reversibility of all microscopic processes is not an obstacle to achieve kinetic gating. It can be achieved either by modulating transition state barriers or by having nonequilibrium reactant and product concentrations or by modulating electron and proton affinities of the microscopic functional states.

As defined by Kim and Hummer (2012), energy transduction efficiency is just the ratio of output to input flux. The flux ratio should be multiplied with the ratio of output to input force to get the free-energy transduction efficiency (see Chapter 3). The force ratio is another efficiency expression that we named the free-energy storage efficiency, and they called it the thermodynamic efficiency. Multiple efficiency maximization simulations produced flux ratio near 1.0 and force ratio from 0.2 to 0.6, that is, the free-energy transduction efficiency between 20 and 60% in the range of membrane voltages from 100 to 200 mV (Kim and Hummer 2012). The peak efficiency of 60% was reached for the mitochondria in state 4 when respiratory control is firmly established.

One unjustified assumption in these simulations is that biological evolution had a goal to reach maximal efficiency. Have molecular machines evolved toward maximal efficiency or not is the fundamental non-resolved question in biological physics (Hoffmann 2016). Up to the present time, there was no attempt to examine for cytochrome oxidase whether the most important charge transfer steps can be optimized from the maximal entropy production requirement for corresponding transitions.

Definitions and explanations from Chapter 7:

AMP and cAMP are abbreviations for, respectively, adenosine monophosphate and cyclic adenosine monophosphate.

An amphipathic chemical compound contains both polar (hydrophilic) and non-polar (hydrophobic) portions in its structure. Polar groups love contacts with water molecules. Amphipathic molecules are phospholipids from biological membranes. Their asymmetric hydrophobicity is distributed lengthwise, with a polar head group at one end and hydrophobic lipid tails at the other end. Peptides in the amphipathic helical conformation have a sidewise asymmetry in hydrophobicity, with polar or charged residues extending from one helix face and hydrophobic residues from another.

CASP, short for Critical Assessment of Structure Prediction, is the biennial challenge in the 3D protein structure prediction initiated by John Moult and other computational biologists in 1994.

To the present day, it remains among biology's grandest challenges—how to determine a protein's 3D shape from its amino-acid sequence. Authors send recently solved protein structures to CASP organization in advance of their public release. CASP then selects up to 100 structures for the world-wide competition among bioinformaticians and other researchers interested in computational biology predictions.

The **cytoskeleton** is a dynamic filamentous network that determines and stabilizes a cell's shape. In eukaryotic cells, it is mainly composed of microfilaments, intermediate filaments, and microtubules.

Microtubules have a major role in forming the cytoskeleton but are highly dynamic. They are polymers composed of two globular proteins, α and β-tubulin.

Photoreceptor cells are neuronal cells with unique design, shape, and function. Rod photoreceptor cells have many mitochondria that cater to the immense free-energy requirements of these cells and a specialized antenna segment in the shape of a rod containing about thousand disk-shaped membranes. Each "disk" side contains roughly 80 thousand rhodopsin molecules. In contrast to other neuronal cells, activated photoreceptor cells react by hyperpolarization to covert an external signal into a traveling wave of electrochemical energy, which the brain perceives as an action potential. The human retina has more than 100 million photoreceptor cells, and each cell has about one billion rhodopsins (Kennedy and Malicki 2009).

Prediction accuracy is the ratio of all correct predictions (positive and negative) to all predictions (correct, false positive, and false-negative).

The **proteome** is the set of all expressed proteins by a genome of examined cells or organisms.

Respiratory complex I is the largest and most complicated enzyme among four respiratory complexes localized in the inner mitochondrial membrane. The enzyme comprises 14 core and at least 30 accessory subunits (polypeptides). It is a marvelous molecular engine proposed to couple electron transport with proton translocation using a piston-like motion of a very long lateral surface helix HL, quite unlike the rotary ATPase motor of the complex V (Efremov and Sazanov 2011, Wirth et al. 2016). Unfortunately, complex I is a significant contributor to the release of harmful oxygen radicals implicated in several degenerative diseases.

Respiratory control in bioenergetics is the respiration rate increase in response to ADP. It is the best measure of mitochondrial function in isolated mitochondria. A high rate of substrate oxidation and ATP synthesis after ADP addition implies low proton leaks and high respiratory control ability of healthy mitochondria (Brand and Nicholls 2011).

Signal peptide at the N-terminus of freshly synthesized protein targets and facilitates entry into a secretory pathway or intracellular organelles. Its middle section consists of predominantly hydrophobic amino acids. The short length of 16 to 30 residues is similar to the span of transmembrane helices.

A **Spin** is an intrinsic form of angular momentum in quantum mechanics. It often refers to the spin quantum number.

Taxol is the anticancer medication that was first isolated from yew trees. It was known already to Caezar how yew tree extract can kill or help during some conditions. His Gallic Wars book described the suicide of Eburonean king Catuvolcus with poisonous yew. The book contains a proud account of Caesar-led total genocide of all Eburones. For our book, it is of interest that taxol stabilizes microtubules and prevents their dissociation and re-grouping during cell proliferation (Steinmetz and Prota 2018). Thus, the stability is useful for a healthy cell only if it is a controlled and transient dynamic stability, no matter how large free energy investment and accompanied dissipation are needed.

References

Aimone, J.B., Li, Y., Lee, S.W., Clemenson, G.D., Deng, W. and Gage, F.H. Regulation and function of adult neurogenesis: From genes to cognition. Physiol. Rev. 94(2014): 991–1026.

Alexander, C.M., Reichsman, F., Hinkes, M.T., Lincecum, J., Becker, K.A., Cumberledge, S. et al. Syndecan-1 is required for Wnt-1-induced mammary tumorigenesis in mice. Nature Genet. 25(2000): 329–332.

Allen, K.N., Entova, S., Ray, L.C. and Imperiali, B. Monotopic membrane proteins join the fold. Trends Biochem. Sci. 44(2019): 7–20.

Aoki, S., Thomas, A., Decaffmeyer, M., Brasseur, R. and Epand, R. The role of proline in the membrane re-entrant helix of caveolin-1. J. Biol. Chem. 285(2010): 33371–33380.

Aoki, S. and Epand, R.M. Caveolin-1 hydrophobic segment peptides insertion into membrane mimetic systems: role of proline residue. Biochim. Biophys. Acta 1818(2012): 12–18.

Aquino, R.S., Teng, Y.H.-F. and Park, P.W. Glycobiology of syndecan-1 in bacterial infections. Biochem. Soc. Trans. 46(2018): 371–377.

Ashcroft, F.K. 1999. Ion Channels and Disease. Academic Press, Cambridge, MA, USA.

Attwell, D. and Laughlin, S.B. An energy budget for signaling in the gray matter of the brain. J. Cereb. Blood Flow Metab. 21(2001): 1133–1145.

Austin, E.D., Ma, L., LeDuc, C., Berman Rosenzweig, E., Borczuk, A., Phillips, J.A. 3rd et al. Whole exome sequencing to identify a novel gene (caveolin-1) associated with human pulmonary arterial hypertension. Circ. Cardiovasc. Genet. 5(2012): 336–343.

Bai, X.-c., Yan, C., Yang, G., Lu, P., Ma, D., Sun, L. et al. An atomic structure of human γ-secretase. Nature 525(2015): 212–217.

Baietti, M.F., Zhang, Z., Mortier, E., Melchior, A., Degeest, G., Geeraerts, A. et al. Syndecan-syntenin-ALIX regulates the biogenesis of exosomes. Nat. Cell Biol. 14(2012): 677–685.

Balali-Mood, K., Bond, P.J. and Sansom, M.S.P. Interaction of monotopic membrane enzymes with a lipid bilayer: A coarse-grained MD simulation study. Biochemistry 48(2009): 2135–2145.

Ball, P. Water as an active constituent in cell biology. Chem. Rev. 108(2008): 74–108.

Banerjee, R., Rudloff, Z., Naylor, C., Yu, M.C. and Gunawardena, S. The presenilin loop region is essential for glycogen synthase kinase 3 β (GSK3β) mediated functions on motor proteins during axonal transport. Hum. Mol. Genet. 27(2018): 2986–3001.

Belyanskaya, L.L., Gehrig, P.M. and Gehring, H. Exposure on cell surface and extensive arginine methylation of Ewing Sarcoma (EWS) protein. J. Biol. Chem. 276(2001): 18681–18687.

Bernfield, M., Kokenyesi, R., Kato, M., Hinkes, M.T., Spring, J., Gallo, R.L. et al. Biology of the syndecans: A family of transmembrane heparan sulfate proteoglycans. Annu. Rev. Cell Biol. 8(1992): 365–393.

Betz, R.C., Schoser, B.G., Kasper, D., Ricker, K., Ramírez, A., Stein, V. et al. Mutations in *CAV3* cause mechanical hyperirritability of skeletal muscle in rippling muscle disease. Nat. Genet. 28(2001): 218–219.

Blomberg, M.R.A. and Siegbahn, P.E.M. The mechanism for proton pumping in cytochrome c oxidase from an electrostatic and quantum chemical perspective. Biochim. Biophys. Acta 1817(2012): 495–505.

Blomberg, M.R.A. The mechanism of oxygen reduction in cytochrome c oxidase and the role of the active site tyrosine. Biochemistry 55(2016): 489–500.

Bloomfield, G. and Kay, R.R. Uses and abuses of macropinocytosis. J. Cell Sci. 129(2016): 2697–2705.

Bourseau-Guilmain, E., Menard, J.A., Lindqvist, E., Indira Chandran, V., Christianson, H.C., CerezoMagaña, M. et al. Hypoxia regulates global membrane protein endocytosis through caveolin-1 in cancer cells. Nat. Commun. 7(2016): 11371. doi: 10.1038/ncomms11371.

Boya, P., Esteban-Martínez, L., Serrano-Puebla, A., Gómez-Sintes, R. and Villarejo-Zori, B. Autophagy in the eye: Development, degeneration, and aging. Prog. Retin. Eye Res. 55(2016): 206–245.

Boyer, P.D. A research journey with ATP synthase. J. Biol. Chem. 277(2002): 39045–39061.

Bracey, M.H., Cravatt, B.F. and Stevens, R.C. Structural commonalities among integral membrane enzymes. FEBS Letters 567(2004): 159–165.

Brand, M.D. and Nicholls, D.G. Assessing mitochondrial dysfunction in cells. Biochem. J. 435(2011): 297–312.

Buchan, D.W.A. and Jones, D.T. The PSIPRED protein analysis workbench: 20 years on. Nucleic Acids Res. 47(2019): W402–W407.

Bucher, D., Hsu, Y.-H., Mouchlis, V.D., Dennis, E.A. and McCammon, J.A. Insertion of the Ca^{2+}-independent phospholipase A_2 into a phospholipid bilayer via coarse-grained and atomistic molecular dynamics simulations. PLoS Comput. Biol. 9(7)(2013): e1003156.

Campostrini, G., Bonzanni, M., Lissoni, A., Bazzini, C., Milanesi, R., Vezzoli, E. et al. The expression of the rare caveolin-3 variant T78M alters cardiac ion channels function and membrane excitability. Cardiovasc. Res. 113(2017): 1256–1265.

Canton, J. Macropinocytosis: New insights into its under appreciated role in innate immune cell surveillance. Front. Immunol. 9(2018): 2286. doi:10.3389/fimmu.2018.02286.

Cao, H., Alston, L., Ruschman, J. and Hegele, R.A. Heterozygous *CAV1* frameshift mutations (MIM 601047) in patients with atypical partial lipodystrophy and hypertriglyceridemia. Lipids Health Dis. 7(2008): 3. doi: 10.1186/1476-511X-7-3.

Caughey, W.S. 1971. Structure—function relationships in cytochrome c oxidase and other hemeproteins. Chapter 12, pp. 248–270. *In*: Dessy, R., Dillard, J. and Taylor, L. (eds.). Advances in Chemistry Vol. 100: Bioinorganic Chemistry, American Chemical Society, USA.

Celus, W., Di Conza, G., Oliveira, A.I., Ehling, M., Costa, B.M., Wenes, M. et al. Loss of caveolin-1 in metastasis-associated macrophages drives lung metastatic growth through increased angiogenesis. Cell Reports 21(2017): 2842–2854.

Chai, F., Li, Y., Liu, K., Li, Q. and Sun, H. Caveolin enhances hepatocellular carcinoma cell metabolism, migration, and invasion *in vitro* via a hexokinase 2-dependent mechanism. J. Cell Physiol. 234(2019): 1937–1946.

Chan, S.I. and Li, P.M. Cytochrome c oxidase: Understanding nature's design of a proton pump. Biochemistry 29(1990): 1–12.

Chang, S.-C., Mulloy, B., Magee, A.I. and Couchman, J.R. Two distinct sites in sonic hedgehog combine for heparan sulfate interactions and cell signaling functions. J. Biol. Chem. 286(2011): 44391–44402.

Chaudhary, N., Gomez, G.A., Howes, M.T., Lo, H.P., McMahon, K-A., Rae, J.A. et al. Endocytic crosstalk: cavins, caveolins, and caveolae regulate clathrin-independent endocytosis. PLoS Biol. 12(2014): e1001832. doi:10.1371/journal.pbio.1001832.

Chatterjee, D., Eckert, C.E., Slavov, C., Saxena, K., Fürtig, B., Sanders, C.R et al. Influence of arrestin on the photodecay of bovine rhodopsin. Angewandte Chemie International Edition 54(2015): 13555–13560.

Cheng, R.K.Y., Fiez-Vandal, C., Schlenker, O., Edman, K., Aggeler, B., Brown, D.G. et al. Structural insight into allosteric modulation of protease-activated receptor 2. Nature 545(2017): 112–115.

Chong, S.J., Low, I.C. and Pervaiz, S. Mitochondrial ROS and involvement of Bcl-2 as a mitochondrial ROS regulator. Mitochondrion 19 Pt A(2014): 39–48.

Chow, B.W., Nuñez, V., Kaplan, L., Granger, A.J., Bistrong, K., Zucker, H.L. et al. Caveolae in CNS arterioles mediate neurovascular coupling. Nature 579(2020): 106–110.

Chronopoulos, A., Thorpe, S.D., Cortes, E., Lachowski, D., Rice, A.J., Mykuliak, V.V. et al. Syndecan-4 tunes cell mechanics by activating the kindlin-integrin-RhoA pathway. Nat. Mater. 19(2020): 669–678.

Cocucci, E., Racchetti, G. and Meldolesi, J. Shedding microvesicles: artefacts no more. Trends Cell Biol. 2(2009): 43–51.

Cohen, A.W., Hnasko, R., Schubert, W. and Lisanti, M.P. Role of caveolae and caveolins in health and disease. Physiol. Rev. 84(2004): 1341–1379.

Corradi, V., Mendez-Villuendas, E., Ingólfsson, H.I., Gu, R.-X., Siuda, I., Melo, M.N. et al. Lipid−protein interactions are unique fingerprints for membrane proteins. ACS Cent. Sci. 4(2018): 709–717.

Csizmok, V., Follis, A.V., Kriwacki, R.W. and Forman-Kay, J.D. Dynamic protein interaction networks and new structural paradigms in signaling. Chem Rev. 116 (2016): 6424–6462.

Čeřovský, V., Hovorka, O., Cvačka, J., Voburka, Z., Bednárová, L., Borovičková, L. et al. Melectin: A novel antimicrobial peptide from the venom of the cleptoparasitic bee *Melecta albifrons*. ChemBioChem. 9(2008): 2815–2821.

David, G. Integral membrane heparan sulfate proteoglycans. FASEB J. 7(1993): 1023–1030.

de Grotthuss, C.J.T. Memoir on the decomposition of water and of the bodies that it holds in solution by means of galvanic electricity. Biochim. Biophys. Acta 1757(2006): 871–875. (This memoir was printed in Rome in 1805).

Deisenhofer, J., Epp, O., Miki, K., Huber, R. and Michel, H. Structure of the protein subunits in the photosynthetic reaction centre of *Rhodopseudomonas viridis* at 3 Å resolution. Nature 318(1985): 618–624.

Deng, Y.F., Huang, Y.Y., Lu, W.S., Huang, Y.H., Xian, J., Wei, H.Q. et al. The caveolin-3 P104L mutation of LGMD-1C leads to disordered glucose metabolism in muscle cells. Biochem. Biophys. Res. Commun. 486(2017): 218–223.

Dobson, L., Reményi, I. and Tusnády, G.E. CCTOP: a consensus constrained topology prediction web server. Nucleic Acids Res. 43(2015): W408–W412.

Doyle, D.A., Cabral, J.M., Pfuetzner, R.A., Kuo, A., Gulbis, J.M., Cohen, S.L. et al. The structure of the potassium channel: molecular basis of K^+ conduction and selectivity. Science 280(1998): 69–77.

Drab, M., Verkade, P., Elger, M., Kasper, M., Lohn, M., Lauterbach, B. et al. Loss of caveolae, vascular dysfunction, and pulmonary defects in caveolin-1 gene-disrupted mice. Science 293(2001): 2449–2452.

Dupuy, A.D. and Engelman, D.M. Protein area occupancy at the center of the red blood cell membrane. Proc. Natl. Acad. Sci. USA 105(2008): 2848–2852.

Earle, W. 1984. Evanescence. Peri-phenomenological Essays. Regnery Gateway, Chicago, IL, USA.

Eisinger, K.R.T., Woolfrey, K.M., Swanson, S.P., Schnell, S.A., Meitzen, J., Dell'Acqua, M. et al. Palmitoylation of caveolin-1 is regulated by the same DHHC acyltransferases that modify steroid hormone receptors. J. Biol. Chem. 293(2018): 15901–15911.

Elfenbein, A. and Simons, M. Auxiliary and autonomous proteoglycan signaling networks. Methods Enzymol. 480(2010): 3–31.

Fagerberg, L., Jonasson, K., von Heijne, G., Uhlén, M. and Berglund, L. Prediction of the human membrane proteome. Proteomics 10(2010): 1141–1149.

Familtseva, A., Jeremic, N. and Tyagi, S.C. Exosomes: cell-created drug delivery systems. Mol. Cell Biochem. 459(2019): 1–6.

Fiedorczuk, K., Letts, J.A., Degliesposti, G., Kaszuba, K., Skehel, M. and Sazanov, L.A. Atomic structure of the entire mammalian mitochondrial complex I. Nature 538(2016): 406–410.

Field, G.D., Sampath, A.P. and Rieke, F. Retinal processing near absolute threshold. From behavior to mechanism. Annu. Rev. Physiol. 67(2005): 491–514.

Fiucci, G., Ravid, D., Reich, R. and Liscovitch, M. Caveolin-1 inhibits anchorage-independent growth, anoikis and invasiveness in MCF-7 human breast cancer cells. Oncogene 21(2002): 2365–2375.

Franke, W.W. Discovering the molecular components of intercellular junctions—A historical view. Cold Spring Harb. Perspect. Biol. 1(2009): a003061. doi: 10.1101/cshperspect.a003061.

Fredriksson, R., Lagerström, M.C., Lundin, L.-G. and Schiöth, H.B. The G-protein-coupled receptors in the human genome form five main families. Phylogenetic analysis, paralogon groups, and fingerprints. Mol. Pharmacol. 63(2003): 1256–1272.

Fukumori, A., Fluhrer, R., Steiner, H. and Haass, C. Three-amino acid spacing of presenilin endoproteolysis suggests a general stepwise cleavage of gamma-secretase-mediated intramembrane proteolysis. J. Neurosci. 30(2010): 7853–7862.

Gao, Y., Hu, H., Ramachandran, S., Erickson, J.W., Cerione, R.A. and Skiniotis, G. Structures of the rhodopsin-transducin complex: Insights into G-protein activation. Mol. Cell 75(2019): 781–790.

Garcia, J., Bagwell, J., Njaine, B., Norman, J., Levic, D.S., Wopat, S. et al. Sheath cell invasion and trans-differentiation repair mechanical damage caused by loss of caveolae in the zebrafish notochord. Curr. Biol. 27(2017): 1982–1989.

Gazzerro, E., Sotgia, F., Bruno, C., Lisanti, M.P. and Minetti, C. Caveolinopathies: from the biology of caveolin-3 to human diseases. Eur. J. Hum. Genet. 18(2010): 137–145.

Gharbaran, R. Advances in the molecular functions of syndecan-1 (SDC1/CD138) in the pathogenesis of malignancies. Criti. Rev. Oncol. Hematol. 94(2015): 1–17.

Golani, G., Ariotti, N., Parton, R.G. and Kozlov, M.M. Membrane curvature and tension control the formation and collapse of caveolar superstructures. Dev. Cell 48(2019): 523–538.

Golde, T.E., Wolfe, M.S. and Greenbaum, D.C. Signal peptide peptidases: A family of intramembrane-cleaving proteases that cleave type 2 transmembrane proteins. Semin. Cell Dev. Biol. 20(2009): 225–230.

Gondelaud, F. and Ricard-Blum, S. Structures and interactions of syndecans. FEBS J. 286(2019): 2994–3007.

Goodsell, D.S. The molecular perspective: Bcl-2 and apoptosis. Stem Cells 20(2002): 355–356.

Goyal, P., Yang, S. and Cui, Q. Microscopic basis for kinetic gating in cytochrome c oxidase: insights from QM/MM analysis. Chem. Sci. 6(2015): 826–841.

Granseth, E., von Heijne, G. and Elofsson, A. A study of the membrane–water interface region of membrane proteins. J. Mol. Biol. 346(2005): 377–385.

Granseth, E. 2010. Dual-topology: one sequence, two topologies. pp. 137–150. *In*: Frshman, D. (ed.). Structural Bioinformatics of Membrane Proteins. Springer-Verlag, Wien, Austria.

Guda, C., Fahy, E. and Subramaniam, S. MITOPRED: a genome-scale method for prediction of nucleus-encoded mitochondrial proteins. Bioinformatics 20(2004): 1785–1794.

Guidotti, G. Membrane proteins. Annu. Rev. Biochem. 41(1972): 731–752.

Gunkel, M., Schöneberg, J., Alkhaldi, W., Irsen, S., Noé, F., Kaupp, U.B. et al. Higher-order architecture of rhodopsin in intact photoreceptors and its implication for phototransduction kinetics. Structure 23(2015): 628–638.

Hardy, J.A. and Higgins, G.A. Alzheimer's disease: the amyloid cascade hypothesis. Science 256(1992): 184–185.

Head, B.P., Patel, H.H. and Insel, P.A. Interaction of membrane/lipid rafts with the cytoskeleton: impact on signaling and function: Membrane/lipid rafts, mediators of cytoskeletal arrangement and cell signaling. Biochim. Biophys. Acta 1838(2014): 532–545.

Heath, J.R., Davis, M.E. and Hood, L. Nanomedicine targets cancer. Sci. Am. 300(2009): 34–41.

Hecht, S., Shlaer, S. and Pirenne, M.H. Energy, quanta, and vision. J. Gen. Physiol. 25(1942): 819–840.

Henderson, R., Baldwin, J.M., Ceska, T.A., Zemlin, F., Beckmann, E. and Downing, K.H. Model for the structure of bacteriorhodopsin based on high-resolution cryo-microscopy. J. Mol. Biol. 213(1990): 899–929.

Henzler-Wildman, K. and Kern, D. Dynamic personalities of proteins. Nature 450(2007): 964–972.

Herst, P.M., Rowe, M.R., Carson, G.M. and Berridge, M.V. Functional mitochondria in health and disease. Front. Endocrinol. 8(2017): 296. doi: 10.3389/fendo.2017.00296.

Hirama, T., Das, R., Yang, Y., Ferguson, C., Won, A., Yip, C.M. et al. Phosphatidylserine dictates the assembly and dynamics of caveolae in the plasma membrane. J. Biol. Chem. 292(2017): 14292–14307.

Hochachka, P.W. 1994. Muscles as Molecular and Metabolic Machines. Boca Raton: CRC Press, FL, USA.

Hoffmann, P.M. How molecular motors extract order from chaos. Rep. Prog. Phys. 79(2016): 032601. doi:10.1088/0034-4885/79/3/032601.

Hoffmann, M., Kleine-Weber, H. and Pöhlmann, S. A multibasic cleavage site in the spike protein of SARS-CoV-2 is essential for infection of human lung cells. Mol. Cell 78(2020): 779–784.

Huertas-Martínez, J., Rello-Varona, S., Herrero-Martín, D., Barrau, I., García-Monclús, S., Sáinz-Jaspeado, M. et al. Caveolin-1 is down-regulated in alveolar rhabdomyosarcomas and negatively regulates tumor growth. Oncotarget 5(2014): 9744–9755.

Iwata, S., Ostermeier, C., Ludwig, B. and Michel, H. Structure at 2.8 Å resolution of cytochrome c oxidase from *Paracoccus denitrificans*. Nature 376(1995): 660–668.

Jackson, M.B. 2006. Molecular and Cellular Biophysics. Cambridge University Press, Cambridge, UK.

Jang, E., Albadawi, H., Watkins, M.T., Edelman, E.R. and Baker, A.B. Syndecan-4 proteoliposomes enhance fibroblast growth factor-2 (FGF-2)–induced proliferation, migration, and neovascularization of ischemic muscle. Proc. Natl. Acad. Sci. USA 109(2012): 1679–1684.

Jang, B., Jung, H., Hong, H. and Oh, E.-S. Syndecan transmembrane domain modulates intracellular signaling by regulating the oligomeric status of the cytoplasmic domain. Cell. Signal. 52(2018): 121–126.

Jeffries, E.P., Di Filippo, M. and Galbiati, F. Failure to reabsorb the primary cilium induces cellular senescence. FASEB J. 33(2019): 4866–4882.

Jiang, X.Z., Luo, K.H. and Ventikos, Y. Principal mode of Syndecan-4 mechanotransduction for the endothelial glycocalyx is a scissor-like dimer motion. Acta Physiol. (Oxf). 228(2020): e13376. doi: 10.1111/apha.13376.

Johnstone, R.M., Adam, M., Hammond, J.R., Orr, L. and Turbide, C. Vesicle formation during reticulocyte maturation. Association of plasma membrane activities with released vesicles (exosomes). J. Biol. Chem. 262(1987): 9412–9420.

Juretić, D., Hendler, R.W. and Westerhoff, H.V. 1988.Variation of efficiency with free energy dissipation in theoretical models of oxidative posphorylation and cytochrome oxidase. pp. 205–212. *In*: Lemasters, J.J., Hackenbrock, C.R., Thurman, R.G. and Westerhoff, H.V. (eds.). Integration of Mitochondrial Function. Springer, Boston, MA, USA. doi.org/10.1007/978-1-4899-2551-0_18.

Juretic, D., Jeroncic, A. and Zucic, D. Sequence analysis of membrane proteins with the web server SPLIT. Croatica Chemica Acta 72(1999): 975–997.

Juretić, D., Zoranić, L. and Zucić, D. Basic charge clusters and predictions of membrane protein topology. J. Chem. Inf. Comput. Sci. 4(2002): 620–632.

Juretić, D., Sonavane, Y., Ilić, N., Gajski, G., Goić-Barišić, I., Tonkić, M. et al. Designed peptide with a flexible central motif from ranatuerins adapts its conformation to bacterial membranes. Biochim. Biophys. Acta 1860(2018): 2655–2668.

Juretić, D. and Simunić, J. Design of α-helical antimicrobial peptides with a high selectivity index. Expert Opin. Drug Discov. 14(2019): 1053–1063.

Juretić, D., Golemac, A., Strand, D.E., Chung, K., Ilić, N., Goić-Barišić, I. et al. The spectrum of design solutions for improving the activity-selectivity product of peptide antibiotics against multidrug-resistant bacteria and prostate cancer PC-3 cells. Molecules 25(2020): 3526. doi: 10.3390/molecules25153526.

Kaila, V.R.I., Verkhovsky, M.I. and Wikstörm, M. Proton-coupled electron transfer in cytochrome oxidase. Chem. Rev. 110(2010): 7062–7081.

Kasuga, K., Kaneko, H., Nishizawa, M., Onodera, O. and Ikeuchi, T. Generation of intracellular domain of insulin receptor tyrosine kinase by γ-secretase. Biochem. Biophys. Res. Commun. 360(2007): 90–96.

Kazanis, I., Lathia, J.D., Vadakkan, T.J., Raborn, E., Wan, R., Mughal, M.R. et al. Quiescence and activation of stem and precursor cell populations in the subependymal zone of the mammalian brain are associated with distinct cellular and extracellular matrix signals. J. Neurosci. 30(2010): 9771–9781.

Käll, L., Krogh, A. and Sonnhammer, E.L.L. A combined transmembrane topology and signal peptide prediction method. J. Mol. Biol. 338(2004): 1027–1036.

Kennedy, B. and Malicki, J. What drives cell morphogenesis: A look inside the vertebrate photoreceptor. Dev. Dynam. 238(2009): 2115–2138.

Kevany, B.K. and Palczewski, K. Phagocytosis of retinal rod and cone photoreceptors. Physiology (Bethesda) 25(2010): 8–15.

Kim, T.S., Heinlein, C., Hackman, R.C. and Nelso, P.S. Phenotypic analysis of mice lacking the *Tmprss2*-encoded protease. Mol. Cell Biol. 26(2006): 965–975.

Kim, C.A., Delépine, M., Boutet, E., El Mourabit, H., Le Lay, S., Meier, M. et al. Association of a homozygous nonsense caveolin-1 mutation with Berardinelli-Seip congenital lipodystrophy. J. Clin. Endocrinol. Metab. 93(2008): 1129–1134.

Kim, Y.C. and Hummer, G. Proton-pumping mechanism of cytochrome c oxidase: a kinetic master-equation approach. Biochim. Biophys. Acta 1817(2012): 526–536.

Kirkham, M., Nixon, S.J., Howes, M.T., Abi-Rached, L., Wakeham, D.E., Hanzal-Bayer M. et al. Evolutionary analysis and molecular dissection of caveola biogenesis. J. Cell Sci. 121(2008): 2075–2086.

Kobayashi, S., Takeshima, K., Park, C.B., Kim, S.C. and Matsuzaki, K. Interactions of the novel antimicrobial peptide buforin 2 with lipid bilayers: proline as a translocation promoting factor. Biochemistry 39(2000): 8648–8654.

Koch, K.-W. and Dell'Orco, D. Protein and signaling networks in vertebrate photoreceptor cells. Front. Mol. Neurosci. 8(2015): 67. doi: 10.3389/fnmol.2015.00067.

Koepke, J., Olkhova, E., Angerer, H., Müller, H., Peng, G. and Michel, H. High resolution crystal structure of paracoccus denitrificans cytochrome c oxidase: New insights into the active site and the proton transfer pathways. Biochim. Biophys. Acta Bioenerg. 1787(2009): 635–645.

Kozic, M., Fox, S.J., Thomas, J.M., Verma, C.S. and Rigden, D.J. Large scale ab initio modeling of structurally uncharacterized antimicrobial peptides reveals known and novel folds. Proteins 86(2018): 548–565.

Krishna, A. and Sengupta, D. Interplay between membrane curvature and cholesterol: Role of palmitoylated caveolin-1. Biophys. J. 116(2019): 69–78.

Kvansakul, M. and Hinds, M.G. Structural biology of the Bcl-2 family and its mimicry by viral proteins. Cell Death Dis. 4(2013): e909. doi:10.1038/cddis.2013.436.

Kyte, J. and Doolittle, R.F. A simple method for displaying the hydropathic character of a protein. J. Mol. Biol. 157(1982): 105–132.

Kwon, M.-J., Park, J., Jang, S., Eom, C.-Y. and Oh, E.-S. The conserved phenylalanine in the transmembrane domain enhances heteromeric interactions of syndecans. J. Biol. Chem. 291(2016): 872–881.

Lamaze, C., Tardif, N., Dewulf, M., Vassilopoulos, S. and Blouin, C.M. The caveolae dress code: structure and signaling. Curr. Opin. Cell Biol. 47(2017): 117–125.

Lambert, C., Mann, S. and Prange, R. Assessment of determinants affecting the dual topology of hepadnaviral large envelope proteins. J. Gen. Virol. 85(2004): 1221–1225.

Landgraf, P., Wahle, P., Pape, H.-C., Gundelfinger, E.D. and Kreutz, M.R. The survival-promoting peptide Y-P30 enhances binding of pleiotrophin to syndecan-2 and -3 and supports its neuritogenic activity. J. Biol. Chem. 283(2008): 25036–25045.

Lane, N. and Martin, W. The energetics of genome complexity. Nature 467(2010): 929–934.

Langleben, D.D., Schroeder, L., Maldjian, J.A., Gur, R.C., McDonald, S., Ragland, J.D. et al. Brain activity during simulated deception: An event-related functional magnetic resonance study. NeuroImage 15(2002): 727–732.

Le Lay, S., Blouin, C.M., Hajduch, E. and Dugail, I. Filling up adipocytes with lipids. Lessons from caveolin-1 deficiency. Biochim. Biophys. Acta 1791(2009): 514–518.

Lee, H., Park, D.S., Razani, B., Russell, R.G., Pestell, R.G. and Lisanti, M.P. Caveolin-1 mutations (P132L and null) and the pathogenesis of breast cancer: caveolin-1 (P132L) behaves in a dominant-negative manner and caveolin-1 (–/–) null mice show mammary epithelial cell hyperplasia. Am. J. Pathol. 161(2002): 1357–1369.

Lee, J.-K., Park, S.-C., Hahm, K.-S. and Park, Y. A helix-PXXP-helix peptide with antibacterial activity without cytotoxicity against MDRPA-infected mice. Biomaterials 35(2014): 1025–1039.

Lee, E., Kim, J.K., Jeon, D., Jeong, K.-W., Shin, A. and Kim, Y. Functional roles of aromatic residues and helices of papiliocin in its antimicrobial and antiinflammatory activities. Sci. Rep. 5(2015): 12048. doi: 10.1038/srep12048.

Lents, N.H. 2018. Human Errors: A Panorama of Our Glitches, from Pointless Bones to Broken Genes. Houghton Mifflin Harcourt, New York, NY, USA.

Lepore, R., Kryshtafovych, A., Alahuhta, M., Veraszto, H.A., Bomble, Y.J., Bufton, J.C. et al. Target highlights in CASP13: Experimental target structures through the eyes of their authors. Proteins 87(2019): 1037–1057.

Léveillard, T. and Sahel, J.-A. Metabolic and redox signaling in the retina. Cell. Mol. Life Sci. 74(2017): 3649–3665.

Li, X., Dang, S., Yan, C., Gong, X., Wang, J. and Shi, Y. Structure of a presenilin family intramembrane aspartate protease. Nature 493(2013): 56–63.

Li, J.-P. and Kusche-Gullberg, M. Heparan sulfate: Biosynthesis, structure, and function. Int. Rev. Cell Mol. Biol. 325(2016): 215–273.

Li, S., Zhang, W. and Han, W. Initial substrate binding of γ-secretase: the role of substrate flexibility. ACS Chem. Neurosci. 8(2017): 1279–1290.

Lightowlers, R.L., Rozanska, A. and Chrzanowska-Lightowlers, Z.M. Mitochondrial protein synthesis: Figuring the fundamentals, complexities and complications, of mammalian mitochondrial translation. FEBS Letters 588(2014): 2496–2503.

Lim, S., Smith, K.R., Lim, S.-T.S., Tian, R., Lu, J. and Tan, M. Regulation of mitochondrial functions by protein phosphorylation and dephosphorylation. Cell Biosci. 6(2016): 25. doi. 10.1186/s13578-016-0089-3.

Liu, X., Zhao, J., Zhang, Y., Ubarretxena-Belandia, I., Forth, S., Lieberman, R.L. et al. Substrate–enzyme interactions in intramembrane proteolysis: γ-secretase as the prototype. Front. Mol. Neurosci. 13(2020): 65. doi: 10.3389/fnmol.2020.00065.

Lomize, M.A., Pogozheva, I.D., Joo, H., Mosberg, H.I. and Lomize, A.L. OPM database and PPM web server: resources for positioning of proteins in membranes. Nucleic Acids Res. 40(2012): D370–D376.

MacKinnon, R. Potassium channels and the atomic basis of selective ion conduction. Biosci. Rep. 24(2004): 75–100.

Mandyam, C.D., Schilling, J.M., Cui, W., Egawa, J., Niesman, I.R., Kellerhals, S.E. et al. Neuron-targeted caveolin-1 improves molecular signaling, plasticity, and behavior dependent on the hippocampus in adult and aged mice. Biol. Psychiatry 81(2017): 101–110.

Manon-Jensen, T., Itoh, Y. and Couchman, J.R. Proteoglycans in health and disease: the multiple roles of syndecan shedding. FEBS J 277(2010): 3876–3889.

Matthews, R.A.J. Tumbling toast, Murphy's Law and the fundamental constants. Eur. J. Phys. 16(1995): 172–176.

Marx, D. Proton transfer 200 years after von Grotthuss: Insights from ab initio simulations. ChemPhysChem. 7(2006): 1848–1870.

McClintock, P.V.E. and Kaufman, I.K. A review of Handbook of Ion Channels, by Jie Zheng and Matthew C. Trudeau. Contemporary Physics 59(2018): 305–307.

Mergia, A. The role of caveolin 1 in HIV infection and pathogenesis. Viruses 9(2017): 129. doi:10.3390/v9060129.

Minetti, C., Sotgia, F., Bruno, C., Scartezzini, P., Broda, P., Bado, M. et al. Mutations in the caveolin-3 gene cause autosomal dominant limb-girdle muscular dystrophy. Nat. Genet. 18(1998): 365–368.

Mitchell, P. Coupling of phosphorylation to electron and hydrogen transfer by a chemi-osmotic type of mechanism. Nature 191(1961): 144–148.

Mohan, J., Morén, B., Larsson, E., Holst, M.R. and Lundmark, R. Cavin3 interacts with cavin1 and caveolin1 to increase surface dynamics of caveolae. J. Cell Sci. 128(2015): 979–991.

Morante-Redolat, J.M. and Porlan, E. Neural stem cell regulation by adhesion molecules within the subependymal niche. Front. Cell Dev. Biol. 7(2019): e102. doi: 10.3389/fcell.2019.00102.

Morgan, M.R., Hamidi, H., Bass, M.B., Warwood, S., Ballestrem, C. and Humphries, M.J. Syndecan-4 phosphorylation is a control point for integrin recycling. Dev. Cell 24(2013): 472–485.

Mundy, D.I., Li, W.P., Luby-Phelps, K. and Anderson, R.G. Caveolin targeting to late endosome/lysosomal membranes is induced by perturbations of lysosomal pH and cholesterol content. Mol. Biol. Cell. 23(2012): 864–880.

Nagle, J.F. and Morowitz, H.J. Molecular mechanisms for proton transport in membranes. Proc. Natl. Acad. Sci. USA 75(1978): 298–302.

Naing, S.-H., Kalyoncu, S., Smalley, D.M., Kim, H., Tao, X., George, J.B. et al. Both positional and chemical variables control *in vitro* proteolytic cleavage of a presenilin ortholog. J. Biol. Chem. 293(2018): 4653–4663.

Neumann, J.R., Dash-Wagh, S., Jack, A., Räk, A., Jüngling, K., Hamad, M.I.K. et al. The primate-specific peptide Y-P30 regulates morphological maturation of neocortical dendritic spines. PLoS ONE 14(2019): e0211151. doi.org/10.1371/journal.pone.0211151.

Newport, T.D., Sansom, M.S.P. and Stansfeld, P.J. The MemProtMD database: a resource for membrane-embedded protein structures and their lipid interactions. Nucleic Acids Res. 47(2019): D390–D397.

Nielsen, A.L., Holm, I.E., Johansen, M., Bonven, B., Jørgensen, P. and Jørgensen, A.L. A new splice variant of glial fibrillary acidic protein, GFAP epsilon, interacts with the presenilin proteins. J. Biol. Chem. 277(2002): 29983–29991.

Nielsen, H., Tsirigos, K.D., Brunak, S. and von Heijne, G. A brief history of protein sorting prediction. Protein J. 38(2019): 200–216.

Okada, T., Sugihara, M., Bondar, A.N., Elstner, M., Entel, P. and Buss, V. The retinal conformation and its environment in rhodopsin in light of a new 2.2 Å crystal structure. J. Mol. Biol. 342(2004): 571–583.

Okawa, H., Sampath, A.P., Laughlin, S.B. and Fain, G.L. ATP consumption by mammalian rod photoreceptors in darkness and in light. Curr. Biol. 18(2008): 1917–1921.

Ozawa, E., Nishino, I. and Nonaka, I. Sarcolemmopathy: Muscular dystrophies with cell membrane defects. Brain Pathol. 11(2001): 218–230.

Palczewski, K., Kumasaka, T., Hori, T., Behnke, C.A., Motoshima, H., Fox, B.A et al. Crystal structure of rhodopsin: A G protein-coupled receptor. Science 289(2000): 739–745.

Pan, B.T. and Johnstone, R.M. Fate of the transferrin receptor during maturation of sheep reticulocytes *in vitro*: selective externalization of the receptor. Cell 33(1983): 967–978.

Panteleev, P.V., Bolosov, I.A., Kalashnikov, A.À., Kokryakov, V.N., Shamova, O.V., Emelianova, A.A. et al. Combined antibacterial effects of goat cathelicidins with different mechanisms of action. Front. Microbiol. 9(2018): 2983. doi: 10.3389/fmicb.2018.02983.

Panza, F., Lozupone, M., Logroscino, G. and Imbimbo, B.P. A critical appraisal of amyloid-β-targeting therapies for Alzheimer disease. Nat. Rev. Neurol. 15(2019): 73–88.

Parton, R.G., Hanzal-Bayer, M. and Hancock, J.F. Biogenesis of caveolae: A structural model for caveolin-induced domain formation. J. Cell Sci. 119(2006): 787–796.

Parton, R.G., Kozlov, M.M. and Ariotti, N. Caveolae and lipid sorting: Shaping the cellular response to stress. J. Cell Biol. 219(2020): e201905071. doi.org/10.1083/jcb.201905071.

Parton, R.G. Caveolae: Structure, function, and relationship to disease. Annu. Rev. Cell Dev. Biol. 34(2018): 111–136.

Paulmann, M., Arnold, T., Linke, D., Özdirekcan, S., Kopp, A., Gutsmann, T. et al. Structure-activity analysis of the dermcidin-derived peptide DCD-1L, an anionic antimicrobial peptide present in human sweat. J. Biol. Chem. 287(2012): 8434–8443.

Pereira, M.M., Santana, M. and Teixeira, M. A novel scenario for the evolution of haem–copper oxygen reductases. Biochim. Biophys. Acta 1505(2001): 185–208.

Peysselon, F., Xue, B., Uversky, V.N. and Ricard-Blum, S. Intrinsic disorder of the extracellular matrix. Mol. BioSyst. 7(2011): 3353–3365.

Pleasant-Jenkins, D., Reese, C., Chinnakkannu, P., Kasiganesan, H., Tourkina, E., Hoffman, S. et al. Reversal of maladaptive fibrosis and compromised ventricular function in the pressure overloaded heart by a caveolin-1 surrogate peptide. Lab. Invest. 97(2017): 370–382.

Pogozheva, I.D. and Lomize, A.L. Evolution and adaptation of single-pass transmembrane proteins. Biochim. Biophys. Acta. 1860(2018): 364–377.

Ray, L.C., Das, D., Entova, S., Lukose, V., Lynch, A.J., Imperiali, B. et al. Membrane association of monotopic phosphoglycosyl transferase underpins function. Nat. Chem. Biol. 14(2018): 538–541.

Razani, B., Woodman, S.E. and Lisanti, M.P. Caveolae: From cell biology to animal physiology. Pharmacol. Rev. 54(2002): 431–467.

Reynolds, S.M., Käll, L., Riffle, M.E., Bilmes, J.A. and Noble, W.S. Transmembrane topology and signal peptide prediction using dynamic Bayesian networks. PLoS Comput. Biol. 4(2008): e1000213. doi:10.1371/journal.pcbi.1000213.

Ricker, K., Moxley, R.T. and Rohkamm, R. Rippling muscle disease. Arch. Neurol. 46(1989): 405–408.

Rieth, M.D., Lee, J. and Glover, K.J. Probing the caveolin-1 P132L mutant: Critical insights into its oligomeric behavior and structure. Biochemistry 51(2012): 3911–3918.

Root, K.T., Plucinsky, S.M. and Glover, K.J. Recent progress in the topology, structure, and oligomerization of caveolin: A building block of caveolae. Curr. Top. Membr. 75(2015): 305–336.

Root, K.T., Julien, J.A. and Glover, K.J. Secondary structure of caveolins: a mini review. Biochem. Soc. Trans. 47(2019): 1489–1498.

Sampson, L.J., Davies, L.D., Barrett-Jolley, R., Standen, N.B. and Dart, C. Angiotensin II-activated protein kinase C targets caveolae to inhibit aortic ATP-sensitive potassium channels. Cardiovasc. Res. 76(2007): 61–70.

Saunders, S., Jalkanen, M., O'Farrell, S. and Bernfield, M. Molecular cloning of syndecan, an integral membrane proteoglycan. J. Cell Biol. 108(1989): 1547–1556.

Schou, K.B., Mogensen, J.B., Morthorst, S.K., Nielsen, B.S., Aleliunaite, A., Serra-Marques, A. et al. KIF13B establishes a CAV1-enriched microdomain at the ciliary transition zone to promote Sonic hedgehog signalling. Nat. Commun. 8(2017): 14177. doi: 10.1038/ncomms14177.

Schrauwen, I., Szelinger, S., Siniard, A.L., Kurdoglu, A., Corneveaux, J.J., Malenica, I. et al. A frame-shift mutation in CAV1 is associated with a severe neonatal progeroid and lipodystrophy syndrome. PLoS One 10(2015): e0131797. doi:10.1371/journal.pone.0131797.

Selkoe, D.J. and Wolfe, M.S. Presenilin: running with scissors in the membrane. Cell 131(2007): 215–221.

Senior, A.W., Evans, R., Jumper, J., Kirkpatrick, J., Sifre, L., Green, T. et al. Improved protein structure prediction using potentials from deep learning. Nature 577(2020): 706–710.

Shajahan, A.N., Wang, A., Decker, M., Minshall, R.D., Liu, M.C. and Clarke, R. Caveolin-1 tyrosine phosphorylation enhances paclitaxel-mediated cytotoxicity. J. Biol. Chem. 282(2007): 5934–5943.

Shao, C., Tian, H., Wang, T., Wang, Z., Chou, S., Shan, A. et al. Central β-turn increases the cell selectivity of imperfectly amphipathic α-helical peptides. Acta Biomater. 69(2018): 243–255.

Shen, J. and Kelleher III, R.J. The presenilin hypothesis of Alzheimer's disease: Evidence for a loss-of-function pathogenic mechanism. Proc. Natl. Acad. Sci. USA 104(2007): 403–409.

Shin, S.Y., Lee, S.-H., Yand, S.-T., Park, E.J., Lee, D.G., Lee, M.K. et al. Antibacterial, antitumor and hemolytic activities of α-helical antibiotic peptide, P18 and its analogs. J. Peptide Res. 58(2001): 504–514.

Sohn, J., Brick, R.M. and Tuan, R.S. From embryonic development to human diseases: The functional role of caveolae/caveolin. Birth Defects Res. C 108(2016): 45–64.

Son, C.Y., Yethiraj, A. and Cui, Q. Cavity hydration dynamics in cytochrome *c* oxidase and functional implications. Proc. Natl. Acad. Sci. USA 114(2017): E8830–E8836.

Sousa, F.L., Thiergart, T., Landan, G., Nelson-Sathi, S., Pereira, I.A.C., Allen, J.F. et al. Early bioenergetic evolution. Phil. Trans. R. Soc. B 368(2013): 20130088. doi.org/10.1098/rstb.2013.0088.

Spillane, J., Kullmann, D.M. and Hanna, M.G. Genetic neurological channelopathies: molecular genetics and clinical phenotypes. J. Neurol. Neurosurg. Psychiatry 87(2016): 37–48.

Stansfeld, P.J., Goose, J.E., Caffrey, M., Carpenter, E.P., Parker, J.L., Newstead, S. et al. MemProtMD: Automated insertion of membrane protein structures into explicit lipid membranes. Structure 23(2015): 1350–1361.

Steinmetz, M.O. and Prota, A.E. Microtubule-targeting agents: Strategies to hijack the cytoskeleton. Trends Cell Biol. 28(2018): 776–792.

Stephen, R., Filipek, S., Palczewski, K. and Sousa, M.C. Ca^{2+}-dependent regulation of phototransduction. Photochem. Photobiol. 84(2008): 903–910.

Sylvester, J.E., Fischel-Ghodsian, N., Mougey, E.B. and O'Brien, T.W. Mitochondrial ribosomal proteins: candidate genes for mitochondrial disease. Genet. Med. 6(2004): 73–80.

Syn, N., Wang, L., Sethi, G., Thiery, J.-P. and Goh, B.-C. Exosome-mediated metastasis: From epithelial–mesenchymal transition to escape from immunosurveillance. Trends Pharmacol. Sci. 37(2016): 606–617.

Thinakaran, G., Borchelt, D.R., Lee, M.K., Slunt, H.H., Spitzer, L., Kim, G. et al. Endoproteolysis of presenilin 1 and accumulation of processed derivatives *in vivo*. Neuron 17(1996): 181–190.

Tomita, T. Probing the structure and function relationships of presenilin by substituted-cysteine accessibility method. Methods Enzymol. 584(2017): 185–205.

Tsukihara, T., Aoyama, H., Yamashita, E., Tomizaki, T., Yamaguchi, H., Shinzawa-Itoh, K. et al. Structures of metal sites of oxidized bovine heart cytochrome c oxidase at 2.8 Å. Science 269(1995): 1069–1074.

Tsukihara, T., Aoyama, H., Yamashita, E., Tomizaki, T., Yamaguchi, H., Shinzawa-Itoh, K. et al. The whole structure of the 13-subunit oxidized cytochrome c oxidase at 2.8 Å. Science 272(1996): 1136–1144.

Tusnády, G.E. and Simon, I. The HMMTOP transmembrane topology prediction server. Bioinformatics 17(2001): 849–850.

Ulmschneider, J.P. and Ulmschneider, M.B. Molecular dynamics simulations are redefining our view of peptides interacting with biological membranes. Acc. Chem. Res. 51(2018): 1106–1116.

Vaidyanathan, R., Reilly, L. and Eckhardt, L.L. Caveolin-3 microdomain: Arrhythmia implications for potassium inward rectifier and cardiac sodium channel. Front. Physiol. 9(2018): 1548. doi: 10.3389/fphys.2018.01548.

Vatta, M., Ackerman, M.J., Ye, B., Makielski, J.C., Ughanze, E.E., Taylor, E.W. et al. Mutant caveolin-3 induces persistent late sodium current and is associated with long-QT syndrome. Circulation 114(2006): 2104–2112.

von Heijne, G. The distribution of positively charged residues in bacterial inner membrane proteins correlates with the trans-membrane topology. EMBO J. 5(1986): 3021–3027.

von Heijne, G. Control of topology and mode of assembly of a polytopic membrane protein by positively charged residues. Nature 341(1989): 456–458.

von Heijne, G. Membrane protein structure prediction. Hydrophobicity analysis and the positive-inside rule. J. Mol. Biol. 225(1992): 487–494.

Walters, R.F.S. and DeGrado, W.F. Helix-packing motifs in membrane proteins. Proc. Natl. Acad. Sci. USA 103(2006): 13658–13663.

Wang, J., Ma, K., Ruan, M., Wang, Y., Li, Y., Fu, Y.V. et al. A novel cecropin B-derived peptide with antibacterial and potential anti-inflammatory properties. Peer J. 6(2018): e5369. doi: 10.7717/peerj.5369.

Weinrauch, Y., Abad, C., Liang, N.S., Lowry, S.F. and Weiss, J. Mobilization of potent plasma bactericidal activity during systemic bacterial challenge. Role of group IIA phospholipase A2. J. Clin. Invest. 102(1998): 633–638.

Westerhoff, H.V., Juretić, D., Hendler, R.W. and Zasloff, M. Magainins and the distruption of membrane-linked free-energy transduction. Proc. Natl. Acad. Sci. USA 86(1989): 6597–6601.

Wikström, M.K. Proton pump coupled to cytochrome c oxidase in mitochondria. Nature 266(1977): 271–273.

Wikström, M., Verkhovsky, M.I. and Hummer, G. Water-gated mechanism of proton translocation by cytochrome c oxidase. Biochim. Biophys. Acta 1604(2003): 61–65.

Wikström, M. and Sharma, V. Proton pumping by cytochrome c oxidase—A 40 year anniversary. Biochim. Biophys. Acta 1859(2018): 692–698.

Wikström, M., Krab, K. and Sharma, V. Oxygen activation and energy conservation by cytochrome c oxidase. Chem. Rev. 118(2018): 2469–2490.

Wilkins, H.M. and Swerdlow, R.H. Amyloid precursor protein processing and bioenergetics. Brain Res. Bull. 133(2017): 71–79.

Williams, T.M., Medina, F., Badano, I., Hazan, R.B., Hutchinson, J., Muller, W.J. et al. Caveolin-1 gene disruption promotes mammary tumorigenesis and dramatically enhances lung metastasis *in vivo*. J. Biol. Chem. 279(2004): 51630–51646.

Wrobel, A.G., Benton, D.J., Xu, P., Roustan, C., Martin, S.R., Rosenthal, P.B. et al. SARS-CoV-2 and bat RaTG13 spike glycoprotein structures inform on virus evolution and furin-cleavage effects. Nat. Struct. Mol. Biol. 27(2020): 763–767.

Xia, D., Watanabe, H., Wu, B., Lee, H.S., Li, Y., Tsvetkov, E. et al. Presenilin-1 knockin mice reveal loss-of-function mechanism for familial Alzheimer's disease. Neuron 85(2015): 967–981.

Xu, J. and Voth, G.A. Computer simulation of explicit proton translocation in cytochrome c oxidase: the D-pathway. Proc. Natl. Acad. Sci. USA 102(2005): 6795–800.

Yao, W., Rose, J.L., Wang, W., Seth, S., Jiang, H., Taguchi, A. et al. Syndecan 1 is a critical mediator of macropinocytosis in pancreatic cancer. Nature 568(2019): 410–414.

Yau, K.-W. and Hardie, R.C. Phototransduction motifs and variations. Cell 139(2009): 246–264.

Yoshikawa, S., Shinzawa-Itoh, K., Nakashima, R., Yaono, R., Yamashita, E., Inoue, N. et al. Redox-coupled crystal structural changes in bovine heart cytochrome c oxidase. Science 280(1998): 1723–1729.

Zakaryan, R.P. and Gehring, H. Identification and characterization of the nuclear localization/retention signal in the EWS proto-oncoprotein. J. Mol. Biol. 363(2006): 27–38.

Zhang, X. and Cote, R.C. cGMP signaling in vertebrate retinal photoreceptor cells. Front. Biosci. 10(2005): 1191–1204.

Zhao, Y.Y., Zhao, Y.D., Mirza, M.K., Huang, J.H., Potula, H.H., Vogel, S.M. et al. Persistent eNOS activation secondary to caveolin-1 deficiency induces pulmonary hypertension in mice and humans through PKG nitration. J. Clin. Invest. 119(2009): 2009–2018.

Zheng, J. and Trudeau, M.C. (eds.). (2015). Handbook of Ion Channels. CRC Press, Boca Raton, FL, USA.

Zhou, Y. and MacKinnon, R. The occupancy of ions in the K^+ selectivity filter: Charge balance and coupling of ion binding to a protein conformational change underlie high conduction rates. J. Mol. Biol. 333(2003): 965–975.

Zong, S., Wu, M., Gu, J., Liu, T., Guo, R. and Yang, M. Structure of the intact 14-subunit human cytochrome c oxidase. Cell Res. 28(2018): 1026–1034.

Zucić, D. and Juretić, D. Precise annotation of transmembrane segments with garlic—a free molecular visualization program. Croatica Chemica Acta 77(2004): 397–401.

Chapter 8

The Maximum Entropy Production

Applications in the Bioenergetics of Bacterial Photosynthesis

8.1 Brief personal introduction

The possibility of high or even maximal entropy production by living beings was my main motivation to switch the scientific career from solid-state theoretical physics to biophysics in the early 1970s because neither biology nor physics had good answers to why life emerged and developed. At present, the "why question" is as often evaded among physicists in favor of a "how" question (Salthe 2004) as it was in 1970 when I started thinking about mysteries of biological evolution in the context of known physical laws. The question "Why" was the favorite way of thinking for my student days' roommate and friend Ante Mudnić, a gifted theoretical physicist from Split, who prematurely died. In due time, he would have asked the same question Lee Smolin asked: "Why are the laws of nature as we find them to be, and not otherwise?" (Smolin 1997). Accidentally, I was not wise or lucky in choosing my first mentors in theoretical physics. At that time, Master of Science degree work lasted two to four years on average in Yugoslavia, almost like the PhD research. Institute Ruđer Bošković, where I worked then, had high scientific standards and the intention to get rid of the students who could not finish it. After my young mentor declined to continue with mentorship, the authorities raised the bar in my case. They imposed the condition that I must come with the research idea, perform the calculations, and publish the paper in some foreign journal, all without having a mentor or any help from my senior colleagues at the Department of Theoretical Physics. I noticed David Pines's statement (1962) that nobody used Feynman's diagrams to calculate **polaron**'s ground-state energy and found which perturbation diagrams have to be summed to do just that (Juretić 1971a). Ergo, Feynman's diagram method saved my scientific carrier in the early 70s when I was left without a mentor for the MSc degree and with no prospect to get the PhD degree in Croatia. My contribution to the polaron topics research was recognized next year as "the most important contribution for the finite radius of convergence of the perturbation series" (Baltz and Birkholz 1972).

When I became interested in the question of how physics can explain the existence and the evolution of life (Juretić 1971b), Ilya Prigogine's publications and books were my main inspiration. He even offered me to join his group for one year, but I had some reservations. When you switch to a different research field, a minimum of two years are needed to get familiar enough with it for achieving satisfactory progress. Hence, after getting a master's degree in theoretical physics, I decided to accept the offer to become a PhD student in the Biophysics Department of Pennsylvania State University, PA, USA. The Department was established by Ernest C. Pollard, the founder and the first president of the US Biophysical Society. By the way, Professor Pollard was an excellent example (among many others mentioned in this book) of renewed physicists who switched their career from physics toward biology and biophysics in the period marked by the Second World War atrocities. Pollard's biophysical insights from his Penn. State Univ. period (Pollard 1965, Keith et al. 1977) are still alive in recent publications dealing with microgravity effects on human cells (Ullrich and Thiel 2020) and with the problem when starvation can end in a different outcome. It can prolong life in the "sleeping beauty" stage (Gray et al. 2004) or induce programmed cell death connected to the increase in intracellular viscosity (De Virgilio 2012). Ernie Pollard had a well-balanced anticommunist attitude and healthy suspicion about students coming from communist countries.

My PhD research started in an even stranger manner. For about 18 months, I was producing millions of yeast mutants without any success in fulfilling the desired goal—to find the mutant, which is not able to synthesize the phosphatidylcholine. Then, one day my mentor Prof. Alec Keith did a crucial experiment that proved this goal to be the mission impossible. His instructions for my research could not be fulfilled because the particular strain of yeast cells that we used was not able to take up the choline supplemented in the growth media. Thus, all cells died when I produced the desired mutation. These cells did not have the alternative pathway for lecithin synthesis and could not be used to study why that phospholipid is essential for eukaryotic cells. We then used the *Neurospora crassa* chol-1 mutant.

The methods and concepts I learned during the PhD study were very useful in life sciences but had nothing to do with the question of how physics can explain the life phenomenon. Nobody in the Biophysics Department was interested in the entropy production of living cells, and my seminars about that topic did not change that situation. Nonetheless, I returned to Croatia as the first scientist with a PhD in biophysics from the USA. At that time, Croatia was the republic inside the communist state of Yugoslavia, de facto under Belgrade's and totalitarian Tito's rule. Most of my acquaintances in the US had no idea that Croatia existed at all and even lesser knowledge about its location in Europe. My attempts to continue with biophysical experiments in Croatia were idealistic and unproductive. One example will illustrate why it was so. About half of the paid-for chemicals and enzymes arrived after the prolonged three years period. At that time (the late 70s), Yougoslavia was attempting to cut down on all nonessential foreign payments. The Playboy magazine, international scientific journals, and biochemicals not produced inside the country were all banned for import by Belgrade's authorities. Some federal republics (Slovenia) resisted Belgrade's orders to cut down on already meager expenses for the scientific community. Croatian communist leaders had all local powers in their hands and no backbone to resist unreasonable new rules. Soon, the situation became worse with the arbitrary power shutdowns in different city neighborhoods. My precious enzymes, three years late in arriving, could not survive at room temperature. It was crazy to carry them around to various city locations where elevators and refrigerators still worked.

The motivation to do some theoretical work revived again in my mind. I published two papers about thermodynamics and the efficiency of photosynthesis (Juretić 1983, 1984). I was not satisfied with these publications, and there was little, if any, response. Most experts were of the opinion that a beam of photons is decidedly nonthermodynamic, with the exception of well-defined thermodynamic functions for photons in blackbody radiation (Hill 1977). My family situation changed for the worse, which also stimulated a desire to seek a new far-away position where scientific research would be easier. Less discrimination against disabled persons was also a desire close to my heart. After the divorce, I was in a bitter mood because my physical disability (weak legs) was used as the main reason for not allowing me and my son to ever spend a night together at my place during his early years. That such generous contacts at my new flat were prescribed by the judge did not help at all. In communist countries, laws always have a selective application—strict obedience for some and the freedom to ignore them for others. Thus, I accepted the offer for another training period in the USA, Bethesda, Washington, D.C., this time around.

I soon learned that Terrel Hill also lived and worked at Bethesda. Since I had a high opinion about his contributions to free energy transduction in biology, I asked to meet him, and he kindly accepted. This single conversation with him in 1986 left a profound impact. After talking about my ideas on how to calculate entropy production for light-activated energy transduction, he remarked:

"My intuition tells me that your proposal can work." He suggested for me to join efforts with his young collaborator Hans Westerhoff. Our joint paper resulted from that collaboration (Juretić and Westerhoff 1987). The conclusion from that paper was still not entirely satisfying to me, namely, that an organism can find a compromise with neither the efficiency nor the free-energy dissipation being optimal for light-driven proton pump operating under far-from-equilibrium conditions.

Upon my second return to Croatia in 1989, the situation was precarious again. Each Yugoslav republic asked for greater freedom as allowed by the constitution, but instead got the internationally supported aggression with the intent of replacing Yugoslavia with Greater Serbia. My job was secure for several years after my transfer from the University of Rijeka to the University of Split, where I accepted the Associate Professor position. Croatia was internationally recognized as an independent country after the heroic and tragic defense of Vukovar. There was little change in the Croatian authorities' mindset, no Poland-like lustration, and most of the communist laws inherited as such. The career advancement in teaching and research required fulfilling bureaucratic quotas in published papers during many four to five years periods with widely different criteria in different research fields. I felt secure enough only after getting the permanent full professorship position in 2001, which is roughly equivalent to tenure. At that time, I was 57 years old, and nobody expected novel discoveries from me.

That was the time when I told myself: "If you really want to fulfill your dream of finding some physical explanation why life exists, it is now or never time." Hence, I devoted three months of my time to modeling photosynthesis and calculating entropy production in charge transfer steps. Nothing else had priority except obligatory routine teaching. I was fortunate and grateful to get some data from Hans Westerhoff and his ex-student Bart van Rotterdam. My younger colleague Paško Županović become interested in that project, and we had a fruitful collaboration for about ten years.

8.2 Entropy production calculations for the simplified models of photosynthesis

During the last several decades, *Rhodobacter sphaeroides* have become the favorite research subject for laboratories interested in how photosynthesis connects light absorption to electron transfer and proton transport. This remarkably versatile bacterium can grow via fermentation or respiration, and it can also obtain energy through photosynthesis. Its photosynthetic machinery is located in membrane invaginations named chromatophores that are convenient for isolation and structural studies of relevant integral membrane proteins. After capturing light by antenna complexes containing bacteriochlorophylls, the **exciton** energy is transferred to the reaction center. The reaction center of *Rhodobacter sphaeroides* belongs to the first transmembrane proteins with successfully resolved 3D structure by X-ray diffraction (Allen et al. 1987, Ermler et al. 1994). The very first one was the reaction center of *Rhodopseudomonas viridis* (Deisenhofer et al. 1984, Deisenhofer and Michel 1989).

We used experimentally determined kinetic constants as published by van Rotterdam in his PhD thesis (van Rotterdam 1998). We simplified his kinetic scheme for the initial photosynthetic steps of the purple photosynthetic Eubacterium *Rhodobacter sphaeroides* to the five-state model (Figure 8.1). We retained only the essential kinetic steps of the an-oxygenic chlorophyll-based photosynthetic cycle. The MaxEP principle was applied to several of the initial irreversible steps leading to active proton transport and recovery of bacteriochlorophyll ground state (Juretić and Županović 2003, 2005). It was an indirect application of that principle because we were not looking for maximal total entropy production.

Membrane proteins performing the steps illustrated in Figure 8.1 are the reaction center, electron transfer proteins, and cytochrome bc1 complex, which couple electron transfer to proton release into the bacterium's periplasmic space. Since periplasmic space is diffusionally well connected (for proton and water diffusion) to extracellular space, the final charge separation step performed by the

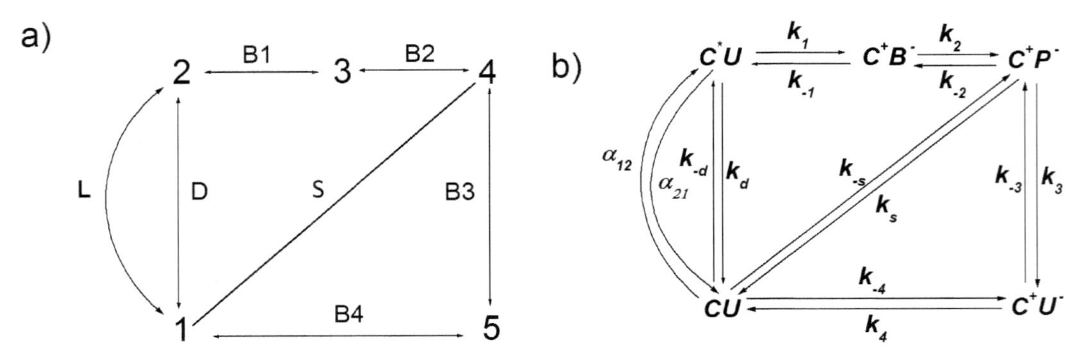

Figure 8.1: a) Bacteriochlorophyll-based 5-state kinetic model for the bacterial photosynthesis with light-activated transition L, non-radiative transition D back to the ground state 1, and relaxation transitions B1–B4 from the excited state 2 to intermediate states 3, 4, and 5, also ending in the ground state 1. The non-productive slip transition S is assumed to connect states 4 and 1. b) States identity and rate constants for all transitions. The following labels are used for different molecules: C and C* for, respectively, ground and excited state of bacteriochlorophyll, U for ubiquinone electron acceptor, B for accessory bacteriochlorophyll, and P for pheophytin electron acceptor. The light-activated transition L connects the CU and excited state C*U with rate constants α_{12} and α_{21}. The excited state can be depopulated through a non-radiative transition D back to the ground state with corresponding kinetic constants k_d and k_{-d}, or through productive relaxation pathway B1 to B4, which includes electron transfer and charge separation through states C^+B^-, C^+P^-, and C^+U^-. The non-productive slip transition S does not produce proton transport. Electron cycling is coupled to proton pumping only in the recovery transition B4.

cytochrome bc1 complex represents active proton transport (pumping) from inside (cytoplasmic) to the outside environment, which creates an electrochemical proton gradient. Each transition from the five-state kinetic scheme has associated forward and backward kinetic constant. Only some transitions are productive in the sense of being required for the generation of the protonmotive force.

Even a simpler kinetic model for bacterial photosynthesis was used by Lavergne and Joilot (1996) with no branched pathway (just one cycle). Their model considers only three different states of a photochemical pigment: ground state, excited state, and oxidized state. The single recovery transition lumps into one all productive transitions in which electron transfer is coupled to proton transfer. It emphasized the fundamental function of the bioenergetic electron transfer chain—the irreversible conversion of redox energy into the electrochemical potential difference of protons. In turn, the protonmotive force drives the phosphorylation process, which is not considered as belonging to the initial charge transfer steps.

The word irreversible implies that dissipation takes place in the bioenergetic electron transfer chain. Lavergne and Joliot (1996) considered the dissipated power of free energy, which is named the dissipation function and defined as entropy production multiplied by the absolute temperature T. In their simple example of a single active force (photon free energy) and a single flux, it is easy to solve the problem of how the flux of free energy can be maximized, presumably the goal of biological evolution. The dissipation function must be maximized to find maximum power that the photochemical converter can deliver. Ross and Calvin (1967) and Knox (1969) have concluded already in 60' that one must maximize the dissipation function to find optimal parameters for photosynthetic free-energy conversion. Lavergne and Joliot (1996) found that maximizing dissipation resulted in a very high optimal **quantum yield** of 0.98 for chlorophyll and 0.97 for bacteriochlorophyll. Reasonable experimental values for light intensity and kinetic constants were used to get this result, which is close to experimentally observed quantum yields (also named photochemical yields). In our modeling experiments II, III, and IV for the 5-state kinetic scheme (Figure 8.1), we used the data provided by van Rotterdam (1998) and three different choices for light intensity and equilibrium constant K(B4) (Table 1 from Juretić and Županović 2003). The photochemical yield was defined as the ratio of driven to driving flux: J(B4)/J(L). All optimizations also resulted in a high optimal photochemical yield in the range from 0.94 to 0.95.

We also used the 3-state scheme for bacterial photosynthesis similar to that of Lavergne and Joliot (1996) and maximized dissipation in productive charge-separation steps. For a different choice of light intensities and radiationless rate constant transition to the ground state, we obtained optimal quantum yield in the range from 0.93 to 0.98, in accord with experiments. Under the supervision of Dr. Robert Niven, Sarah Hall applied our 3-state kinetic scheme to plant photosynthesis with much more complex photosystem I and II located in thylakoid membranes of chloroplasts. The chlorophyll photocycle included only the transition of radiative energy from the ground state for photosystem I or II to excited state, electron transfer to acceptors, and recovery step to ground state coupled to proton transport. All steps are kinetically reversible but with the dominance of forward reactions. The Mathlab code was constructed for complete optimization of all rate constants. The calculated optimal quantum yield of 0.985 was close to our result. The reaction time (the inverse of the rate constant) for the first productive electron transfer step was close to 1ps, similar to observed values for PS I and PS II (Ke 1996a,b).

The complexity of photosynthesis networks, either bacterial or plant-based, has been so much reduced in all of these modeling studies that the similarity of observed and optimized values is not conclusive evidence that looking for the maximal entropy production in certain productive pathways is a worthwhile approach. Hence, one can take the recurse of analyzing the simplest photosynthetic system still capable of light-activated proton transport against its electrochemical gradient. It is the bacteriorhodopsin photocycle (see Chapter 12). In our most recent simulations of bacteriorhodopsin photocycle, we also looked for maximal dissipation in enzymatic transitions and obtained the same result of very high quantum yield (Juretić et al. 2019b). All of these optimization results are robust with respect to change in numerical estimates for used parameters. The basic reason for little sensitivity is very high photon energy hν in comparison to thermal energy k_BT (k_B is the Boltzmann constant).

In natural photosynthetic systems, high yield implies a large departure from thermodynamic equilibrium. Evolutionary optimization is for high quantum yield, high dissipation in dark reactions, and low usable chemical energy. The optimization by evolutionary algorithms reproduced the observed high quantum yields from 0.95 to 0.98 for *Blastochloris viridis* and *Rhodobacter sphaeroides* (Fingerhut et al. 2008). These authors concluded that artificial systems for light harvesting and charge separation in the initial reaction steps could increase the free energy conversion efficiency regarding natural biological systems without a significant decrease in a high quantum yield. The conclusion is in accord with the experimental confirmation of predicted free energy change through the high power input arbitrarily far from equilibrium (Liphardt et al. 2002). Interestingly, just the opposite argument has been recently presented by Robert Jennings and coauthors. They considered high quantum efficiency (the other name for quantum yield) as the main indication for minimal dissipation in initial photosynthetic steps (see Chapter 5).

In the meantime, 20 years old insightful contribution of Lavergne and Joliot to BTOL (biophysics textbook online, 2000) has been mostly forgotten. Its internet link is difficult to find, and when found, it is seen to the dismay of a reader that all equations from that paper are corrupted with nonsense symbols. It is a good reason to recapitulate some of their findings concerning the thermodynamics of the excited states of photosynthesis. When excited, the pigment molecule loses excess vibrational and electronic energy, meaning that it dissipates heat despite the extreme rapidity of relaxation in less than the picosecond time frame. However, pigments are arranged in close proximity, thus forming the antenna complex, which prevents the decay back to the ground state and facilitates the excitation transfer to the reaction center. Excited pigment from the reaction center is conveniently located close to an electron acceptor and donor pair (the redox couple). This initiates electron transfer and converts the part of photon free-energy into the redox potential difference, that is, into an electrochemical form.

Actual concentrations of pigments in the excited and ground state can be different for a factor of 10^{10} or even more in favor of a ground state. The occupation probability of excited and ground state must be taken into account when estimating the energy involved in this electrochemical process. It means that free-energy available from a photon is considerably smaller than its energy hν due to an entropic decrease of the available free energy. For open-circuit condition, when the bacteriochlorophyll pigment does not interact with membrane proteins (excluding the pathway for energy utilization), the photon free energy is only about 70% of its energy hν under bright daylight illumination. An equivalent Carnot machine with no work output has a Carnot yield of about 0.70. When the photochemical circuit is under load, we can define photochemistry's quantum yield as the flux yield. It is the flux of utilized electrons over flux of absorbed photons: $\phi = J/I$, where J and I are, respectively, electron and photon flux. The affinity of the system at the molecular scale or difference in standard potentials between excited and ground state is then:

$$A = A_C + k_B T \cdot \ln(1 - \phi) \tag{8.1}$$

where A_C denotes the maximal Carnot yield for the open circuit affinity.

Naturally, the photochemical yield ϕ and dissipated power Φ depend on the illumination intensity I: $\Phi = JA = \phi AI$. When illumination intensity and temperature are both kept constant, it is still possible to find the maximal value of dissipated power and corresponding optimal values of quantum yield ϕ and affinity transfer efficiency A/A_C (Juretić and Županović 2003). From power per unit surface maximization, Lavergne and Joliot (2000) even predicted the optimal antenna size.

8.3 The Maximal Transitional Entropy Production (MTEP) theorem and its applications

The optimal values of rate constants can also be found from the maximal dissipation requirement in the corresponding transition. Together with my former student Željana Bonačić Lošić, I derived the proof (theorem) in 2007, stating that maximal entropy production can be found for each enzymatic transition. This proof confirmed that we were on the right track in the Juretić and Županović publication (2003). Proof details were published later (Dobovišek et al. 2011, Juretić et al. 2019b). Algebraic derivations are lengthy and will not be reproduced here. I derived it first graphically as an extension of Terrel Hill's method, which required the introduction of a special diagram class: the diagrams that did not contain the chosen i-j transition. The theorem used the trade-off between affinity decrease and flux increase for increasing values of a forward rate constant for the i→j transition. The analysis always leads to the conclusion that maximal transitional entropy production value must exist for any considered transition between functional states i and j.

We required fixed equilibrium constants, cyclic turnover of enzymes, and a steady state condition for validity and applications of the **MTEP theorem**. Stijn Bruers used the phrase "partial steady state MaxEP" (Bruers 2007) to distinguish the application of MTEP condition to reactions catalyzed by enzymes (Juretić and Županović 2003, Dewar et al. 2006) from better-known applications of the MaxEP principle. He concluded that Paltridge's climate model (1979) also belongs to the same class of partial steady state validations of MaxEP because of common features: steady state and only partial entropy production maximization. By the way, partial MaxEP is probably a better term than "partial steady state," a phrase with the implication that some chemical components in a reaction system violate the stationarity condition. The unconvincing similarity among atmospheric compartments and enzyme functional states also did not help maintain the proposed name "partial steady state MaxEP" in the research literature.

For the MTEP theorem application, one must decide which transitions between functional states are interesting for optimization. Initially, overall entropy production can be dissected into all possible contributions from state-to-state transitions. Separate MTEP optimization for each transition will single out those transition-events in the kinetic scheme that can produce the largest increase in the overall entropy production. Usually, these are catalytic or charge transfer steps that are rate-controlling for the turnover and productive in terms of free energy conversion into a biologically useful form. In the case of our 5-state model for the turnover of *Rhodobacter sphaeroides* reaction center, the sum of maximal partial entropy productions in the productive

B1-B2-B3-B4 transitions (Figure 8.1a) is between 80 and 90% of the total entropy production for all of the tested models (Juretić and Županović 2003). These transitions involve indispensable charge transfer and charge separation steps for the conversion of driving to a driven force-flux couple. Out of these four transitions, the last two are more important because of a) greater contribution to total entropy production, and b) less well known kinetic parameters. Accordingly, we have chosen an iterative self-consistent optimization of forward kinetic constants in B3 and B4 transitions (Figure 8.1a) from the requirement that entropy production is maximized simultaneously in both of them. The optimal values of forward rate constants were then found numerically in the 5-state model. The iteration to the stable optimal values was very fast and robust regarding a wide range of initial values. Due to many simplifications introduced in all of these models, an exact agreement can not be expected among optimized and measured values. Still, an approximate agreement can be easily achieved, and that is beyond anything which Prigogine's minimum entropy production theorem can deliver in the same research field (Prigogine 1967).

For example, with a choice of $\alpha_{12} = 100\ s^{-1}$, an optimal recovery rate constant $k_4 = 254\ s^{-1}$ corresponds to the reaction time of 3.9 ms and maximal entropy production in the B4 transition (Figure 8.1). The recovery step is rate-limiting for the photocycle turnover time. It follows that the measured minimum turnover time of 2 ms for the photosynthetic reaction center of *Rhodobacter sphaeroides* (Osvath and Maroti 1997) is comparable to our optimal k_4 value obtained from the MTEP theorem application. For 25 times lower light intensity, when $\alpha_{12} = 4\ (s^{-1})$, our optimization model 5-state III (Table 1 from Juretić and Županović 2003) produces optimal $k_4 = 10.1\ (s^{-1})$. This optimal value is in good agreement with observed $k_{recovery} = 8.6\ (s^{-1})$ published by many different experimental teams and collected together in the van Rotterdam thesis (1988, page 55, Table 2). In the optimization experiment 5-state II at higher light intensity ($\alpha_{12} = 100\ (s^{-1})$) (Table 1 from Juretić and Županović 2003), the overall entropy production is distributed as 2.92, 0.33, 0.80, 0.10, 0.02, 13.55, and 2.06 kJ $mol^{-1}K^{-1}s^{-1}$ among transitions L, D, S, B1, B2, B3, and B4, respectively. Irreversible transitions coupling electron transport to proton pumping (B3 and B4) contributed 79% of the total entropy production.

8.4 The backpressure effect, optimal and maximal thermodynamic efficiency

Despite its simplicity, our model and its optimization exhibit the backpressure regulation of proton pumping by the protonmotive force. When pmf, as driven secondary force, is increased, the decrease in the optimal values follows for recovery rate constant k_4, the proton extrusion, and entropy production contributions from the productive pathway. On the other hand, we found the optimal values increased for the occupancy of the charged ubiquinone state C^+U^-, and overall efficiency η of free-energy transduction. Thus, the MTEP application was able to reproduce the first demonstration of a backpressure effect by the protonmotive force on the steady-state operation of the reaction centers (van Rotterdam et al. 2001). For photosystem I reaction center complexes incorporated in liposomes, Pennisi et al. (2010) also observed and modeled protonmotive force's backpressure effect. In modeling, they used a set of electrodiffusion equations combined with a proton pump with backpressure for transmembrane potential. In all cases, the backpressure effect is very sensitive to membrane permeability to protons—a nice confirmation of Mitchell's chemiosmotic theory.

Since the efficiency of free-energy transduction (see Chapter 3, section 3.1, and Chapter 9, section 9.2, for efficiency definitions) increases with higher protonmotive force, one can ask: is there some maximal efficiency value that can be reached, and how high is it? The answer to this question may tell us something about frequent proposals that biological evolution's main goal is to maximize efficiency. Firstly, we must distinguish maximal from optimal η values. It is easy to find both of them when some forward rate constant is varied. The MTEP theorem application's output is the optimal efficiency, which in general does not coincide with the η_{max} value. For our 5-state model, these two values are about 5% different from each other in various optimization cases. When pmf is fixed at $X_{out} = -18.55$ kJ/mol (corresponding to –196 mV membrane potential), an optimal efficiency of 17.7% is close to the maximal energy transduction efficiency of 18.4% for the MTEP optimization subject to that constraint. These values can be directly compared with observed overall efficiency in the reconstituted system because the same definition has been used, namely that thermodynamic efficiency equals the product of force ratio and the flow ratio (van Rotterdam et al. 2001; see Chapter 3 for efficiency definitions). The flow ratio was discussed in section 8.2 as quantum or photochemical yield, and the conclusion was that proton per photon stochiometry is in the range from 0.93 to 0.98. The thermodynamic force ratio of 18% reported by van Rotterdam et al. (2001) is reduced to 17% when multiplied with the photochemical yield equal to 0.95. Thus, the estimate for the output power divided by the input power does not exceed 18%, in excellent agreement with the results obtained from MTEP optimizations.

In theoretical calculations, we can increase the electrochemical proton gradient as much as we like, with the aim to increase the thermodynamic efficiency up to its theoretical maximum. In a laboratory, all possible constraints and restrictions come into play. Van Rotterdam et al. (2002) worked with the photosynthetic reaction center (PSRC) isolated from *Rhodobacter sphaeroides* and reconstituted into small unilamellar liposomes. They were perplexed why the PSRC is not more efficient. After noticing that free energy of about 90 kJ/mol contained in photons is reduced to about 20 kJ/mol (or 0.2 V) for protonmotive force, they speculated that protonmotive force above 0.2 V is dangerous. Due to the extreme thinness of biological membranes, membrane potentials significantly higher than 0.2 V (closer to 0.3 V) correspond to such a strong electric field that it exceeds the breakdown voltage for those membranes. The cytoplasmic lipid membrane of *Rhodobacter sphaeroides* cannot withstand a transmembrane voltage substantially greater than 200 mV (Geyer and Helms 2006). These ideas, shared by Roy Caplan too (1982), were a good introduction to a more general concept of "sacrifice-of-efficiency-for-protection" by Rutherford et al. (2012). The optimal value of free-energy transduction efficiency of about 20% is, at the same time, the maximal value that could be reached from *in-silico* experiments using optimization based on the MTEP theorem. We obtained this result by assuming that the highest innocuous secondary force of approximately 200 mV has been reached. It appears that during the first steps of bacterial photosynthesis, about 80% of photon input power is dissipated as heat.

As a theoretical exercise, we can examine what happens when pmf is increased beyond biologically imposed limits. A gradual decrease in proton pumping with increased backpressure is not enough to prevent efficiency increase as output to input force ratio is increased. At approaching the protonmotive force of about 900 mV, the transduction efficiency passes through the maximum value of almost 83%, with the precipitous decline to zero value for the pmf of 980 mV (Juretić and Županović 2005). Net proton flux also vanishes in that far-from-equilibrium steady state, which we have discussed in Chapter 3, section 3.3. It is the forward static head state when free energy storage is maximal, but the shutdown occurs in output power production. The analogy with the near-equilibrium static head state is not complete because minimal overall entropy production is not reached when net driven flux vanishes. Nonetheless, the total entropy production is about 10^{12} times smaller for the vanishing proton flux condition in comparison to physiological circumstances when the protonmotive force is four times smaller (Juretić and Županović 2005). As a rough estimate, it is the best indication that productive proton transport is responsible for two inseparable consequences of being alive: enormously increased dissipation and an optimal degree of free energy transduction into self-serving cellular pathways.

The fragility of biological membranes is not the only reason why biological evolution never attained steady states with a high enough protonmotive force to have negligible dissipation and the energy conversion efficiency in the range from 70 to 80%. Transmembrane proton transport is coupled to electron transfer between protein located donors and acceptors of electrons (Kaila et al. 2010, Wikström and Sharma 2018). After Wikström discovered that cytochrome c oxidases from different biological sources act as proton pumps, it became clear that physical and chemical mechanisms should be clarified about the question of how electron-proton coupling is achieved. Low dielectric constant ε = 3–6 in protein interior yields electric fields of over billion volts per meter for electron transfer near the active site of cytochrome c oxidase, more than enough to drive electron-coupled proton expulsion against the transmembrane electric field component of observed protonmotive force (Kaila et al. 2010). A similar situation holds for other redox-driven proton pumps. Different stoichiometries and identities of membrane proton pumps produce a similar maximum pmf of circa 200 mV (Blomberg and Siegbahn 2014). There are some curious exceptions for bacterial and archeon species (Ferguson and Ingledew 2008) that are using the oxidation of ferrous ions by oxygen at acid pH to generate the pmf of over 250 mV, but such highly charged conditions are created mostly by the pH difference across the membrane, while membrane potential is low or even positive inside, thus harmless for the integrity of lipid membrane.

Are there some biological or thermodynamic constraints acting as the evolutionary selection pressure for optimizing redox potentials in the respiratory chain at values needed to generate pmf of circa 200 mV (Bergdoll et al. 2016)? One possibility is that redox potential differences in rate-limiting proton-pumping catalytic steps should be high enough, but not higher than those required for driving the proton transport against a maximum pmf of ca. 200–220 mV without causing the release of toxic compounds: superoxide, hydrogen peroxide, and hydroxyl radical. Another possibility is that relatively low energy transduction efficiency is an evolutionary adaptation to the low light intensities in the sub-surface habitat of purple bacteria providing protection against over-illumination (Sener et al. 2016). Finally, we can assume that increases in information entropy (MaxENT principle) and overall entropy production are favored by the thermodynamics of far-from-equilibrium systems endowed with regulative and adaptive capabilities. Therefore, one would expect that the distribution of overall redox potential over a large number of proton-consuming and proton-transporting catalytic steps is a natural development, provided that compatibility is maintained with biologically optimal pmf for each particular species.

8.5 Benefits of increased dissipation for the thermodynamics of photosynthesis

Since mainstream research about thermodynamics' application to bioenergetics of photosynthesis is still subject to controversy about relevant entropy production principles, let us briefly enumerate additional advantages of using some versions of the MaxEP principle instead of the proposed minimum dissipation principles. In our example of using the MTEP version of the MaxEP requirement in modeling bacterial photosynthesis, we obtained the distribution of optimized kinetic constant values over many orders of magnitude, starting for initial picosecond transitions to the final (recovery) millisecond recovery step. It is well known and repeatedly observed hallmark of all light-activated biological cycles. We proposed that a productive charge-separation pathway was optimized during biological evolution for maximal contribution to overall entropy production because it led both to the biological gain of maximal power transfer and to the increased entropy gain for the universe. This proposal brings biological evolution in complete synergy with thermodynamic evolution: the former accelerates the latter.

Seamless unification of kinetics and far-from-equilibrium thermodynamics is another benefit. The possibility to incorporate kinetic details characteristic for each particular case of energy conversion empowers optimized models to make realistic predictions. These predictions can be tested in experiments or have been already verified by ubiquitous observations without realizing that some common cause exists. Of no less importance is our observation that the highly nonlinear force-flux relationship has a huge advantage over the linear mode of operation.

As is usual in science, one new result opens many further questions. The most important of those is the question about the relationship of Darwinian evolution and selection with evolution and selection directed by above mentioned physical principles. We assumed and proved that appropriate modifications of Kirchhoff's laws, MaxENT, and MaxEP principles are relevant for biological networks' present-day operation. Hence, it is likely that the same principles worked hand in hand with biological evolution and selection through more than three billion years of cellular evolution and even before in the much shorter period of chemical evolution leading to the rise of first cells. Such an insight, grounded in a much more extensive comparison of experimental data with modeling predictions, should resolve the old problem of why biological and thermodynamic evolution are not going in the same direction: less complexity, higher entropy, and more disorder. The opposite seems to happen during a long enough period of biological evolution.

The mistake commonly made from the living-structure-centric point of view is neglecting what happens to the environment during living cells' activity. It has been argued that entropy increase in the universe is higher in the presence than in the absence of life (Ulanowicz and Hannon 1987), and that efficiency of photon free energy conversion into biomass is at most around 10%, which means that the entropy increase in the overall environment is many times larger than structural entropy decrease due to synthesis of additional biomacromolecules. Free energy dissipation in the joint system, composed of cells and their environment, accelerates when cells are metabolically active. This effect can be traced to the activities of cellular enzymes and bioactive peptides, which serve as gates or catalysts for coupling thermodynamic to biological evolution.

Previously, physicists did not have the proper tools to study the conservation or optimization of macromolecular function, and so it was not clear at all how to extract some optimal structural characteristics selected as the consequence of functional optimization. Based on the work of Jaynes (1957a,b, 1980, 2003) (Figure 8.2) and Dewar (2003, 2005) in statistical mechanics and information theory, the maximum entropy principle and maximum entropy production principle are seen today to provide a statistical interpretation of evolution as a natural selection of the most probable system behavior under given constraints. In biology, these principles might indeed seem incompatible with macromolecular evolution, which is traditionally viewed as leading to low probability structures (low configurational entropy). However, our results suggested that these principles can be applied to the probabilities of macromolecular functional states and the optimization of fluxes among these states. In our recent publications (Dobovišek et al. 2011, 2014, Juretić et al. 2019a,b), we argue that the MaxEnt principle and MTEP tool are good candidates for extracting optimized functional, kinetic, thermodynamic, and structural information. Shannon's entropy (see Chapter 3 for the definition) is associated with the probability distribution of functional states. We must emphasize that a maximal Shannon's state entropy does not contradict the low entropy associated with a macromolecular structure. It is just very condensed information about numbers, identity, and the probability of the most important enzyme functional states. More decisive information is how these

Figure 8.2: Edwin Thompson Jaynes at Stanford in 1960 (G. Larry Bretthorst, Washington University in St. Louis, released this picture to the public domain on Aug 30, 2007 (https://commons.wikimedia.org/wiki/File:ETJaynes1.jpg). He kindly confirmed on October 28, 2020, that the picture could be used in this book.

states are connected with "flows" and "forces." With known chemical affinities and corresponding flows in each transition between directly connected functional states, the entropy production for that catalytic transition can also be calculated. The distribution of entropy production contributions in all catalytic transitions helps in finding rate-limiting steps in which directed nanocurrents of electrons, protons, and atoms are essential for catalysis.

We cannot expect metabolism to be at the same time maximally efficient and minimally wasteful of free energy. For instance, futile proton circuits produced by mitochondria in brown adipose tissue serve mainly to increase entropy production and maintain nonshivering thermogenesis by producing heat. It is a useful ability for small and young mammals to keep them warm, maybe even crucial for mammals' evolutionary success (Cannon and Nedergaard 2004). Yet, it is impossible to explain from the point of view that the fitness arrow of biological evolution points toward minimal entropy production (Sabater 2006).

Major photosynthetic producers of organic material on Earth are plants, green algae, and cyanobacteria. In a common fashion to purple photosynthetic bacteria, all of these species use integral membrane proteins named reaction centers. The operating spectral window for these reaction centers is quite narrow, from 680 to 700 nm. Ron Milo asked why this window was chosen during evolution (Milo 2009). Milo calculated the "overpotential" used for the free-energy transformation process through charge separation and wasted as heat per photon, that is, not harvested as chemical energy. The assumption that overpotential can't be made smaller than it is represents another example of a novel viewpoint that allows waste heat or entropy production to dictate the value of the reaction center excitation energy. When energy costs accounting for charge separation are taken into consideration, an optimal window for efficiency dependence on wavelength is found close to the operating spectral window from 680 to 700 nm. Efficiency is maximal at about 710 nm, but it is a broad maximum in the range from 680 to 720 nm. Milo suggested that the fitness function for the selection process during about three billion years of oxygenic photosynthesis was energy harvest maximization constrained with a fixed minimally required overpotential. This proposal does not correspond to any physical principle, such as MinEP or MaxEP principle, as a selection principle. Instead, it gives equal importance to wasted heat and harvested chemical energy, constraining and channeling biological evolution and selection. Like other authors, Milo also calculated overall photosynthetic efficiency to be no more than 10% to 15% at its maximal value. Thus, one can equally well claim that real evolutionary constraints, function goals, or fitness functions are close to maximal wasted power and minimally required harnessed power.

8.6 Biotechnological applications

The scientific mind has a natural tendency to improve on nature's design to suit our needs. The reaction centers with light-harvesting complex 1 (RC-LH1) from *Rhodobacter spheroides* can power a low-consumption light-emitting diode (LED) display for a short time. Still, it is challenging to orient these complexes in protein multilayers with satisfactory stability and thickness (Ravi et al. 2019). Constructed RC-containing power cells by these authors generated a photovoltage as high as 0.45 V upon illumination. Their biohybrid photovoltaics carry out solar energy harvesting, energy conversion, and energy storage in the integrated architecture: organic polypeptides and chromophores sandwiched between inorganic semiconductor electrodes. PufX-deficient RC-LH1 used by Ravi et al. (2019) are not capable of light energy conversion *in vivo* but are fully functional and have a higher heat tolerance in biohybrid devices for photocurrent generation (Liu et al. 2018). PufX polypeptide profoundly affects membrane morphology and its bioenergetic functions *in vivo* (Siebert 2004) because it associates both with light-harvesting polypeptides and reaction center polypeptides.

A gain of a new function in artificial devices has been accompanied by the loss of wild-type function in living bacteria. A similar design development happened over and over again during biological evolution when genetic changes produced beneficial new functions with some cost to previous protein functions. After discussing how deleting the gene for the production of PufX protein increases the number of LH pigments, Liu et al. (2018) concluded that the bacterium could be used as a factory for the production of proteins specifically tailored to suit a particular biohybrid application. These two applications are proof of the concept that higher photovoltages and photocurrents can be reached than the physiological values attained by *Rhodobacter spheroides*. Higher efficiencies can also be reached when energy conversion is limited to the first step—the conversion of the light energy into the electron current.

8.7 Feedback from research papers citing our 2003 contribution to the thermodynamics of photosynthesis

The passage of time often helps to clarify scientific questions about the meaning and merits of published research, but there are many exceptions, and I feel that our Juretić and Županović (2003) article is one of them. Most of the authors who cited it lumped it together with some other contributions to support a thesis (whatever it was) promoted by these authors. To Robert Niven, it was of interest whether the maximization of entropy production could be used to predict the steady-state behavior of various reaction mechanisms (2010a) and biochemical processes (2010b). He concluded that the MaxEP principle "has been successfully applied—in a heuristic sense." Zhu et al. (2013) echoed Niven (2010a) in putting together modeling results for MaxEP application to ecosystems (Meysman and Bruers 2007, 2010) with our optimizations of simple photosynthetic models. Belkin et al. (2015) lumped together Swenson's (1997) paper about human ecology and our modeling of bacterial photosynthesis under a common proposition: "it is plausible to hypothesize that MEPP (MaxEP) might be one of the main guiding principles for not only the evolution of biological species, but their appearance on Earth due to the strongly nonequilibrium energy distribution created by the Sun." Martyushev and Seleznev (2015) described all our papers in this field as "the series of papers dedicated to the use of MEPP (MaxEP) for determining the optimal characteristics of enzymes." Bras (2015) presented our 2003 contribution together with climate modeling by Paltridge (1975) and Kleidon et al. (2010) as evidence for the usefulness of the MaxEP principle in describing transport phenomena far from equilibrium. It did not help that Garth Paltridge was still looking in 1975 for an appropriate physical principle and used the phrase "minimum entropy exchange" for his 1975 contribution to climate modeling. Skene (2017) cited the Juretić and Županović 2005 contribution (where essentially the same results were presented as in Juretić and Županović 2003 article) among many other examples of how MaxEP "has been applied in a wide range of non-equilibrium systems." A curious claim has been recently put in the review paper by Arango-Restrepo et al. (2019) that "the MEP (MaxEP) principle has been observed in biochemical processes" when they included our 2003 reference while attempting to enumerate all instances when that principle has been "observed." Leyva et al. (2019) mentioned it, among other references, to support "the possibility that the phenomenon of life itself and consequently, the underlying order, are tuned or optimized in these systems to fulfill a determined set of specific criteria or rules." The comment is connected with their rather unique research question on how to find the optimal size of the prebiotic vesicles from entropic considerations.

The same general trend is present in specific short comments about Juretić and Županović 2003 paper (Christen 2006, Moroz 2008, Niven 2009, Wang and Bras 2009, Martin and Horvath 2013, Bradford 2013, Delgado-Bonal and Martín-Torres 2015). These authors presented our results as support for the application of the MaxEP principle in biology while dealing with seemingly unrelated research topics—from electrical circuits to the Mars climate. They did not notice that we never looked for or found maximal total entropy production. However, we did point out that "unconstrained force contributes a negative term to total entropy production" as the consequence of LeChatelier–Braun principle (Kubo 1976). We also emphasized that our optimization procedure "does not produce maximal total entropy production neither in the productive pathway nor in the whole system." We identified the major contributors to the total entropy production (charge-transfer and charge separation steps) and found the optimal value of the total entropy production in a large number of different photosynthetic models. Martyushev and Seleznev (2006) and Wang et al. (2007) repeated our claim "that photosynthetic proton pumps operate close to the maximum entropy production mode," but did not distinguish between maximum entropy production mode in the productive pathway and the absence of a maximum in total entropy production. This confusion about the meaning of our first highly cited paper on that topic is partly our fault because we did not provide complete theoretical justification for the EP maximization in chosen transitions and instead invoked the MaxEP principle as the theoretical background.

Andrej Dobovišek from Slovenia had long discussions with me about this topic during the initial phase of the Croatia-Slovenia collaborative project in 2007 dealing with the MaxEP principle application to cellular biochemical networks. I have shown to him how maximal total entropy production can be found in the three-state model for a cyclic kinetic scheme, something that Terrel Hill already considered (page 77 from his 1977 book). I derived this result in 2004 (Figure 8.3) and presented it at several international conferences in the following years (Juretić et al. 2005, 2006). The attractive feature was that maximal information entropy could also be found, thus bringing together the MaxENT and MayEP principle in the same application for finding optimal kinetic constants. Less attractive was an arbitrary but obligatory requirement that the product of all kinetic constants in a chosen direction must remain constant. It was the only possibility to force the model toward exhibiting a maximum in total entropy production. This additional constraint requires that the sum of all activation energies must remain constant in a chosen cycle direction no matter how kinetic constants are changed. Dobovišek et al. used this additional constraint but interpreted it first as energy conservation (2017) and subsequently as free-energy conservation

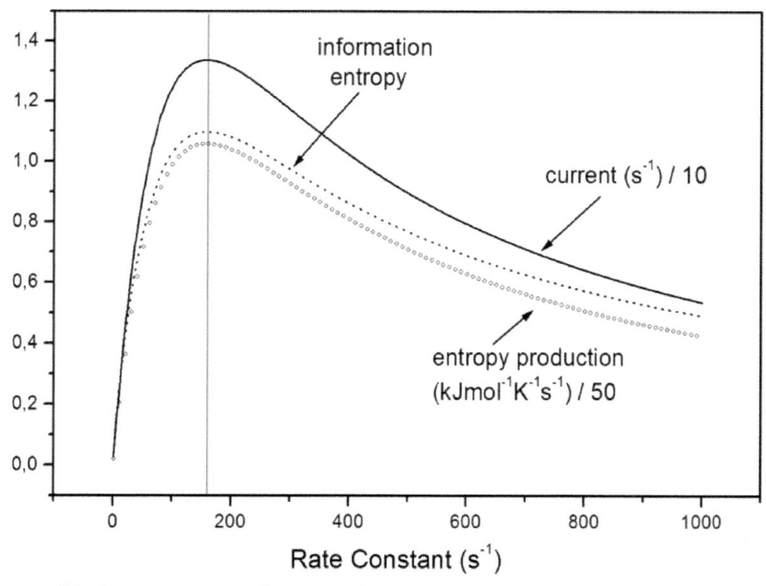

Figure 8.3: The reversible three-state scheme for Michaelis-Menten enzyme kinetics has been optimized in this educational example of using the MaxEP principle to find optimal parameters. Two out of three forward rate constants are varied until the same optimal value of 159 s^{-1} (thin vertical line) is found in the far from equilibrium steady state (defined with an arbitrary choice of X = 74 kJ/mol) with maximal total entropy production. The optimization constrictions were that a) the product of all forward constants is fixed, and b) the reverse rate constants have the inverse value of forward rate constants. The third forward constant is determined from the given value of the thermodynamic force X. The variation of two forward rate constants is then equivalent to a variation of free energy changes for each transition with the constraint that the sum of all activation energies does not change in the forward direction. The stable steady-state reached during the iteration of rate constants corresponds to the equal spacing of basic free energy levels and equal state probabilities $p_i = 1/3$ (i = 1, 2, 3).

(2018). They then found optimal parameters in the case of a two-state kinetic model for glucose isomerase. The model produced unrealistic equal probability for considered functional states under steady-state conditions. We recently examined how the optimization for maximal total entropy production works in the case of the three-state kinetic model for the PC1 lactamase (see Chapter 10). It turned out that predicted rate constants are very different from their known experimental values. Worse, the MaxEP requirement for total EP gave a lower σ_{tot} value compared with calculated σ_{tot} when observed rate constants are used (Juretić et al. 2019a, Table 2, last column).

Albarrán-Zavala and Angulo-Brown (2007) first praised our model for photosynthesis as being "very complete" and then gave it an equal weight as the minimum entropy production in the treatment of photosynthesis (Andriesse and Hollestelle 2001). Ozturk (2012) used a verbatim repetition of identical sentences in his publication. It is an unfair comparison of the alleged MinEP theorem application to photosynthesis with our results. We concluded in Chapter 5 that Andriesse and Hollestelle's application of minimum entropy production requirement for photosynthesis is nothing else but a mathematical error and cannot be compared with our results regardless of interpretation.

Monokrousos et al. (2011) presented our 2003 results as the support for the general evolution criterion (Glansdorff and Prigogine 1964). It is not so simple. The cited criterion refers to a minimum of generalized entropy production, and it reduces to the theorem of minimum entropy production in the linear region. We are not the only ones who expressed doubts about Prigogine's theorem applicability to biochemical energy conversion. Prigogine's collaborator René Lefever recently expressed the opinion that "living matter is not as sparing of dissipation as the theorem of minimum entropy production seemed to suggest" (Lefever 2018).

The variety of interpretations did not stop there. Petela (2008) described the contribution of Jurević and Županović (misspelling my surname) as the support for the claim that the photosynthesis process occurs according to the Second Law of Thermodynamics. I assume that he recalled a great quote from Sir Artur Eddington, who summarized the importance of the second law (Eddington 1928): "If someone points out to you that your pet theory of the universe is in disagreement with Maxwell's equations—then so much the worse for Maxwell's equations. . . But if your theory is found to be against the second law of thermodynamics I can give you no hope; there is nothing for it but to collapse in deepest humiliation." Surprisingly, some authors questioned the validity of applying the Second Law of Thermodynamics to photosynthesis and took issue with the possibility of associating free energy to photons absorbed by chlorophylls (Jennings et al. 2005, 2006).

As already mentioned in Chapter 5, other experts were quick to find errors in the Jennings and coauthors' reasoning that plant photochemistry may function with negative entropy production (equivalent to the claim that the Second Law is violated). Hence, we shall briefly mention here only the second issue about photons free energy. Jennings et al. (2005) cited Parson's (1978) results as the reason why the concept of chemical potential cannot be applied to photosynthetic systems. The opposite view has been amply discussed and defended (Westerhoff and van Dam 1987, Meszéna and Westerhoff 1999, Meszéna et al. 2000). The photosynthetic efficiency for energy conversion does deviate from the Carnot efficiency but is, in fact, lower than the Carnot's efficiency (Zhang and Choi 2018). Besides, William Parson was quite explicit in his 2007 paper with Robert Knox (Knox and Parson 2007), when he refuted the claim by Jennings et al. (2005) that the change in a pigment's free energy on excitation is a) approximately equal to the photon energy, and (b) "in agreement with" Parson's earlier paper (Parson 1978).

The example used by Meszéna and Westerhoff (1999) is for a photon with a 700 nm wavelength that carries an energy of 1.7 eV. When it is used for photosynthesis, it's chemical potential is not higher than 1.25 eV. The different procedures employed by Parson in 1978 led to the same result—not more than about 70% of a photon's energy is available to extract useful work that could be obtained from the system. It is not enough to perform splitting a water molecule. At pH 6, this crucial step of photosynthesis requires 1.3 eV for each of the four electrons involved. Thus, water splitting by a single photon is impossible because a photon delivers free energy, rather than energy (Meszéna and Westerhoff 1999). The cited conclusion invoked the bioenergetics of photosynthesis to answer the question of why biological evolution had to develop a complex electron transport chain before oxygen atmosphere and higher life forms (plants and animals) could arise (Sagan et al. 1993, Westerhoff et al. 1997).

Before ending this chapter, I would like to mention one question opened by Juretić and Županović 2003 paper that nobody tackled afterward. It is the question of the MTEP theorem application to a cyclic set of chemical reactions catalyzed by several functionally connected enzymes. An unfortunate lack of interest for the independent verification of the MTEP theorem for an arbitrary transition between functional states of a single enzyme is the most likely reason why the theorem's generalization was not explored for cyclic reaction pathways catalyzed by multiple enzymes. Our results from the 2003–2019 period suggest the desirability of a theoretical justification for optimizing the rate-limiting step or steps in biological cycles through the maximum entropy production requirement in these transitions.

Definitions and explanations from Chapter 8:

Exciton is an electrically neutral quasiparticle consisting of an electron bound to a hole by the electrostatic force.

MTEP theorem is the geometric and analytic proof for the hypothesis that maximal entropy production can be found for any chosen transition between enzyme functional states. A steady state is assumed, and the constancy of equilibrium constants for all transitions. The theorem uses the tradeoff between transition affinity and flux. A forward kinetic constant or corresponding activation energy is usually varied. The theorem's proof is more complex than Prigogine's proof for the minimal entropy production theorem in the static head steady state. Presently, the theorem's validity is restricted to the enzyme's functional states. It is not restricted to close to equilibrium situations nor linear force-flux relationships. The proof uses the extension of the Terrel Hill "Free energy transduction in biology" framework (Hill 1977) with the same assumptions and restrictions on top of those stated above. The word "theorem," usually used in mathematics, may seem out of place in enzymology, bioenergetics, or experimental physics. However, the word has a long tradition in mathematical and theoretical physics. When experts accept theorem correctness, it can become the foundation for generalizations and wide-ranging applications as one of the physical principles. Historically speaking, entropy production theorems had an essential role in developing irreversible thermodynamics (Saha et al. 2009). As far as I am aware, nobody tried to examine the correctness, falsehood, in-built restrictions, or possible generalizations for the MTEP theorem.

The **polaron concept** describes how an electron in a dielectric crystal disturbs surrounding atoms and moves along together with atoms' nonequilibrium distribution. The phenomenon reduces mobility and increases the electron's effective mass, thus warranting the polaron inclusion among quasiparticles.

Quantum yield in the context of initial photosynthetic steps is the ratio of the produced flux (of electrons) to the absorbed flux of photons. The alternative names are photochemical yield, flux yield, and quantum efficiency.

References

Albarrán-Zavala, E. and Angulo-Brown, F. A simple thermodynamic analysis of photosynthesis. Entropy 9(2007): 152–168.

Allen, J.P., Feher, G., Yeates, T.O., Komiya, H. and Rees, D.C. Structure of the reaction center from *Rhodobacter sphaeroides* R-26: the protein subunits. Proc. Natl. Acad. Sci. USA 84(1987): 6162–6166.

Andriesse, C.D. and Hollestelle, M.J. Minimum entropy production in photosynthesis. Biophys. Chem. 90(2001): 249–253.

Arango-Restrepo, A., Barragán, D. and Rubi, J.M. Self-assembling outside equilibrium: emergence of structures mediated by dissipation. Phys. Chem. Chem. Phys. 21(2019): 17475–17493.

Baltz, R. and Birkholz, U. 1972. Polaronen. pp. 233–341. *In*: Madelung, O. (ed.). Festkörperprobleme 12. Advances in Solid State Physics, vol 12. Springer, Berlin, Germany. https://doi.org/10.1007/BFb0107703.

Belkin, A., Hubler, A. and Bezryadin, A. Self-assembled wiggling nano-structures and the principle of maximum entropy production. Sci. Rep. 5(2015): 8323. doi: 10.1038/srep08323.

Bergdoll, L., ten Brink, F., Nitschke, W., Picot, D. and Baymann, F. From low- to high-potential bioenergetic chains: Thermodynamic constraints of Q-cycle function. Biochim. Biophys. Acta 1857(2016): 1569–1579.

Blomberg, M.R. and Siegbahn, P.E. Proton pumping in cytochrome c oxidase: energetic requirements and the role of two proton channels. Biochim. Biophys. Acta. 1837(2014): 1165–1177.

Bradford, R.A.W. An investigation into the maximum entropy production principle in chaotic Rayleigh–Bénard convection. Physica A 392(2013): 6273–6283.

Bras, R.L. Complexity and organization in hydrology: A personal view. Water Resour. Res. 51(2015): 6532–6548.

Bruers, S. A discussion on maximum entropy production and information theory. J. Phys. A: Math. Theor. 40(2007): 7441–7450.

Cannon, B. and Nedergaard, J. Brown adipose tissue: Function and physiological significance. Physiol. Rev. 84(2004): 277–359.

Caplan, S.R. 1982. pp. 431–441. *In*: Oosawa, F. (ed.). Dynamic Aspects of Biopolyelectrolytes and Biomembranes, Kodansha, Tokyo, Japan.

Christen, T. Application of the maximum entropy production principle to electrical systems. J. Phys. D Appl. Phys. 39(2006): 4497–4503.

De Virgilio, C. The essence of yeast quiescence. FEMS Microbiol. Rev. 36(2012): 306–339.

Deisenhofer, J., Epp, O., Miki, K., Huber, R. and Michel, H. X-ray structure analysis of a membrane protein complex. Electron density map at 3 Å resolution and a model of the chromophores of the photosynthetic reaction center from *Rhodopseudomonas viridis*. J. Mol. Biol. 180(1984): 385–398.

Deisenhofer, J. and Michel, H. Nobel lecture: The photosynthetic reaction center from the purple bacterium *Rhodopseudomonas viridis*. EMBO J. 8(1989): 2149–2170.

Delgado-Bonal, A. and Martín-Torres, F.J. Evaluation of the atmospheric chemical entropy production of Mars. Entropy 17(2015): 5047–5062.

Dewar, R.C. Information theory explanation of the fluctuation theorem, maximum entropy production and self-organized criticality in non-equilibrium stationary states. J. Phys. A: Math. Gen. 36(2003): 631–641.

Dewar, R.C. Maximum entropy production and the fluctuation theorem. J. Phys. A: Math. Gen. 38(2005): L371–L381.

Dewar, R., Juretić, D. and Županović, P. The functional design of the rotary enzyme ATP synthase is consistent with maximum entropy production. Chem. Phys. Lett. 430(2006): 177–182.

Dobovišek, A., Županović, P., Brumen, M., Bonačić Lošić, Ž., Kuić, D. and Juretić, D. Enzyme kinetics and the maximum entropy production principle. Biophys. Chem. 154(2011): 49–55.

Dobovišek, A., Županović, P., Brumen, M. and Juretić, D. 2014. Maximum entropy production and maximum Shannon entropy as germane principles for the evolution of enzyme kinetics. pp. 361–382. *In*: Dewar, R.C., Lineweaver, C.H., Niven, R.K. and Regenauer-Lieb K. (eds.). Beyond the Second Law. Springer-Verlag, Berlin, Heidelberg, Germany.

Dobovišek, A., Vitas, M., Brumen, M. and Fajmut, A. Energy conservation and maximal entropy production in enzyme reactions. Biosystems 158(2017): 47–56.

Dobovišek, A., Markovič, R., Brumen, M. and Fajmut, A. The maximum entropy production and maximum Shannon information entropy in enzyme kinetics. Physica A 496(2018): 220–232.

Eddington, A.S. (1928 and 1948). The Nature of the Physical World. Macmillan, New York, NY, USA.

Ermler, U., Fritzsch, G., Buchanan, S.K. and Michel, H. Structure of the photosynthetic reaction center from *Rhodobacter sphaeroides* at 2.65 Å resolution: cofactors and protein-cofactor interactions. Structure 2(1994): 925–936.

Ferguson, S.J. and Ingledew, W.J. Energetic problems faced by micro-organisms growing or surviving on parsimonious energy sources and at acidic pH: I. *Acidithiobacillus ferrooxidans* as a paradigm. Biochim. Biophys. Acta 1777(2008): 1471–1479.

Fingerhut, B.P., Zinth, W. and de Vivie-Riedle, R. Design criteria for optimal photosynthetic energy conversion. Chem. Phys. Lett. 466(2008): 209–213.

Geyer, T. and Helms, V. Reconstruction of a kinetic model of the chromatophore vesicles from *Rhodobacter sphaeroides*. Biophys. J. 91(2006): 927–937.

Glansdorff, P. and Prigogine, I. On a general evolution criterion in macroscopic physics. Physica (Amsterdam) 30(1964): 351–374.

Gray, J.V., Petsko, G.A., Johnston, G.C., Ringe, D., Singer, R.A. and Werner-Washburne, M. "Sleeping beauty": Quiescence in *Saccharomyces cerevisiae*. Microbiol. Mol. Biol. Rev. 68(2004): 187–206.

Hill, T.L. 1977. Free Energy Transduction in Biology. The Steady-State Kinetic and Thermodynamic Formalism. Academic Press, New York, NY, USA.
Jaynes, E.T. Information theory and statistical mechanics. Phys. Rev. 106(1957a): 620–630.
Jaynes, E.T. Information theory and statistical mechanics II. Phys. Rev. 108(1957b): 171–190.
Jaynes, E.T. The minimum entropy production principle. Ann. Rev. Phys. Chem. 31(1980): 579–601.
Jaynes, E.T. 2003. Probability Theory: The Logic of Science. Cambridge University Press, Cambridge, UK.
Jennings, R.C., Engelmann, E., Garlaschi, F., Casazza, A.P. and Zucchelli, G. Photosynthesis and negative entropy production. Biochim. Biophys. Acta 1709(2005): 251–255.
Jennings, R.C., Casazza, A.P., Belgio, E., Garlaschi, F.M. and Zucchelli, G. Reply to "Commentary on photosynthesis and negative entropy production by Jennings and coworkers" by J. Lavergne. Biochim. Biophys. Acta 1757(2006): 1460–1462.
Juretić, D. A diagrammatic approach to the polaron ground-state energy. Phys. Stat. Sol. (b) 47(1971a): K49–K51.
Juretić, D. About entropy production of biological systems. Periodicum biologorum 73(1971b): A9. Presented at the First Biophysical Meeting of Scientists from Yugoslavia, Krapinske Toplice 1970.
Juretić, D. The thermodynamic and kinetic limits on the process of free energy storage by photosynthetic systems. Croat. Chem. Acta 56(1983): 383–387.
Juretić, D. Efficiency of free energy transfer and entropy production in photosynthetic systems. J. Theor. Biol. 106(1984): 315–327.
Juretić, D. and Westerhoff, H.V. Variation of efficiency with free-energy dissipation in models of biological energy transduction. Biophys. Chem. 28(1987): 21–34.
Juretić, D. and Županović, P. Photosynthetic models with maximum entropy production in irreversible charge transfer steps. J. Comp. Biol. Chem. 27(2003): 541–553.
Juretić, D. and Županović, P. 2005. The free-energy transduction and entropy production in initial photosynthetic reactions. pp. 161–171. *In*: Kleidon, A. and Lorenz, R.D. (eds.). Non-equilibrium Thermodynamics and the Production of Entropy: Life, Earth and Beyond. Springer, Berlin, Germany.
Juretić, D., Županović, P. and Botrić, S. 2005. The maximum entropy production principle as the guideline for predicting evolution of complex systems. Math/Chem/Comp_2005 conference presentation, Dubrovnik, Croatia.
Juretić, D., Županović, P. and Dewar, R. 2006. Evolution and selection in physics and in biology. From Solid State to BioPhysics III conference presentation, Cavtat (Dubrovnik), Croatia.
Juretić, D., Bonačić Lošić, Ž., Kuić, D., Simunić, J. and Dobovišek, A. The maximum entropy production requirement for proton transfers enhances catalytic efficiency for β-lactamases. Biophys. Chem. 244(2019a): 11–21.
Juretić, D., Simunić, J. and Bonačić Lošić, Ž. Maximum entropy production theorem for transitions between enzyme functional states and its applications. Entropy 21(2019b): 743. doi:10.3390/e21080743.
Kaila, V.R.I., Verkhovsky, M.I. and Wikström, M. Proton-coupled electron transfer in cytochrome oxidase. Chem. Rev. 110(2010): 7062–7081.
Ke, B. 1996a. The stable primary electron acceptor QA and the secondary electron acceptor QB. pp. 289–304. *In*: Photosynthesis: Photobiochemistry and Photobiophysics, Vol. 10, Advances in Photosynthesis. Kluwer Academic Publishers, Dordrecht, The Netherlands.
Ke, B. 1996b. Photosystem I—Introduction. pp. 419–430. *In*: Photosynthesis: Photobiochemistry and Photobiophysics, Vol. 10, Advances in Photosynthesis, Kluwer Academic Publishers, Dordrecht, The Netherlands.
Keith, A.D., Pollard, E.C. and Snipes, W. Inositol-less death in yeast results in a simultaneous increase in intracellular viscosity. Biophys. J. 17(1977): 205–212.
Kleidon, A., Malhi, Y. and Cox, P.M. Maximum entropy production in ecological environmental system: Applications and implications. Philos. Trans. R. Soc. B 365(2010): 1295–1455.
Knox, R.S. Thermodynamics and the primary processes of photosynthesis. Biophys. J. 9(1969): 1351–1362.
Knox, R.S. and Parson, W.W. On "Entropy consumption in primary photosynthesis" by Jennings et al. Biochim. Biophys. Acta 1767(2007): 1198–1199.
Kubo, R. 1976. Thermodynamics. North Holland, Amsterdam, The Netherlands.
Lavergne, J. and Joliot, P. Dissipation in bioenergetic electron transfer chains. Photosynth. Res. 48(1996): 127–138.
Lavergne, J. and Jolliot, P. 2000. Thermodynamics of the excited states of photosynthesis. BTOL-Bioenergetic. www.biophysics.org. Copyright by Lavergne and Jolliot, 2000. Cramer, W.A. Biophysical Society, USA.
Lefever, R. The rehabilitation of irreversible processes and dissipative structures' 50th anniversary. Phil. Trans. R. Soc. A 376(2018): 20170365. doi:10.1098/rsta.2017.0365.
Leyva, Y., Martin, O., Perez, N., Suarez-Lezcano, J. and Fundora-Pozo, M. The optimal size of protocells from simple entropic considerations. Eur. Biophys. J. 48(2019): 277–283.
Liphardt, J., Dumont, S., Smith, S.B., Tinoco Jr., I. and Bustamante, C. Equilibrium information from nonequilibrium measurements in an experimental test of Jarzynski's equality. Science 296(2002): 1832–1835.

Liu, J., Friebe, V.M., Swainsbury, D.J.K., Crouch, L.I., Szabo, D.A., Frese, R.N. et al. Engineered photoproteins that give rise to photosynthetically-incompetent bacteria are effective as photovoltaic materials for biohybrid photoelectrochemical cells. Faraday Discuss. 207(2018): 307–327.

Martin, O. and Horvath, J.E. Biological evolution of replicator systems: Towards a quantitative approach. Orig. Life Evol. Biosph. 43(2013): 151–160.

Martyushev, L.M. and Seleznev, V.D. Maximum entropy production principle in physics, chemistry and biology. Phys. Rep. 426(2006): 1–45.

Martyushev, L. and Seleznev, V. Maximum entropy production: application to crystal growth and chemical kinetics. Curr. Opin. Chem. Eng. 7(2015): 23–31.

Meszéna, G. and Westerhoff, H.V. Non-equilibrium thermodynamics of light absorption. J. Phys. A: Math. Gen. 32(1999): 301–311.

Meszéna, G., Westerhoff, H.V. and Somsen, O. Reply to Comment on 'Non-equilibrium thermodynamics of light absorption'. J. Phys. A: Math. Gen. 33(2000): 1301–1303.

Meysman, F.J.R. and Bruers, S. A thermodynamic perspective on food webs: Quantifying entropy production within detrital-based ecosystems. J. Theor. Biol. 249(2007): 124–139.

Meysman, F.J.R. and Bruers, S. Ecosystem functioning and maximum entropy production: a quantitative test of hypotheses. Philos. Trans. R. Soc. Lond. B Biol. Sci. May 12; 365(2010): 1405–1416.

Milo, R. What governs the reaction center excitation wavelength of photosystems I and II? Photosynth. Res. 101(2009): 59–67.

Monokrousos, A., Bottaro, A., Brandt, L., Di Vita, A. and Henningson, D.S. Nonequilibrium thermodynamics and the optimal path to turbulence in shear flows. Phys. Rev. Lett. 106(2011): 134502. doi: 10.1103/physrevlett.106.134502.

Moroz, A. On a variational formulation of the maximum energy dissipation principle for non-equilibrium chemical thermodynamics. Chem. Phys. Lett. 457(2008): 448–452.

Niven, R.K. Steady state of a dissipative flow-controlled system and the maximum entropy production principle. Phys. Rev. E Stat. Nonlin. Soft Matter Phys. 80(2009): 021113. doi: 10.1103/PhysRevE.80.021113.

Niven, R.K. Simultaneous extrema in the entropy production for steady-state fluid flow in parallel pipes. J. Non-Equilib. Thermodyn. 35(2010a): 347–378.

Niven, R.K. Minimization of a free-energy-like potential for non-equilibrium flow systems at steady state. Phil. Trans. R. Soc. B 365(2010b): 1323–1331.

Osvath, S. and Maroti, P. Coupling of cytochrome and quinone turnovers in the photocycle of reaction centers from the photosynthetic bacterium *Rhodobacter sphaeroides*. Biophys. J. 73(1997): 972–982.

Ozturk, M. An exergy analysis of biological energy conversion. Energy Sources, Part A 34(2012): 1974–1983.

Paltridge, G.W. Global dynamics and climate—A system of minimum entropy exchange. Q. J. R. Meteorol. Soc. 101(1975): 475–484.

Paltridge, G.W. Climate and thermodynamic systems of maximum dissipation. Nature 279(1979): 630–631.

Parson, W.W. Thermodynamics of the primary reactions of photosynthesis. Photochem. Photobiol. 28(1978): 389–393.

Pennisi, C.P., Greenbaum, E. and Yoshida, K. Analysis of light-induced transmembrane ion gradients and membrane potential in photosystem I proteoliposomes. Biophys. Chem. 146(2010): 13–24.

Petela, R. An approach to the exergy analysis of photosynthesis. Solar Energy 82(2008): 311–328.

Pines, D. 1962. Pines contribution. pp. 155–170. *In*: Kuper, C.G. and Whitfield, G.D. (eds.). Polarons and Excitons. Plenum Press, New York, NY, USA.

Pollard, E.C. Theoretical studies on living systems in the absence of mechanical stress. J. Theor. Biol. 8(1965): 113–123.

Prigogine, I. 1967. Introduction to Thermodynamics of Irreversible Processes. Interscience, New York, NY, USA.

Ravi, S.K., Rawding, P., Elshahawy, A.M., Huang, K., Sun, W., Zhao, F. et al. Photosynthetic apparatus of *Rhodobacter sphaeroides* exhibits prolonged charge storage. Nat. Commun. 10(2019): 902. doi: 10.1038/s41467-019-08817-7.

Ross, R.T. and Calvin, M. Thermodynamics of light emission and free-energy storage in photosynthesis. Biophys. J. 7(1967): 595–614.

Rutherford, A.W., Osyczka, A. and Rappaport, F. Back-reactions, short-circuits, leaks and other energy wasteful reactions in biological electron transfer: Redox tuning to survive life in O_2. FEBS Letters 586(2012): 603–616.

Sabater, B. Are organisms committed to lower their rates of entropy production? Possible relevance to evolution of the Prigogine theorem and the ergodic hypothesis. BioSystems 83(2006): 10–17.

Sagan, C., Thompson, W.R., Carlson, R., Gurnett, D. and Hord, C. A search for life on Earth from the Galileo spacecraft. Nature 365(1993): 715–721.

Saha, A., Lahiri, S. and Jayannavar, A.M. Entropy production theorems and some consequences. Phys. Rev. E 80(2009): 011117. doi:10.1103/physreve.80.011117.

Salthe, S.N. The spontaneous origin of new levels in a scalar hierarchy. Entropy 6(2004): 327–343.

Sener, M., Strumpfer, J., Singharoy, A., Hunter, C.H. and Schulten, K. Overall energy conversion efficiency of a photosynthetic vesicle. eLife 5(2016): e09541. doi: 10.7554/eLife.09541.

Siebert, C.A., Qian, P., Fotiadis, D., Engel, A., Hunter, C.N. and Bullough, P.A. Molecular architecture of photosynthetic membranes in *Rhodobacter sphaeroides*: the role of PufX. EMBO J. 23(2004): 690–700.

Skene, K.R. Thermodynamics, ecology and evolutionary biology: A bridge over troubled water or common ground? Acta Oecologica 85(2017): 116–125.

Smolin, L. 1997. The Life of the Cosmos. Oxford University Press, Oxford, UK.

Swenson, R. Autocatakinetics, evolution, and the law of maximum entropy production: A principled foundation toward the study of human ecology. Adv. Hum. Ecol. 6(1997): 1–46.

Ulanowicz, R.E. and Hannon, B.M. Life and the production of entropy. Proc. R. Soc. Lond. 232(1987): 181–192.

Ullrich, O. and Thiel, C.S. 2020. Cellular and molecular responses to gravitational force-triggered stress in cells of the immune system. Chapter 17. pp. 301–326. *In*: Choukér, A. (ed.). Stress Challenges and Immunity in Space: From Mechanisms to Monitoring and Preventive Strategies. Springer Nature, Switzerland.

van Rotterdam, B.J. 1998. Control of light-induced electron transfer in bacterial photosynthesis (PhD Thesis). Universiteit van Amsterdam, Amsterdam, The Netherlands.

Van Rotterdam, B.J., Westerhoff, H.V., Visschers, R.W., Bloch, D.A., Hellingwerf, K.J., Jones, M.R.et al. Pumping capacity of bacterial reaction centers and backpressure regulation of energy transduction. Eur. J. Biochem. 268(2001): 958–970.

van Rotterdam, B.J., Crielaard, W., van Stokkum, I.H.M., Hellingwerf, K.J. and Westerhoff, H.V. Simplicity in complexity: the photosynthetic reaction center performs as a simple 0.2 V battery. FEBS Lett. 510(2002): 105–107.

Wang, J., Bras, R.L., Lerdau, M. and Salvucci, G.D. A maximum hypothesis of transpiration. J. Geophys. Res. 112(2007): G03010. doi:10.1029/2006JG000255.

Wang, J. and Bras, R.L. A model of surface heat fluxes based on the theory of maximum entropy production. Water Resour. Res. 45(2009): W11422. doi:10.1029/2009WR007900.

Westerhoff, H.V. and van Dam, K. 1987. Thermodynamics and Control of Biological Free Energy Transduction. Elsevier. Amsterdam, The Netherlands.

Westerhoff, H.V., Crielaard, W. and Hellingwerf, K.J. 1997. Global bioenergetics. pp. 57–94. *In*: Gräber, P., Milazzo, G. and Walz, D. (eds.). Bioenergetics. Birkhäuser Basel, Basel, Switzerland.

Wikström, M. and Sharma, V. Proton pumping by cytochrome c oxidase—A 40 year anniversary. Biochim. Biophys. Acta 1859(2018): 652–698.

Zhang, H.I. and Choi, M.Y. Generalized formulation of free energy and application to photosynthesis. Physica A 493(2018): 125–134.

Zhu, Y., Song, J., Xu, Z., Sun, J., Zhang, Y., Li, Y. et al. Development of thermodynamic optimum searching (TOS) to improve the prediction accuracy of flux balance analysis. Biotechnol. Bioeng. 110(2013): 914–923.

CHAPTER 9

Coupling Thermodynamics with Biological Evolution through Bioenergetics

9.1 Fundamental principles

Some previously expressed thoughts will be repeated in this chapter because we wish to formulate a consistent general picture before going into free energy transduction details for specific enzymes. The recent book by Smith and Morowitz (2016) emphasized the importance of the phase transition paradigm for the emergence and the nature of the living state, which attracted the attention of physicists interested in the life-origin issues. The authors did not consider entropy production changes connected to sudden structural changes or catalytic events. A similar viewpoint, challenging the biological systems' tendency to maximize energy dissipation and the fundamental role of irreversible thermodynamics in life processes, has been expressed by other authors too (Pross and Pascal 2017). We shall present some arguments for the opposite viewpoint in this chapter.

Equilibrium thermodynamics, with its main concepts and applications, belongs to the foundations of physics. General relativity and quantum mechanics also belong to these foundations. For nearly a century, leading physicists published many inconclusive attempts to unite general relativity theory and quantum mechanics without ever abandoning the main achievements of equilibrium thermodynamics. All of them had a common consolidated viewpoint about the suspect scholarship of research results contradicting the Second Law of Thermodynamics.

A different situation applies to theories describing the time evolution and characterization of nonequilibrium states. It is an active research field full of controversy with a little consensus except in the domain of near-equilibrium thermodynamics with linear force-flux relationships, where Nobel Prize winner Lars Onsager laid the groundwork in the early 30s (see Chapter 1). The maximum entropy production principle is just one important contribution in this field. Physicists involved in that research are very much satisfied that it can be applied for near-equilibrium situations as well—always taking into account boundary conditions and conservation laws (Županović and Juretić 2004, Županović et al. 2004, Christen 2006). We used the **mesh currents** concept (that already incorporates Kirchhoff's current law) to show that steady-state current distribution in electrical networks obeys the maximum entropy production principle, which is, in this case, equivalent to Kirchhoff's loop law (Županović et al. 2004). The energy conservation assumption was critical in that derivation. Christen (2006) did not use mesh currents and confirmed that the steady-state maximizes the total entropy production rate in a simple example of two parallel constant resistors. The same example served to illustrate the superiority of the MaxEP principle compared to Prigogine's minimum entropy production principle. Although this example is straightforward, the considered system is firmly in the domain of linear irreversible thermodynamics (validity of Ohm's law)—the domain where the MinEP theorem has been derived, and the MinEP principle formulated.

Far-from-equilibrium thermodynamics is a much more complex research field leading to a variety of new concepts and applications at the forefront of physics (Montefusco et al. 2015). When considering electrical systems, the MaxEP principle's advantage is that it can be applied to nonlinear current-voltage relations such as electric arcs. The Steenbeck principle from 1932 gives good predictions of arc properties in different cases. It can be deduced from the MaxEP principle, which is much more general than the Steenbeck principle (Christen 2006). It is one of the many examples when a particular minimum or least dissipation principle perfectly corresponds to the MaxEP principle. To wit, the Steenbeck's principle states that the voltage drop and corresponding heat power at fixed current is minimized in real gas discharges. One must take into account the current stabilizing external resistor to consider the total system for the MaxEP application (Weber et al. 2015).

The MaxEP principle is equivalent to Onsager's principle of the least dissipation of energy (Ziman 1956, Gyarmati 1969, Županović et al. 2005, Verhás 2014). Confusingly opposite character of the extremal value (minimal or maximal) is just the consequence of different definitions for the considered dissipation function. Ziman explained it very clearly (Ziman 1956). Besides Onsager's principle, that has been firmly established for about 90 years (see Chapters 1 and 5), he cited even older Lord Rayleigh' theorem (1896), and Jeans theorem (1923) involving the dissipation function. Perhaps the simplest formulation of the theorem states that the distribution of currents J_k in a network of resistances $L_{k,k'}$ with given applied electromotive forces X_k is such that the expression

$$\Sigma L_{k,k'} J_k J_{k'} - 2\, \Sigma J_k X_k$$

is a minimum (Jeans 1923). This expression corresponds to internal entropy production (first term), and external entropy production multiplied with the factor –2 (second term). Thus, from a physical point of view, the steady-state current distribution makes the entropy production a maximum in the network with linear-response constituents (Jeans 1923, Onsager's 1931 papers, Kohler 1948, Ziman 1956, Županović et al. 2004).

There is no contradiction with Prigogine's minimal entropy production theorem, nor with the corresponding principle. Some critics pointed out that the minimal entropy production principle was never formulated correctly, and that it must be sharply distinguished from the minimum entropy production theorem. They even claimed that the minimal entropy production principle is a mathematical error (see Chapter 5). Others found fault with the derivation of the Prigogine's theorem because the stationary state was not correctly taken into account (Bertola and Cafaro 2008).

Polettini (2011) repeated and confirmed our results about the maximum entropy production principle for a steady-state electrical network (Županović et al. 2004), adding that it is nothing but a restatement of Onsager's least dissipation principle. We cannot agree with the last statement of Polettini's (2011) section C, named Županović and coworkers. It reads: "The MAXEP principle (MaxEP in our notation) does not imply that "currents in a linear planar network arrange themselves so as to achieve the state of maximum entropy production." That is due to the minimum entropy production principle." The fascination with the minimum principles, despite all evidence to the contrary in the case of entropy production, is a major bone of contention among researchers. It is probably due to profoundly ingrained energy minimization procedures in physics-curriculum education. However, there is nothing strange in the free energy achieving its minimum through the maximum energy dissipation (Moroz 2008). The possibility emerged that some version of the maximum entropy production principle can be applied in the linear and nonlinear ranges.

Interestingly, the MaxEP principle itself acquires well defined general validity within the steepest-entropy-ascent principle purported to be valid both in quantum and classical world (Beretta 2014, Beretta et al. 2017). This ambitious approach was motivated by the search for the nonequilibrium thermodynamics theory, which is fully compatible with the Second Law of Thermodynamics, representing its extension and generalization as one of the fundamental principles for all of physics from subatomic to astronomical domains. It is essentially a variational problem on how to maximize the entropy production rate subject to all constraints. In theoretical physics, it is known as the problem of constructing the Lagrangian and the action functional as the integral over Lagrangian. The variational principles for deriving Newton's equation of motion and Maxwell's equations of electromagnetism have used the corresponding Lagrangians and the least action principle as an extremely elegant common formalism from which all of mechanics and electromagnetism can be derived.

Richard Feynman learned very early in his life how powerful is the least action principle. That inspirational knowledge transfer from his high school teacher was probably instrumental for the formulation of quantum mechanics in terms of the least action principle (Feynman et al. 1966, 2006, Phillips 2013). Feynman's path integral formulation of quantum mechanics generalizes the classical mechanics' action principle (Feynman and Hibbs 1965). The path integral method involves a rather strange concept of summing all possible histories of the system from the initial to the final state, and famous Feynman diagrams greatly facilitate that job. In his well-known textbook, "The Feynman Lectures on Physics," one lecture is intended to be for "entertainment." It is the lecture about the action principle. Two remarks from that lecture (Vol 2, lecture 19), added as a note after the lecture, are of interest to us:

a) "if currents are made to go through a piece of material obeying Ohm's law, the currents distribute themselves inside the piece so that the rate at which heat is generated is as little as possible," and
b) " the entropy developed per second by collisions is as small as possible." (Feynman et al. 1966, 2006; www.feynmanlectures.caltech.edu/II_19.html).

Feynman was not infallible, as he readily admitted. When a lady student complained about not receiving any points from her teacher, although she faithfully followed one particular statement from his textbook, Feynman answered: "...You should, in science, believe logic and arguments, carefully drawn, and not authorities... I am not sure how I did it, but I goofed. And you goofed, too, for believing me." (From the New Millenium Edition Preface of The Feynman Lectures on Physics). In remarks a) and b), Feynman followed Maxwell's (1954) and Jeans' (1923) derivations of Kirchhoff's laws for dc circuits by variational methods. Van Baak (1999) and Perez (2000) were inspired by these nuggets of Feynman's wisdom to publish their variational and thermodynamic interpretation of Kirchhoff's loop theorem. It appears that neither Feynman in 1961 nor Van Bakk and Perez at the turn of the century were aware of Kohler's contribution in 1948 and Ziman's results from 1956 and 1960 mentioned above and in Chapter 5. Both students and physics teachers should relish Feynman's advice against uncritical belief in authorities, as we mentioned above.

Even stranger concept of simultaneous time evolution in the past and the future has been used to construct the Lagrangian for the proposed application of action principle to the system's irreversible evolution (Glavatskiy 2015). For dissipative systems, the stochastic least action principle is fast becoming the cornerstone (Wang 2006, Seifert 2012). Robert Endres (2017) recently showed that maximal entropy production is a consequence of the least-action principle applied to dissipative multistable systems. He speculated that the high complexity state is more likely as long as a significantly larger entropy production offsets the extra cost from the entropy reduction due to complexity. When this phase-transition to higher complexity and higher entropy production is repeated in astronomically high numbers every half hour (approximate time required for some bacterial cells to divide in favorable circumstances), the final result is a tight coupling between biological and thermodynamic evolution: the former accelerates the latter in accord with the Second Law of Thermodynamics.

It is not an entirely new development of a new research paradigm. Davis proposed that prebiotic and biological evolution follows a path of least action (Davis 1996, 1998). Roderick Dewar used in 2003 the action integral alongside irreversible paths to formulate the maximum entropy production principle (Dewar 2003) in agreement with the Jaynes maximization procedure (Jaynes 1980).

The present challenge in the origin-of-life research is to study how quasi-steady states with high or optimal entropy production were naturally selected in the prebiotic-like environments as the consequence of the stochastic least action principle (see Chapter 13 and Chapter 16, section 16.18).

A high level of free-energy transduction cannot be divorced from the emergence of new (dissipative) structures and a high entropy production level in macroscopic systems. It is important to recall that a high level of entropy production is a convenient measure not only for a distance from equilibrium but also for an acceleration of evolution. It does not matter if we shall label such an evolution as physical, thermodynamic, or biological evolution. Accelerated evolution usually means a shorter lifetime for the system of interest. An average star lives for millions or billions of years, while our life is measured in decades. Curiously the conversion factor of 10^5 to 10^7 is similar for a star lifetime compared to our lifespan and for the mean free-energy transduction intensity of our mitochondria compared to a star's free-energy transduction. Both show the same inverse relationship: longer life means lower entropy production and the other way around. Common to living in a fast or a slow lane is the low efficiency of free energy transduction. Different is a fascinating life's ability to speed up or slow down dissipation close to theoretical upper or lower limits (see Chapter 16, sections 10 and 11).

9.2 Efficiency of biological processes

For convenience, we shall repeat here the definitions from Chapter 3 pertaining to efficiency. When speaking about biological processes' efficiency, it is important to distinguish free-energy storage efficiency from free-energy transduction efficiency. These two efficiencies are connected. Energy storage efficiency is the ratio of output to the input force. This ratio is already contained as the $-X_2/X_1$ expression from Chapter 3 equation for the free-energy transduction efficiency: $\eta = -\frac{J_2X_2}{J_1X_1}$. We have seen in the case of linear relationships between forces and fluxes that maximal energy storage efficiency is obtained when maximal output force X_2 is achieved in the static head steady state, but the output flux J_2 vanishes in that case ($J_2 = 0$). In a linear case, it is not possible to achieve at the same time a maximal free-energy storage efficiency and non-zero free-energy transduction efficiency. There is no trade-off between maximal energy storage and non-vanishing energy conversion. The same principle holds for nonlinear energy conversions. For example, the living cell cannot convert free energy into a more suitable form while storing a maximum possible amount of free energy for future needs. It is like an impossibility to eat the cake and save it at the same time for tomorrow's delight. If storage efficiency is maximal, the conversion efficiency is precisely zero.

We have learned from bioenergetics the nature of the most important output force produced and maintained during photosynthesis and respiration. The protonmotive force is the output force, while either photosynthesis or respiration provides the driving force. The photon free energy is used to perform charge separation during photosynthesis. Life diversification, the emergence of multicellular organisms, and accelerated biological evolution became possible when life learned how to use the photon free energy. Only a small percentage of photon free energy suffices to create an optimal (but not the maximal) protonmotive force. The major part of proton free-energy is dissipated, that is, exported as useless heat to the environment. However, the protonmotive force and nonzero free-energy conversion efficiency cannot be maintained without the continuous destruction of free energy packages. An optimal electrochemical proton gradient is then converted into ATP synthesis. All subsequent uphill biochemical reactions require ATP hydrolysis. In the case of an active bacterial cell, the cell is effectively dead in seconds after the ATP synthesis stops. ATP molecules are never stored in the cells, and synthesis-hydrolysis cycles, needing ATP hydrolysis, take place at the breakneck pace.

Regulatory mechanisms maintaining the protonmotive force at a safe upper limit of around 200–250 mV must have been developed very early during life evolution. The biochemical composition of cells is such that an active life is possible only far from thermodynamic equilibrium and far from the static head state. Then, entropy production is closer to maximal than minimal values. It is higher from the static head value for more than 12 orders of magnitude in our kinetic model for bacterial photosynthesis (Juretić and Županović 2005; see Chapter 12 of this book too).

Life enjoys living only in the case when free energy transduction efficiency is not zero. When it is close to zero for us, we are dead skeletons or in a deep freeze without revival possibility, but some organisms can be resurrected from such a state. All viruses outside cells or bacterial viruses (phages) outside bacteria are essentially dead crystals. They become alive again when their DNA or RNA enters the cell and trick it into producing new viruses or phages. Some bacteria are so hardy that their spores can be revived after spending 40 to 250 million years in amber or salt crystals (Price 2000). It is not exactly the "Jurassic Park" revival example but close enough when we accept the molecular biology conclusions about all living beings' fundamental sameness. Other bacteria, such as *Deinococcus radiodurans*, can be revived after being desiccated or irradiated so much that their entire DNA has been broken down (Krisko and Radman 2013). Some small animals, such as tardigrades (water bears), can even endure being in a vacuum near absolute zero temperature or liquid nitrogen for a prolonged period (Møbjerg et al. 2011).

Resurrection plants are unusual for their ability for a selective slowdown of metabolic processes in response to changing ecological circumstances. These charismatic botanical taxa have long mesmerized and inspired writers and scientists with their potential to come back to life after they seemingly perished. For instance, resurrection plants come to flower again in the rain after being completely desiccated and irradiated in a desert for years (Griffiths et al. 2014). Interestingly, at least one of these amazing plants, the *Anastatica hierochuntica* (True Rose of Jericho), avoids oxidative damage by coupling highly active antioxidant systems to excess free-energy dissipation mechanisms (Eshel et al. 2017). Some frogs can be revived after being frozen during winter. I shall leave it to readers a sweet joy to discover how frogs can survive their death from total freezing. There are many more such resurrection examples, which all testify to life robustness and revival ability. None explains what happens afterward. Did free-energy transduction efficiency reach some optimal or maximal level compatible with new restrictions, which are beneficial for growth, multiplication, and diversification?

Before attempting to answer this question, we must first realize that biochemical processes important for life are as a rule nonlinear. For physicists and engineers, this is a considerable complication. Onsager's theory of linear irreversible thermodynamics and Prigogine's theorem about minimal entropy production in the static head state no longer hold in the non-linear far-from-equilibrium domain. Ohm's law and Kirchhoff's laws for electrical currents are also restricted to the linear relationship between forces and fluxes. Before we published our papers about this topic in 2003 and 2004, most other scientists were convinced that minimal entropy production principle certainly regulates what happens in the linear regime. In contrast, it was expected that the maximum entropy production principle might hold only for some specific highly nonlinear cases of flux-force relationships. Our research contributions in the 2003–2019 period added some enzymes and some bioenergetic systems for which the maximal transitional entropy production theorem (Juretić et al. 2019) can be applied. The theorem is valid for steady states of nonlinear flux-force relationships arbitrarily far from equilibrium (see Chapter 8). We proved that maximizing entropy production in selected productive steps of simplified bioenergetic systems led to the order of magnitude agreement of <u>optimal</u> kinetic constants and <u>optimal</u> performance parameters with measured kinetic constants and performance parameters.

9.3 Power transfer in the nonlinear domain

As the universal physical law, we are sure that energy conservation holds for nonlinear biochemical circuits. But can we construct the analogs of Kirchhoff's laws for nonlinear circuits? Rate constants and macromolecular states can change during evolution and even during the biochemical circuit operation, complicating the matter. The solution is to find if the stable steady-state exists, which can be described by using the MaxEP principle. When all external forces are fixed (for instance, the photon free energy, and protonmotive force), the cell conditions are analogous to an electrical network with all electromotive forces fixed. The analogs of Kirchhoff's laws and energy conservation condition holds for biochemical circuits as well (Juretić and Županović 2003). We realized that greater flexibility of biochemical networks with nonlinear flux-force relationships could be considered as an advantage instead of considering it just as an additional complication (Juretić and Westerhoff 1987, Juretić et al. 1988, Juretić 1992, Juretić and Županović 2003, Juretić et al. 2019).

We examined in these papers nonlinear free-energy transduction models with the slip in proton pumping (Westerhoff and van Dam 1987). We used the six-state model to simulate the behavior of a redox-linked proton pump (the cytochrome c oxidase) and several different kinetic schemes to simulate the light-driven proton pump (the bacteriorhodopsin). Kinetic parameters become more critical further away from equilibrium. In the case of steady-state kinetics, the diagrammatic method can be used to study free-energy-transducing systems (Hill 1977, Westerhoff and van Dam 1987). Incomplete coupling of proton pumps can be due to proton leakage and slip. The slip is catalyzed by less than a perfect mechanism of membrane-embedded proton pumps. It is easily recognized in the corresponding diagram as the only line common to driving and driven cycle. For instance, it is the transition line D, which separates the input cycle for the absorption of photon free-energy from the output cycle for protons pumping across a membrane against an electrochemical potential difference for protons (Figure 8.1a from Chapter 8 and Figure 12.1a from Chapter 12).

The generalization of Kirchhoff's voltage law leads to an input-free-energy difference as the driving force in the driving cycle. Similarly, the output free-energy difference is the secondary force produced in the driven cycle. The output flow becomes equal to zero, and free-energy transduction efficiency vanishes when the output secondary force is maximal, making it possible to define the generalization of the static head state in the nonequilibrium region (see Chapter 3). In general, minimal entropy production does not coincide with the static head state condition in the nonlinear domain. Far-from-equilibrium entropy production may be minimal away from the static head state where net output flux vanishes or multiple minima can be found (Juretić and Westerhoff 1987). Clearly, Prigogine's theorem and the MinEP principle do not apply. Far-from-equilibrium entropy production can increase with increased free-energy transduction efficiency. However, the concomitant increase in energy transduction efficiency and entropy production is not unique to nonlinear relations between flows and fluxes. For a linear response, one can also find the range of output to input force ratios with a joint increase or decrease of efficiency and dissipation (Figure 1 from Juretić and Westerhoff 1987; Figure 2 from Nath 2019). In any case, when output to input force ratio decreases from its maximal value, the "back-pressure" of an output force is released, and the transduction efficiency increases with increased dissipation, as can be seen from Figure 12.3 (Chapter 12) - another evidence that widely accepted wisdom of the inverse relationship between dissipation and efficiency is often wrong for free-energy transduction in bioenergetics!

In addition to the "forward" static head state, an effective degree of coupling can also be defined in the nonlinear region (Juretić 1992). It can be expressed in terms of slip coefficients and forces for two force-flux pairs (see Chapter 3). We have shown that the maximal free energy storage is always higher in a nonlinear than in the linear case. Claims to the contrary (Stucki et al. 1983) have been corrected by Roy Caplan (Pietrobon and Caplan 1985). In that sense, the nonlinear mode of loose coupling with slippage is more efficient than free-energy conversion in the linear domain. In the presence of two proton pumps, primary pump such as cytochrome c oxidase, and secondary pump, such as F_0F_1-ATPase, the increased slippage in the primary proton pump leads to the higher flux control coefficient and higher control exerted by the F_0F_1-ATPase (Juretić 1992).

We also realized that free-energy transduction is closely related to power transfer and asked what would happen if irreversible transitions involved in the power transfer are optimized for rate constants so that associated entropy productions are maximal. In a simplified bacterial photosynthetic model, a stable steady-state was obtained with the optimal efficiency of free energy transduction being close to experimentally measured values (Van Rotterdam 1998, Juretić and Županović 2003). We got the same satisfactory agreement with experiments when we used different kinetic model and different macromolecule in the case of bacteriorhodopsin. We repeated modeling bacteriorhodopsin in 2019 by using an updated kinetic scheme and confirmed the usefulness of looking for distribution of dissipations for all transitions of the productive pathway leading to charge separation when our maximal transitional entropy production (MTEP) theorem is used (Juretić et al. 2019). In all cases of photosynthetic models, the nonlinear relationship between fluxes and forces turned out to be a huge advantage over working in the linear regime. Optimal power transfer was around 90% (Juretić and Županović 2003), much higher than the maximal power transfer of 50% in linear circuits (Kong 1995).

The take-home point is that the prejudice about the desirability of maximal free-energy conversion efficiency or maximal free-energy storage efficiency of living systems must be given up. What counts for life is very high power transfer through all the hierarchy of irreversible energy transduction steps. The intimate connection between the entropy production and power transfer makes it possible to use the MTEP theorem to find optimal and high power transfer modes. The continuous free-energy input from our Sun is a practically infinite energy source for the biosphere. That is why the advantage of the nonlinear mode in its superior capability to transfer and dissipate large amounts of free-energy is more important during evolution than its disadvantage in terms of limited overall energy conversion efficiency (usually below 20%), and relatively low free-energy storage efficiency of less than 10%. Thus, entropy production by biological macromolecules, organelles, and cells cannot be entirely dismissed as the "lost work" (Jennings et al. 2020), nor can it be named the "useful work" (Martyushev and Seleznev 2015).

Notice that we have reversed the traditional insight into what is important for a photosynthetic cell in its interaction with the environment. Some physicists claimed that there is no entropy production during initial photosynthetic steps or that photosynthesis is associated with negative entropy production (see Chapter 5). These calculations were criticized, and their results were rejected by other scientists without any clear effect. The belief that entropy production must be very low, if not exactly zero, during initial photosynthetic steps, remained firmly entrenched in mainstream thinking. It is not surprising given the quantum nature of photon absorption, exciton emergence, and transfer to photosynthetic protein center or its active site. However, the contrast with highly intensive free-energy transduction due to photosynthesis is still a glaring contradiction to such a viewpoint. Physicists like to forget about active-site chemistry connected to electron and proton transport—the main cause for a high entropy production in the initial photosynthetic steps. How can anyone neglect that bacterial and chloroplast photosynthesis exhibit much more intensive free-energy transduction that equivalent volume of the Sun, a well-known fact from bioenergetics (see Introduction)? Free-energy transduction and concomitant power transfer during respiration and photosynthesis are a hundred thousand to ten million times more intensive compared to an equivalent volume of an average star like our Sun (Metzner 1985).

What are the advantages for us, bacteria and green plants supporting a high level of entropy production? We all produce a large outflow of entropy in the form of thermal radiation (infrared light) and some heating of our environment. No ill consequences occurred until our civilization got addicted to fossil fuel burning. Benefits were a seemingly small percentage of free-energy storage in the form of protein, DNA, RNA, carbohydrate, and lipid synthesis and a rich pattern of metabolic fluxes. For instance, less than 10% of available free-energy is incorporated into biomass by fast-growing bacteria. Microcalorimetric measurements of the growth process in microorganisms by Bermudez and Wagensberg (1986) and many other authors demonstrated that the major thermodynamic process is a large entropy outflow from the growing cells. All microorganisms prefer exponential growth when supplied with their favorite organic compounds and needed minerals (Forrest and Walker 1971). Then, from the point of view of physics, the free-energy converted into chemical bonds is not significant compared to the free-energy change from catabolism. When this is done in continuous culture experiments, a steady-state is established with zero entropy change. Despite large entropy outflow, the steady-state is maintained due to high free-energy input in the form of continuously supplied organic compounds and high entropy production needed to maintain metabolic fluxes. Everything is perfectly balanced by artificial means.

At a global level of our biosphere, the balance is maintained mainly by microscopic and macroscopic photosynthetic species (plants) subject to numerous growth restrictions. Did natural growth restrictions result in minimal dissipation, as many researchers expected mainly due to Prigogine's theorem of minimum entropy production, his high reputation, and very successful popularization activity? It attracted a lot of attention from biologists who did not grasp the meaning of extending the theorem's application in the form of the "principle" far outside from its validity domain (Martyushev 2013). Instead, life's presence led to an additional increase in entropy production (Ulanowicz and Hannon 1987). Synergy was established between thermodynamic evolution, which is presumably leading toward the equilibrium state, and biological evolution leading to an ever greater distance from thermodynamic equilibrium. How is this possible?

9.4 Evolution-coupling hypothesis

Our answer to this conundrum was to postulate the "evolution coupling" hypothesis (Juretić and Županović 2005, Juretić et al. 2019). According to this hypothesis, the biological evolution accelerates thermodynamic evolution, and both evolution types are harmoniously connected to the entropy production. Why do we consider the entropy production increase during natural evolution as a valuable insight? Acceleration of thermodynamic evolution can only lead to the entropy production increase, independently of how it was achieved, by photosynthesis or other means of dissipating the free energy gradients. Accordingly, the evolution-coupling hypothesis can be generalized to hold for any open system driven by some external force to maintain a certain minimal (optimal) distance from the equilibrium state and also from the static head state. These two sorts of well-known evolutions, biological and thermodynamic, are not opposed to each other but exist (where life exists) in a synergistic relationship. The arrow of evolution in physics and in biology aims toward the same target. It is a very promising development for physicists, that common statistical interpretation of evolution can be proposed in such different disciplines as physics and biology. In agreement with Martyushev (2013), we can propose that approaching the maximum entropy production within given constraints "governs" all known evolutionary processes of nonequilibrium systems and that it "guides" prebiotic and biological evolution as well.

Evolutionary biology is not a new research field, and similar hypotheses are close to a hundred years old. Alfred Lotka carefully examined the question of when and how the Darwinian concept of natural selection can be considered the consequence of a physical principle or law of evolution (Lotka 1922a,b). To state his words: "evolution proceeds in such direction as to make the total energy flux through the system maximum compatible with constraints." He also added that the physical quantity, which tends toward a maximum, has the dimension of power. Howard T. Odum named Lotka's law the maximum power principle (Odum and Odum 1976, Tilley 2004). Other authors attempted a critical reassessment of Lotka's maximum power principle (Sciubba 2011), noticing that it can be reformulated under Ziegler's maximum entropy production paradigm (Ziegler and Wehrli 1987) when some additional conditions are met. Lotka repeatedly mentioned the optimal use of the flow of available energy. He stressed the need for biological evolution to include some combination of optimal efficiency and maximum dissipation principle. The calculations presented in earlier chapters show how optimal efficiency can be derived directly from the MTEP theorem, thus avoiding the cumbersome search for joint optimization of efficiency and entropy production.

The reaction steps involved in the creation of the electrochemical proton gradient are similar in photosynthesis and respiration. The optimization for the maximal entropy production leads to the realistic optimal efficiency values and high information entropy. Therefore, it is likely that the whole bioenergetics, as the energy conversion by living beings, can be optimized, and probably was optimized during biological evolution for reaching the maximal possible entropy production and high information entropy given external and internal constraints. The steady-state of optimal/maximal entropy production in different bioenergetic processes is likely to be a tightly regulated state. Any sudden and substantial change in entropy production rate is either quickly repaired or leads to a catastrophic relaxation toward the equilibrium state and cellular death.

We believe that the driving force behind the evolution of life is the generalization of the Second Law: every system, either organic or inorganic, evolves as fast as possible. We can consider the entropy production as the best measure for optimal distance from the equilibrium state and also for the evolution speed degree. Maximal entropy production by major processes in bioenergetics (when all constraints are properly taken into account) means that life increases the speed of evolution in every system (on every planet) where it arises or finds itself. Biological evolution accelerates thermodynamic evolution. Life exists in the universe because it speeds up the evolution of the universe.

Darwinian evolution of life and the origin of first cells in pre-biological evolution are completely compatible with these two powerful thermodynamic principles: the maximum entropy principle of Jaynes and the maximum entropy production principle of Dewar and Ziegler (Martyushev and Seleznev 2006). Even the problems that Darwinian evolution cannot handle, such as the question of how first cells on the Earth arose, are, in principle, solvable by using these two evolution principles from physics.

What saves us from thermodynamic death is a spatial separation between entropy producing entities (living organisms) and planetary entropy sinks. Increasing the entropy of oceans, atmosphere, and the universe beyond Earth did not matter to life. It was even beneficial according to the Gaia hypothesis (Lovelock and Margulis 1974, Pujol 2002), which postulates that biosphere evolution is intimately connected with the evolution of life's physical environment. Life succeeded in accelerating its evolution by establishing a positive feedback loop between biological and thermodynamic evolution. A positive feedback loop speeds up both evolutions. Of course, life was always inventive to find new ways to channel the input power into those dissipative metabolic pathways where electrochemical rather than only thermal free-energy conversion can occur.

There is something common to all living systems which have a high rate of entropy production. These are topologically closed systems that are at the same time open to the exchange of energy and matter (atoms, ions, molecules) with their surroundings. Each living system capable of intensive free energy transduction has easily defined boundaries. Regardless of a system nature, be it mitochondria, chloroplast, prokaryotic or eukaryotic cell, its topologically closed boundary is a lipid membrane bilayer with incorporated membrane proteins. One of the many membrane functions is tightly connecting high entropy production with a fast exchange of energy and matter through the external environment. For instance, brown fat cells' mitochondria keep the whole body warm by accelerating proton currents through some of their integral membrane proteins. In this example, an optimal free-energy transduction efficiency and high entropy production are maintained at the same time to prevent death from freezing. Orderly proton leaks promote increased disorder channeled away as heat. The lesson for us is that the emergence of ever more complex islands of order in biology is the condition for allowing a higher level of exported disorder. In other words, increased order is not the goal of universal evolution, but it is likely to be the best means of how universal evolution can speed up the increase in disorder.

Unfortunately, our words from everyday speech, order and disorder, are not the best connection with, respectively, entropy decrease and entropy increase. For instance, gravity increases entropy while creating wonderful and highly ordered galactic, stellar, and planetary entities. It should not surprise us when some physicists recently realized that gravity is an entropic force (Verlinde 2011). We have already described the hydrophobicity (Chapter 2) as the entropic force of extraordinary importance for folding biological macromolecules into their complex structures suitable for performing equally complex functions. A truly amazing diversity of shapes and functions for eukaryotic membranes have all been achieved through biological evolution by fine regulation of hydrophobicity, fluidity, phase transitions, phase separations, and other physical properties and forces with a major influence of the hydrophobic force. To us, it looks like a given situation in a quasi-steady state, which changes very little with passage of time because biological evolution is a mostly slow process taking many generations to produce significant differences. It is easy to forget that stability of far from equilibrium state can be maintained only in the case when some external forcing with free-energy input is used to maintain a high level of entropy production. Biological evolution has not done anything to reverse the thermodynamic evolution, but on the contrary biological evolution accelerates thermodynamic evolution. This statement is the essence of the evolution-coupling hypothesis.

The bioenergetics of organelles and cells is not the only example of the evolution-coupling hypothesis in action. It is a good example of how life acts as the catalytic agent in the universe, which speeds up evolution of universe. It is also the example of the autocatalysis, a self-sustaining set of biochemical reactions, including replication and multiplication of living organisms. If not contained within deep gravity wells and restricted within the nursing water-rich environment, life would spread like an explosion through the whole space that can feed it with enough energetic photons. Within our planetary system, it is certainly possible that human civilization can spread life to many planets and their moons, but it is also possible that simple life forms already exist in deep oceans hidden under Europe's ice cover and some other Jupiter's and Saturn's moons. There are no photons to be used there as the free-energy input. If life originated near hydrothermal vents at the bottom of first Earth oceans, as many scientists speculate, it had to overcome the same obstacle. It must have used some other free energy sources instead of photons. Fortunately, the very first prokaryotes, named archaea, have their descendants living today, and their capabilities to extract free-energy from unlikely sources and in extreme environments are truly amazing. This very complex story's common thread is how life developed the ability to speed up chemical reactions for many orders of magnitude. It directs us to consider another efficiency, catalytic efficiency, which distinguishes biochemistry from organic chemistry when we study enzymes' catalytic efficiency.

Are there some specific examples of entropy-production-driven reaction networks capable of increasing catalytic efficiency? We considered several such cases in this book, all for enzymes that can increase the catalytic rate from billion to billion×billion times. In the example of β-**lactamase**, more evolved enzymes, that is, enzymes more distant from a putative common ancestor, have increased catalytic efficiency associated with increased entropy production (see Chapter 10). How do we explain the almost linear-like relationship between evolutionary distance and reaction rate, between evolutionary distance and entropy production, and between reaction rate and entropy production? In our viewpoint, even the least evolved enzymes can initiate a chain reaction of nanoscale electron or proton currents when stimulated by proper substrate, which acts as a trigger and as a donor of charged particles. Nanocurrents within any single enzyme are extremely weak but still impossible to divorce from entropy production, which is associated with the enzymatic cycle leading to product release and enzyme return to its former trigger-happy state. Rare beneficial mutations were preserved if they increased the reaction rate and the turnover rate for enzyme cycling, because of the competitive advantage such mutations offered to the organism. It is natural then to expect that biological evolution led to increased entropy production of organisms. Indeed, mammals and birds have considerably greater entropy production than lizards, fishes, and other cold-blooded species. They are also quicker, smarter, and more flexible in spreading over a great variety of habitats. Just as more evolved enzymes are better in connecting biological to thermodynamic evolution, so are more evolved species better in coupling their own evolution to thermodynamic evolution. We could not win over or reverse thermodynamic evolution. However, it turned out that it is possible to speed it up so that a small percentage of energy conversion is directed toward power transfer satisfying all life needs. Is ordering coupled to dissipation the phenomenon unique to life?

9.5 Self-ordering and self-organization as a natural system response to increased dissipation

Self-ordering phenomenon such as crystallization does not involve any function. On the contrary, water to ice crystallization is a perfectly natural mechanism for how all cellular functions can be destroyed when ice crystals start to grow until they break up cellular membranes. Some species, named psychrophiles, are resistant to freezing and thawing processes (Price 2000). One of many different methods nature repeatedly invented to enable the survival below the water freezing temperature is the evolution of proteins able to attach themselves to some ice crystal surfaces in such a way as to achieve the inhibition of ice crystal growth (Bar Dolev et al. 2016). Antifreeze proteins and their peptide or polymer mimics have tremendous application potential ranging from improving the texture of ice cream products, frozen food storage, cryosurgery, cryopreservation (Mitchell and Gibson 2015, Biggs et al. 2017), to brave attempts to maintain frozen bits of rapidly diminishing biodiversity (Browne et al. 2019).

We can first ask the question is crystal growth connected in any way with dissipation? In 1990, Adrian Hill found that crystalization follows the selection rule governed by entropy production rate per unit area of different growth morphologies. The crystal form selected at any driving force is that which, in its formation, releases entropy from the free energy gradient at the highest rate (Hill 1990). His research results are in accord with previous findings by Ziegler (1963), Sawada (1981, 1984), and Shimizu and Sawada (1983). Some other papers published in the Russian language did not attract proper attention, as pointed out by Martiouchev and Seleznev (2000) (in his subsequent papers, English spelling is Martyushev for the first author). Instead, some authors sharply criticized both MinEP and MaxEP extremal principles (Andresen et al. 1984, Hunt et al. 1987). One of these authors subsequently formulated a version of the MinEP principle (Anderson and Gordon 1994) and even proposed that maximal efficiency is “an attractive design principle for biological systems” (Anderson et al. 2002). We criticized attempts to connect maximal efficiency to minimal entropy production in Chapter 3 and this chapter. However, the criticism does not apply to the approach of Anderson et al. (2002) because they defined maximal efficiency in the sense of minimal overall entropy production without allowance for the quantitative relationship between these two quantities.

Crystal growth and growth of other inanimate dissipative structures are often described in terms of self-organization. That notion was both criticized (Abel and Trevors 2006) and supported in numerous examples starting from Glansdorff and Prigogine book in 1971 to chapters authored by a large number of scientists in the books published by Kleidon and Lorentz (2005) and Dewar et al. (2014). Martyushev and Chervontseva pointed out in 2009 that the loss of the morphological stability during crystalization or coagulation can lead to abrupt or continuous growth rate changes corresponding, respectively, to nonequilibrium phase transitions of first or second order. Their report contains the majority of all previous relevant results and proposes the entropy production as the nonequilibrium analog of the potentials used in equilibrium thermodynamics. Ben-Jacob and Garik published beautiful photographs of dendritic crystal growth in Nature’s periodical (1990), suggesting that the dynamically selected morphology is the fastest-growing one. These authors stressed the crucial difference between equilibrium and nonequilibrium selection criterion. The selected phase at equilibrium minimizes the free energy when state variables are known irrespective of the system’s prior history. Nonequilibrium growth processes are time-dependent because they depend on the prior history of the system. They concluded that the corresponding selection principle and theoretical prediction of growth morphologies must exist and proposed that entropy production is dominant in selecting growth morphologies far from equilibrium.

We can infer that the maximum entropy production paradigm is more general than the phase-transition paradigm advanced by Smith and Morowitz (2016). These authors expressed their dislike for the entropy production expression by putting these words into parenthesis “entropy production” and mentioning minimum or maximum entropy production only as a small print footnote (page 507 from their book). Consequently, they did not examine the connection between nonequilibrium phase transitions, entropy production, membrane bioenergetics, the chemiosmotic theory, and metabolic power optimization. The book also did not mention the seminal contributions of Alfred Lotka, Hans Ziegler, John Ziman, Terrel Hill, and many other researchers who studied how flows, dissipation, and power output are associated in the linear and nonlinear domain. Instead, Smith and Morowitz proposed that "produced entropy" in diffusing systems should be named the "equilibration entropy," and used the phrase "equilibrium entropy production" in opposition to Prigogine's definition of entropy production (see the first three equations in Chapter 3). Also neglected by them were all valiant attempts to answer the question posed by Paul Ehrenfest more than 100 years ago (in 1911): Is there some function which, like the entropy in the equilibrium state of an isolated system, achieves its extreme value in the stationary nonequilibrium state? We believe that the proper choice of the maximum entropy production principle, together with all constraints and restrictions imposed upon a considered open system, goes a long way toward a physical understanding of how bioenergetics connects the origin and nature of life to its wider environment.

Initially, the MaxEP principle was used theoretically and for experimental verification only in the special cases: the selection of solutions at **bifurcation** points, sudden changes in crystal growth morphologies, and thermomechanics. The generality of S-shaped growth curves (sigmoidal shape curves with a slow rise, fast linear-like-growth, and slow saturation growth afterward) suggested that the principle also holds for growth under restricted supply of "nutrients" of inanimate systems and living cells without sharp nonequilibrium transitions (Martyushev and Axelrod 2003, Bejan and Lorente 2011). What will be the outcome depends very much on the considered time-frame. For very short space-time intervals, basic physical principles restrict everything to the strange world of time-symmetric quantum mechanics devoid of evolution as we know it in our macroscopic world. For small periods, evolution in accord with the MaxEP principle becomes possible outside equilibrium, and for a long period decreasing entropy production becomes a dominant feature (Martyushev 2013). This pattern is recapitulated in every individual's life: increasing entropy production initially, maximal EP reached early in the childhood, and slowly decreasing entropy production until death, which leads to precipitous EP decline (Aoki 1991). Likewise, Ichiro Aoki concluded that the pattern of age-related entropy production changes per unit surface (metabolic heat production) is similar for individuals from other species regarded as "mini-ecosystem" and for any ecosystem regarded as a "superorganism." For all species, metabolic heat production peaked early with a subsequent gradual decline over the lifetime.

The emergence of self-organized criticality (threshold dynamics) is reproducible for macroscopic systems under the constraint of a slow driving flux and fixed external force. Roderick Dewar pointed out in 2003 that for such systems MaxEP principle corresponds to maximal flux in accord with Sawada proposal in 1981, Per Bak's ideas on power-law distribution of cluster sizes and time scales (Bak et al. 1987, Bak 1996), and Henrik Jensen's ideas on self-organized criticality (Jensen 1998). In his contributions, Bak remarked that only a scattered handful of oddballs were working on understanding life itself, which is the most interesting of all problems. Remarkably, his study of sandpiles inspired his question: Does biological evolution operate at the self-organized critical state?

Connecting this chapter to introductory thoughts, we can question why artificial boundaries should be erected between thermodynamic and biological evolution? No matter how dissipative evolution is named, it is always connected to some self-ordering and self-organization. The connection is of the feedback type—the emergence of order accelerates the evolution and vice versa. Contrary to a pervasive viewpoint among biologists, the Darwinian selection is not sufficient to understand everything about evolution. When considered only in the light of evolving cells and species, the study of evolution suffers from a living-structure-centric attitude, an analog of also pervasive **anthropocentric** view. This paradigm should be changed to take into account how bioenergetic processes bridge life and the universe, and if a handful of oddballs must devote their life to do it, it is worthwhile, in my opinion.

9.6 Game of life as a side reaction

When everything is accounted in terms of free-energy changes, we have seen that the energy invested in the construction of organic molecules, the constituents of even the most primitive cell, pales compared with free-energy dissipation associated with the same biosynthetic processes. In their insightful papers about the origin of bioenergetics, Nick Lane, William Martin, and Michael Russel (Martin and Russel 2003, Lane et al. 2010, Lane and Martin 2012, Lane 2017) observed that "life is not so much a reaction as a side reaction of the cell's core bioenergetic process." Since side reaction is usually described as the less important reaction driven by the main reaction, is it too much to say that free-energy dissipation or entropy production is the cause rather than the consequence of life? Admittedly, this statement can be regarded as a drastic change in paradigm, but let us consider some of its ramifications. Firstly, universal evolution surely invented and created dissipative structures that increased entropy production before life originated. It may well be the consequence of the Second Law generalization that the most probable relaxation pathway toward the thermodynamic equilibrium is the fastest one. I am aware this is still a controversial statement, just one out of many proposed Second Law generalizations that consider how fast systems dissipate free-energy gradients. Nevertheless, it gives us hope that life exists elsewhere in the universe as one of the myriad means nature found to speed up its thermodynamic evolution. Secondly, this change in paradigm elegantly solves the controversy whether viruses and phages are alive or not. These particles are almost nude nucleic acids with some minimal protein or protein-lipid envelope. Viruses are completely inert outside living cells. They are dead in the same sense as all cellular macromolecules. At the same time, none of these structures can arise nor can be revived in beneficial circumstances without a major investment of free-energy dissipation with subsequent biochemical reactions.

I would like very much that viruses' resurrection is not possible even when they enter host cells. In that case, I would not suffer my whole adult life from the consequences of childhood poliomyelitis. However, the resurrection of seemingly dead structures like viruses and desiccated seeds, and activation of barely alive parasitic organisms, are just some of the many means biological evolution found for regulating entropy production level from very low to a very high level required by vigorous growth. Therefore, at the end of this chapter, we should again approach the question of how life couples its evolution to thermodynamic evolution. When claimed that it is maximal entropy production when all constraints are taken into account, this can mean such rigorous constraints that very low entropy production is maintained. There are obviously some extreme situations when life must die to have any chance of being resurrected. Death and life are much more intimately connected than usually considered. It is entirely possible that life virtuosity excels at its amazing regulatory capabilities for controlling entropy production levels according to its needs. Lost control can result in an irreversible runaway entropy production process until all life-supporting structures disappear. We can say that a real game of life is building safety valves against external changes to maintain the precarious balance against too high and too low entropy production. The pool sizes of key enzymes, ions, reactants, and products are often extremely small in a cell. Flexible and fast regulation is needed to ensure the metabolic homeostasis and continuous fluxes during variable supply and demand so that pool sizes can be kept "half full and half empty" (Knuesting and Scheibe 2018). It is intentional if this statement brings you the mental association with the tradeoff condition often mentioned in this book when discussing the entropy production.

Definitions and explanations from Chapter 9:

The **Anthropocentric viewpoint** interprets everything in the world according to human values. It boils down to a morally questionable attitude to attach the price in whatever currency we use to undisturbed micro-ecology and developed land with human infrastructures after the destruction of the natural habitat is completed at that location. Development and "progress" always win for economists, politicians, mass media, and all of those who value economic growth at the expense of natural ecological services.

Bifurcation term describes the discontinuity of evolution when a small change in the bifurcation parameter causes a sudden topological change in the considered system.

Lactamase enzymes hydrolyze and inactivate antibiotic molecules belonging to the class of β-lactam antibiotics (penicillins, cephalosporins, carbapenems, and others). The enzymes provide the antibiotic resistance capability to bacteria producing them. The competition among bacteria is an ancient affair, and so is the confrontation of molds with bacteria. Therefore, we should not be surprised that bacteria developed enzymes with antibiotic-fighting ability a billion years before our chemists synthesized the first penicillin-type antibiotics. We merely helped the widespread growth of antibiotic-resistant bacterial strains by the indiscriminate antibiotics usage.

Mesh currents are virtual circulation currents introduced by electrical engineers to simplify the electric circuit analysis using Kirchhoff's voltage law. One mesh current is introduced for each mesh (cycle) from a planar electric network containing voltage sources and resistors.

References

Abel, D.L. and Trevors, J.T. Self-organization vs. self-ordering events in life-origin models. Phys. Life Rev. 3(2006): 211–228.

Andresen, B., Zimmermann, E.C. and Ross, J. Objections to a proposal on the rate of entropy production in systems far from equilibrium. J. Chem. Phys. 81(1984): 4676–4677.

Andresen, B. and Gordon, J.M. Constant thermodynamic speed for minimizing entropy production in thermodynamic processes and simulated annealing. Phys. Rev. E 50(1994): 4346–4351.

Andresen, B., Shiner, J.S. and Uehlinger, D.E. Allometric scaling and maximum efficiency in physiological eigen time. Proc. Natl. Acad. Sci. USA 99(2002): 5822–5824.

Aoki, I. Entropy principle for human development, growth and aging. J. Theor. Biol. 150(1991): 215–223.

Bak, P., Tang, C. and Wiesenfeld, K. Self-organized criticality. An explanation of 1/f noise. Phys. Rev. Lett. 59(1987): 381–384.

Bak, P. 1996. How Nature Works: The Science of Self-Organized Criticality. Springer-Verlag, New York, USA.

Bar Dolev, M., Braslavsky, I. and Davies, P.L. Ice-binding proteins and their function. Annu. Rev. Biochem. 85(2016): 515–542.

Bejan, A. and Lorente, S. The constructal law origin of the logistics S curve. J. Appl. Phys. 110(2011): 024901. doi: 10.1063/1.3606555.

Ben-Jacob, E. and Garik, P. The formation of patterns in non-equilibrium growth. Nature 343(1990): 523–530.

Beretta, GP. Steepest entropy ascent model for far-nonequilibrium thermodynamics: Unified implementation of the maximum entropy production principle. Phys. Rev. E 90(2014): 042113. doi: 10.1103/PhysRevE.90.042113.

Beretta, G.P., Al-Abbasi, O. and von Spakovsky, M.R. Steepest-entropy-ascent nonequilibrium quantum thermodynamic framework to model chemical reaction rates at an atomistic level. Phys. Rev. E 95(2017). doi: 10.1103/physreve.95.042139.

Bermudez, J. and Wagensberg, J. On the entropy production in microbiological stationary states. J. Theor. Biol. 122(1986): 347–358.

Bertola, V. and Cafaro, E. A critical analysis of the minimum entropy production theorem and its application to heat and fluid flow. Int. J. Heat Mass Transf. 51(2008): 1907–1912.

Biggs, C.I., Bailey, T.L., Graham, B., Stubbs, C.D., Fayter, A. and Gibson, M.I. Polymer mimics of biomacromolecular antifreezes. Nat. Commun. 8(2017): 1546. doi: 10.1038/s41467-017-01421-7.

Browne, R.K., Silla, A.J., Upton, R., Della-Togna, G., Marcec-Greaves, R., Shishova, N.V. et al. Sperm collection and storage for the sustainable management of amphibian biodiversity. Theriogenology 133(2019): 187e200. doi.org/10.1016/j.theriogenology.2019.03.035.

Christen, T. Application of the maximum entropy production principle to electrical systems. J. Phys. D Appl. Phys. 39(2006): 4497–4503.

Davis, B.K. A theory of evolution that includes prebiotic self-organization and episodic species formation. Bull. Math. Biol. 58(1996): 65–97.

Davis, B.K. The forces driving molecular evolution. Prog. Biophys. Mol. Biol. 69(1998): 83–150.

Dewar, R.C. Information theory explanation of the fluctuation theorem, maximum entropy production and self-organized criticality in non-equilibrium stationary states. J. Phys. A: Math. Gen. 36(2003): 631–641.

Dewar, R.C., Lineweaver, C.H. and Regenauer-Lieb, K. (eds.). 2014. Beyond the Second Law. Entropy Production of Non-equilibrium systems. Springer-Verlag, Berlin, Germany.

Ehrenfest, P. and Ehrenfest, T. 1911. Begriffliche Grundlagen der statistischen Affassung in der Mechanik. pp. 3–90. *In*: Klein, F. and Müller, C. (eds.). Encyclopädie der mathematischen Wissenschaften mit Einschluβ ihrer Anwendungen, Band IV, 2. Teil. Leipzig: Teubner, Germany. Translated as: Ehrenfest, P. and Ehrenfest, T. 1959. The Conceptual Foundations of the Statistical Approach in Mechanics. New York: Cornell University Press, New York, NY, USA.

Endres, R.G. Entropy production selects nonequilibrium states in multistable systems. Sci. Rep. 7(2017): 14437. doi: 10.1038/s41598-017-14485-8.

Eshel, G., Shaked, R., Kazachkova, Y., Khan, A., Eppel A., Cisneros, A. et al. *Anastatica hierochuntica*, an *Arabidopsis* desert relative, is tolerant to multiple abiotic stresses and exhibits species-specific and common stress tolerance strategies with its halophytic relative, *Eutrema* (*Thellungiella*) *salsugineum*. Front. Plant Sci. 7(2017): 1992. doi: 10.3389/fpls.2016.01992.

Feynman, R.P. and Hibbs, A.R. 1965. Quantum Mechanics and Path Integrals. McGraw-Hill, New York, NY, USA.

Feynman, R.P., Leighton, R.B. and Sands, M. (1966, 2006). Feynman lectures on physics, Vol. 2, Lecture 19. Addison-Wesley, San Francisco, CA, USA.

Forrest, W.W. and Walker, D.J. The generation and utilization of energy during growth. Adv. Microb. Physiol. 5(1971): 213–274.

Glansdorff, P. and Prigogine, I. 1971. Thermodynamics Theory of Structure, Stability and Fluctations. Willey-Interscience, New York, NY, USA.

Glavatskiy, K.S. Lagrangian formulation of irreversible thermodynamics and the second law of thermodynamics. J. Chem. Phys. 142(2015): 204106. doi.org/10.1063/1.4921558.

Griffiths, C.A., Gaff, D.F. and Neale, A.D. Drying without senescence in resurrection plants. Front. Plant Sci. 5(2014). doi: 10.3389/fpls.2014.00036.

Gyarmati, I. On the “Governing principle of dissipative processes” and its extension to non-linear problems. Annalen der Physik 23(1969): 353–378.

Hill, T.L. 1977. Free Energy Transduction in Biology. The Steady-State Kinetic and Thermodynamic Formalism. Academic Press, New York, USA.

Hill, A. Entropy production as the selection rule between different growth morphologies. Nature 348(1990): 426–428.

Hunt, K.L.C., Hunt, P.M. and Ross, J. Dissipation in steady states of chemical systems and deviations from minimal entropy production. Physica 147A(1987): 48–60.

Jaynes, E.T. The minimum entropy production principle. Ann. Rev. Phys. Chem. 31(1980): 579–601.

Jeans, J.H. 1923. The Mathematical Theory of Electricity and Magnetism. 4th Ed. Cambridge University Press, Cambridge, UK.

Jennings, R.C., Belgio, E. and Zucchelli, G. Does maximal entropy production play a role in the evolution of biological complexity? A biological point of view. Rend. Lincei. Sci. Fis. Nat. 31(2020): 259–268.

Jensen, H.J. 1998. Self-Organized Criticality. Emergent Complex Behaviour in Physical and Biological Systems. Cambridge University Press, Cambridge, UK.

Juretić, D. and Westerhoff, H.V. Variation of efficiency with free-energy dissipation in models of biological energy transduction. Biophys. Chem. 28(1987): 21–34.

Juretić, D., Hendler, R.W. and Westerhoff, H.V. 1988. Variation of efficiency with free energy dissipation in theoretical models of oxidative posphorylation and cytochrome oxidase. pp. 205–212. *In*: Lemasters, J.J., Hackenbrock, C.R., Thurman, R.G. and Westerhoff, H.V. (eds.). Integration of Mitochondrial Function. Springer, Boston, MA, USA. doi: https://doi.org/10.1007/978-1-4899-2551-0_18.

Juretić, D. Membrane free-energy converters: The benefits of intrinsic uncoupling and nonlinearity. Acta Pharmaceutica 42(1992): 373–376.

Juretić, D. and Županović, P. Photosynthetic models with maximum entropy production in irreversible charge transfer steps. J. Comp. Biol. Chem. 27(2003): 541–553.

Juretić, D. and Županović, P. 2005. The free-energy transduction and entropy production in initial photosynthetic reactions. pp. 161–171. *In*: Kleidon, A. and Lorenz, R.D. (eds.). Non-equilibrium Thermodynamics and the Production of Entropy: Life, Earth and Beyond, Springer, Berlin, Germany.

Juretić, D., Simunić, J. and Bonačić Lošić, Ž. Maximum entropy production theorem for transitions between enzyme functional states and its applications. Entropy 21(2019): 743. doi: 10.3390/e21080743.

Kleidon, A. and Lorentz, R.D. (eds.). 2005. Non-equilibrium Thermodyamics and the Production of Entropy. Life, Earth, and Beyond. Springer-Verlag, Berlin, Germany.

Knuesting, J. and Scheibe, R. Small molecules govern thiol redox switches. Trends Plant. Sci. 23(2018): 769–782.

Kohler, M. Behandlung von Nichtgleichgewichtsvorgängen mit Hilfe eines Extremalprinzips. Z. Phys. 124(1948): 772–789.

Kong, C.S. A general maximum power transfer theorem. IEEE Trans. Educ. 38(1995): 296–198.
Krisko, A. and Radman, M. Biology of extreme radiation resistance: the way of *Deinococcus radiodurans*. Cold Spring Harb. Perspect. Biol. 5(2013): a012765. doi: 10.1101/cshperspect.a012765.
Lane, N., Allen, J.F. and Martin, W. How did LUCA make a living? Chemiosmosis in the origin of life. BioEssays 32(2010): 271–280.
Lane, N. and Martin, W.F. The origin of membrane bioenergetics. Cell 151(2012): 1406–1416.
Lane, N. Proton gradients at the origin of life. Bioessays 39(2017). doi: 10.1002/bies.201600217.
Lovelock, J.E. and Margulis, L. Atmospheric homeostasis by and for the biosphere—the Gaia hypothesis. Tellus 26(1974): 2–10.
Lovelock, J. 1988. The Ages of Gaia. Norton, New York, NY, USA.
Lotka, A.J. Contribution to the energetics of evolution. Proc. Natl. Acad. Sci. USA 8(1922a): 147–151.
Lotka, A.J. Natural selection as a physical principle. Proc. Natl. Acad. Sci. USA 8(1922b): 151–154.
Martin, W. and Russell, M.J. On the origins of cells: a hypothesis for the evolutionary transitions from abiotic geochemistry to chemoautotrophic prokaryotes, and from prokaryotes to nucleated cells. Phil. Trans. R. Soc. Lond. B 358(2003): 59–85.
Martiouchev, L.M. and Seleznev, V.D. Maximum-entropy production principle as a criterion for the morphological-phase selection in the crystallization process. Doklady Physics 45(2000): 129–131. Translated from Doklady Akademii Nauk 371(2000): 466–468.
Martiouchev, L.M. and Axelrod, E.G. From dendrites and S-shaped growth curves to the maximum entropy production principle. JETP Letters 78(2003): 476–479. From: Pisma v Zhurnal Eksperimentalnoi i Teoreticheskoi Fiziki 78(2003): 948–951.
Martyushev, L.M. and Seleznev, V.D. Maximum entropy production principle in physics, chemistry and biology. Phys. Rep. 426(2006): 1–45.
Martyushev, L.M. and Chervontseva, E.A. On the problem of the metastable region at morphological instability. Phys. Lett. A 373(2009): 4206–4213.
Martyushev, L.M. Entropy and entropy production: Old misconceptions and new breakthroughs. Entropy 15(2013): 1152–1170.
Martyushev, L. and Seleznev, V. Maximum entropy production: application to crystal growth and chemical kinetics. Curr. Opin. Chem. Eng. 7(2015): 23–31.
Maxwell, J.C. 1954. A Treatise on Electricity and Magnetism, Vol. 2. Dover, New York, NY, USA.
Metzner, H. Bioelectrochemistry of photosynthesis: a theoretical approach. Bioelectrochem. Bioenerg. 13(1984): 183–190.
Mitchell, D.E. and Gibson, M.I. Latent ice recrystallization inhibition activity in nonantifreeze proteins: Ca^{2+}-activated plant lectins and cation-activated antimicrobial peptides. Biomacromolecules 16(2015): 3411–3416.
Montefusco, A., Consonni, F. and Beretta, G.P. Essential equivalence of the general equation for the nonequilibrium reversible-irreversible coupling (GENERIC) and steepest-entropy-ascent models of dissipation for nonequilibrium thermodynamics. Phys. Rev. E 91(2015): 042138. doi: 10.1103/PhysRevE.91.042138.
Moroz, A. On a variational formulation of the maximum energy dissipation principle for non-equilibrium chemical thermodynamics. Chem. Phys. Lett. 457(2008): 448–452.
Møbjerg, N., Halberg, K.A., Jørgensen, A., Persson, D., Bjørn, M., Ramløv, H. et al. Survival in extreme environments—on the current knowledge of adaptations in tardigrades. Acta Physiol. 202(2011): 409–420.
Nath, S. Coupling in ATP synthesis: Test of thermodynamic consistency and formulation in terms of the principle of least action. Chem. Phys. Lett. 723(2019): 118–122.
Odum, H.T. and Odum, E.C. 1976. Energy Basis for Man and Nature. McGraw-Hill Book Company, New York, NY, USA.
Pérez, J.-P. Thermodynamical interpretation of the variational Maxwell theorem in dc circuits. Am. J. Phys. 68(2000): 860–863.
Phillips, R. In retrospect: The Feynman lectures on physics. Nature 504(2013): 30–31.
Pietrobon, D. and Caplan, S.R. Flow-force relationships for a six-state proton pump model: intrinsic uncoupling, kinetic equivalence of input and output forces, and domain of approximate linearity. Biochemistry 24(1985): 5764–5776.
Polettini, M. Macroscopic constraints for the minimum entropy production principle. Phys. Rev. E 84(2011): 051117. doi: 10.1103/PhysRevE.84.051117.
Price, P.B. A habitat for psychrophiles in deep Antarctic ice. Proc. Natl. Acad. Sci. USA 97(2000): 1247–1251.
Pross, A. and Pascal, R. How and why kinetics, thermodynamics, and chemistry induce the logic of biological evolution. Beilstein J. Org. Chem. 13(2017): 665–674.
Pujol, T. The Consequence of maximum thermodynamic efficiency in Daisyworld. J. Theor. Biol. 217(2002): 53–60.
Rayleigh, Lord. 1896. The Theory of Sound, Vol. I., 2nd ed. MacMillan and Co. Ltd., London, GB.

Sawada, Y. A thermodynamic variational principle in nonlinear non-equilibrium phenomena. Prog. Theor. Phys. 66(1981): 68–76.

Sawada, Y. A thermodynamic variational principle in nonlinear systems far from equilibrium. J. Stat. Phys. 34(1984): 1039–1045.

Sciubba, E. What did Lotka really say? A critical reassessment of the "maximum power principle". Ecol. Modell. 222(2011): 1347–1353.

Seifert, U. Stochastic thermodynamics, fluctuation theorems and molecular machines. Rep. Prog. Phys. 75(2012): 126001. doi: 10.1088/0034-4885/75/12/126001.

Shimizu, H. and Sawada, Y. Relative stability among metastable steady state structures in chemical reaction systems. J. Chem. Phys. 79(1983): 3828–3835.

Smith, E. and Morowitz, H.J. 2016. The Origin and Nature of Life on Earth. The Emergence of the Fourth Geosphere. Cambridge Univ. Press, Cambridge, UK.

Steenbeck, M. Energetik der Gasentladungen. Z. Phys. 33(1932): 809–815.

Stucki, J., Compiani, M. and Caplan, S.R. Efficiency of energy conversion in model biological pumps. Optimization by linear nonequlibrium thermodynamic relations. Biophys. Chem. 18(1983): 101–109.

Tilley, D.R. Howard T. Odum's contribution to the laws of energy. Ecol. Modell. 178(2004): 121–125.

Ulanowicz, R.E. and Hannon, B.M. Life and the production of entropy. Proc. R. Soc. Lond. 232(1987): 181–192.

Van Baak, D.A. Variational alternatives to Kirchhoff's loop theorem in dc circuits. Am. J. Phys. 67(1999): 36–44.

Van Rotterdam, B. 1998. Control of Light-Induced Electron Transfer in Bacterial Photosynthesis, PhD thesis, University of Amsterdam. Amsterdam, The Netherlands.

Verhás, J. Gyarmati's variational principle of dissipative processes. Entropy 16(2014): 2362–2383.

Verlinde, E.P. On the origin of gravity and the laws of Newton. J. High Energy Phys. (2011): 029, doi: 10.1007/JHEP04(2011)029.

Wang, Q.A. Maximum entropy change and least action principle for nonequilibrium systems. Astrophys. Space Sci. 305(2006): 273–281.

Weber, N., Galindo, V., Stefani, F. and Weier, T. The Tayler instability at low magnetic Prandtl numbers: between chiral symmetry breaking and helicity oscillations. New J. Phys. 17(2015): 113013. doi: 10.1088/1367-2630/17/11/113013.

Westerhoff, H.V. and van Dam, K. 1987. Thermodynamics and Control of Biological Free Energy Transduction. Elsevier, Amsterdam, The Netherlands.

Ziegler, H. 1963. Some extremum principles in irreversible thermodynamics with application to continuum mechanics. Chapter 2. *In*: Sneddon, I.N. and Hill, R. (eds.). Progress in Solid Mechanics, Volume 4. North-Holland, Amsterdam, The Netherlands,

Ziegler, H. and Wehrli, C. On a principle of maximal rate of entropy production. J. Non-Eq. Thermodyn. 12(1987): 229–243.

Ziman, J.M. The general variational principle of transport theory. Can. J. Phys. 34(1956): 1256–1273.

Županović, P., Juretić, D. and Botrić, S. Kirchhoff's loop law and the maximum entropy production principle. Phys. Rev. E 70(2004): 056108. doi: 10.1103/PhysRevE.70.056108.

Županović, P. and Juretić, D. The chemical cycle kinetics close to the equilibrium state and electrical circuit analogy. Croat. Chem. Acta 77(2004): 561–571.

Županović, P., Juretić, D. and Botrić, S. On the equivalence between Onsager's principle of the least dissipation of energy and maximum entropy production principle. Conduction of heat in an anisotropic crystal. FIZIKA A 14(2005): 89–96.

CHAPTER 10

Perfect Enzymes, According to Biochemists

10.1 Enhancing the reaction rates

Acceleration of catalysis rates may well have been the first sign of chemical evolution preceding the first cells' formation. The comparison of some chemical reactions leads to an enormous enhancement factor of up to 10^{19} for the reaction rate when the enzyme is present (Wolfenden and Snider 2001). The mysteries of enzyme catalysis are far from being completely solved. It is unknown how natural evolution increased enzyme performance parameters (see the section *Catalytic efficiency increase...* from the Introduction chapter for the definition of catalytic efficiency and other quantitative tools for measuring enzyme's performance). Enzymologists and biologists have become experts during the past several decades on the reverse engineering goal of diminishing or destroying the catalytic efficiency of enzymes interesting to them. In combination with advanced crystallographic methods for the study of the 3D protein structures, this gradual development illuminated not only which amino acid residues are the most important for catalytic function, but also why their space orientation and mutual contacts contribute to the enhancement of reaction rates. It turned out that the substrate, which enters into a reaction, is often also the trigger for subtle changes within **the enzyme active site**. Altogether, these fast changes bring about the transformation of the substrate into product or products. It all happens in milliseconds or even shorter periods, while left to itself, a substrate would take millions of years to transform into a product spontaneously. Also, when enzymes are involved, the reaction conditions are quite mild, often at room temperature, and all reactants and enzymes are usually so harmless that we can eat or drink them without any ill effect. That cannot be claimed for many organic chemistry reactants used by our industry.

How did blind biological evolution invent these miracles of enzyme catalysis? This mystery is even more challenging to solve. The present-day methods from the new science of bioinformatics allow us to go deeper in the past when multicellular life did not yet exist, nor cells like our cells (eukaryotic cells). It means reconstructing past events, which happened billions of years before the inferred age (about 560 million years) of the oldest found fossils of the first animals (Bobrovskiy et al. 2018, Evans et al. 2020). Bioinformaticians and evolution biologists discovered that basically, the same housekeeping enzymes existed already two and three billion years ago. They may have been present in some simpler primitive form close to four billion years ago. It appears that a combination of chemical and biological evolution solved the secret of how one can enhance the reaction rate 10^{19}-fold already so early in Earth's history. This remarkable rate enhancement happened concomitant with the emergence of first cells during a relatively short geological time of several hundred million years near the end of the Hadean period about 3.9 billion years ago. At that time, humans would not be able to survive the harsh surface conditions of early Earth for longer than several minutes.

Enhancement of reaction rate or catalytic efficiency is much more challenging to achieve by biochemists than the inhibition of reaction rates, despite all insights gained about the main obstacles impeding the creation of more efficient synthetic enzymes. It is, therefore, a surprising and potentially useful prediction of our calculations that more efficient enzymes are tightly associated with higher overall dissipation (Bonačić Lošić et al. 2017, Juretić et al. 2019a,b). Those superior enzymes, for which biochemists used the colorful name "perfect enzymes" (Newton et al. 2015, Sharma and Guptasarma 2015), are predicted by us to achieve even higher catalytic efficiency when they manage to enter into an optimal catalytic mode characterized with maximal entropy production for crucial catalytic steps. For instance, we described how "perfect enzymes" can be improved by maximum entropy production requirement for the rate-determining catalytic step or steps (Bonačić Lošić et al. 2017, Juretić et al. 2019a). Proving that it is possible in practice, Sharma and Guptasarma achieved in their experiments (2015) a substantial increase in the catalytic efficiency of already "perfect" triosephosphate isomerase and used even more colorful name, the "super-perfect" enzymes. They did not consider entropy production.

10.2 Coupling enzyme kinetics to dissipation

Dissipation and entropy production terms are seldom used when describing enzyme kinetics. It is a reasonable attitude considering that both terms describe an increase in entropy or conversion of free energy into useless energy. Minimal dissipation and minimal entropy production are often assumed in bioenergetics and enzyme kinetics and even elevated as lofty principles enslaving all of the known life. Coming from Nobel Prize winners in physics and chemistry, these principles appealed greatly to biochemists and molecular biologists in their attempts to find a firm foundation for their expectations that life does not tolerate any waste. Physicists, interested in biology, were well aware that increased efficiency can be realized by decreasing free-energy dissipation. Thus, they also liked the idea that some general principles about minimal entropy production exist and govern biological macromolecules and living structures' evolution. We mentioned before (See Chapters 5–9) the critique of Ross and others (Hunt et al. 1987, Ross and Vlad 2005), stating that Prigogine's principle of minimal entropy production is based on a mathematical error. However, there are many different formulations for the alternative principle of maximum entropy production, all more or less controversial, and none accepted as the mainstay in any natural science.

I shall firstly propose in this book to look instead for common optimal solutions nature found during the evolution of inanimate and living matter. Secondly, I shall examine the microscopic details of how evolution is made apparent from the physicists' viewpoint. The emergence of nanocurrents can be taken as the revealing sign of evolution in action accompanied by increased entropy production. The particles involved in nanocurrents can be electrons, protons, atoms, molecules, dust particles, or anything else, a tell-tale sign that order appeared together with a disorder in a small part of the universe, which we examine. We used here the word disorder as a rather crude description for entropy increase. We assume that microscopic currents, like macroscopic currents, are irreversible events that cannot be dissociated from entropy production in the sense of entropy production definition we used earlier (see Introduction). Nanocurrents do not appear in the absence of driving forces, nor alone as a single nanocurent strictly limited to a particular nanospace. The sum of all nanocurrents forms separate or interacting macroscopic currents. All are caused by certain gradients or imbalances in free energy distribution. The assumption is that nature has a tendency to dissipate energy gradients and does not care if it can be best done by establishing local areas of increased order if the overall result is entropy increase. The remaining question then is if nature will rather choose a faster or slower way to perform this feat, having in mind that faster means greater and slower means smaller entropy production.

In previous chapters, we have already mentioned several examples of optimizing simplified bioenergetic systems, protein nanomotors, or enzyme kinetic cycles, to examine if observed kinetic and performance parameters can be approximately reproduced. We used the MTEP theorem I formulated and proved with my former student Željana Bonačić Lošić in the first decade of this century (Bonačić Lošić et al. 2017; see Chapter 8, section 8.3). This theorem states that maximal entropy production can be found for each transition between enzyme functional states. The theorem validity can be extended to bioenergetic systems consisting of many enzymes. Encouragingly, during our modeling experiments, we found a similar hierarchy as measured for widely different relaxation constants during the first irreversible steps in bacterial photosynthesis (Juretić and Županović 2003). We also reproduced from the MTEP requirement the measured value of approximately 20% for optimal light-conversion efficiency. We did not consider the following steps: how the proton electrochemical gradient is converted into ATP synthesis and how ATP/ADP gradient is converted into work for all cellular processes. These later photosynthetic stages decreased the maximal efficiency of photosynthesis to less than 10%. The ATP-synthase rotatory nanomotor performs the first next step of using the protonmotive force for ATP synthesis. It is very efficient (Turina et al. 2016). Its essential kinetic features can be reconstructed (see next chapter) from the maximal entropy production requirement for the crucial conversion of proton nanocurrent into elastic torsional strain and ADP and inorganic phosphate fusion (Dewar et al. 2006).

A caveat should be mentioned here concerning the restrictions used in all simulations and optimizations we performed. The essential restriction is the steady-state approximation in terms of small systems thermodynamics, as formulated in the Terrel L. Hill book from 1977 (see Chapter 3). After more than 40 years, his method of understanding and describing bioenergetic transformation problems can still be used to advantage when building further upon it or expanding accessible bridges among physics, bioenergetics, enzyme kinetics, biological evolution, and other life sciences. The steady-state approximation implies the constancy of all gradients, all driving forces, and all substrate and product nonequilibrium concentrations. It is a realistic approximation for cells, organs, and bodies' general tendency to maintain the homeostasis. For example, the multiple regulatory feedback mechanisms ensure the stable environment of healthy organs and cells in the human body. Other organisms experience transitions between dormant and active periods, both with nearly constant but different parameters. Single bacteria, archaea, or eukaryotic cells in the planktonic state (when floating in water solution) have more difficulty ensuring near constancy of their cytoplasmic environment and employ different mechanisms to cope with external changes. In the flow reactors for growing bacteria or yeast cells, we can ensure the constancy of external constraints by a constant supply of growth substrates and continuous removal of products. Biochemists routinely use the pseudo-steady-state assumption built in the Michaelis-Menten kinetics for more than one century, and this kinetic description has been adopted in many other research fields as well. The additional restriction in all of our simulations is the constancy of all equilibrium constants.

10.3 The MTEP theorem tool for the identification of rate-limiting transitions

Of lately, we changed our goal in simulations from reproducing observed kinetic constants to finding rate-limiting transitions between enzyme functional states (Juretić et al. 2019b). For rate-limiting transitions, the maximal entropy production requirement (the MTEP theorem) leads to predicting higher values for optimal kinetic constants, a higher overall entropy production, and higher catalytic efficiency at the same time (when compared with measured or inferred values). This goal required dissecting all possible intermediate or slippage steps in the catalytic cycle concerning their contribution to overall entropy production. In modeling studies, we used our maximum entropy production theorem for enzyme functional transitions, either separately for each transition or iteratively until the joint maximum has been reached for two or more subsequent transitions.

Triosephosphate isomerase (TIM) is an essential enzyme for efficient free-energy transduction in glycolysis. It performs substrate (dihydroxyacetone phosphate) to product (D-glyceraldehyde-3-

phosphate) conversion 10^9 times faster when TIM is present in solution. The simulation study for the triosephosphate isomerase is presented in Figure 10.1. It assumes the four-state reversible kinetic scheme with functional states: enzyme E(1), the enzyme-substrate complex ES(2), an intermediate complex EZ(3), and the enzyme-product complex EP(4) (Bonačić Lošić et al. 2017). All eight kinetic constants have been estimated for known substrate and product concentrations (Knowles and Albery 1977) and subsequently used for different optimization approaches (Petterson 1992, Marín-Sanguino and Torres 2002). Our approach left one of four forward kinetic constant free to change without changing the equilibrium constants. We looked for its optimal value when maximal entropy production is reached in a chosen transition. Thus, it was the MTEP theorem's application separately to all of four transitions among enzyme functional states. It revealed that product release (step four from the four-state catalytic cycle) is the rate-limiting step. Only the maximal entropy production requirement for the last catalytic step increased the entropy production for that and all other steps. In addition to the roughly 30% increase of the total entropy production and flux, the same increase occurred for the catalytic efficiency (top black parts of each histogram column in Figure 10.1). The turnover rate k_{cat} increased 37%.

We performed all of our simulations using a thermodynamic optimization requirement that can be applied to any enzyme with known functionally important states. Thus, our method is conceptually more straightforward than other optimization schemes used to increase the enzyme catalytic efficiency. The favorite function for maximization is usually the enzyme activity expressed as the rate of product synthesis. This function keeps increasing without limit if some reasonable constraints are not imposed in the form of inequalities for rate constants (Heinrich et al. 1991, Marin-Sanguino and Torres 2002). Among a multitude of proposed optimization principles for biochemical reaction systems, the majority assume various supposed evolutionary goals formulated as "objective" functions (Heinrich et al. 1991, de Leon et al. 2008). None of the flux, thermodynamic

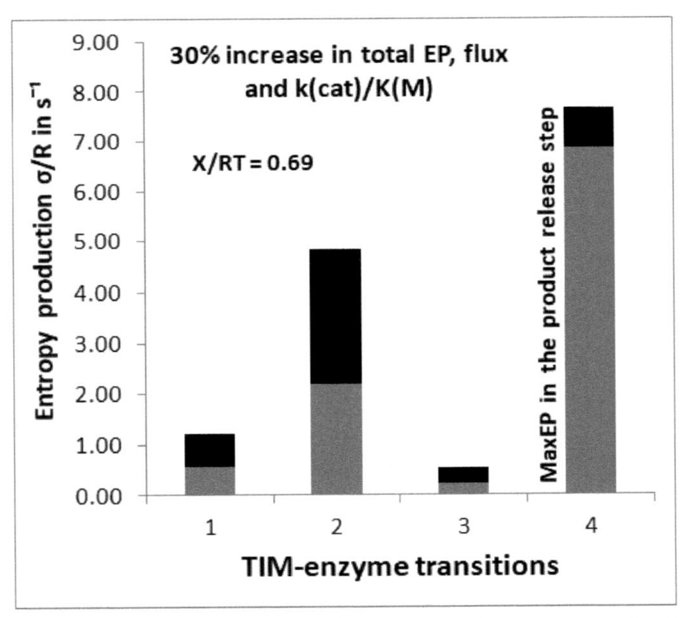

Figure 10.1: The maximal entropy production requirement for the product release step four in the triosephosphate catalytic cycle produces a 30% increase in the total entropy production, the catalytic rate (flux), and the catalytic efficiency for that "perfect enzyme." The entropy production increase is shown (in black) for each of four transitions (12, 23, 34, 41 labeled, respectively, 1, 2, 3, and 4) between enzyme functional states E(1), ES(2), EZ(3), and EP(4). The 30% increase in total entropy production is obtained after summing all four partial EP increases produced by the MTEP optimization for the last product release step. Flux J and catalytic efficiency k_{cat}/K_M values exhibited a similar 30% increase regarding the measured values (in gray) after the same optimization for the product release kinetics. For details, see Bonačić Lošić et al. (2017).

efficiency, output power, growth rate, yield maximization, or other maximizations of hypothetical goals for metabolic networks or biological growth has a firm foundation in the mathematical proof and physical principle independent of what looks desirable to us from a biological or evolutionary perspective. That is not a secure basis for finding the optimization principle or principles leading to the rationalization of structural design, functioning, and possible improvements of present-day biochemical systems. The MTEP theorem application removes the need to impose ill-defined upper limits for rate constants and streamline the optimization procedure. Entropy production is well defined and easily calculated physical function. The optimal values of all biological parameters of interest can be obtained by leaning on the proof that maximal EP must exist for all transitions between functionally important states. Our optimization approach shows that the reaction rate for the triosephosphate isomerase "perfect enzyme" can be considerably improved. Besides, our method demonstrates that entropy production maximization in the product release step leads to performance improvements.

What is so special about the product release step? The catalysis rate in biochemistry is often limited by the product release step (Albery and Knowles 1976). In the case of triosephosphate isomerase, the loop-6 opening leads to the product (D-glyceraldehyde-3-phosphate) release (Malabalan et al. 2010, Wierenga et al. 2010). This conformational change consists of directed fast movements of many atoms over long distances in every single enzyme's microworld. It should not be surprising that entropy is produced by such flow of atoms in the activated enzymes. There is no current or even nanocurrent disconnected from entropy production at room temperature.

What is so special about the triosephosphate isomerase enzyme (TIM enzyme) to deserve the attention of readers interested in bioenergetics? Bioinformatics helped to put the TIM enzyme firmly among the most ancient of housekeeping enzymes. The glycolysis derived ATP synthesis does not happen without the TIM enzyme. It is just as essential (if not more essential) for cancer cells, as for our healthy cells, for yeast cells as for lowly bacteria or archaea cells. Biochemists are well aware of the prominent TIM role in gluconeogenesis, pentose phosphate pathway, and lipid metabolism. A single amino acid substitution in TIM's polypeptide chain leads to death at an early age (Daar et al. 1986).

10.4 Entropy production and evolutionary distances

Enzyme evolution is our new topic. **Phylogenetic** tree analysis tools are already so advanced in bioinformatics that evolutionary distances can be found from inferred common ancestor enzyme. Our goal is to compare evolutionary distances to a common ancestor with changes in catalytic efficiency and overall entropy production. It turned out in the case of beta-lactamases that enzyme performance parameters increased together with total entropy production and together with the greater evolutionary distance from a common ancestor enzyme (Juretić et al. 2019a). This observation is in line with our expectations that biological evolution goes hand in hand with thermodynamic evolution. Each improvement of enzyme performance parameters is handsomely paid with greatly increased entropy production. We gained the impression that accelerated thermodynamic evolution through increased entropy production is either the cause or consequence of performance improvements achieved during biological evolution. Furthermore, it is easy to predict that additional gains in enzyme performance parameters can occur after using the maximal entropy production requirement for identified rate-limiting substrate to product transformations.

We shall now present recently published details about simulation results with beta-lactamase enzymes. We used measurements performed by Christensen and co-authors (1990) to compare *S. aureus* PC1, *E. coli* RTEM, and *B. cereus* β-lactamase 1 by using the same reversible Michaelis-Menten scheme (Figure 10.2) for the catalytic cycle of all β-lactamases. We estimated reverse kinetic constants, which were not determined in their experiments, from kinetic modeling, and also made sure that all concentrations were approximately constant at the time (0.12 s after substrate addition) when measurements were performed.

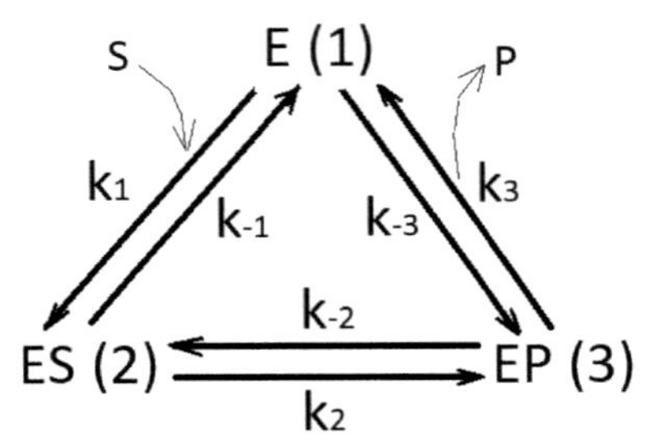

Figure 10.2: Reversible Michaelis-Menten cyclic kinetic scheme with the forward and backward (superscript "–") kinetic constants. Enzyme E (state 1), substrate S, enzyme-substrate complex ES (state 2), product P, and enzyme product complex EP (state 3) are assumed to maintain nonequilibrium steady-state concentrations by continuous removal of product and inflow of substrate. Modified Figure 1a from Juretić et al. (2019b).

Figure 10.3 illustrates an almost linear increase of catalytic constant and overall entropy production with increased evolutionary distance (gray and black squares, respectively). An additional increase of these parameters (circles) occurs after the optimization of substrate acylation with maximal entropy production requirement for that proton-shuttle step. A similar increase in performance parameters with increased evolutionary distance occurred after the MTEP optimization of the second proton transfer step leading to the product release (Juretić et al. 2019a,b). Thus, the optimization of proton transfer steps during the evolution of lactamases achieved the joint increase in catalytic efficiency and overall entropy production. The observation can be used to obtain an additional enzyme efficiency increase beyond and above the best nature's accomplishments. At present, we are not aware of mutations leading to the increase in catalytic constants when all of the important catalytic constants are measured. It is likely, however, that such mutations would represent an additional increase in evolutionary distance. Due to the lack of such data, we could not verify this proposition. We assumed that evolutionary distances did not change after optimizations,

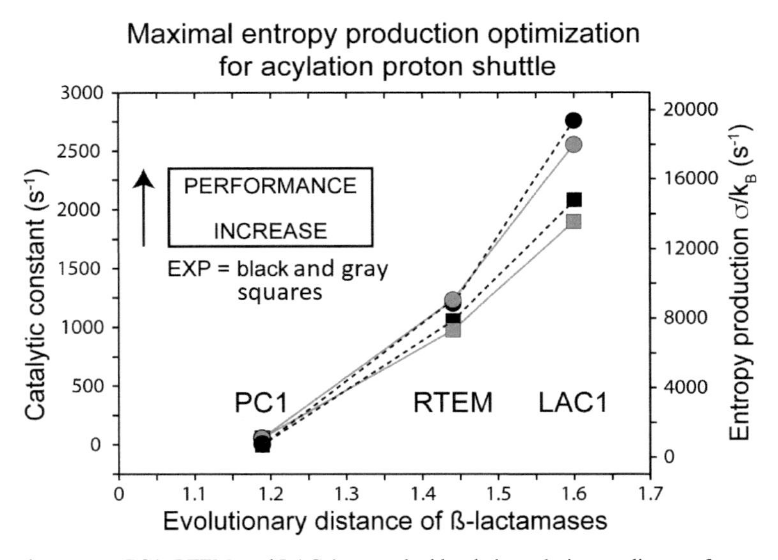

Figure 10.3: Beta-lactamases PC1, RTEM, and LAC 1 are ranked by their evolutionary distance from a common ancestor (x-axis) using the phylogenetic tree of β-lactamase sequences. The catalytic constant or turnover number (left y-axis, gray squares) and overall entropy production (right y-axis, black squares) are calculated from known and estimated kinetic constants. The catalytic constant (gray circles) increased for more evolved lactamases together with overall entropy production increase (black circles) when maximal entropy production requirement is imposed for the ES ↔ EP proton transfer acylation step (the second catalytic step). See Juretić et al. (2019a) for the calculation details. All points have been re-calculated for this book after correcting the original calculations for evolutionary distances.

the assumption which may have contributed to the slight deformation of linear relationship for calculated points.

The 3D plot is more appropriate to represent calculation results after separate optimizations in each of the three transitions: substrate binding E+S ↔ ES, acylation ES ↔ EP, and deacylation step with product release EP ↔ E+P. The acylation and deacylation transitions involve the transport of several protons. The catalysis performed by β-lactamases would not be possible without creating the proton nanocurrent. Vectorial proton current comprises crucial amino acids from the enzyme active site. Proton transfers also involve the substrate as the trigger for opening proton current flow, transition state changes, and even a water molecule close to the active site. As seen from the 3D plot (Figure 10.4), only the optimization of proton transfer steps increases the catalytic constant and entropy production for all three lactamases (blue and green lines, symbols, and arrows). Maximal entropy production for the substrate-binding step leads to a decrease in overall entropy production.

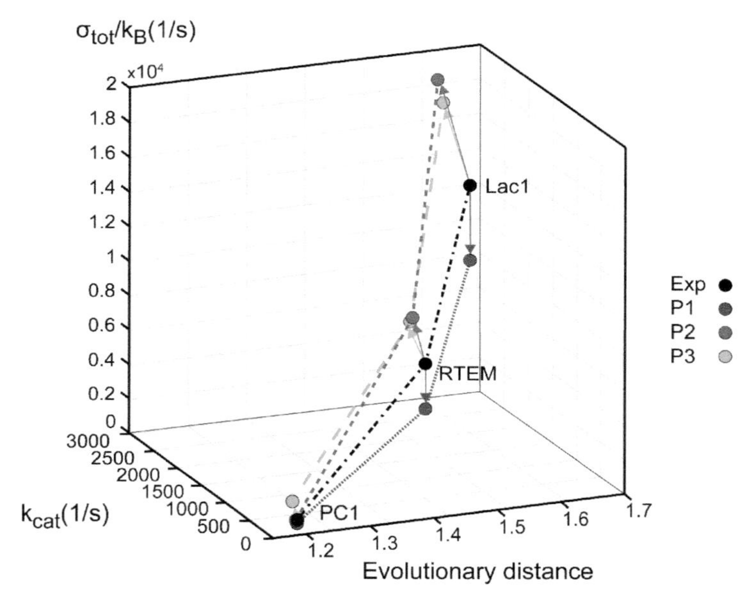

Figure 10.4: For three β-lactamases, PC1, RTEM, and LAC 1, increased evolutionary distance from a common ancestor (at y-axis) is associated with increased enzyme activity (the turnover number k_{cat} at x-axis) and increased overall entropy production (at z-axis). The coordinates for these three enzymes represent values calculated from experiments and kinetic modeling (black), optimized from MTEP requirement in the E+S ↔ ES transition (red, P1 symbols), in the ES ↔ EP transition (blue, P2 symbols), and the EP ↔ E+P transition (green, P3 symbols). Only the optimization of proton nanocurrents for maximal dissipation (ES ↔ EP and EP ↔ E+P transitions involving proton transfer) results in better performance parameters from those measured in experiments. After Juretić et al. (2019b). Figure 10.4 points have been re-calculated in Juretić et al. (2019b) after correcting the original calculations for evolutionary distances from Juretić et al. (2019a).

10.5 Iterative entropy production maximizations for both proton transfer steps

What about simultaneous maximizations for maximum entropy production in both proton transfer steps? It would be the P2P3 optimization (see symbols from Figure 10.4). Other combined optimizations, the P1P2, and P1P3 did not involve all proton transfers. After P1P2 and P1P3 optimizations, the performance parameters did not increase compared to values calculated from experimentally measured and estimated kinetic constants. We performed these calculations iteratively due to the complexity of solving analytically joint maximizations. The result (Figure 10.5) illustrates that significant additional increase over experimentally measured values

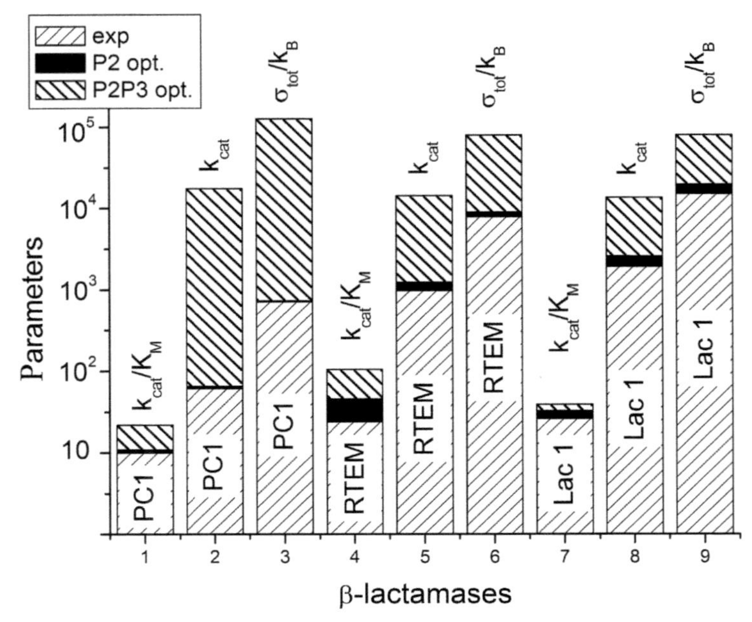

Figure 10.5: Increase in performance parameters: catalytic efficiency k_{cat}/K_M (s·µM)$^{-1}$ and catalytic constant k_{cat} (s^{-1}) when first proton transfer is optimized (the P2 optimization) and when both proton transfer steps are simultaneously optimized (the P2P3 optimization). Notice that the y-axis is logarithmic, meaning that all increases (left-leaning stripes at the top of each column) are for one to two orders of magnitude larger compared to experimental values (right-leaning bottom stripes in each column). At this scale, the performance increases illustrated in Figure 10.3 (the P2 optimization) reduce to a black horizontal line. Increases in overall entropy production σ_{tot}/k_B (s^{-1}) mirror the increases in turnover number k_{cat} and the increase in evolutionary distance (only roughly represented here as higher x-axis values). After Juretić et al. (2019b).

can be achieved after the P2P3 optimization for a) specificity constant k_{cat}/K_M, b) catalytic constant k_{cat}, and c) entropy production σ_{tot}/k_B (K_M is Michaelis-constant and k_B is the Boltzmann's constant).

At the logarithmic scale that we are using for the y-axis of Figure 10.5, the catalytic constant (turnover number) for all three β-lactamases achieves almost the same optimal value, which is likely to be the maximal possible value that can be produced by natural or directed evolution. Actual calculated values range from 13400 (s^{-1}) for β-lactamase 1 from *B. cereus* to 17500 (s^{-1}) for PC1 β-lactamase from *S. aureus*. The results indicate a vast but limited and similar latent evolution potential for all three β-lactamases. β-lactamases have been long considered as perfect enzymes in colorful terminology used by Knowles and Albery (1977), that is, enzymes that have reached the end of their natural evolutionary development. Indeed, wild-type β-lactamases have a high turnover number with an excellent substrate, such as benzylpenicillin. Their k_{cat} was determined to be 61, 975, and 1905 s^{-1}, respectively, for the least evolved PC1, RTEM, and the most evolved LAC1, when calculated using measurements obtained under the same steady-state conditions (Juretić et al. 2019a). Our simulations' take-home message is the theoretical possibility that human ingenuity can improve even the "perfect enzymes."

10.6 Life had enough time to explore how order can develop by increasing disorder

The performance parameters introduced above indicate how increased order, efficiency, and complexity developed during evolution, but we have to be careful how these terms are defined. A few words about terminology are in order, even if we repeat the introductory chapter's explanations.

Biochemists are using the word catalytic efficiency for specificity constant k_{cat}/K_M, but sometimes also as a common name for both performance parameters: catalytic constant k_{cat} (also named the turnover number), and specificity constant k_{cat}/K_M. From the presented results (Figures 10.3. 10.4, and 10.5), it is clear that the increase in overall entropy production is associated with the increased evolutionary distance and performance improvements in k_{cat} and k_{cat}/K_M. With just one enzyme example for the concomitant increase of efficiency and entropy production during evolution, it is not possible to extract any general conclusion. Still, we can formulate the hypothesis that this is so for some other enzymes too, because it can stimulate research if coupled efficiency and entropy production increase has not been previously recognized but is nevertheless common to enzymes' evolution.

The β-lactamase is a specific example in which organismal survival depends on the hydrolytic activity of a single enzyme (Tomatis et al. 2008). However, other mechanisms and enzymes, crucial to cell or organism survival, also exist, leaving an open field for testing the proposed hypothesis about the joint increase of dissipation and catalytic efficiency. Physicists may not like this hypothesis at all, because it goes against their prejudice that efficiency is the very opposite of converting free energy into useless or dissipated energy, which is manifested as entropy production. The confusion of using the same terminology, that is, efficiency, with different meanings in different fields, complicates the matter. In physics, efficiency is the ratio of energy output to energy input or the ratio of work/power output to work/power input. Thus, one must be very careful to define efficiency in mathematical terms before attempting to interpret its meaning. For instance, in the static head state, the energy storage efficiency (the ratio of energy output to energy input) is maximal, but the free-energy transduction efficiency (the ratio of power output to power input) vanishes. The entropy production is then minimal, as stated by Prigogine's theorem for linear flux-force relationships.

As we repeatedly mentioned before, this is not the situation which can be tolerated by living organisms. Still, it is the situation familiar to physicists when they associate maximal efficiency (of energy storage) with minimal energy dissipation. Physicists are then suspicious toward claims that increased order can arise from increased disorder. The reverse of this statement (disorder arising from order) is raised at the level of most general universal law, the Second Law of Thermodynamics, which surely must hold for living organisms as well. Natural evolution toward greater complexity, or greater order, or increased enzyme activity and catalytic efficiency, is then somewhat of a mystery. However, we should be humble enough to realize that life had at least 3.5 billion years more than we had to explore how order can develop from disorder without breaking the Second Law of Thermodynamics. We should also be humble enough to realize that through all of 3.5 billion years of biological evolution, life has been operating within far from equilibrium conditions, mostly unexplored by physicists due to their complexity and difficulty of extracting any general conclusions.

Physicists are not even sure how to define thermodynamic functions in the case of far-from-equilibrium states, much less the relationships among them. We should stick with giants in the physics of evolution, who had insights of such sweeping generality that it can be applied to the physics of living beings as well. These are Josiah Willard Gibbs, Edwin T. Jaynes, and Terrell Leslie Hill in that order. Molecular details of how these tasty protein nanomachines, which we named enzymes, work are incomparably better known today than at the time of their most creative research period. What we learned during the past several decades is truly amazing, but still far the realm of complete understanding, predicting, and directing the evolution of enzymes and bioenergetic systems. It would be useful to predict evolvability or ultimate limits to evolution for certain enzyme classes, as we did for β-lactamases. Quite another question is: can such limits be reached at all and how?

Nobody knows how to predict which amino acid substitution would increase enzyme activity. We may have to wait for the creation of artificial intelligence minds better than our minds to achieve substantial progress in this research field. Francis Arnold, the Nobel prize winner for 2018, has shown to us just how difficult that task is. Each step forward in this direction asked first for several steps backward. At the end, when bacteria achieved the desired outcome in controlled conditions, it was still difficult, if not impossible, to say why some mutation was crucial even if spatially distant from the enzyme active site. To be fair, the common goal of directed evolution is to force some bacterial enzymes to gain a new ability; in other words, to accept different substrates and produce different products, something that we need, not the bacteria accustomed to living in the same environment.

Definitions and explanations from Chapter 10:

The **enzyme active site** is the region of an enzyme where substrates bind and undergo transport or chemical transformation. Substrates can be ions, small organic molecules, or even macromolecules (nucleic acids or proteins), while enzymes can be quite complex protein structures containing many polypeptides with multiple active sites.

Phylogenetics in biology evaluates evolutionary relationships among heritable traits, such as protein or DNA sequences.

References

Albery, W.J. and Knowles, J.R. Evolution of enzyme function and the development of catalytic efficiency. Biochemistry 15(1976): 5631–5640.

Arnold, F.H. Directed evolution: Bringing new chemistry to life. Angew. Chem. Int. Ed. 57(2018): 4143–4148.

Bobrovskiy, I., Hope, J.M., Ivantsov, A., Nettersheim, B.J., Hallmann, C. and Brocks, J.J. Ancient steroids establish the Ediacaran fossil *Dickinsonia* as one of the earliest animals. Science 361(2018): 1246–1249.

Bonašić Lošić, Ž., Donđivić, T. and Juretić, D. Is the catalytic activity of triosephosphate isomerase fully optimized? An investigation based on maximization of entropy production. J. Biol. Phys. 43(2017): 69–86.

Christensen, H., Martin, M.T. and Waley, G. β-lactamases as fully efficient enzymes. Determination of all the rate constants in the acyl-enzyme mechanism. Biochem. J. 266(1990): 853–861.

Daar, I.O., Artymuik, P.J., Phillips, D.C. and Maquat, L.E. Human triose-phosphate isomerase deficiency: a single amino acid substitution leads in a thermolabile enzyme. Proc. Natl. Acad. Sci. USA 83(1986): 7903–7907.

de León, M.P., Cancela, H. and Acerenza, L. A strategy to calculate the patterns of nutrient consumption by microorganisms applying a two-level optimisation principle to reconstructed metabolic networks. J. Biol. Phys. 34(2008): 73–90.

Dewar, R., Juretić, D. and Županović, P. The functional design of the rotary enzyme ATP synthase is consistent with maximum entropy production. Chem. Phys. Lett. 430(2006): 177–182.

Evans, S.D., Hughes, I.V., Gehling, J.G. and Droser, M.L. Discovery of the oldest bilaterian from the Ediacaran of South Australia. Proc. Natl. Acad. Sci. USA (2020): 202001045. doi: 10.1073/pnas.2001045117.

Heinrich, R., Schuster, S. and Holzhütter, H.-G. Mathematical analysis of enzymic reaction systems using optimization principles. Eur. J. Biochem. 201(1991): 1–21.

Hunt, K.L.C., Hunt, P.M. and Ross, J. Dissipation in steady states of chemical systems and deviations from minimal entropy production. Physica 147A(1987): 48–60.

Juretić, D. and Županović, P. Photosynthetic models with maximum entropy production in irreversible charge transfer steps. J. Comp. Biol. Chem. 27(2003): 541–553.

Juretić, D., Bonačić Lošić, Ž., Kuić, D., Simunić, J. and Dobovišek, A. The maximum entropy production requirement for proton transfers enhances catalytic efficiency for β-lactamases. Biophys. Chem. 244(2019a): 11–21.

Juretić, D., Simunić, J. and Bonačić Lošić, Ž. Maximum entropy production theorem for transitions between enzyme functional states and its applications. Entropy 21(2019b): 743. doi: 10.3390/e21080743.

Knowles, J.R. and Albery, W.J. Perfection in enzyme catalysis: the energetics of triosephosphate isomerase. Acc. Chem. Res. 10(1977): 105–111.

Malabalan, M.M., Amyes, T.L. and Richard, J.P. A role for flexible loops in enzyme catalysis. Curr. Opin. Struct. Biol. 20(2010): 702–710.

Marín-Sanguino, A. and Torres, N. Modeling, steady state analysis and optimization of the catalytic efficiency of the triosephosphate isomerase. Bull. Math. Biol. 64(2002): 301–326.

Newton, M.S., Arcus, V.L. and Patrick, W.M. Rapid bursts and slow declines: on the possible evolutionary trajectories of enzymes. J. R. Soc. Interface 12(2015): 20150036. http://dx.doi.org/10.1098/rsif.2015.0036.

Pettersson, G. Evolutionary optimization of the catalytic efficiency of enzymes. Eur. J. Biochem. 206(1992): 289–295.

Ross, J. and Vlad, M.O. Exact solutions for the entropy production rate of several irreversible processes. J. Phys. Chem. A 109(2005): 10607–10612.

Sharma, P. and Guptasarma, P. "Super-perfect" enzymes: Structural stabilities and activities of recombinant triose phosphate isomerases from *Pyrococcus furiosus* and *Thermococcus onnurineus* produced in *Escherichia coli*. Biochem. Biophys. Res. Commun. 460(2015): 753–758.

Tomatis, P.E., Fabiane, S.M., Simona F., Carloni, P., Sutton, B.J. and Vila, A.J. Adaptive protein evolution grants organismal fitness by improving catalysis and flexibility. Proc. Natl. Acad. Sci. USA 105(2008): 20605–20610.

Turina, P., Petersen, J. and Gräber, P. Thermodynamics of proton transport coupled ATP synthesis. Biochim. Biophys. Acta 1857(2016): 653–664.

Wierenga, R.K., Kapetaniou, E.G. and Venkatesan, R. Triophosphate isomerase: a highly evolved biocatalyst. Cell. Mol. Life. Sci. 67(2010): 3961–3982.

Wolfenden, R. and Snider, M.J. The depth of chemical time and the power of enzymes as catalysts. Acc. Chem. Res. 34(2001): 938–945.

Chapter 11

ATP Synthase Molecular Motor

11.1 What is unique about ATP synthase and ATP turnover?

A key enzyme for cell respiration, the ATP synthase (known as F_0F_1-ATPase), is also the key enzyme for photosynthesis (Walker 2013, Junge and Nelson 2015). The ATP synthase funnels both photosynthesis and respiration into ATP synthesis. It is the smallest known biological nanomachine. The ATP synthase proton-driven molecular motor is a quite unique protein designed by biological selection as an early evolutionary pinnacle with a singular capability to convert the protonmotive force into the biologically useful energy of a covalent bond. Today, it is widely distributed in all known living beings, from archaea and bacteria to plants and animals. ATP synthases from photosynthetic organisms produce the major part of ATP molecules on our planet. This protein is so universally conserved across life that rotary catalysis, as its unique feature, may have evolved even before the last universal common ancestor of life (LUCA) collected DNA, RNA, and ribosomes inside the protective membrane-lipid barrier (Lane 2017).

The ATP synthesis performed by ATP synthase enables living cells to use myriad intelligent ways (also designed by natural evolution) to couple ATP hydrolysis to almost all uphill works, such as signaling, regulation, movements, detection, replication, and operation of all other cellular nanomotors. It is also the essential first step to synthesizing and recycling of macromolecules needed by our cells and organs. When converted into mass units, a single day ATP turnover of about 60 kg by an adult human is similar to her body mass. In microorganisms, the synthesis of one gram of cell mass requires the hydrolysis of 20 to 100 grams of ATP (Stouthamer 1977). The turnover time for ATP molecules in *Escherichia coli* is only about 0.2 seconds (Chapman and Atkinson 1977). The ATP contribution to phosphorylation and signaling in human cells is so crucial that about a hundred thousand phosphorylation sites exist in the human proteome (Kamerlin et al. 2013). The ATP synthase complexes in all of our mitochondria are the main suppliers of our ATP molecules. The embodiment of ATP synthase in tight-coupling membrane impermeable to protons is the first prerequisite for its work in the direction of ATP synthesis when ADP concentration is in large excess over ATP concentration. It is a far-from-equilibrium condition that enables bioenergetic membranes to produce ATP molecules.

ATP/ADP ratio and inside versus outside proton concentrations must be both far from equilibrium for ATP synthesis to occur. The adenylate energy charge

$$AEC = (ATP + \tfrac{1}{2}ADP)/(ATP+ADP+AMP)$$

is sometimes used to indicate the metabolic activity in ATP synthesis and hydrolysis (Ramaiah et al. 1964, Kieft and Rosacker 1991). For microbial cells, AEC higher than 0.8 reflects an active community. The ATP synthase can also work in the direction of ATP hydrolysis when high-enough concentration gradients are not maintained any more during some pathological conditions. An AEC value less than 0.5 or 0.4 can indicate the presence of dead or moribund cells (Chapman et al.

1971). AEC value is a crude tool invented by Atkinson in 1964 (Ramaiah et al. 1964). High AEC values above 0.9 are typical for healthy human erythrocytes (Chapman et al. 1971)—cells lacking mitochondria and all other internal organelles, thus proving that glycolysis (fermentation) is enough to produce a highly nonequlibrium distribution of ATP, ADP, and AMP.

Two functional domains of F_0F_1-ATPase, a membrane-intrinsic rotary motor F_0, and a membrane-extrinsic catalytic head F_1, are joined together to form that integral membrane protein. The connector subunits form the central stalk and the peripheral stalk. The central stalk rotates together with the F_0 motor. Polypeptides from the peripheral stalk compose the stator, which is bound both to the membrane-located F_0 domain and to the F_1 head domain. The free energy transduction process performed by ATP synthase consists of several energy conversion steps. It starts with the conversion of electrochemical proton gradient into the rotation of oligomeric c polypeptides from the membrane F_0 motor. The elastic torque is transmitted to the chemical F_1 motor. The F_1 subunit of ATP synthase is responsible for rotary catalysis leading to ATP synthesis and release. Proton-driven rotation of the c-ring in bare F_0 is very fast (about 1000 revolutions per second). It slows down to about 100 revolutions per second when F_0 and F_1 are coupled together in the holoenzyme. ADP molecule and inorganic phosphate P_i must be squeezed together in the absence of water molecules to create the ATP molecule. The dynamic equilibrium is established for the reaction ADP + $P_i \rightleftarrows$ ATP, which takes place at the open state configuration of the active site. The binding of ADP and P_i to the open state configuration happens spontaneously, but the release of the synthesized ATP molecule requires additional free energy transduction. Due to this process's importance, a great number of laboratories examined the details of ATP synthase structure and dynamics of ATP synthesis during the last several decades.

11.2 Being praised for errors

My book on bioenergetics (in Croatian language, Juretić 1997) depicted the naïve scheme of how ATPase stator and rotor fit into the membrane environment. I put it proudly on the cover page. Several months after book publication, I learned about the 1997 Nobel Prize in Chemistry for elucidating how the ATP synthase catalyzes the ATP formation (Boyer 1998, Walker 1998). It turned out that the arrangement of F_0 polypeptides is just the opposite of my cover page picture (Figure 11.1). The central rotary shaft is not composed of a and b polypeptides but of c_1 and c_2 polypeptides. The rotor is not rotating in the clockwise direction but in the counterclockwise direction during ATP hydrolysis (Walker 2013). In the cover picture's legend, I mentioned the clockwise direction of rotation during ATP synthesis, and that was the correct choice when the rotation is viewed from the F_0 toward F_1.

My isolation in Croatia during aggression on disarmed Croatia, when even the single remaining road toward Zagreb was under constant artillery fire, contributed to my ignorance about the valiant efforts of Nobel Prize winners Paul D. Boyer (2002) and John E. Walker (2013). The contributing factor was also that paid for international scientific journals never arrived at the University of Split. Zagreb had more pressing priorities than to worry about scientists in Split. Early internet connections were slowly introduced at the University of Split, and this situation of isolation from the world's scientific community gradually improved only after my book was published. I had the idea that something should survive as a help for young Croatian scientists out of my additional training in biophysics, bioinformatics, and bioenergetics at NIH (USA). Croatian researchers were very slow to enter in bioenergetics, biophysics, and bioinformatics, at the time when these exciting interdisciplinary fields connecting biology to other sciences were already booming in the USA. After more than 20 years, this situation improved at Zagreb research institutions, especially in bioinformatics, mainly due to private initiatives, while centralized Ministry of Science bureaucracy remained firmly entrenched into the obsolete 19th-century division of sciences with mostly lip service to interdisciplinarity. When presenting the just-published book (Juretić 1997), theoretical chemist Dr. Ante Graovac praised my good intuition for putting the proposed (erroneous) scheme

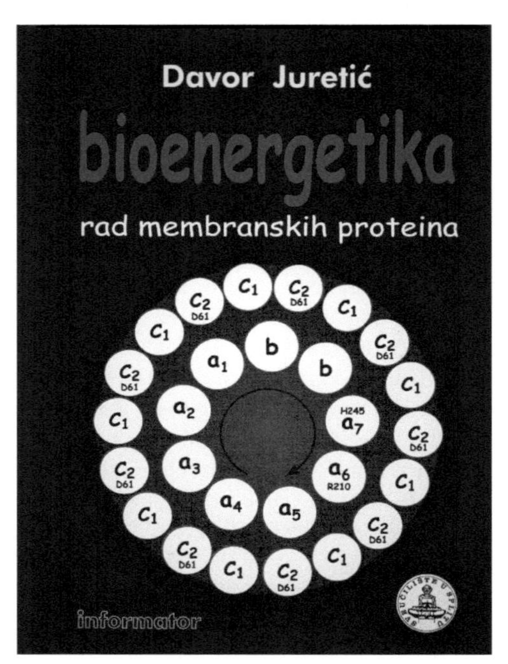

Figure 11.1: The cover page of my book: "Bioenergetics: Work of Membrane Proteins" (Juretić 1997).

for the cover page F_0 motor. At that time, close to the end of the year 1997, he learned about Nobel Prize awarded to Paul Boyer and John Walker. As you can imagine, I had mixed fillings.

11.3 Questions about ATP synthase mechanism of action shaped the bioenergetics

To better understand the historical role of ATP molecules and ATP synthase in the emergence of bioenergetics, it is instructive to see how a scientist can recover from error with Nobel Prize winners being no exception (Allchin 2002). I do not mean errors in methodology. There is no excuse for using flawed experimental or theoretical methods. I mean that more in-depth understanding can be gained when we become aware of salient unwarranted assumptions that entered into our theoretical framework. In our mind, we are all theoreticians prone to construct lovable theoretical edifices that would explain to our satisfaction those events or images that caught our attention. Scientific theories have been all subjected to rigorous selection, often involving hot disputes, but none escaping the reality check in experiments designed to prove them wrong. Thus, all scientific theories are more or less provisional. In our age of greatly accelerated scientific accomplishments, it is beneficial for all involved to be aware of a crucial role of strong emotions, intuitions, and contradictory conservative-revolutionary requirements that enter into each ground-breaking scientific research. Scientific thinking is balanced at the razor's edge of the highest respect for earlier fruits of scientific creativity and audacious willingness to solve still unsolved problems even if it requires a revolutionary paradigm shift. What makes life interesting for a scientist is a regular, almost lawful rule that each solved mystery generates new and wonderful worlds to explore. Following these thoughts, let us briefly trace the evolution of bioenergetics and an ever better understanding of how ATP synthase works as a marvelous molecular engine.

There were four contending theories about the mechanism of oxidative phosphorylation in the period from the middle of the previous century to the 1970s: the chemical hypothesis of Slater (1953), the hypothesis about proton activity in a lipid phase of Williams (1961), the chemiosmotic hypothesis of Mitchell (1966), and conformational coupling hypothesis of Boyer (Boyer et al.

1966, Boyer 1974). Rotary catalysis of ATP synthase, outlined in the previous section, was only the climax-discovery after experimental evidence overcame the passionate wars among scientists about the mechanism of F_0F_1-ATP synthase activity (Gresser et al. 1982, Prebble 2019). Opposite opinions were finally joined together by Nobelist Paul D. Boyer, who was able to disregard the exaggerations of his own pet theories and to incorporate the essential aspects of Peter Mitchell's chemiosmotic theory (Boyer et al. 1977, Boyer 1998, 2002, Prebble 2013).

11.4 The Martian scientists

Boyer was par excellence enzymologist, biochemist, and molecular biologist who published significant advancements in understanding biochemical problems connected with some 25 different enzymes before tackling ATP synthase mysteries (Boyer 2002). In the enzymology field, he was mainly interested in the mechanism of action. There were some common traits in his long career with that of Terrel Hill. Terrel was the first to come with the idea that the phrase "molecular biology" may be attractive to interdisciplinary-minded scientists at the forefront of modern biology who were interested in using recently developed tools and concepts from physics. Thus, he established the very first Institute of Molecular Biology. The Institute was located at the University of Oregon. Terrel declined to lead the Institute. Instead, he appointed Aaron Novick as the Founding Director (Chamberlin 2015).

After participation in the atomic bomb building, Aaron Novick decided to move from physics and chemistry into biology (University of Oregon Archives, Aaron Novick collection). Exactly how this happened is an attractive example of deep interconnections between physics and biology, war and peace, visionary thinking pinnacles, and down to earth scientific drudgery. Leo Szilard asked Novick if he would be interested in joining him in 'an adventure in biology.' Leo Szilard was The Martian scientist (I shall leave it to readers a joy to discover which Hungarian scientists were aliens from Mars as Leo jokingly suggested). Aaron immediately accepted his offer to take the risky and drastic change in his scientific career. Physicists were well aware that Szilard was in a class of its own, even among those considered geniuses (Lanouette and Silard 2018). To support this claim, I can mention several tidbits. Szilard was the first mortal to win in the struggle with **Maxwell's demon** from 1922 to 1929 and, in the process, became the founding father of informatics. He also won the strategic battle in the struggle with Hitler's demons by hiding his concepts from 1933 about neutron-induced chain reaction until his famous letter (signed by Albert Einstein) to President Franklin D. Roosevelt. The letter initiated the Manhattan project and the construction of the first nuclear reactors and atomic bombs. The government decided to use atomic bombs against civilians over Szilard and other scientists' protests and petitions. Finally, Szilard won the struggle against his bladder cancer by directing medical doctors how to use the cobalt 60 therapy treatment he designed himself. There are numerous examples of other scientists getting Nobel Prize for discoveries and concepts he invented, published, and often patented. Szilard did not have time, funds, or patience to pursue single-mindedly all of his brain children toward experimental verifications. Amazingly, his contributions to biology were also ground-breaking, although achieved late in his career.

Aaron Novick, Terrel Hill, and Paul Boyer all had the PhD in physical chemistry or chemistry, an appreciation for interdisciplinary research, and the inclination to use physical and chemical methods and concepts in biology. Throughout his long career, Paul Boyer often used heavy or radioactive isotopes of oxygen, phosphorus, and carbon: ^{18}O, ^{32}P, and ^{14}C. In the 99th year of his life, Paul Boyer had an amazingly youthful look and an obvious interest and love for life. His photo from the Wikipedia page seems to be telling us: "I still have all my marbles." It is the sentence Terrel Hill used in the last email to a friend when he too was in the tenth decade of his life (from the biographical memoir about Terrel published by Ralf Chamberlin in 2015, see Introduction). The nearly 80 years long marriage of Paul Boyer and Lyda Whicker was similar in the extraordinary duration to Terrel's Hill 71 years long marriage to Laura Eta Gano. The loving support of a woman does wonders for a man. When I was 15, a thought occurred to me that my future wife is just born in my native town (Split). We married in Split when I was 58, and she was indeed 15 years younger, the descendant from another of Split's noble families (my mother Sonja Juretić, née Grisogono, was a descendent of at least 800 years old Grisogono nobles, while Ljiljana de Nutrizio's noble ancestors had equally deep roots in Dalmatia). She is a great woman with a pure and noble heart, well aware of all my weaknesses, but always standing by my side.

11.5 The learning curve includes recovering from errors and adding educated insight

After accepting the professor position at the University of California Los Angeles (UCLA) in 1963, Paul Boyer soon became the founding director of the Molecular Biology Institute at UCLA. He also established the interdisciplinary and interdepartmental PhD program at UCLA and edited or established several influential biochemical journals. All these activities and the supportive role of his family, students, and colleagues contributed to Boyer's ultimate success in understanding the "Splendid molecular machine" of ATP synthase (Boyer 1997). However, there was something even more important. I would say that it was his persistence, open-mindedness, flexibility, educated intuition, and willingness to come with new concepts even in his older age, which ultimately led to the highest recognition for Boyer's contribution.

The persistence is nicely reflected in Boyer's words: "One of my favorite sayings is that most of what you accomplish in research is the coal that you mine while looking for diamonds" (Boyer 2002). For instance, Boyer was convinced in 1961 that investigations with ^{32}P "had hit pay dirt." A long-sought "energy-rich" intermediate of oxidative phosphorylation (see Chapter 1) was tentatively identified by his group (Peter and Boyer 1963). Further studies did not confirm it. The first recognition of a phosphohistidine in biochemical systems was a worthwhile discovery, but unrelated to the oxidative phosphorylation problem. Again in Boyer's words: "In Olympic analogy, we were reaching for a gold but were fortunate to have obtained a bronze" (Boyer 2002).

Boyer's research interests and priorities evolved in synchrony with career development and an ever-larger number of bright collaborators he adroitly inspired. One of his favorite projects at UCLA was to solve the mystery of how *Escherichia coli* active transport is coupled to oxidative phosphorylation. This time around, he was focused on the role of a bioenergetic membrane (the cytoplasmic membrane in the case of *E. coli*), but refused to consider the chemiosmotic theory of Peter Mitchell. Still, he maintained his willingness to stick his neck out by proposing new concepts. His new concept was energy transduction from the oxidation of substrates to form a high energy membrane state (Klein and Boyer 1972).

What exactly is this energized state was an open question. Two conclusions from that paper were prominent: a) that *E. coli* can use respiration to drive active transport of amino acids, carbohydrates, and potassium ions without ATP participation, and b) that the uncoupler of oxidative phosphorylation, 2,4-dinitrophenol, inhibits the active transport. Protons were never mentioned in that paper, nor the fact that the uncoupler 2,4-DNP is a proton ionophore that can shuttle protons across the membrane. Boyer did not believe that proton migration can lead directly to ATP formation, as Mitchell proposed. Hence, at that time, he rejected the whole concept of the chemiosmotic coupling mechanism and the crucial role of the electrochemical proton gradient in cellular bioenergetics. For his taste, the chemiosmotic hypothesis was not specific enough to explain the formation of chemical bonds. He preferred to consider different manners of how energy can be captured in protein conformational changes. In retrospect, Boyer correctly rejected the direct role of protons in the ATP synthesis mechanism but erred in being a stubborn holdout in regard to Mitchell's concept of protonmotive force and energy-linked proton translocation. When cast in the chemiosmotic framework, the experimental results coming from his and Harold's laboratory (Harold 1974) in the early 1970s helped dismiss the "caloric catastrophe" question (Minkoff and Damadian 1973a).

11.6 The caloric catastrophe question and Nobel Prizes

The caloric catastrophe idea was a colorful way of stating that solute pump models are thermodynamically untenable if the active transport of small molecules and ions across the cytoplasmic membrane is dependent only on ATP for energy (Minkoff and Damadian 1973b). These authors did not correct the omission to consider the protonmotive force, nor did they propose any other clear cut explanation for the alleged breakdown of thermodynamic laws. The well-known textbook on bioenergetics by Nicholls and Ferguson (2002, 2013) does not mention the "caloric catastrophe" idea at all, not even in a historical context. This whole idea did nothing to help Raymond Damadian in his intensive lobbying to get Nobel Prize. His open support for creationism also did not help (Kinley 2015), but his childhood desire to help millions of cancer patients was fulfilled to his satisfaction. He certainly deserved the Nobel Prize for his discovery and applications of magnetic resonance imaging in medicine (Kauffman 2014). As a physician, Damadian was not afraid to build upon previous discoveries of many gifted physicists. When subjected to the noninvasive magnetic resonance imaging to detect some problems in our body, we can all be thankful that by some miracle, Damadian's father escaped the wholesale genocide performed over Armenian people in 1915.

Many physicists who discovered and applied nuclear magnetic resonance in physics and chemistry (the foundation for Damadian contributions to medicine) were Jewish. They escaped even greater genocide performed by Hitler against Jews during the Second World War. While preparing the holocaust, Adolf Hitler arrogantly proclaimed to Nobelist Max Planck: "If the dismissal of Jewish scientists means the annihilation of contemporary German science, then we shall do without science for a few years" (Beyerchen 1977). The most recent example of doing without science is the mismanagement of Dr. Li Wenliang's early warning about the coronavirus COVID-19 pandemic.

Another reason for this short digression about Damadian is to put Nobel Prizes in a proper perspective. The Nobel Prize committee is not infallible. For instance, António Egas Moniz got the Prize in Physiology or Medicine (in 1949) for the lobotomy variant of psychosurgery, which is nowadays recognized as ethically unacceptable. But Moniz's discovery of cerebral imaging (angiography) already in 1927 (Artico et al. 2017) had similarly seminal significance in the medical diagnostics as the magnetic resonance imaging technique. Thus, it deserved the Nobel Prize. Numerous other scientists fully deserved the Nobel Prize but never obtained it.

11.7 First breakthrough in solving the mechanism of proton-driven ATP synthesis

The 1972 paper by Klein and Boyer was the last instance when Boyer omitted to mention the chemiosmotic hypothesis. In his defense, he cited four reviews during the 1967–1974 period in the influential *Annual Review of Biochemistry* journal with negative assessments of Mitchell's proposal that proton transport could drive ATP synthesis (Boyer 2002). Boyer served as Editor or Associate Editor of that journal from 1965 to 1988 (Allchin 2002). Being well aware of Mitchell's work (through long-lasting personal correspondence with him too), Boyer started to think in early 1972 how active transport, charge separation, and proton downhill transport can be converted and captured into conformational changes of membrane proteins. He addressed the mechanism of action question for ATP-synthase. Since the mitochondrial F_1-ATP domain can be conventionally isolated to study its structure and activity, Boyer first focused on the "simpler" question about possible catalytic site cooperativity in ATP hydrolysis of isolated F_1-ATPase. With his capable group of students and colleagues, he plunged ahead in a study about this topic but mentioned in his recollections: "Occasionally in biochemical research one encounters a property of a system that seems designed to confuse and thwart the researcher" (Boyer 2002).

After discoveries enter into textbooks, it is hard to realize how dramatic and far from obvious were these crucial insights when first presented to students and contemporary scientists. Words and phrases describing these insights are currently presented as known facts in textbooks. But a terminology had to be invented by investigators as the last recourse when all older terms proved to be inadequate. For instance, after discovering or designing novel peptide antibiotics, I had to invent a new vocabulary for them (adepantins, trichoplaxins, kiadins, flexampin). The alternative was to increase an already enormous number of abbreviations used in biological sciences (see section 15.4 from Chapter 15 concerning how peptide antibiotics can inhibit bioenergetic energy conversions and ATP synthesis). By the way, the characterization "enormous" is appropriate for recently discovered huge bacteriophages (Al-Shayeb et al. 2020), so that new phage groups were named after the words for "big" in the languages of the paper coauthors: from Mahaphage (Sanskrit) to Biggiephage (Australian). Huge phages blur the gap between life and non-life with predicted ribosome subunits and F-type H^+-transporting ATPase subunit b (Al-Shayeb et al. 2020, supplementary Table 5B). Philip Bell noticed how some giant viruses build "viral factory" compartments inside prokaryotic cells, not unlike the eukaryotic nucleus. He proposed a viral origin of the eukaryotic nucleus (Bell 2020). It is the relevant topic of the present chapter and the origin of life problem (see Chapter 13).

Discovery insights can come at any occasion, even during sleeping or completely unrelated conversations, because the unconscious brain of a dedicated scientist constantly mulls the matters over and attempts to solve the troubling problem. In Paul Boyer's case, he was bored by some scientific presentation that he listened to in 1972, lost interest in understanding it, but was too polite to leave the conference hall. An epiphany suddenly occurred to him, a change in paradigm in his previous thinking that free energy is mostly used to form a covalent bond between ADP and inorganic phosphate. The experimental observations he had in mind could also be explained with the idea that the major use of captured energy is invested in the conformational change of the catalytic site that causes ATP release. Tightly bound ADP is easily phosphorylated in the presence of high P_i concentration, but energy input is needed for the ATP release, as we already mentioned in the section 11.1 (see also the end of the last section from this chapter).

Some other researchers come independently to the same conclusion. Boyer realized that this new concept is worthy of publication in a highly regarded scientific journal. When one such journal declined to publish his paper, he did something that only the best researchers do. He sent the same article to a higher quality journal where it was duly published (Boyer et al. 1973) and subsequently provided exhaustive additional experimental details (Cross and Boyer 1975). A novel concept required a new name, as we mentioned above. Boyer realized that energy-linked binding changes are unique for the intact ATP synthase and isolated F_1-ATPase. He used the phrase "The Binding Change Mechanism" for the first time in his 1979 contribution (Boyer 1979).

11.8 Rotary catalysis: The second breakthrough concept by Paul Boyer

The second revolutionary concept proposed by Boyer did not require any new words or phrases. Old knowledge from classical mechanics of rotary motors sufficed when combined with the binding change mechanism and Mitchell's chemiosmotic hypothesis. Assuming that any enzyme, and in particular the ATP synthase, acts as a rotary nanomotor was a bizarre notion to biochemists in the 1970s. The plausibility of only one biological rotary motor was proposed at that time—the motor that enables bacteria to swim by moving their flagella (Berg and Anderson 1973). Proton flux drives the flagellar motor in accordance with the chemiosmotic mechanism (Manson et al. 1977, Mitchell 1979). Peter Mitchell published the 1985 paper in which he suggested the concept of proton-driven molecular rotation for the F_0F_1-ATPase similar to that of bacterial flagellar rotator motor (Mitchell 1985). He proposed that the γ subunit acts as the rotating proton gate. The Nobel Prize may have distracted him from publishing this concept earlier.

Boyer's students and closest associates were initially skeptical when Boyer presented his hypothesis about rotary catalysis. However, they quickly realized that the proposed unification from biochemistry, physics, and physiology (ion transport) is very alluring when focused on experimental tests on how this unique enzyme works. It is almost equally inciting for truly ambitious students to prove that their great professor erred as it is to prove that he or she was on the right track. The best professors are thrilled when proven wrong, just as Nobel laureate Christian de Duve said: "I would be delighted if someone should one day carry out the experiments I suggest, were it only to show my ideas wrong" (de Duve 2003). Professor Boyer was on the right track this time. Imagination and the ability to recognize important and relevant contributions of other research groups were equally essential ingredients in the professor's thinking as persistence to achieve the desired breakthrough in the face of previous failures. Boyer's group soon published experimental results about this topic as an indication that rotary catalysis is possible (Boyer and Kohlbrenner 1981, Gresser et al. 1982).

11.9 Direct experimental evidence for rotary catalysis by other research groups

The outstanding contributions of Efraim Racker and Yasuo Kagawa were eye-opening for Boyer. These researchers realized that F_0F_1-ATPase and most other proteins from thermophilic bacterium are highly resistant to harsh physicochemical treatments by denaturing agents. Renaturation and reconstitution in maximally simplified artificial membrane systems were possible without any loss in catalytic function. The coupling of protons flow to ATP synthesis or hydrolysis was so good that an electrochemical proton gradient of about 300 mV was reached when driven with ATP hydrolysis. The conclusion was quite clear already in the early 1970s from the Racker's group: only the subunits of thermophilic F_0F_1-ATPase are required for coupling proton translocation to ATP hydrolysis (Kagawa et al. 1973). Kagawa group studies with ATP-synthase subunits isolated from thermophilic bacteria and reconstituted into artificial vesicles found additional connections between specific subunits and their functions (Yoshida et al. 1977). The thermophilic F_0 subunit was confirmed to be the proton-conducting pathway across the membrane. The subunit β from the F_1 part was responsible for ATP hydrolytic activity. The proposed model of ATP synthase structure (Yoshida et al. 1977) had a circular distribution of α and β subunits with a centrally positioned γ subunit drawn as a dotted circle connected both to F_0 and to F_1. Boyer's inspiration was that the γ subunit revolution was the only way it can influence the catalysis in the outer ring of three β subunits.

The rotational catalysis mechanism appeared likely after high-resolution structural data became available for F_1-ATPase from Walker's laboratory (Abrahams et al. 1994). It had to be interpreted in the light of numerous experiments performed by Boyer's group and his former students from 1981 to 1995 (Boyer and Kohlbrenner 1981, Gresser et al. 1982, Kandpal and Boyer 1987, Duncan et al. 1995). Remaining doubts disappeared after the laboratories of Yoshida and Kinoshita in Japan published the visual confirmation that ATP hydrolysis drives the rotation of the central rotor formed by the F_1-ATPase γ subunit (Noji et al. 1997).

At present, it is easy to find hundreds of colorful pictures and animation movies illustrating how all ATPase polypeptides are assembled and the dynamics of their stator and rotor complexes. They are all involved in an intricately choreographed dance. The stator connects F_0 and F_1 as a long peripheral stalk, which prevents unproductive F_1 rotation. Central rotor stalk dynamics also causes peripheral stalk dynamics during the rotation period. In other words, the stator is not completely static. That was predicted earlier (Blum et al. 2001) and nicely illustrated in recent ATP synthase movies composed together from high-resolution cryo-electron microscopy data acquisition (Stewart et al. 2012).

11.10 Optimization of the transitions' state parameters

Our approach was different. We asked the question about the evolutionary optimization of that marvelous biomolecular machine. Was it partially due to the tendency to reach high or even maximal entropy production in some catalytic steps? Is it possible at all to achieve high efficiency and a maximal entropy production at the same time? To answer these questions, we considered just the

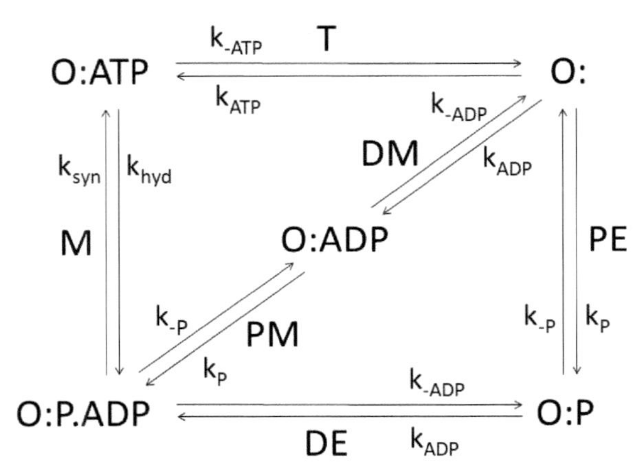

Figure 11.2: The kinetic model of the ATPase F_1 subunit. The open state with a symbol O: can bind inorganic phosphate P_i to become the O:P state. It can also bind the ADP molecule to become the O:ADP state. When O:P state binds the ADP molecule, it becomes the O:P.ADP state. Only O:P.ADP state can be transformed into O:ATP state in the M transition. These two states are connected with ATP synthesis and ATP hydrolysis rate constants k_{syn} and k_{hyd}, respectively. ATP molecule is released in the T transition. Other transitions between open enzyme states are PE, DE, PM, and DM. This figure is similar but not identical to our recently published Figure 5 from Juretić et al. (2019b). We corrected here the subscript labels for kinetic constants in transitions DM and DE, which are, of course, ADP subscripts instead of ATP subscripts as erroneously entered into Figure 5 from Juretić et al. (2019b).

essential open states of the catalytic subunit F_1 and transitions connecting these states (Figure 11.2). The M and T transitions are crucial for ATP synthesis and release from the active site of the F_1 head. In the plant cells, the release of newly synthesized ATP molecules happens in the cytoplasm, where it is needed to perform all cellular works. Three ATP molecules are synthesized and released per revolution.

All rate constants are known or estimated from experiments in the kinetic model presented in Figure 11.2. I was fortunate to discover the gem of the paper about transition-state kinetics and storage of elastic energy by chloroplast ATPase (CF_0F_1-ATP synthase) published by Pänke and Rumberg in 1999. It connected the parameters measured by them with a biophysical model of how ATPase catalysis works. It does not matter that the kinetic model was not presented as Hill's diagram (Hill 1977) by these authors (Pänke and Rumberg 1999). It was easy to transform their kinetic scheme for transitions among open states into Hill's diagram (Figure 11.2) and to calculate all state probabilities p_i, currents J_i, affinities A_i, and transitional entropy productions P_i after following the now-classic Terrel Hill procedure (Hill 1977) for steady-state free energy conversion by biological macromolecules.

By using this kinetic model, we examined (Dewar et al. 2006) if it is possible to find optimal parameters when the Maximum Entropy Production Principle (MaxEP) and Maximum Shannon' Information Entropy (MaxENT) requirement are jointly used. The MaxEP principle was used in the form of the maximum transitional entropy production theorem (MTEP) (Chapters 8 and 10). I first verified that the calculated ATP synthesis flux J_M fits well the experimental data kindly provided by Pänke and Rumberg (Figure 11.3). With this task done, I applied the MTEP theorem, which I described latter in 2011 (Dobovišek et al. 2011) and in 2019 (Juretić et al. 2019b). My collaborators included information entropy maximization in a joint optimization procedure for which we found to produce optimal kinetic parameters (Dewar et al. 2006) very similar to observed parameters by Pänke and Rumberg (1999). Remarkably, when transition state parameter κ is varied, the single joint maximum is found for information entropy and the entropy production P_M in the M transition (Figure 11.4) responsible for ATP synthesis or hydrolysis (Figure 11.3).

The optimization for the transition state parameters in the M transition (Figure 11.2) does not ensure that overall entropy production is maximal under given constraints. These parameters are,

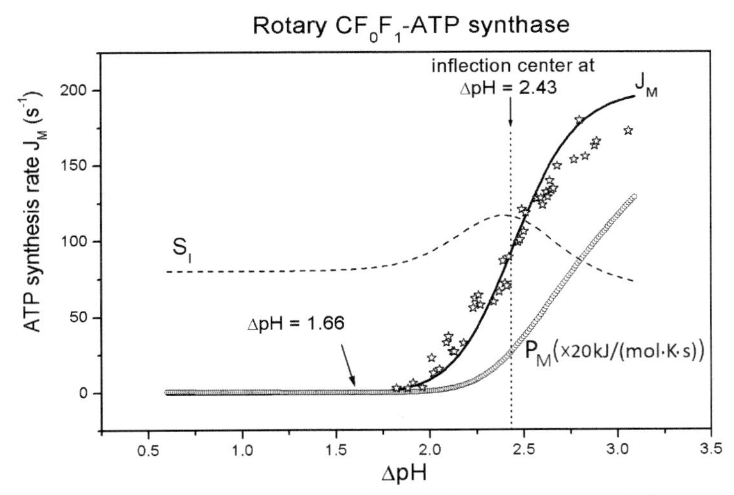

Figure 11.3: Dependence of steady-state parameter S_I (Shannon's information entropy), flux J_M, and partial entropy production P_M on pH difference $\Delta\mu_H{}^+$, when ADP and ATP concentrations are such that net ATP synthesis is favored for the transition M which couples proton gradient to the rotation of the central shaft, elastic energy storage, and ATP synthesis. S_I and P_M values are multiplied, respectively, with the factor 100 and 20 to render corresponding curves equally visible as the J_M ($\Delta\mu_H{}^+$) curve. The **inflection point** of the J_M ($\Delta\mu_H{}^+$) curve at 2.43 ΔpH is very close to maximal S_I at ΔpH = 2.4 when the best distribution of state probabilities enables rapid regulation of J_M in nearly linear far-from-equilibrium flux-force relationship. I prepared this figure before Roderick Dewar created and published with us a similar picture (Dewar et al. 2006; Figure 2b).

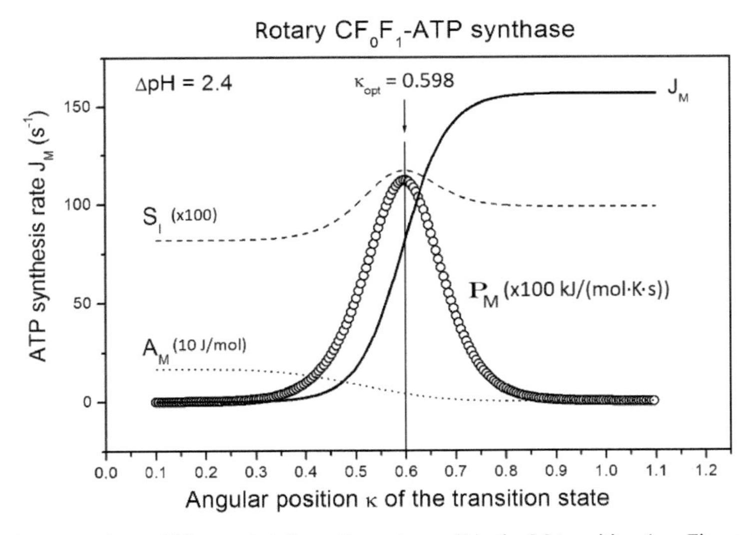

Figure 11.4: Steady-state values of Shannon's information entropy (S_I), the M-transition (see Figure 11.2) flux J_M, and entropy production P_M for different relative angular position κ of the transition state when the protonmotive driving force ΔpH ($\Delta\mu_H{}^+$) is fixed at 2.4 (equivalent to 13.4 kJ/mol). Coincidence maximum is found for S_I and P_M for the catalytic dwell at κ = 0.598. S_I and P_M values are both multiplied with the factor 100 to render corresponding curves equally visible as the J_M dependence on κ. The Pänke and Rumborg experiments (1999) estimated that κ = 0.6. The inflection point of the calculated J_M (κ) curve for the ATP synthesis rate is also very close to κ = 0.6, enabling maximal sensitivity to angular position changes. I prepared this figure before Roderick Dewar created and published with us a similar picture (Dewar et al. 2006; Figure 2a).

however, in agreement with empirical estimates about the optimal angular position for the ATP-binding transition (Pänke and Rumberg 1999, Dewar et al. 2006). After optimization, the greatest contribution to overall entropy production comes from the recovery step: the T transition (from

O:ATP to O:) when ATP is released (Juretić et al. 2019b). The second highest contribution comes from the M step. Together, these two transitions contribute 93% of total entropy production and the highest flux corresponding to the conversion of proton transport into rotary catalysis for ATP synthesis and ATP release (Table 11.1). An analogy of Kirchhoff's current law in steady-state enzyme cycling requires the identity of fluxes in M and T transitions, in DM and PM transitions, and in PE and DE transitions. Also, the sum of branched fluxes must be equal to the ATP synthesis and release flux.

Table 11.1: The steady-state distribution of fluxes, state probabilities, and transitional entropy productions after ATP synthase optimization for MaxENT and MTEP. The MTEP requirement is applied to the ATP synthesis transition (M).

State probabilities		Transition entropy productions ($kJmol^{-1}K^{-1}s^{-1}$)		Transition fluxes (s^{-1})	
$p_{O:}$	0.04	σ_T	3.175	J_T	81.07
$p_{O:ATP}$	0.30	σ_M	1.117	J_M	81.07
$p_{O:ADP}$	0.49	σ_{DM}	0.158	J_{DM}	72.45
$p_{O:P}$	0.01	σ_{PM}	0.121	J_{PM}	72.45
$p_{O:P.ADP}$	0.16	σ_{PE}	0.026	J_{PE}	8.62
		σ_{DE}	0.007	J_{DE}	8.62
Σp_i	1.00	σ_{tot}	4.604	Kirchhoff's law: $J_T = J_M = (J_{DM} = J_{PM}) + (J_{PE} = J_{DE})$	

11.11 MTEP theorem application and rate-determining catalytic steps

Biochemists are well aware that the last step in any catalytic cycle is likely to be the rate-determining step. It is usually the slowest catalytic step too. When it is speeded up to high values allowed by the MTEP theorem, this is seen as an increase in productive flux and overall entropy production increase. The increase in catalytic efficiency also occurs for single-cycle enzymatic schemes. For example, the MTEP optimizations of another evolutionary perfected molecular engine, the triosephosphate isomerase, also revealed that the most significant contribution to overall entropy production comes from the rate-limiting recovery step (Chapter 10 and Bonačić Lošić et al. 2017).

The sigmoidal shape of the ATP synthesis flux is measured and predicted using the transition state theory (Pänke and Rumberg 1999, Dewar et al. 2006). It is quite clear from Figures 11.3 and 11.4. The sigmoidal enzyme kinetics is often the sign of cooperativity and metabolic regulation (see Chapter 9, section 9.5, and Chapter 14, section 14.1). The inflection point of a sigmoidal reaction rate curve is the point of maximal sensitivity to variable chemical potential. At this point, the ATP synthesis rate is maximally sensitive to the electrochemical proton gradient. It can hardly be accidental that this point is also a) the point of maximal information entropy, b) the point of maximal entropy production, and c) the point of the best far-from-equilibrium linearity in the force-flux relationship. Optimal metabolic control is achieved for maximal entropy production. This conclusion cannot be just one isolated case for one free-energy transforming enzyme, and it certainly deserves additional investigations. It would be indeed strange if natural evolution did not take advantage of achieving better metabolic regulation and of increasing the efficiency of its "working horses" (enzymes) by speeding up the entropy increase of the universe. In our review papers (Dobovišek et al. 2014, Juretić et al. 2019b) and in Chapters 8, 10, and 12 from this book, a reader can find other examples of MTEP-theorem-based optimization for enzyme kinetics. It resulted in accelerating rate-limiting step or steps with the concomitant increases in catalytic efficiency, free-energy transduction efficiency, output flux, and overall entropy production.

To sum up, in the case of ATP synthase nanomotor working in the direction of ATP synthesis, MaxENT and MTEP requirements predicted an optimal angular position for the ATP-binding transition, which was close to the experimental value (Dewar et al. 2006). The application of the

MaxENT principle led to the realization that the optimal steady-state is associated with the inflection point in the curve of ATP synthesis rate *versus* the protonmotive force. That is also a special steady state when metabolic control reaches maximal possible reaction speed and high optimal free-energy conversion efficiency. Another novel insight from Dewar et al. 2006 paper is the realization that maximal possible entropy export from living cells, compatible with given constraints, is achieved spontaneously as the most probable outcome of functional cellular networks that are already in the maximal information entropy state. Altogether, our results in modeling chloroplast ATP synthase suggested a statistical interpretation of ATP synthase function's evolutionary optimization and led to a realistic description of the ATP synthase rotary motor action.

The vacuolar, or V-ATPase, is also a rotary motor similar to F_0F_1-ATPase, except that it works exclusively to pump protons (McGuire et al. 2017). The other difference is V-ATPase's regulated trafficking among cellular organelles (needing acidification to properly function) and cell surface membrane. The reconstruction of free-energy changes during enzyme cycling among its functional states led to a seemingly opposite conclusion regarding our analysis, namely that binding affinity changes driven by ATP hydrolysis serve to prevent dissipation and channel the chemical energy into mechanical work (Singharoy et al. 2019). The authors added that in contrast, ATP hydrolysis rapidly dissipates the chemical energy in an aqueous environment. Their rationale looks reasonable. A cycle of transitions at protein-protein interfaces takes place over the millisecond timescale, considerably slower than the picoseconds scale of energy dispersal and dissipation in water solution. However, this picture neglects that the ATP molecule is stable at room temperature (25°C) in buffered aqueous solution between pH 6.8 and 7.4. Often repeated statements about ATP's low stability in water (Bonora et al. 2012) refer to ATP in a cytoplasmic environment. ATP environment in an aqueous solution at the same pH is different because it is devoid of numerous atoms, molecules, and enzymes normally present in the cytoplasm and capable of accelerating the ATP hydrolysis rate. In the absence of enzymatic catalysis, there is only a glacially slow conversion of half ATP molecules into ADP and phosphate ions after about six years (the $t_{1/2}$ half-life) (Stockbridge and Wolfenden 2009, Stockbridge et al. 2010, Wolfenden 2011). The dissipation because of hydrolysis is fast in water or alcohol (methanol), but it is rare in the absence of some ATPase enzyme. Paradoxically, the ATPase enzyme is both increasing and decreasing the dissipation rate. The increase is due to the high frequency of enzyme-substrate encounters leading to greatly enhanced ATP hydrolysis. The decrease is due to the slow channeling of chemical energy into secondary work or current pathways. The channeling capability for dissipation pathways is inherent in all ATPases because of its double-gain aspect: faster entropy increase in the universe, and increased power for biological needs.

11.12 Tangled impact after the publication of thermodynamic optimization for ATP synthase

Roderick Dewar carefully repeated my calculations presented in Figures 11.3 and 11.4. He gave it better theoretical background with the unification of MaxEnt and MaxEP principles for the crucial ATP synthase transition, the transition coupling proton transport to rotary catalysis for ATP synthesis (Dewar et al. 2006). The majority of subsequent papers citing our work lumped it together with other publications supporting the application of the MaxEP principle to thermodynamically coupled biological systems (Demirel 2014, Skene 2017, Nath 2019a,b). Some introduced the confusing claim that our paper reviewed the minimum entropy production method (Moroz and Wimpenny 2011), although minimum EP was mentioned only in introductory sentences laying out the background for Dewar et al. 2006 paper. Others mentioned it as the showcase for predictability potential of the MaxEP theory (Whitfield 2007, Koch and Britton 2018) in the analysis of enzymatic transitions connected to metabolic networks (Martin and Horvath 2013), or as a claim that ATP synthase evolved in accordance with the MaxEP principle (Jia et al. 2012, Bradford 2013). Some have seen the similarity of Paltridge's (1979) and our approach in maximizing only partial entropy production (Bruers 2007; see section 8.3). None attempted to extend our calculations for more realistic ATP synthase models with updated values for measured kinetic constants and transition state parameters (activation enthalpy and entropy).

The most detailed appraisal of Martyushev and Seleznev (2015) also has nothing to do with the ATP design and evolutionary optimization. They mentioned how Melkikh and Seleznev (2012) defined useful work and pointed out that our choice for the maximization procedure corresponds to the maximal useful work: "...the maximization procedure used by the authors corresponds to the maximum useful work. This logic is especially apparent when considering the enzymes of ATP synthesis. An enzyme of ATP synthesis actually performs the chemical work to deviate intracellular composition from the equilibrium value using the ATP, ADP + P reaction." Martyushev and Seleznev (2015) pointed out in their opinion paper that MaxEP principle is "quite rarely used in chemical kinetics." The examples they considered are all presented as MaxEP applications for determining optimal characteristics of enzymes. Since cited references (Juretić and Županović 2003, Dewar et al. 2006, Dobovišek et al. 2011) are for papers I initiated, with mostly my calculations and writing contribution, it is educational in the context of this chapter to focus on what was missed by Martyushev and Seleznev's (2015) interpretation.

First, the "useful work" in the case of F_0F_1-ATPase can be defined in at least three possible ways with the same goal of finding the maximal possible efficiency of energy conversion (Kulish et al. 2016). The subjective choice of what can be regarded as a useful work arises because this enzyme consists of two motors, the membrane proton-transporting F_0 motor and the F_1 motor responsible for the rotatory catalysis. Both motors can perform several free-energy transduction steps. The insight about the structure-action relationship has been just the opposite for F_0 and F_1 motors—the first was well understood even before detailed structural information became available (Elston et al. 1998), while the second one remained somewhat enigmatic despite the early availability of structural information. For instance, the maximal efficiency of the F_1 motor is around 0.01% when useful work per unit time is calculated as the product of average angular speed and the counter-torque in its active regime of predominant ATP hydrolysis (Kulish et al. 2016). However, mechanical work (rotation) against torque is not the useful outcome of F_1 motor activity in the direction of ATP hydrolysis. The useful outcome for some bacterial cells under anaerobic conditions is the establishment of the pH gradient across the membrane. But *in vivo*, the operation of ATP synthases from chloroplast, mitochondria, and most bacteria under aerobic conditions are inhibited from working in the direction of ATP hydrolysis (Walker 2013). When protonmotive force drops too much under some pathophysiological conditions (myocardial ischemia, for instance), the ATP hydrolysis of F_0F_1-ATPase may occur to restore the electrochemical proton gradient (Nesci et al. 2015). The enzyme can also be transformed into an energy-dissipating structure that performs maximal ATP hydrolysis not coupled to a generation of a proton gradient (Lippe et al. 2019). Some neurodegenerative diseases and cell death can follow. If the proton gradient equivalent to 100 mV across the membrane is established by ATP synthase, the maximal efficiency for this useful work can reach 75% (Kulish et al. 2016). In the normal mode of ATP synthesis, the F_0 motor drives the F_1 motor by converting the protonmotive force into the useful work of ATP synthesis. The efficiency for that energy conversion ranges from 60% (Elston et al. 1998) to 70% (von Ballmoos et al. 2009). In Dewar et al. publication (2006), we clearly stated that chosen experimentally determined parameters are relevant only for net positive ATP synthesis. Thus, the ATP hydrolysis mode and corresponding useful work of active proton transport performed by ATP synthase are irrelevant for discussing our 2006 paper and for considering the energy transduction of that enzyme in physiological conditions.

Second, the efficiency of energy conversion has a simple relationship to useful work. When the driving force is kept at a constant value (the condition we always used in our modeling experiments), useful work is proportional to efficiency. Consequently, the search for maximal useful work is then equivalent to the search for maximal free energy transduction efficiency. In our 2003–2019 publications about enzymes' optimizations, we explicitly rejected the maximal efficiency as the selection or optimization criterion. In our first review about principles guiding the evolution of enzyme kinetics (Dobovišek et al. 2014), we pointed out that we do not consider maximal efficiency as an important objective for natural selection. We considered physical principles, such as MaxEP, as a better and more fundamental guide for finding optimal kinetic parameters. Optimization for maximal efficiency and maximal useful work was never performed in any of our publications. We always calculated optimal (not maximal) values for efficiency and the kinetic parameters of interest, but these optimal values corresponded to the maximal value for information entropy and the maximal value for entropy production associated with the enzymatic transition leading to ATP synthesis in the case of ATP synthase. Arguments from Martyushev and Seleznev (2015) and invoked Melkikh and Seleznev's (2012) paper have mispresented our optimization procedure for ATP synthesis mode as maximal useful work calculations giving maximal efficiency for ATP hydrolysis mode operation of that proton pump. As discussed in the previous paragraph, the efficiency values from 86% to 100% derived from maximal useful work calculations by Melkikh and Seleznev (2012) are excessively high even when estimated for less physiologically relevant ATP hydrolysis mode of ATP synthase. Our modeling result is the optimal efficiency of 0.69 for free energy storage in the form of the E_{out}/E_{in} ratio and for the optimal efficiency of free-energy transduction (Dewar et al. 2006). We followed the approach of Pänke and Rumberg (1999) in lumping together the proton transport with ATP synthesis or hydrolysis in a single conformational transition between enzyme functional states (the transition M from Figure 11.2) directly connected to ATP release (the transition T from Figure 11.2) without any loss of flux through some branched pathway. There is no "slip" between M and T transition. The transport, mechanical, and chemical step are assumed to be tightly coupled in the M to T transitions. In the steady-state condition, currents J_M and J_T are always equal, as are the efficiencies of energy storage and energy conversion. It follows that our optimal efficiency expression from Dewar et al. (2006) is the upper limit because we omitted all sources of proton leaks and presumed an ideal conversion of primary proton flux into secondary ATP synthesis and ATP release flux.

Third, the reversible transition of substrates (ADP and P_i) to product (ATP) is certainly associated with substantial dissipation and cannot be interpreted solely as useful work. This misunderstanding from Martyushev and Seleznev's (2015) paper stems from different possibilities to dissect total entropy production into contributing terms. When clear separation is possible between primary (driving) and secondary (driven) force-flux couples, the entropy production of the system can be expressed as $TP = J_1X_1 + J_2X_2$ (see Chapter 3), and the second contribution to total entropy production can be negative. The efficiency of free-energy transduction is then defined as $\eta = -(J_2X_2)/(J_1X_1)$. The energy transduction can occur when the positive term (index 1) is larger than the negative term (index 2). The numerator J_2X_2 in the efficiency expression can be regarded as output power or useful work per unit time. In our publications, we used much more detailed accounting of all contributions to system entropy production as the starting point for analysis and subsequent optimizations. For all kinetic schemes of enzyme catalytic activity, it is well known that steady-state condition makes it possible to associate well-defined entropy production for each transition between enzyme functional states (Hill 1977). It does not matter how far is the distance from equilibrium and how nonlinear is the relationship between affinities and fluxes for a considered transition. Furthermore, we derived the proof, or theorem, if you like, that transitional entropy production has a well-defined maximum irrespective of which transition is examined (Dobovišek et al. 2011, Juretić et al. 2019b). In general, maximal dissipations are widely different for various catalytic steps. There are rate-limiting transitions with major contributions to P_{tot} that are obvious targets for optimizations. There is no such thing as equality of all transitions either in their contribution to total entropy production or in the influence the chosen transition has on the increase of total entropy production when transitional entropy production is maximized. On the other hand, all catalytic steps have some common features. In the enzymatic cycle, for instance, there may be a few of many transitions among important states, but it is not possible to assign free energy transduction to a single step (transition) in a cycle. The free energy transfer is an indivisible property of the entire cycle (Hill and Eisenberg 1981, Kamp et al. 1988). "Useful work" in the form of energy transfer between small molecules cannot be associated with the ES ↔ EP transition, as claimed by Martyushev and Seleznev (2015). It is the microscopic transition between two excited states of the enzyme complexed with a ligand. It often involves directional movements of water molecules, electrons, protons, and amino acids, but all constrained within nano-dimensions. Macroscopic changes (including work) can occur only after the completion of the catalytic cycle.

The point of view that there is some crucial part of the cycle where free energy transfer occurs was commonplace among biochemists, but not among biophysicists who studied enzyme catalysis, nor for that matter, in Peter Mitchell's publications (see, for instance, Mitchell 1973). Our kinetic model for the ATP synthase catalytic cycle is the special artificial case when intermediate steps in rotatory and elastic power transmission have been all connected into the M transition toward enzyme-bound ATP. However, when ADP and ATP are bound to the F_1 part of ATP synthase, the equilibrium constant of the ATP/ADP couple is close to 1.0, a marked difference from the equilibrium constant of approximately 10^{-8} in free solution. Boyer et al. (1973) concluded that large free energy investment is required to release ATP from the active site (the step T in our notation, Figure 11.2). When corresponding entropy productions are calculated after optimization for the optimal angular position of the transition state (Figure 11.4), the contribution of the ATP release step to total entropy production (T transition) is about three times larger than that of the ATP synthesis step (Table 11.1). In any case, it is incorrect to interpret the dissipative loss from some part of the catalytic cycle as "useful work" (Kulish et al. 2016).

Fourth, we did not seek the maximum substrate to product conversion rate in any of the above-cited research papers. We predicted the reaction rate from the optimal values of kinetic constants. It is not the research topic dealing with ground-state free energy changes, notwithstanding how these changes are interpreted. The activation free energy for each catalytic step is of crucial importance and how it changes under the influence of an enzyme. Hence, in our modeling of ATP synthesis by ATPase, we used the transition state theory because the optimization of reaction rates must deal with excited state parameters. Transition states are rate-determining for enzyme-catalyzed reactions (Richard 2019). Instead of calculating the product of reaction rates in the clockwise and counterclockwise direction of each reaction cycle, we could have calculated the sum of activation free energies, activation enthalpies, and activation entropies in each direction. It would be useful for finding the analogy of Kirchhoff's laws for biochemical circuits. Biological networks dissipate energy in a way similar to electric circuits (Fang and Wang 2020). The product of reaction rate ratios (forward to backward) is equal to the product of equilibrium constants around the cycle, which is proportional to the sum of thermodynamic forces acting in that cycle (Hill 1977). Information is needed about energy transfer and useful work (if any) performed by the enzyme. A single catalytic step does not provide such information, but there are other reasons why it is important to study how each catalytic step can be controlled by speeding it up or slowing it down. Nature's way to do it is by rare mutations causing beneficial amino acid substitutions in the protein structure. Selected for mutations are usually those that simultaneously increase entropy production and catalytic efficiency of more evolved enzymes (Juretić et al. 2019a). Finding the evolutionary potential of any given enzyme is of obvious theoretical and practical interest.

Fifth, a large number of measured rate constant values incorporated in our modeling is more an advantage than a weakness. It allows for comparison among predicted optimal values and observed values when all known constraints are taken into account. A much more frequent case is that some kinetic constants have not been measured. It impedes the analysis of free energy conversion. The values of reverse rate constants are mostly missing in kinetic schemes for enzymes performing one or more irreversible catalytic steps in the forward direction. When these values are estimated from the maximum transition entropy production (MTEP) theorem between enzyme functional states, then the rate-limiting steps can be identified, and the nature of nanocurrents inferred that are essential in each particular case of free-energy conversion (Juretić et al. 2019b). Imposed additional constraint about equality of some kinetic constants was present only in the Dobovišek et al. (2011) paper. That assumption was not present in any other of our research papers cited by Martyushev and Seleznev (2015). Also, we removed the equality restriction as unnecessary in the subsequent paper (Juretić et al. 2019a), dealing with the optimization of the same enzyme (lactamase). Thus, we did reduce the number of restrictions, as Russian authors suggested.

A caveat should be mentioned here about the relationship of the MTEP theorem and the MaxEP principle. The MTEP theorem helps in finding the maximal possible increase of enzyme-driven entropy production, but it is not equivalent to the MaxEP principle. The MaxEP principle has well-known restrictions for a local relationship between the cause and response, small-time intervals, and small volume elements (Martyushev 2013, Martyushev and Seleznev 2015). Biological selection and evolution are not restrained within these bounds. For selection to occur for some animal species, about ten generations are needed. That is an approximate time-frame for rare beneficial mutations to become established in the population. It is not a small time-interval, no matter how it is measured. However, mutations are mostly random events, initially well confined within a small volume and small time-intervals. Quantum thermodynamics would be needed to connect such atomic-level events to the appropriate modification of the maximum entropy production principle. Even so, there is no need to enter the strange world of quantum mechanics in the formulation, proof, or MTEP theorem applications. It does not have the power of principle but can still help to analyze how nature improved enzymes or how this can be done using biotechnological methods. Mirroring natural ways and means how evolution improved enzyme performance is usually a good choice for testing if trial and error feedback cycles can be greatly speeded up by directed evolution in a laboratory (Arnold 2018, Verma and Salunke 2018).

11.13 Our conjecture and concluding thoughts

After reading over the previous paragraph, we still do not know how to apply the maximum entropy production principle to enzyme-focused thermodynamics. The MTEP theorem does not imply the correctness or incorrectness of MaxEP in that field. One can take the educated guess that Hill's cycle fluxes (Hill 1977) would bridge the gap to the MaxEP principle in analogy to mesh currents from electrical engineering (Županović et al. 2004). Instead of asking what is the optimal value of the forward kinetic constant for a considered transition between enzyme functional states (resulting in the maximal EP for that transition), we can ask what the optimal value is for a cycle flux that would produce the maximal EP for that cycle among enzyme functional states? The question implies the existence of maximal EP for each cycle. That is not a proper analogy to mesh current's use together with energy conservation law to get Kirchhoff's loop law (the current distribution in a network) from the maximum entropy production requirement. The difficulty lies in the highly nonlinear flux-force relationships for the enzyme's internal network, which does not become less nonlinear after choosing cycle fluxes and cycle forces instead of operational fluxes and forces. Still, we used generalized Kirchhoff's laws for nonlinear situations throughout this book. A better analogy would be the conjecture that the conditional extremum of total entropy production exists for each enzyme and should be looked for by using the cycle currents as variables and well known variational technique of Lagrange's multipliers to take into account the conservation laws. In the Michaelis-Menten cyclic kinetic scheme (Figure 10.2), the cycle flux is identical to each one of three transition fluxes in the steady state. The maximum in total entropy production exists only if the sum of activation energies does not change (in either direction) during variations of rate constants (Figure 8.3). However, that is not the energy conservation constraint. But that simple and widely used scheme for enzyme kinetics remains the testing ground for the conjecture that the MaxEP requirement describes a theoretical optimal steady-state for each enzyme when all constraints are properly taken into account.

To summarize this chapter, let us see what are the insights of ATP synthase optimization when physical selection principles guide that procedure. The focus on the information entropy increase for the whole kinetic cycle has been equally important as the focus on dissipation increase for the crucial catalytic step connecting the proton transport to rotary motion, elastic strain, and conformational changes leading to ATP synthesis. Obtained optimal values of excited state parameters are in good agreement with their values deduced from experiments. To paraphrase Weber et al. (2015), we have ten years earlier addressed the fundamental question linking bioenergetics and nonequilibrium physics: can the evolved dissipative pathways that facilitate biomolecular function be identified by their extent of entropy production? We have found that dominant entropy-production transitions are inseparably coupled to nonequilibrium conformational switches enabling ATP synthesis and release.

Definitions and explanations from Chapter 11:

Maxwell's demon refers to a famous thought experiment described in 1867 and 1871 by James Clerk Maxwell—one of the most gifted physicists that ever lived (Maxwell 1871, Knott 1911). He devised the contraption of two gas-filled chambers isolated from the environment, but not from each other. A "demon" capable of violating the second law of thermodynamics performed only one "demonical" activity. He would open the door exclusively to fast molecules to pass from one gas-filled compartment to another. That would lead to one chamber warming up and another cooling down without any external influence. Entropy would decrease from the maximal value prescribed by the second law, thus violating it. Demon somehow acquired an exciting life of its own during the last 150 years. It still helps physics teachers to explain the statistical nature of the second law and why the demon does not have the power to break the second law. Remarkably, the research about the demon branched into many unexpected directions, from quantum physics to informatics and physics of life (Leff and Rex 1990).

The inflection point is the point at which the curvature of a function changes its sign. For increasing variable values, the function can change its shape from convex to concave at the inflection point. It happens at the point of the most rapid function's increase—the increase or decrease of the variable value causes the maximal possible change of the function's value.

References

Abrahams, J.P., Leslie, A.G.W., Lutter, R. and Walker, J.E. Structure at 2.8 Å resolution of F_1-ATPase from bovine heart mitochondria. Nature 370(1994): 621–628.

Al-Shayeb, B., Sachdeva, R., Chen, L.-X., Ward, F., Munk, P., Devoto, A. et al. Clades of huge phages from across Earth's ecosystems. Nature 578(2020): 425–431.

Allchin, D. To err and win a Nobel Prize: Paul Boyer, ATP synthase and the emergence of bioenergetics. J. Hist. Biol. 35(2002): 149–172.

Arnold, F.H. Directed evolution: Bringing new chemistry to life. Angew. Chem. Int. Ed. 57(2018): 4143–4148.

Artico, M., Spoletini, M., Fumagalli, L., Biagioni, F., Ryskalin, L., Fornai, F. et al. Egas Moniz: 90 years (1927–2017) from cerebral angiography. Front. Neuroanat. 11(2017): 81. doi: 10.3389/fnana.2017.00081.

Bell, P.J.L. Evidence supporting a viral origin of the eukaryotic nucleus. Virus Research 289(2020): 198168. doi: 10.1016/j.virusres.2020.198168.

Berg, H.C. and Anderson, R.A. Bacteria swim by rotating their flagellar filaments. Nature 245(1973): 380–382.

Beyerchen, A.D. 1977. Scientists Under Hitler: Politics and the Physics Community in the Third Reich, Yale University Press, New Haven, CT, USA.

Blum, D.J., Ko, Y.H., Hong, S., Rini, D.A. and Pedersen, P.L. ATP synthase motor components: Proposal and animation of two dynamic models for stator function. Biochem. Biophys. Res. Commun. 287(2001): 801–807.

Bonašić Lošić, Ž., Donđivić, T. and Juretić, D. Is the catalytic activity of triosephosphate isomerase fully optimized? An investigation based on maximization of entropy production. J. Biol. Phys. 43(2017): 69–86.

Bonora, M., Patergnani, S., Rimessi, A., De Marchi, E., Suski, J.M., Bononi, A. et al. ATP synthesis and storage. Purinergic Signal. 8(2012): 343–357.

Boyer, P.D., Bieber, L.L., Mitchell, R.A. and Szabolcsi, G. The apparent independence of the phosphorylation and water formation reactions from the oxidation reactions of oxidative phosphorylation. J. Biol. Chem. 241(1966): 5384–5390.

Boyer, P.D., Cross, R.L. and Momsen, W. A new concept for energy coupling in oxidative phosphorylation based on a molecular explanation of the oxygen exchange reactions. Proc. Natl. Acad. Sci. USA 70(1973): 2837–2839.

Boyer, P.D. 1974. Conformational coupling in biological energy transductions. pp. 289–301. *In*: Ernster, L. et al. (eds.). Dynamics of Energy-Transducing Membranes. Elsevier, Amsterdam, The Netherlands.

Boyer, P.D., Chance, B., Ernster, L., Mitchell, P., Racker, E. and Slater, E.C. Oxidative phosphorylation and photophosphorylation. Annu. Rev. Biochem. 46(1977): 955–1026. CONTAINS: Boyer, P.D. Coupling mechanisms in capture, transmission and use of energy. Ann. Rev. Biochem. 46(1977): 955–966.

Boyer, P.D. 1979. pp. 461–479. *In*: Lee, C.P., Schatz, G. and Ernster, L. (eds.). Membrane Bioenergetics. Addison-Wesley, Reading, MA, USA.

Boyer, P.D. and Kohlbrenner, W.E. 1981. pp. 231–240. *In*: Selman, B. and Selman-Reiner, S. (eds.). Energy Coupling in Photosynthesis. Elsevier Science Publishing Co., New York. NY, USA.

Boyer, P.D. The ATP synthase—a splendid molecular machine. Annu. Rev. Biochem. 66(1997): 717–749.

Boyer, P.D. Energy, life, and ATP (Nobel Lecture). Angew. Chem. Int. Ed. Engl. 37(1998): 2296–2307.

Boyer, P.D. A research journey with ATP synthase. J. Biol. Chem. 277(2002): 39045–39061.

Bradford, R.A.W. An investigation into the maximum entropy production principle in chaotic Rayleigh–Bénard convection. Physica A 392(2013): 6273–6283.

Bruers, J. A discussion on maximum entropy production and information theory. J. Phys. A: Math. Theor. 40(2007): 7441–7450.

Chamberlin, R.V. 2015. Terrel L. Hill 1917–2014. National Academy of Sciences Biographical Memoirs. www.nasonline.org/memoirs.

Chapman, A.G., Fall, L. and Atkinson, D.E. Adenylate energy charge in *Escherichia coli* during growth and starvation. J. Bacteriol. 108(1971): 1072–1086.

Chapman, A.G. and Atkinson, D.E. Adenine nucleotide concentrations and turnover rates. Their correlation with biological activity in bacteria and yeast. Adv. Microb. Physiol. 15(1977): 253–306.

Cross, R.L. and Boyer, P.D. Rapid labeling of adenosine triphosphate by phosphorus-32-labeled inorganic phosphate and the exchange of phosphate oxygens as related to conformational coupling in oxidative phosphorylation. Biochemistry 14(1975): 392–398.

de Duve, C. A research proposal on the origin of life. Orig. Life Evol. Biosph. 33(2003): 559–574.

Demirel, Y. Information in biological systems and the fluctuation theorem. Entropy 16(2014): 1931–1948.

Dewar, R., Juretić, D. and Županović, P. The functional design of the rotary enzyme ATP synthase is consistent with maximum entropy production. Chem. Phys. Lett. 430(2006): 177–182.

Dobovišek, A., Županović, P., Brumen, M., Bonačić Lošić, Ž., Kuić, D. and Juretić, D. Enzyme kinetics and the maximum entropy production principle. Biophys. Chem. 154(2011): 49–55.

Dobovišek, A., Županović, P., Brumen, M. and Juretić, D. 2014. Maximum entropy production and maximum Shannon entropy as germane principles for the evolution of enzyme kinetics. pp. 361–382. *In*: Dewar, R.C., Lineweaver, C.H., Niven, R.K. and Regenauer-Lieb K. (eds.). Beyond the Second Law. Springer-Verlag, Berlin, Heidelberg, Germany.

Duncan, T.M., Bulygin, V.V., Zhou, Y., Hutcheon, M.L. and Cross, R.L. Rotation of subunits during catalysis by *Escherichia coli* F1-ATPase. Proc. Natl. Acad. Sci. USA 92(1995): 10964–10968.

Elston, T., Wang, H. and Oster, G. Energy transduction in ATP synthase. Nature 391(1998): 510–513.

Fang, X. and Wang, J. Nonequilibrium thermodynamics in cell biology: Extending equilibrium formalism to cover living systems. Annu. Rev. Biophys. 49(2020): 227–246.

Gresser, M.J., Myers, J.A. and Boyer, P.D. Catalytic site cooperativity of beef heart mitochondrial F1 adenosine triphosphatase. J. Biol. Chem. 257(1982): 12030–12038.

Harold, F.M. Chemiosmotic interpretation of active transport in bacteria. Ann. N. Y. Acad. Sci. 227(1974): 297–311.

Hill, T.L. 1977. Free Energy Transduction in Biology: The Steady State Kinetic and Thermodynamic Formalism. Academic Press, New York, NY, USA.

Hill, T.L. and Eisenberg, E. Can free energy transduction be localized at some crucial part of the enzymatic cycle? Q. Rev. Biophys. 14(1981): 463–511.

Jia, H., Liggins, J.R. and Chow, W.S. Acclimation of leaves to low light produces large grana: the origin of the predominant attractive force at work. Phil. Trans. R. Soc. B 367(2012): 3494–3502.

Junge, W. and Nelson, N. ATP synthase. Annu. Rev. Biochem. 84(2015): 631–657.

Juretić, D. 1997. Bioenergetics: Work of Membrane Proteins. Informator, Zagreb, Croatia (in Croatian).

Juretić, D. and Županović, P. Photosynthetic models with maximum entropy production in irreversible charge transfer steps. J. Comp. Biol. Chem. 27(2003): 541–553.

Juretić, D., Bonačić Lošić, Ž., Kuić, D., Simunić, J. and Dobovišek, A. The maximum entropy production requirement for proton transfers enhances catalytic efficiency for β-lactamases. Biophys. Chem. 244(2019a): 11–21.

Juretić, D., Simunić, J. and Bonačić Lošić, Ž. Maximum entropy production theorem for transitions between enzyme functional states and its applications. Entropy 21(2019b): 743. doi: 10.3390/e21080743.

Kagawa, Y., Kandrach, A. and Racker, E. Partial resolution of the enzymes catalyzing oxidative phosphorylation. XXVI. Specificity of phospholipids required for energy transfer reactions. J. Biol. Chem. 248(1973): 676–684.

Kamerlin, S.C.L., Sharma, P.K., Prasad, R.B. and Warshel, A. Why nature really chose phosphate. Q. Rev. Biophys. 46(2013): 1–132. doi: 10.1017/S0033583512000157.

Kamp, F., Welch, G.R. and Westerhoff, H.V. Energy coupling and Hill cycles in enzymatic processes. Cell Biophys. 12(1988): 201–236.

Kandpal, R.P. and Boyer, P.D. *Escherichia coli* F1 ATPase is reversibly inhibited by intra- and intersubunit crosslinking: an approach to assess rotational catalysis. Biochim. Biophys. Acta 890(1987): 97–105.

Kauffman, G. Nobel Prize for MRI imaging denied to Raymond V. Damadian a decade ago. Chem. Educator 19(2014): 73–90.

Kieft, T.L. and Rosacker, L.L. Application of respiration- and adenylate-based soil microbiological assays to deep subsurface terrestrial sediments. Soil Biol. Biochem. 23(1991): 563–568.

Kinley, J. 2015. Gifted Mind: The Dr. Raymond Damadian Story, Inventor of the MRI. Master Books, Green Forest, AR, USA.

Klein, W.L. and Boyer, P.D. Energization of active transport by *Escherichia coli*. J. Biol. Chem. 247(1972): 7257–7265.

Knott, C.G. 1911. Life and Scientific Work of Peter Guthrie Tait. Cambridge University Press, Cambridge, UK. http://www.archive.org/details/lifescientificwo00knotuoft.

Koch, L.G. and Britton, S.L. Theoretical and biological evaluation of the link between low exercise capacity and disease risk. Cold Spring Harb. Perspect. Med. 8(2018): a029868. doi: 10.1101/cshperspect.a029868.

Kulish, O., Wright, A.D. and Terentjev, E.M. F_1 rotary motor of ATP synthase is driven by the torsionally-asymmetric drive shaft. Sci. Rep. 6(2016): 28180. doi: 10.1038/srep28180.

Leff, H.S. and Rex, A.F. 1990. Maxwell's Demon: Entropy, Information, Computing. Princeton University Press, Princeton, NJ, USA.

Lane, N. Proton gradients at the origin of life. Bioessays 39(2017): 1600217. doi: 10.1002/bies.201600217.

Lanouette, W. and Silard, B. 2018. Genius in the Shadows: A Biography of Leo Szilard: The Man Behind The Bomb. Skyhorse Publishing, New York, NY.

Lippe, G., Coluccino, G., Zancani, M., Baratta, W. and Crusiz, P. Mitochondrial F-ATP synthase and its transition into an energy-dissipating molecular machine. Oxid. Med. Cell Longev. 2019(2019): 8743257. doi: 10.1155/2019/8743257.

Manson, M.D., Tedesco, P., Berg, H.C., Harold, F.M. and Van der Drift, C. A protonmotive force drives bacterial flagella. Proc. Natl. Acad. Sci. USA 74(1977): 3060–3064.

Martin, O. and Horvath, J.E. Biological evolution of replicator systems: Towards a quantitative approach. Orig. Life Evol. Biosph. 43(2013): 151–160.

Martyushev, L.M. Entropy and entropy production: Old misconceptions and new breakthroughs. Entropy 15(2013): 1152–1170.

Martyushev, L. and Seleznev, V. Maximum entropy production: application to crystal growth and chemical kinetics. Curr. Opin. Chem. Eng. 7(2015): 23–31.

Maxwell, J.C. 1871. Theory of Heat. Cambridge University Press, Cambridge, UK. https://doi.org/10.1017/CBO9781139057943.

McGuire, C., Stransky, L., Cotter, K. and Forgac, M. Regulation of V-ATPase activity. Front. Biosci. (Landmark Ed). 22(2017): 609–622.

Melkikh, A.V. and Seleznev, V.D. Mechanisms and models of the active transport of ions and the transformation of energy in intracellular compartments. Prog. Biophys. Mol. Biol. 109(2012): 33–57.

Minkoff, L. and Damadian, R. Caloric catastrophe. Biophys. J. 13(1973a): 167–178.

Minkoff, L. and Damadian, R. Energy requirements of bacterial ion exchange. Ann. N. Y. Acad. Sci. 204(1973b): 249–260.

Mitchell, P. Chemiosmotic coupling in oxidative and photosynthetic phosphorylation. Biol. Rev. 41(1966): 445–502.

Mitchell, P. 1973. Chemiosmotic coupling in energy transduction: a logical development of biochemical knowledge. pp. 5–24. *In*: Avery, J. (ed.). Membrane Structure and Mechanisms of Biological Energy Transduction. Plenum Press, London, UK.

Mitchell, P. Keilin's respiratory chain concept and its chemiosmotic consequences. Science 206(1979): 1148–1159.

Mitchell, P. Molecular mechanics of protonmotive F_0F_1 ATPases. Rolling well and turnstile hypothesis. FEBS Lett. 182(1985): 1–7.

Moroz, A. and Wimpenny, D.I. On the variational framework employing optimal control for biochemical thermodynamics. Chem. Phys. 380(2011): 77–85.

Nath, S. Entropy production and its application to the coupled nonequilibrium processes of ATP Synthesis. Entropy 21(2019a): 746. doi: 10.3390/e21080746.

Nath, S. Coupling in ATP synthesis: Test of thermodynamic consistency and formulation in terms of the principle of least action. Chem. Phys. Lett. 723(2019b): 118–122.
Nesci, S., Trombetti, F., Ventrella, V. and Pagliarani, A. Opposite rotation directions in the synthesis and hydrolysis of ATP by the ATP synthase: hints from *a* subunit asymmetry. J. Membr. Biol. 248(2015): 163–169.
Nicholls, D.G. and Ferguson, S.J. 2002. Bioenergetics 3. Academic Press, London, UK.
Nicholls, D.G. and Ferguson, S.J. 2013. Bioenergetics 4. Academic Press, London, UK.
Noji, H., Yasuda, R., Yoshida, M. and Kinosita, K. Jr. Direct observation of the rotation of F_1-ATPase. Nature 386(1997): 299–302.
Pänke, O. and Rumberg, B. Kinetic modeling of rotary CF_0F_1-ATP synthase: storage of elastic energy during energy transduction. Biochim. Biophys. Acta 1412(1999): 118–128.
Peter, J.B. and Boyer, P.D. The formation of bound phosphohistidine from adenosine triphosphate-P^{32} in mitochondria. J. Biol. Chem. 238(1963): 1180–1182.
Prebble, J.N. Contrasting approaches to a biological problem: Paul Boyer, Peter Mitchell and the mechanism of the ATP synthase, 1961–1985. J. Hist. Biol. 46(2013): 699–737.
Prebble, J.N. 2019. Searching for a Mechanism: A History of Cell Bioenergetics. Oxford Univ. Press, New York, NY.
Ramaiah, A., Hathaway, J.A. and Atkinson, D.E. Adenylate as a metabolic regulator. Effect on yeast phosphofructokinase kinetics. J. Biol. Chem. 239(1964): 3619–3622.
Richard, J.P. Protein flexibility and stiffness enable efficient enzymatic catalysis. J. Am. Chem. Soc. 141(2019): 3320–3331.
Singharoy, A., Chipot, C., Ekimoto, T., Suzuki, K., Ikeguchi, M., Yamato, I. et al. Rotational mechanism model of the bacterial V_1 motor based on structural and computational analyses. Front. Physiol. 10(2019): 46. doi: 10.3389/fphys.2019.00046.
Skene, K.R. Thermodynamics, ecology and evolutionary biology: A bridge over troubled water or common ground? Acta Oecol. 85(2017): 116–125.
Slater, E.C. Mechanism of phosphorylation in the respiratory chain. Nature 172(1953): 975–978.
Stewart, A.G., Lee, L.K., Donohoe, M., Chaston, J.J. and Stock, D. The dynamic stator stalk of rotary ATPases. Nat. Commun. 3(2012): 687. doi: 10.1038/ncomms1693.
Stockbridge, R.B. and Wolfenden, R. The intrinsic reactivity of ATP and the catalytic proficiencies of kinases acting on glucose, *N*-acetylgalactosamine, and homoserine. J. Biol. Chem. 284(2009): 22747–22757.
Stockbridge, R.B., Lewis Jr., C.A., Yuan, Y. and Wolfenden, R. Impact of temperature on the time required for the establishment of primordial biochemistry, and for the evolution of enzymes. Proc. Natl. Acad. Sci. USA 207(2010): 22102–22105.
Stouthamer, A.H. 1977. Energetic aspects of the growth of microorganisms. pp. 285–315. *In*: Haddock, B.A. and Hamilton, W.A. (eds.). Microbial Energetics. Cambridge University Press, London, UK.
Verma, S. and Salunke, D.M. Directed evolution—bringing the power of evolution to the laboratory: 2018 Nobel Prize in Chemistry. Current Science 115(2018): 1627–1630.
von Ballmoos, C., Wiedenmann, A. and Dimroth, P. Essentials for ATP synthesis by F_1F_0 ATP synthases. Annu. Rev. Biochem. 78(2009): 649–672.
Walker, J.E. ATP synthesis by rotary catalysis (Nobel lecture). Angew. Chem. Int. Ed. 37(1998): 2308–2319.
Walker, J.E. The ATP synthase: the understood, the uncertain and the unknown. Biochem. Soc. Trans. 41(2013): 1–16.
Weber, J.K., Shukla, D. and Pande, V.S. Heat dissipation guides activation in signaling proteins. Proc. Natl. Acad. Sci. USA 112(2015): 10377–10382.
Williams, R.J.P. Possible functions of chains of catalysts. J. Theor. Biol. 1(1961): 1–17.
Whitfield, J. Survival of the likeliest? PLoS Biol. 5(2007): e142. doi: 10.1371/journal.pbio.0050142.
Wolfenden, R. Benchmark reaction rates, the stability of biological molecules in water, and the evolution of catalytic power in enzymes. Annu. Rev. Biochem. 80(2011): 645–667.
Yoshida, M., Okamoto, H., Sone, N., Hirata, H. and Kagawa, Y. Reconstitution of thermostable ATPase capable of energy coupling from its purified subunits. Proc. Natl. Acad. Sci. USA 74(1977): 936–940.
Županović, P., Juretić, D. and Botrić, S. Kirchhoff's loop law and the maximum entropy production principle. Phys. Rev. E 70(2004): 056108. doi: 10.1103/PhysRevE.70.056108.

CHAPTER 12

Bacteriorhodopsin

Light-harvesting Movie Star

12.1 Bacteriorhodopsin light cycle and its optimization

In this chapter, we shall first closely follow our recent paper about the MTEP theorem and its application to transitions between bacteriorhodopsin spectroscopic states (Juretić et al. 2019). For us, absorption of a photon can trigger an eye-brain communication we call vision. For certain bacteria, a photon triggers a protein quake, charge separation, the electric field build-up, and photosynthesis. Although separated by a billion years of evolutionary gap, the same protein type, and the same chromophore are responsible for both outcomes.

Bacteriorhodopsin is an integral membrane protein with seven membrane-spanning helices, one of them covalently connected to the retinal chromophore via Schiff base. The Schiff base is attached to the lysine side chain approximately in the middle of the 7th helix (helix G) at the central position halfway between the cytoplasmic and extracellular membrane surface. When acting as a photon detector in human rod cells, the protein is named rhodopsin (see Chapter 7). When performing the first step of photosynthesis for *Halobacterium salinarium*, the protein is called bacteriorhodopsin.

Bacteriorhodopsin (bR) is the most straightforward solution nature found for a light-activated proton pump, which can easily perform active transport of protons against the electrochemical proton gradient (Wickstrand et al. 2015). We used the available data from the recent literature to construct the bR photocycle containing dark transition and light-activated transitions involved in proton pumping (Figure 12.1). This light-harvesting proton pump is activated by photon absorption in such a way that its retinal chromophore isomerizes from all-*trans* conformation to a 13-*cis* conformation (Figure 12.2). During the last two decades, structural biologists used synchrotron radiation as a powerful tool in the research field of macromolecular X-ray crystallography. Recently developed **X-ray free-electron lasers** enabled still another jump in focusing the X-ray brilliance at a desired minuscule volume of an illuminated molecule. Instantaneous sample destruction is an unavoidable consequence of the extremely strong X-ray laser beam. Researchers had to develop methods for getting diffraction images before sample vaporization. This trick was named "diffraction before destruction." It consisted of using the 10-femtosecond (10^{-15} s)-long radiation pulses in such a short time range that X-ray-induced radiation damage to a crystallographic structure is not yet seen (Nango et al. 2016). The combination of Swiss Free Electron Laser equipment and the synchrotron radiation technology from Swiss Light Source also enabled the creation of movies by assembling the snapshots of protein movements from time-resolved serial crystallography (Weinert et al. 2019). The bacteriorhodopsin molecular machine became the movie star in the group of Prof. So Iwata from Japan, Dr. Jörg Standfuss from Switzerland, and in other laboratories (Wickstrand et al. 2015, Nango et al. 2016, Nogly et al. 2018, Weinert et al. 2019, Wickstrand et al. 2019). The bR was already destined to become the movie star at the start of this century (Kühlbrandt 2000). It was

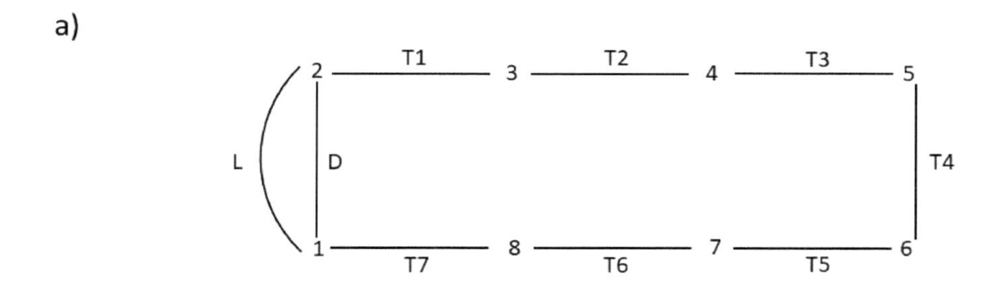

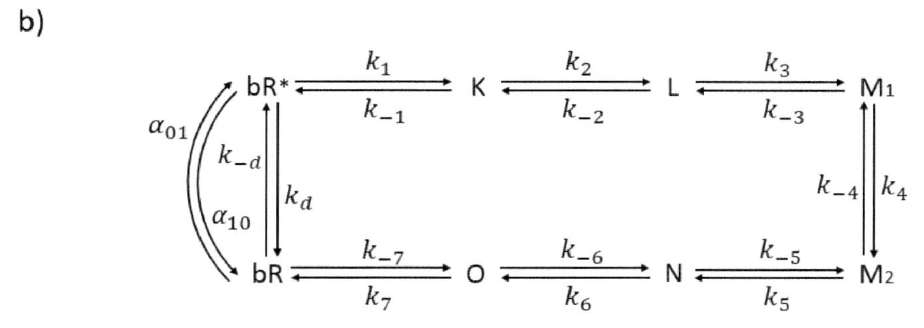

Figure 12.1: The 8-state kinetic model for the bacteriorhodopsin photocycle with a) capital letters for transitions between spectroscopic states one to eight, and with b) capital letters for spectroscopic states and small letters for corresponding rate constants. We assumed the clockwise direction of enzyme cycling, the existence of excited state bR*, and two pathways connecting bR to bR*: light-activated transition L and dark transition D. Rate constants in the L transition are explained in Juretić and Županović (2003). Their meaning is the same as in Figure 8.1b, only the index convention is different (α_{01} instead of α_{12}, α_{10} instead of α_{21}). Very large equilibrium constants are assumed for T1 and T7 transitions since observed light-activated retinal isomerization (T1) and recovery transition to the ground state (T7) are essentially unidirectional irreversible changes. Extremely fast femtosecond and picosecond conformational transitions of the retinal chromophore (T1) are followed by protein conformational dynamics leading to the proton release events in microseconds (T2-T3-T4) and the millisecond proton uptake from the cytoplasmic space (T5-T6-T7) (Weinert et al. 2019). After Juretić et al. (2019).

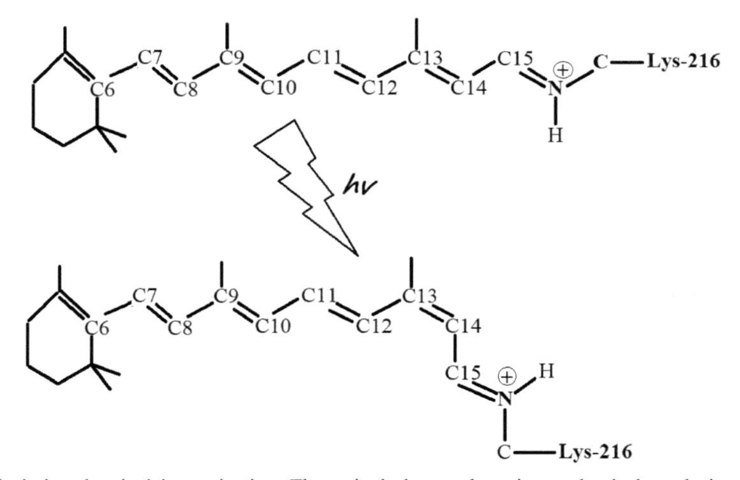

Figure 12.2: Light-induced retinal isomerization. The retinal chromophore is covalently bound via a protonated Schiff base to lysine-216 from bacteriorhodopsin. It undergoes ultrafast isomerization in about four picoseconds following photon absorption. The transition T1 (Figure 12.1) of all-*trans* excited electronic bR* state to 13-*cis* retinal configuration of spectroscopic state K opens the pathway for deprotonation of the Schiff base, the protonation of Asp-85, and ultimately to $L \leftrightarrow M_1 \leftrightarrow M_2$ transitions for proton exit to the extracytoplasmic space. Proton transfer from inside to the outside of the cell stores a certain amount of photon free energy into the bacterial electric field and transmembrane pH gradient despite the major part being dissipated in the photocycle. Stored energy is primarily used to power the ATP-synthase.

natural development after the discovery of purple membrane (PM) from *Halobacterium halobium* (an older name for *Halobacterium salinarium*) in 1967 (Stoeckenius and Rowen 1967), and after the recognition of light-driven proton pumping function for densely packed bR molecules from PM (Oesterhelt 1972, Oesterhelt and Stoeckenius 1973).

Older results about bacteriorhodopsin photocycle were extended and much better associated (Figure 12.1) with specific conformational changes. Ultrafast retinal isomerization can be followed in femtoseconds and picoseconds. It is an essentially irreversible process creating I(460), J(625), and K(590) spectral intermediates (Figure 1 from Wickstrand et al. 2019). These states are local energy minimums, each with stepwise lower energy from the excited state bR*. A significant part of excitation energy is dissipated already after passage to the twisted 13-cis isomer K state lasting only about 3–4 picoseconds (Nogly et al. 2018). However, the protein microenvironment prevented even greater energy dissipation in solution. Measurements of the isomerization efficiency from the all-trans to 13-cis reaction produced the consensus value of approximately 65%, that is, a high quantum yield (Tittor and Oesterhelt 1990, Ernst et al. 2014). Rhodopsin (see Chapter 7) achieves an even higher quantum yield of 67% for active photoproduct. For retinal in solution, the quantum yield of 13-*cis* isomer is only about 6%, that is, ten times lower (Hamm et al. 1996).

We can conclude that opsins of rhodopsin and bacteriorhodopsin have been honed through biological evolution for stereospecific and coherent vibrational motions to ensure the unidirectionality of isomerization, producing the high quantum yield for the proton-gating spectral intermediates. After retinal isomerization (subsumed bR*→I→J→K transitions as the overall bR*↔ K transition in Figure 12.1), the photocycle proceeds via spectroscopic states K(590), L(550), M_1(410), and M_2(410) with a large movement of key amino acid residues, water molecules, and alpha-helices (Nogly et al. 2018, Wickstrand et al. 2019). Light-induced conformational change of retinal triggers the conformational changes and polarity changes in the nearby protein microenvironment. The central protein location deep inside the membrane ceases to be the hydrophobic barrier. The chain of three water molecules forms, thus supporting the Grotthuss mechanism of fast proton transport in water wire (Cukierman 2006) and proton release. All these movements are precisely synchronized in time and space to cause a proton release from Schiff base toward Asp-85 and extracellular space in microseconds (Figure 12.1a, transitions T2, T3, T4). The T4 step for the M_1↔M_2 transition (Figure 12.1) represents the switch between the Schiff base's accessibility to the cytoplasmic and extracellular side. The molecular dynamics study of initial transitions in the bacteriorhodopsin photocycle pointed toward the retinal Schiff base dipole rotation weakening its hydrophobic bond. Thus, the gate is opened for proton-hopping directional movement toward nearby water molecules serving as the bridge to the primary proton acceptor Asp-85 (Ernst et al. 2014). Large structural rearrangements also characterize recovery transitions to spectroscopic states N(560), O(640), and bR(570) (see Figure 12.1 transitions T5, T6, and T7). These dynamical events open a transient proton wire by positioning amino acid residues and water molecules to funnel the proton from the cytoplasmic side to the Asp-96. It allows the Schiff base to accept the proton from Asp-96 so that the photocycle can be repeated (Weinert et al. 2019).

Exported proton does not have to be identical to the cytoplasmic proton accepted from Asp-96. What counts is that the free-energy transduction machinery of bacteriorhodopsin saves just enough of photon free energy to power the ATP-synthase by a created electrochemical proton gradient. This feat would be impossible in the absence of a topologically closed membrane impermeable to protons except through controlled channels formed by integral membrane proteins.

The excess energy protein received after photon absorption disperses away from the active site at 2 km/s. It is faster than the speed of sound in water (1.5 km/s) and similar to primary seismic waves' speed after an earthquake. When translated into nm/ps, the excitation wave speed of 2 nm/ps is more than enough to support recent observations of about 10 angstroms fast movements in the microworld of excited bacteriorhodopsin (Weinert et al. 2019). Energy dispersal and dissipation are facilitated by collective motions of polar residues and water molecules (movie 2 from Nogly et al. 2018). The theory of proteinquakes proposes that absorbed strain energy is released through waves, channeling dissipation in the form of phonons and collective structural deformations (Ansari et al. 1985).

In its natural membrane environment, bacteriorhodopsin dissipates more than 85% of photon free energy. One can ask, why such a high quantum yield and low efficiency of light power conversion to protonmotive power? We have seen that converting protonmotive power into ATP synthesis by an ATP-synthase is a more efficient process. Higher energy conversion efficiency can be easily achieved if the stronger electric field is created, that is, greater than a minimal field of about 130 mV, which is needed to put into rotation the ATP-synthase rotary motor for producing ATP. Photon free energy for photons with a wavelength of about 570 nm, which bacteriorhodopsin prefers to absorb, is quite high and more than enough to create a much stronger electric field. The problem with too strong an electric field is that it will cause a dielectric breakdown of a plasma membrane and cellular death. *H. salinarium* can develop a maximal electric field of about 280 mV (Michel and Oesterhelt 1980). Still, assuming that bacteriorhodopsins can be incorporated in much more robust artificial membranes, we can examine in simulations the cases when weak, strong, and super-strong secondary force is developed corresponding to the membrane potential of –195, –278, and –1185 mV, respectively. Corresponding values for developed secondary force X_{sec} is expressed in kJ/mol as, respectively, –18.84, –26.86, and –123 kJ/mol (Figure 12.3). The first value of 195 mV is quite common for the membrane potential of bacteria, archaea, and mitochondria, and it is identical to one we used in earlier simulations (Juretić and Županović 2003). The second value of –278 mV was also used by us earlier (Dobovišek et al. 2014), similar to the maximal measured value for membrane potential established by *H. Salinarium*. The third and highest value of membrane potential, equal to –1.185 V, corresponds to the maximal efficiency of free energy conversion, which is slightly higher than 70%.

We used estimated kinetic and thermodynamic parameters data for bR (van Stokkum and Lozier 2002, Nango et al. 2016) and performed the simulations designed to answer several questions:

a) Which transition step, out of seven Ti steps (Figure 12.1a), is associated with the greatest entropy production?
b) What is the rate-limiting step among all Ti transitions involved in a complex interplay of retinal, protein atoms, and water molecule movements, resulting in proton pumping, charge separation, and the creation of the protonmotive force (pmf) (Bondar et al. 2008)?
c) When the MTEP theorem is used to optimize each transition, is there a single catalytic step for which photochemical quantum yield, the efficiency of free energy conversion, and total entropy production all exhibit increased optimal values with respect to values obtained without optimization?

In order to apply irreversible thermodynamics to the initial photon absorption step, we introduced the excited state bR*, light-activated transition L from the ground to excited state, and non-radiative transition D back to the ground state. We used our extension of Hill's formalism (Hill 1977) to the light-absorbing systems (Juretić and Županović 2003, Dobovišek et al. 2014). Proton transfer and charge separation take place in the productive T1 to T7 pathway. In a reversible model of van Stokkum and Lozier (2002), all thermodynamic and kinetic parameters (equilibrium constant, forward and reverse kinetic constants) have been estimated for T2 to T6 transitions. The $\tau = 4$ ps time-constant estimate by Nango et al. (2016) was used to calculate the forward constant k_1 as $k_1 = 2.5 \times 10^{11}\ s^{-1}$. With a choice of equilibrium constant $K_7 = 2 \times 10^7$ for recovery transition, the remaining constants can be easily calculated from the requirement that the product of all equilibrium constants must be equal to $\exp(X_{sec}/k_BT)$. The X_{sec} is developed secondary force in the productive charge separation cycle: L →T1→→T7→L (Figure 12.1a) (Hill 1977). We have chosen the kinetic constant k_d for non-radiative D transition as $k_d = 10^8\ s^{-1}$ and the light-absorption rate $\alpha_{01} = 100\ s^{-1}$ by following our choice for modeling bacteriorhodopsin photocycle with the system being at room

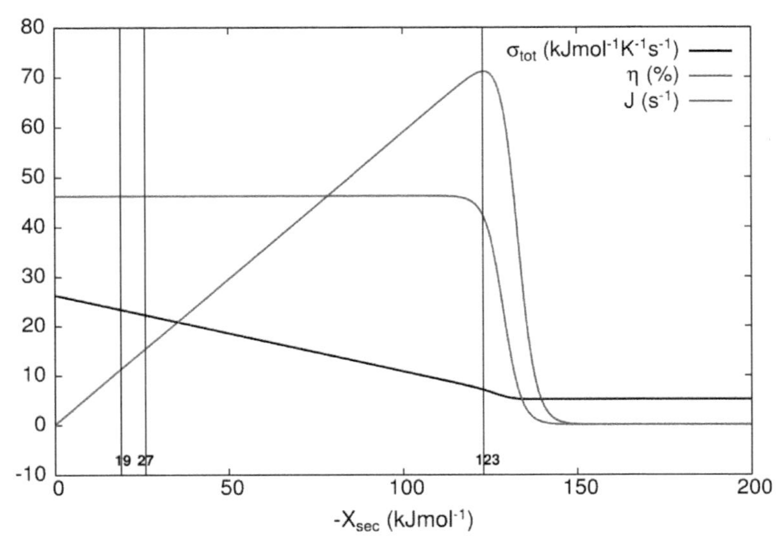

Figure 12.3: Dependence of the overall entropy production σ_{tot}, efficiency η, and proton flux J on secondary force X_{sec} for bacteriorhodopsin photocycle at room temperature T = 298.16 K. The vertical lines represent cutoffs at secondary force values of –18.84, –26.86, and –123 kJ/mol, respectively, corresponding to membrane potential values of –195, –278 and –1185 mV. The secondary force value of –123 kJ/mol corresponds to maximal efficiency η. See Juretić et al. (2019) for details.

temperature T = 298.16 K (Figure 12.1b) (Juretić and Županović 2003). Equilibrium constants in the light cycle L-D are found as $K_L = \exp(h\nu/k_BT_R)$ and $K_D = \exp(h\nu/k_BT)$, where T_R is an effective temperature, which is higher than T and increases with increased light absorption rate α_{01} and increased light intensity $I = J_L$. This is the same approach we used earlier (Juretić and Županović 2003). Fixed forward rate constants in the light cycle L-D, the light-absorption rate α_{01} = 100 s^{-1} and relaxation rate $k_d = 10^8$ s^{-1} in dark conditions to ground state bR, are also the same for presented results (Figure 12.3) as in Juretić and Županović (2003). The main difference concerning that earlier study is the higher number of considered states (eight instead of five) and the omission of the slip transition in the productive pathway after we took into account the updated consensus about light-activated functional and structural bR transitions (Nango et al. 2016, Weinert et al. 2019, Wickstrand et al. 2019).

According to van Stokkum and Lozier (2002), when T1 transition (very fast relaxation from excited state bR* to K_{590} spectroscopic state) is not considered, the major part of the free energy is dissipated in the recovery T7 step (O→bR). It ensures the vectorial transport of protons from the cytoplasm to the external space. Our choice of the equilibrium constant for the recovery step (see the previous paragraph) is in accord with the observed unidirectionality of proton transport. The accumulation of intermediate states then does not occur regardless of how strong is the proton back pressure (the pmf value). The calculations of entropy production for each step in the charge separation pathway confirmed that there is no single “protonation switch” solely responsible for the vectorial transport (van Stokkum and Lozier 2002). When all transitions are included, the ranking of transitional entropy productions is: $\sigma_1 > \sigma_7 > \sigma_L > \sigma_5 > \sigma_6 > \sigma_2 = \sigma_3 > \sigma_4 > \sigma_D$ in the physiological range of membrane potentials (delineated by two vertical lines from Figure 12.3 at –19 and –27 kJ/mol). At the maximal free-energy transduction efficiency (reached at the third vertical line from Figure 12.3 for X_{sec} = –123 kJ/mol), the recovery transition T7 becomes responsible for the highest contribution to overall entropy production (Juretić et al. 2019).

We also investigated how proton flux J, entropy production σ_{tot}, and efficiency η vary as the secondary force X_{sec} varies. The secondary force is the electrochemical proton gradient created by light-activated proton pumping across the topologically closed membrane. As it is seen from

Figure 12.3, flux stays almost constant, and entropy production decreases as the efficiency increases from zero to its maximal value when one varies secondary force from zero to –123 kJ/mol. After additional secondary force increases, flux and efficiency fall to the zero or near-zero values, and entropy production decreases to a finite, almost constant low value.

The MTEP applications and corresponding optimization results did not change much the distribution of transitional entropy productions. MTEP application to the recovery step (T7 transition) for the kinetic model of the bacteriorhodopsin photocycle leads to a moderate increase in the output flux and overall entropy production compared to values obtained without optimization (Juretić et al. 2019). The increase in turnover number, flux, and catalysis rate is in accord with our results described in previous chapters, that single out the last catalytic step as an essential determinant for the whole free-energy transduction process. We note that state probability p_2 of the excited state bR* is very small because the transition from state 2 to state 3 is very fast with large forward kinetic constant k_1 and correspondingly large equilibrium constant. The thermal relaxation to the ground state bR is also fast for the D transition.

12.2 How perfect is photon free-energy conversion by bacteriorhodopsin?

Asking the question of why photosynthetic bacteria developed the light-activated protein pump for expelling protons prompts us to connect this chapter to the previous one about ATP synthase. What would happen if bR and ATPase are incorporated with proper orientation in some topologically closed membrane (natural or artificial)? By proper orientation, we mean that the catalytic F_1 head of ATPase is oriented toward the interior (cytoplasm for *H. salinarium* cells) while bR pumps the protons toward exterior space. Of course, this has been already achieved by *H. salinarium* cytoplasmic membrane with the vitally important result of closing the proton circuit in such a way that ATP synthesis occurs upon illumination. The chemiosmotic hypothesis crossed the threshold to confirmed theory in 1974 when Racker and Stoeckenius incorporated bR and ATP-synthase in artificial phospholipid vesicles and observed light-activated ATP synthesis (Racker and Stoeckenius 1974). By the way, the role of electrochemical proton gradient changes from being the secondary force for bacteriorhodopsin to being the primary force for ATP synthase operating in the ATP synthesis mode.

From this chapter and Chapter 10 about "perfect enzymes," we see that the relationship between entropy production and efficiency change when secondary force can arise. The catalytic efficiency definition (Chapter 10) differs from the free-energy transduction efficiency (Chapter 3). For only one force-flux couple, increased flux is tightly connected to increased catalytic efficiency and increased entropy production. However, free-energy transduction efficiency cannot be defined for a single force-flux couple (Juretić et al. 2019). For more than one force-flux couple, the increase in secondary force leads to the rise in free-energy storage efficiency (Chapter 3) and the increase in overall free-energy-transduction efficiency. A moderate decrease of total entropy production also occurs (Figure 12.3) (see Figure 6 from Juretić and Županović 2003, as the same result when different kinetic scheme is analyzed). The secondary force increase causes the proton back pressure when the secondary force is the electrochemical proton gradient produced by photosynthesis or respiration. In theory, a high proton back pressure brings the system close to the far-from-equilibrium nonlinear analogy of the static head state. The current of expelled protons becomes equal to the current of backward flowing protons. With zero net proton current, the free-energy transduction efficiency vanishes despite reaching maximum in free-energy storage efficiency. Going in a reverse direction, decreasing the secondary force (from the static head value) brings about the concomitant increase in entropy production and free-energy transduction efficiency, as we already noticed in 1987 (Juretić and Westerhoff 1987).

In practice, the hierarchy of free-energy transduction steps is never blocked in some analogy of the static head state with close to minimal entropy production and vanishing transduction efficiency. The charge separation acrosss the membrane is responsible only for creating the protonmotive force as the initial secondary force. Slower subsequent steps transform each secondary force into the driving force at the next levels of free-energy converters. Life can hardly be imagined without a spreading cascade of free-energy converters irrespective of the cell type being bacterial, algal, animal, or plant cells. What comes to mind are fascinating slow-motion photographs of lighting bolts spreading their tendrils toward the ground as they seek the least resistance paths.

To be more specific, when a downstream force-flux couple is activated, such as the ATP-synthase, this can lead toward increased free-energy transduction efficiency. An additional hierarchy of downstream free-energy converters, such as voltage-gated channels, can move optimal transduction efficiency into the physiological range of membrane potentials and pH gradients. Thus, metabolically more complex energy transduction patterns tend to increase energy transduction efficiency. Some optimal state is reached when all enzymes or channels are suitably optimized by biological evolution.

There are obvious limits to increasing secondary force and free-energy transduction efficiency when too much of a bonanza stops or even destroys everything. There must also be some limits to the secondary force decrease when it becomes too weak to drive numerous downstream free energy converters. Entropy production must not decrease too much and must not increase too much, and the same holds for energy conversion efficiency. In other words, these fundamental physical and functional (biological) characteristics of energy conversion have become tightly regulated and optimized during biological evolution. If this is so, we would expect that MTEP optimization (with all constraints taken into account) cannot change much already optimal values of these parameters inferred from measurements. We assume that biological evolution used almost all available options to speed up the enzyme evolution toward a more efficient dissipation of free energy gradients. Indeed, only small increases of entropy production, net proton flux, and energy conversion efficiency can be achieved after MTEP optimization for the last (recovery step) in the bacteriorhodopsin photocycle (Table 12.1). MTEP optimization did not increase proton flux or associated dissipation when performed separately for the other six productive transitions between bR functional (spectroscopic) states. The possibility remains that a significant increase in performance parameters can be gained by joint optimizations in two or more proton-hopping transitions. We achieved that goal in joint MTEP optimizations for several subsequent photocycle transitions leading to active transport of protons, firstly for the bacterial photosynthesis model, and secondly for simpler 5-state bacteriorhodopsin kinetic scheme (Juretić and Županović 2003).

The meaning of all Table 12.1 parameters is explained in that paper, in Juretić et al. 2019 paper, and next to the asterisk sign below this table. Still, the photochemical yield deserves an additional explanation to avoid the confusion with the quantum yield mentioned earlier. Various photochemical or quantum yields can be defined, measured, and calculated. In the case of bacteriorhodopsin, the photon energy is used to convert all-*trans* retinal to 13-*cis* twisted conformation, photon flux J_L into proton flux $J = J_T$, and open circuit affinity A (maximal steady-state affinity of a pigment) into the voltage drop on a load A_{oc}. Corresponding yields are respectively: moles of 13-cis retinal-protein complex formed/moles of absorbed quanta at $\lambda = 570$ nm, the ratio J(T)/J(L) of net flux J_T through M states over the rate J_L of light absorption, and affinity transfer efficiency A/A_{oc}. Almost 100% photochemical yield J_T/J_L, obtained without or with MTEP optimization for the bacteriorhodopsin modeling (Table 12.1), is in accord with optimal yields of 0.98 and 0.97 obtained by modeling the bacterial photosynthesis when the pigment is, respectively, chlorophyll or bacteriochlorophyll (Lavergne and Joliot 1996). Experimental measurements also confirmed that at least 90% of bR cycles through the formation of M spectral intermediates (Govindjee et al. 1990). In their review paper, Ernst et al. (2014) stated that the selectivity of isomerization is 100% for bR. At present, it appears that the bacteriorhodopsin is better perfected molecular machine through natural evolution than any other we already considered in this book. Billion years of evolution ensured that the flux ratio for all-trans to 13-cis isomerization efficiency increased from low values in solution (Hamm et al. 1996) to maximally possible (close to 100%) in a protective protein shell. However, we should mention that close to 100% flux conversion efficiency cannot be used as an argument for minimal

Table 12.1: Kinetic models of the bacteriorhodopsin photocycle without and with MTEP optimization in the recovery T7 transition (Juretić et al. 2019).

	$X_{sec} = -18.84$ kJmol^{-1}		$X_{sec} = -26.86$ kJmol^{-1}		$X_{sec} = -123$ kJmol^{-1}	
Parameters*	No optimization	T7 optimization	No optimization	T7 optimization	No optimization	T7 optimization
k_7 (s^{-1})	700	1750	700	1750	700	1670
σ_L (kJmol^{-1}K^{-1}s^{-1})	2.1	2.5	2.1	2.5	0.7	0.8
σ_D (kJmol^{-1}K^{-1}s^{-1})	$9.8 \cdot 10^{-3}$	$1.2 \cdot 10^{-2}$	$9.8 \cdot 10^{-3}$	$1.2 \cdot 10^{-2}$	0.5	0.5
σ_1 (kJmol^{-1}K^{-1}s^{-1})	14.7	16.3	13.4	14.9	$7.3 \cdot 10^{-3}$	$9.0 \cdot 10^{-3}$
σ_2 (kJmol^{-1}K^{-1}s^{-1})	$2.5 \cdot 10^{-3}$	$2.9 \cdot 10^{-3}$	$2.5 \cdot 10^{-3}$	$2.9 \cdot 10^{-3}$	$2.3 \cdot 10^{-3}$	$2.7 \cdot 10^{-3}$
σ_3 (kJmol^{-1}K^{-1}s^{-1})	$2.5 \cdot 10^{-2}$	$2.9 \cdot 10^{-2}$	$2.5 \cdot 10^{-2}$	$2.9 \cdot 10^{-2}$	$2.3 \cdot 10^{-2}$	$2.7 \cdot 10^{-2}$
σ_4 (kJmol^{-1}K^{-1}s^{-1})	$4.9 \cdot 10^{-2}$	$5.7 \cdot 10^{-2}$	$4.9 \cdot 10^{-2}$	$5.7 \cdot 10^{-2}$	$4.4 \cdot 10^{-2}$	$5.2 \cdot 10^{-2}$
σ_5 (kJmol^{-1}K^{-1}s^{-1})	0.5	0.6	0.5	0.6	0.4	0.6
σ_6 (kJmol^{-1}K^{-1}s^{-1})	0.4	0.7	0.4	0.7	0.3	0.6
σ_7 (kJmol^{-1}K^{-1}s^{-1})	5.7	5.9	5.7	5.9	5.1	5.3
σ_{tot} (kJmol^{-1}K^{-1}s^{-1})	23.4	25.9	22.2	24.5	7.1	7.8
J (s^{-1})	46.2	51.1	46.2	51.1	42.1	46.5
η (%)	11.1	11.1	15.8	15.8	71.1	71.2
J/J_L (%)	99.96	99.96	99.96	99.96	98.08	98.28
A/A_{oc}(%)	91.95	91.95	91.95	91.95	97.33	97.20
S	1.28	1.20	1.28	1.20	1.22	1.14
p_1	0.46	0.51	0.46	0.51	0.51	0.55
p_2	$1.9 \cdot 10^{-10}$	$2.1 \cdot 10^{-9}$	$1.9 \cdot 10^{-10}$	$2.1 \cdot 10^{-9}$	$7.8 \cdot 10^{-9}$	$2.1 \cdot 10^{-9}$
p_3	0.02	0.02	0.02	0.02	0.02	0.02
p_4	0.06	0.07	0.06	0.07	0.06	0.06
p_5	0.10	0.11	0.10	0.11	0.09	0.10
p_6	0.13	0.14	0.13	0.14	0.12	0.13
p_7	0.15	0.13	0.15	0.13	0.14	0.12
p_8	0.07	0.03	0.07	0.03	0.06	0.03

* Recovery kinetic constant k_7, entropy productions σ_i, flux $J = J(T) = J_T$ in the productive cycle, free-energy transduction efficiency η, entropy S, and state probabilities p_i for three representative values of secondary force X_{sec} in 8-state kinetic models of the bacteriorhodopsin photocycle without and with MTEP optimization in the recovery T7 transition (see Figure 12.1). A/A_{oc} is affinity transfer efficiency, and J/J_L is the photochemical yield (quantum efficiency) (Juretić and Županović 2003).

dissipation (Dioumaev 2001). Rather, it is an argument for enviable evolutionary accomplishment to create the force field by exquisitely fine-tuned atomic movements during dissipative space-time events triggered by an initial excitation.

The MTEP optimization reproduced the same hierarchical order of spectroscopic states, structural dynamics, lifetimes, and kinetic constants that led to the functional activity of bacteriorhodopsin and bacterial photosynthesis during biological evolution (Juretić and Županović 2003). To paraphrase Annila and Kuismanen (2006), nature's quest to abolish free-energy gradients by energy dispersal, as quickly as possible, leads naturally to the hierarchically organized energy transduction machinery, just as it emerged in the evolutionary process. Different formulations of the maximum entropy production principle provide the rationale of how various evolutionary processes emerge with hierarchical organization, while the MinEP principle does not.

12.3 Spin-off projects including bacteriorhodopsin and other bR-like proteins

Recent spin-off projects that arose from bR, rhodopsin, and other bR-like proteins are too numerous to describe comprehensively. I shall mention just a few of them. Thanks to bR, the dreams about constructing energetically independent artificial cells have approached the realization. Berhanu et al. (2019) described the giant unilamellar vesicle (GUV) encapsulating the cell-free translation system (named PURE, Shimizu et al. 2001) and the artificial organelle with membrane incorporated bR and F_0F_1-ATPsynthase. It is a much more complex arrangement than those of Racker and Stoeckenius from 1974 that we described above. Present authors demonstrated that their artificial cell GUV could synthesize some of its proteins when illuminated. This positive feedback loop achieved a moderate enhancement of the ATP photosynthesis rate. Although restricted to photosynthesis of only two proteins (bR and F_0), the Berhanu et al. (2019) artificial photosynthetic cell can supply sufficient free energy for operating gene expression. Did these authors construct the artificial **autotrophic** cell or contributed to understanding how first proto-cells learned to use sunlight as an energy source?

The question of how cells first become self-sustaining, as photoautotrophs or chemolithotrophs, is still very much contested (Deamer and Weber 2010). The construction of artificial autotrophic cells would be an important milestone in learning the minimal requirements for producing food from inorganic sources (water, carbon dioxide, and light energy in the case of photoautotrophs) (Berhanu et al. 2019). As is usual in science practice, the research team leader and the last author of the paper, Professor Yutetsu Kuruma at Tokyo Institute of Technology, provided the gist of their achievements for the public release. He was careful to emphasize the importance of constructing a complex encapsulated membrane structure with just the right amount of cholesterol to control spontaneous membrane integration of proteins and to inhibit proton leaks. In his words, "the cell membrane is the most important aspect of forming the cell." The authors of Berhanu et al. 2019 paper claimed that their research is pertinent to how primordial cells gained the ability to drive primitive metabolism. It is the origin of life question that will be discussed in the next chapter.

One optoelectronic application is a recent finding that bacteriorhodopsin improves the performance of solar cells (Das et al. 2019). The integration of bR molecules between perovskite and TiO_2 enhanced the charge transport upon light excitation. Energy conversion efficiency increased from 14.6% to 17%. According to the authors, this bio-inspired hybrid system, containing artificial and biological materials, has the potential to surpass either inorganic or organic molecules in transforming the light energy into an electrical current. It will not be an easy job because perovskite solar cells' energy conversion efficiency has been boosted to reach 23.2% (Jiang et al. 2019).

Nanophotocatalysis is another way to capture and store solar energy. A similar hybrid system of bR assembled on inorganic Pt/TiO_2 surface is efficient in hydrogen generation when exposed to light irradiation (Balasubramanian et al. 2013). Methanol is used as the electron donor to scavenge the photochemically produced holes. The group leader for this research, Dr. Elena Rozhkova at the Center for Nanoscale Materials, Argonne National Laboratory, pointed out that hydrogen generation is not the natural function of bacteriorhodopsin. Platinum nanocatalyst is used for proton reduction chemistry to convert the light-driven proton flux by bR into molecular hydrogen. Bacteriorhodopsin also corrected the titanium dioxide's inability to react with visible light. A particularly impressive result is 25 times rise in the photocatalytic efficiency under white light illumination compared with the monochromatic green light. This resilient protein (Hampp 2000) offers a new path to green hydrogen fuel at a wide range of demanding conditions (high salt concentration, elevated temperatures, low and high pH, dry conditions, short to long illumination wavelength) without environmental pollution. When hydrogen is combined with oxygen in a fuel cell, the benign output is water, while electricity and heat are desired products. Potential commercial developments are promising due to the low-cost biotechnological production of the purple membrane (from *Halobacterium salinarium*) with a tightly integrated two-dimensional lattice of uniformly oriented bR molecules. Rozhkova et al. recently patented this hybrid system for light-driven hydrogen generation (Rozhkova et al. US patent 10,220,378 B2 from 5 March 2019).

My bachelor thesis from 1968 was about plasmon excitations in metals. Why do I mention it here? Because surface plasmons can enhance the photocurrent generation by bR and speed up its photocycle almost 100 times (Yen et al. 2011). The authors of this Nano Letters paper discovered how to control silver nanoparticles synthesis to achieve the optimal size and shape for the spectral overlap of surface plasmon resonance with the blue light-absorbing M-intermediate. The photon's absorption in the wavelength region from 350 to 550 nm (blue light) introduced the bypass in the bR photocycle. The protein then does not pass any more through slow M↔N, N↔O, and O→bR recovery transitions (T5 to T7 transitions from Figure 12.1) lasting about 15 ms. Instead, rapid reprotonation and reisomerization of the retinal occurs through only 200 ns of the direct transition from M to bR state. It was necessary to use proton-selective Nafion membranes, bacteriorhodopsin, and optimized silver nanoparticles to achieve this nanotechnology feat. Nafion is the brand name for the fluoropolymer proton conductor discovered about 50 years ago by Walter Grot of DuPont (Grot 2011). Li et al. (2018) describe this and many other bR-based bioelectronic devices. When oriented bacteriorhodopsins are incorporated in robust bioelectronic devices, high efficiency can be reached with an additional bonus of smaller entropy production, that is, considerably less heating with a minimal decrease in produced proton current. Li et al. commented (2018) that bR has excellent potential in medically related studies on how to construct retinal prostheses. They had in mind the individuals with retinal diseases such as retinitis pigmentosa or age-related macular degeneration. It is not easy to build implants that would replace dead cells and restore eyesight. Researchers Nicole Wagner and Jordan Greco from Farmington-based company LambdaVision (U.S.A.) took the initiative to organize the manufacture of retinal implants at the International Space Station (ISS). The gravity absence helps in the formation of multiple homogenous layers containing bacteriorhodopsin (NASA proposal 19-1-H8.01-3699). During the next few years, we shall learn if their literally sky-high project was worthwhile or not.

The same researchers participated in examining how the bR photocycle kinetics can be manipulated by constructing the hybrid system containing quantum dots and purple membrane with the bacteriorhodopsin A103C mutant (Wax et al. 2019). The research field of quantum dots (QDs) was initiated by Alexsey Ekimov in the former Soviet Union and Louis Brus in the United States, who discovered that striking color changes of semiconductor nanoparticles depend on their size (Ekimov and Onushchenko 1982, Rossetti et al. 1983, Brus 1983). In biotechnological applications, QDs have some advantages over organic fluorophores. When used as fluorescent reporters, QDs can be one to two orders of magnitude brighter and more stable than traditional organic dies. Quantum physics explains why their superior properties depend on their size (Brus 1983). The electronic properties of QDs are intermediate between discrete molecules and bulk materials. When QD absorbs the photon with sufficient energy, an exciton can form. It is an electrically neutral quasiparticle subject to quantum confinement in the QD with a similar diameter to its wavelength. The quantum-confinement effect is so basic that almost any ordered structure of nanoscale size (including peptides) will emit a bright light when illuminated (Hauser and Zhang 2010, Rogach 2014). Graphene QDs can be fabricated from several coal types with a high yield (Ye et al. 2013). In semiconductors, quantum confinement occurs when the particle diameter is about 10 nm. Excitons are then trapped into a single point that is many orders of magnitude smaller than the period mark at the end of this sentence. Thus, Mark Reed used the name "quantum dot" (Reed 1993).

It was only to be expected that sooner or later, some research groups would construct hybrid systems by tagging suitable QD to the natural or modified 2D purple membrane with tightly packed bacteriorhodopsins (Li et al. 2007, Rakovich et al. 2010). The size of QDs is similar to the size of proteins such as bacteriorhodopsin. Hybrid bR-QD nanosystems opened a plethora of potential applications. Early attempts to obtain the bR-based bioelectronic application were in the field of optical memories (Hampp 2000). It is possible to channel into a branched pathway the same recovery transition T7 that we found to be critical for the photocycle optimization. The branched pathway $P_1(525) \leftrightarrow P_2(445) \leftrightarrow Q(380)$ is activated from the O(640) state by red-light absorption during the lifetime of the O intermediate (Wise et al. 2002, Wax et al. 2019). The Q state is not present as a natural functional state of the bacteriorhodopsin photocycle (Greco et al. 2013). It was discovered by Popp et al. (1993) as the long-lived state that is thermally stable for years. The Q state can be converted to the bR ground state when protein is illuminated with high energy photons (~ 380 nm). The stability of the blue-shifted Q photoproduct points toward potential applications of bR-QD systems (see the previous paragraphs) for writing, reading, and erasing information, long-term memory storage, optical computing, pattern recognition systems, robotic vision, and volumetric optical memories (Stuart et al. 2002, Li et al. 2018, Wax et al. 2019). From our analysis, it follows that the MTEP optimization for the T1 transition can increase the O state occupancy and efficiency of a branched recovery pathway. It suggests that joint optimizations of T1 and T7 transitions (see Figure 12.1) may be the best option to increase all performance parameters.

Site-directed mutagenesis is one way to optimize the bR photocycle for increased efficiency of light-driven proton pumping. At present, it is challenging to predict amino acid substitutions, which can achieve that goal. However, the conjugation of the wild-type bR with QD already resulted in increased efficiency through Förster resonance energy transfer (Förster 1946, Rakovich et al. 2010). Wild-type bR does not contain any Cys residue. With the bR mutant A103C, mentioned above, the conjugation with QDs is even better because cysteine serves as the capping agent by forming cysteine-to-QD thiol linkage (Wax et al. 2019). Controlling the QD/protein molar ratio is anything but trivial matter so that both a chosen protein and QD must be modified for optimal conjugation (Wax and Zhao 2019). Fortunately, the A103C substitution is located at the cytoplasmic loop region of the bR and does not impair the proton translocation. The bR can be considered an inverted membrane protein with crucial polar residues buried deep into its interior and deep inside the purple membrane bilayer. Hence, its active site is well protected.

There is some concern about the toxicity of bR-QD systems in potential biomedical applications such as bio-imaging probes and sensors. Cadmium is toxic heavy metal, but it is well protected when present only in QD's core region. Gold is non-toxic, while silver may even have beneficial antimicrobial activity. Krivenkov et al. (2019) have recently fabricated QDs with the CdSe core and ZnS/CdS/ZnS multilayer shell. They found that an optimal bR-to-QD molar ratio of 3:1 enables the interaction of a single five nm-sized QD with a naturally arranged trimer of bR molecules in the purple membrane. Cadmium is protected from leakage by the ZnS outer QD envelope. The efficiency of light conversion by the hybrid system was substantially higher than for the purple membrane alone. Elena Rozhkova group constructed the synthetic nano-bio hybrid system with bR molecules that did not contain cadmium, but for the different purpose mentioned earlier: toward hydrogen generation from water (Wang et al. 2017).

Of course, environmentally friendly nanotechnology is not a human invention. When celebrating the molecule of the month, the bacteriorhodopsin, scientists from the Protein Data Bank, Educational Portal, mentioned four different kinds of bacterial rhodopsins. This statement was published by PDB in 2002 (doi:10.2210/rcsb_pdb/mom_2002_3). Scientists were then only beginning to grasp the nature's billion years love affair with the ability of bR-like molecules to channel the dissipation of photon energy packages toward different needs of prokaryotic cells, eukaryotic cells, and even some giant viruses (Yutin and Koonin 2012, Pusharov et al. 2018). We don't know yet why or how some viruses tweak the host cell metabolism toward rhodopsin production. Possibly, such viruses endow cells with an additional phototropic potential that is beneficial for viral life-cycle. Finkel et al. (2013) estimated that about 50% of marine prokaryotic microorganisms harbor a rhodopsin gene annotated as type-1 microbial rhodopsins: proteorhodopsin, xanthorhodopsin, and bacteriorhodopsin. These genetic products are far cheaper to produce and maintain, and about three times more abundant than light-harvesting photochemical reaction centers, but incapable of carbon fixation.

Proteorhodopsins were first identified in the group of γ-proteobacteria. The wide distribution and abundance of proteorhodopsins in numerous microorganisms provided another association of their name with the god Proteus from Greek mythology, who could change his appearance (Bamann et al. 2014). Global Ocean Survey metagenome recovered almost 4000 proteorhodopsin sequences already in 2007 (Rusch et al. 2007). Proteorhodopsin's energy conversion rate on surface waters amounts to about 10^{13} W. This enormous energy harvesting rate is roughly equivalent to the fossil fuel consumption rate by humanity (Govorunova et al. 2017). It appears that all proteorhodopsins have the light-activated proton-pumping ability, just like bR, and contribute to cell fitness under starvation conditions, but most of them are likely to have the photoreceptor, photosensory, phototrophic, and phototaxis function (Bamann et al. 2014, Brown 2014). Sensory (photodetector) functions are indicated by very slow photocycle or weak proton transport. An additional chromophore, named salinixanthin, serves as a natural quantum dot for efficient photon-energy transfer to xanthorhodopsin that was found in eubacterium *Salinibacter ruber* (Balashov and Lanayi 2007).

Present rough division of labors and functions delineate over seven thousand different rhodopsins involved in a) bioenergetics (bR, proteorhodopsins, halorhodopsins, sodium-pumping rhodopsins, enzymerhodopsins), b) sensory reception (proteorhodopsins, regulation of photosynthesis, histidine kinase rhodopsins), and c) channel activity (cation channelrhodopsins, anion channelrhodopsins) (Govorunova et al. 2017). Just one subfamily, the channelrhodopsins, served to establish the optogenetics as a new neuroscience discipline. Research projects from optogenetics are currently going on in over a thousand laboratories all over the world (Ernst et al. 2014, Grote et al. 2014, Sineshchekov et al. 2017, Kuhne et al. 2019). Channelrhodopsins are algal light-activated and light-gated passive ion channels. They are used as tools to control the excitation by the light of well-defined nerve cell subpopulations. They show promise, too, as the optogenetic therapy for alleviating abnormally repetitive behavior (Burguière et al. 2013) and for the fabrication of earlier mentioned retinal prostheses (Sahel and Roska 2013).

Human needs for whatever application always require a dedicated goal-oriented optimization with a deep understanding of structure-dynamics-function relationships as the first prerequisite. Fortunately, nature has provided us with many bR-like structures to pick up what we need. Even the chromophore, the retinal, is used in a completely different manner by rhodopsin (see Chapter 7) and bR. Light-activated retinal is converted from 11-*cis* to all-*trans* configuration in rhodopsin, which is almost the opposite of all-*trans* to 13-*cis* conformational change in bR (Figure 12.2). A single amino acid substitution at position 105 in proteorhodopsin converts it from the green-light-absorbing molecule, suitable for bacteria dwelling on the marine surface, to the blue-light-absorbing molecule for deep sea living bacteria (Ernst et al. 2014). Replacing the crucial Asp-85 residue from bR with threonine converts bR into a halorhodopsin-like chloride pump (Sasaki et al. 1995), the proof that ion specificity for the active-transport function is determined by the interaction between retinal and amino acid at the sequence position 85. Similarly, the pathway to the future rational design of optogenetic tools depends on the biophysical approach to gain a detailed understanding of how channelrhodopsins work (Kuhne et al. 2019).

Human technology for sustainable growth has a vital need for high-efficiency light-activated energy converters. Naturally evolved bioenergetic devices do not need energy conversion efficiency higher than 10 to 20%, but we do. The bR efficiency ranges from 11% to 71%. However, the fragility of natural membranes composed of polar lipids does not allow higher than about 16% energy conversion efficiency (Table 12.1). When enough robust artificial bioenergetic membranes are developed, it will become possible to reach an overall photosynthetic conversion efficiency of about 50% if bacteriorhodopsin-like proton pumps are used. The take-home lesson is that the wisdom of biological evolution should never be underestimated. Still, human ingenuity can undoubtedly improve over naturally designed nanomachines. We are not restricted by universal genetic code to use only the materials and design solutions compatible with that ancient genetic fossil. In the next chapter, we shall explore if bioenergetics is even more ancient than the genetic code.

Definitions and explanations from Chapter 12:

Autotrophic cells produce organic molecules by using radiation energy quanta (photons) or inorganic chemical reactions.

Nanophotocatalysis refers to the light-activated catalytic activity of nanostructured photocatalysts. It is a new field of physical chemistry that combines nanotechnology with known photocatalytic methods to harness solar energy for catalyzing various chemical reactions.

X-ray free-electron laser employs relativistic electrons from a synchrotron light source. The wavelength of the radiation can be tuned until X-rays are produced. The method can be used to perform time-resolved crystallography experiments.

References

Annila, A. and Kuismanen, E. Natural hierarchy emerges from energy dispersal. BioSystems 95(2009): 227–233.

Ansari, A., Berendzen, J., Bowne, S.F., Frauenfelder, H., Iben, I.E., Sauke, T.B. et al. Protein states and proteinquakes. Proc. Natl. Acad. Sci. USA 82(1985): 5000–5004.

Balashov, S.P. and Lanyi, J.K. Xanthorhodopsin: Proton pump with a carotenoid antenna. Cell Mol. Life Sci. 64(2007): 2323–1328.

Balasubramanian, S., Wang, P., Schaller, R.D., Rajh, T. and Rozhkova, E.A. High-performance bioassisted nanophotocatalyst for hydrogen production. Nano Letters 13(2013): 3365. doi: 10.1021/nl4016655.

Bamann, C., Bamberg, E., Wachtveitl, J. and Glaubitz, C. Proteorhodopsin. Biochim. Biophys. Acta 1837(2014): 614–625.

Berhanu, S., Ueda, T. and Kuruma, Y. Artificial photosynthetic cell producing energy for protein synthesis. Nat. Commun. 10(2019): 1325. doi: 10.1038/s41467-019-09147-4.

Bondar, A.-N., Baudry, J., Suhai, S., Fischer, S. and Smith, J.C. Key role of active-site water molecules in bacteriorhodopsin proton-transfer reactions. J. Phys. Chem. B 112(2008): 14729–14741.

Brown, L.S. Eubacterial rhodopsins—Unique photosensors and diverse ion pumps. Biochim. Biophys. Acta 1837(2014): 553–561.

Brus, L.E. A simple model for the ionization potential, electron affinity, and aqueous redox potentials of small semiconductor crystallites. J. Chem. Phys. 79(1983): 5566–5571.

Burguière, E., Monteiro, P., Feng, G. and Graybiel, A.M. Optogenetic stimulation of lateral orbitofronto-striatal pathway suppresses compulsive behaviors. Science 340(2013): 1243–1246.

Cukierman, S. Et tu, Grotthuss! and other unfinished stories. Biochim. Biophys. Acta 1757(2006): 876–885.

Das, S., Wu, C., Song, Z., Hou, Y., Koch, R., Somasundaran, P. et al. Bacteriorhodopsin enhances efficiency of perovskite solar cells. ACS Appl. Mater. Interfaces 11(2019): 30728–30734.

Deamer, D. and Weber, A.L. Bioenergetics and life's origins. Cold Spring Harb. Perspect. Biol. 2(2010): a004929. doi: 10.1101/cshperspect.a004929.

Dioumaev, A.K. Infrared methods for monitoring the protonation state of carboxylic amino acids in the photocycle of bacteriorhodopsin. Biochemistry (Moscow) 66(2001): 1269–1276.

Dobovišek, A., Županović, P., Brumen, M. and Juretić, D. 2014. Maximum entropy production and maximum Shannon entropy as germane principles for the evolution of enzyme kinetics. pp. 361–382. *In*: Dewar, R.C., Lineweaver, C.H., Niven, R.K. and Regenauer-Lieb K. (eds.). Beyond the Second Law. Springer-Verlag, Berlin, Heidelberg, Germany.

Ekimov, A.I. and Onushchenko, A.A. Quantum-size effects in optical spectra of semiconductor microcrystals. Sov. Phys. Semicond. 16(1982): 775–778.

Ernst, O.P., Lodowski, D.T., Elstner, M., Hegemann, P., Brown, L.S. and Kandori, H. Microbial and animal rhodopsins: structures, functions, and molecular mechanisms. Chem. Rev. 114(2014): 126–163.

Finkel, O.M., Béjŕ, O. and Belkin, S. Global abundance of microbial rhodopsins. ISME J. 7(2013): 448–451.

Förster, T. Energiewanderung und Fluoreszenz. Naturwissenschaften 6(1946): 166–175.

Govindjee, R., Balashov, S.P. and Ebrey, T.G. Quantum efficiency of the photochemical cycle of bacteriorhodopsin. Biophys. J. 58(1990): 597–608.

Govorunova, E.G., Sineshchekov, O.A., Li, H. and Spudich, J.L. Microbial rhodopsins: Diversity, mechanisms, and optogenetic applications. Annu. Rev. Biochem. 86(2017): 845–872.

Greco, J.A., Wagner, N.L. and Birge, R.R. Fourier transform holographic associative processors based on bacteriorhodopsin. Int. Journ. Uncov. Comput. 8(2013): 433–457.

Grot, W. 2011. Fluorinated Ionomers (Second Edition). Elsevier Inc., New York, NY, USA.

Grote, M., Engelhard, M. and Hegemann, P. Of ion pumps, sensors and channels—Perspectives on microbial rhodopsins between science and history. Biochim. Biophys. Acta 1837(2014): 533–545.

Hamm, P., Zurek, M., Röschinger, T., Patzelt, H., Oesterhelt, D. and Zinth, W. Femtosecond spectroscopy of the photoisomerisation of the protonated Schiff base of all-trans retinal. Chem. Phys. Lett. 263(1996): 613–621.

Hampp, N. Bacteriorhodopsin as a photochromic retinal protein for optical memories. Chem. Rev. 100(2000): 1755–1776.

Hauser, C.A.E. and Zhang, S. Peptides as biological semiconductors. Nature 468(2010): 516–517.

Hazelbauer, G.L., Falke, J.J. and Parkinson, J.S. Bacterial chemoreceptors: high-performance signaling in networked arrays. Trends Biochem. Sci. 33(2008): 9–19.

Hill, T.L. 1977. Free Energy Transduction in Biology. The Steady State Kinetic and Thermodynamic Formalism. Academic Press, New York, NY, USA.

Jiang, Q., Zhao, Y., Zhang, X., Yang, X., Chen, Y., Chu, Z. et al. Surface passivation of perovskite film for efficient solar cells. Nat. Photonics 13(2019): 460–466.

Juretić, D. and Westerhoff, H.V. Variation of efficiency with free-energy dissipation in models of biological energy transduction. Biophys. Chem. 28(1987): 21–34.

Juretić, D. and Županović, P. Photosynthetic models with maximum entropy production in irreversible charge transfer steps. J. Comp. Biol. Chem. 27(2003): 541–553.

Juretić, D., Simunić, J. and Bonačić Lošić, Ž. Maximum entropy production theorem for transitions between enzyme functional states and its applications. Entropy 21(2019): 743. doi: 10.3390/e21080743.

Krivenkov, V., Samokhvalov, P. and Nabiev, I. Remarkably enhanced photoelectrical efficiency of bacteriorhodopsin in quantum dot—Purple membrane complexes under two-photon excitation. Biosensors and Bioelectronics 137(2019): 117–122.

Kuhne, J., Vierock, J., Tennigkeit, S.A., Dreier, M.A., Wietek, J., Petersen, D. et al. Unifying photocycle model for light adaptation and temporal evolution of cation conductance in channelrhodopsin-2. Proc. Natl. Acad. Sci. USA 116(2019): 9380–9389.

Kühlbrandt, W. Bacteriorhodopsin—the movie. Nature 406(2000): 569–570.

Lavergne, J. and Joliot, P. Dissipation in bioenergetic electron transfer chains. Photosynthesis Research 48(1996): 127–138.

Li, R., Li, C.M., Bao, H. and Bao, Q. Stationary current generated from photocycle of a hybrid bacteriorhodopsin/quantum dot bionanosystem. Appl. Phys. Lett. 91(2007): 22390. doi:/10.1063/1.2801521.

Li, Y.T., Tian, Y., Tian, H., Tu, T., Gou, G.-Y., Wang, Q. et al. A review on bacteriorhodopsin-based bioelectronic devices. Sensors (Basel) 18(2018): E1368. doi: 10.3390/s18051368.

Michel, H. and Oesterhelt, D. Three-dimensional crystals of membrane proteins: bacteriorhodopsin. Proc. Natl. Acad. Sci. USA 77(1980): 1283–1285.

Nango, E., Royant, A., Kubo, M., Nakane, T., Wickstrand, C., Kimura, T. et al. A three-dimensional movie of structural changes in bacteriorhodopsin. Science 354(2016): 1552–1557.

Nogly, P., Weinert, T., James, D., Carbajo, S., Ozerov, D., Furrer, A. et al. Retinal isomerization in bacteriorhodopsin captured by a femtosecond x-ray laser. Science 361(2018): pii: eaat0094. doi: 10.1126/science.aat0094.

Oesterhelt, D. The purple membrane from *Halobacterium halobium*. Hoppe-Seyler's Z. Physiol. Chem. 353(1972): 1554–1555.

Oesterhelt. D. and Stoeckenius, W. Functions of a new photoreceptor membrane. Proc. Natl. Acad. Sci. USA 70(1973): 2853–2857.

Popp, A., Wolperdinger, M., Hampp, N., Brauchle, C. and Oesterhelt, D. Photochemical conversion of the 0-intermediate to 9-cis-retinal-containing products in bacteriorhodopsin films. Biophys. J. 65(1993): 1449–1459.

Pushkarev, A., Inoue, K., Larom, S., Flores-Uribe, J., Singh, M., Konno, M. et al. A distinct abundant group of microbial rhodopsins discovered using functional metagenomic. Nature 558(2018): 595–599.

Racker, E. and Stoeckenius, W. Reconstitution of purple membrane vesicles catalysing light-driven proton uptake and adenosine triphosphate formation. J. Biol. Chem. 249(1974): 662–663.

Rakovich, A., Sukhanova, A., Bouchonville, N., Lukashev, E., Oleinikov, V., Artemyev, M. et al. Resonance energy transfer improves the biological function of bacteriorhodopsin within a hybrid material built from purple membranes and semiconductor quantum dots. Nano Lett. 10(2010): 2640–2648.

Reed, M.A. Quantum dots. Sci. Am. 268(1993): 118–123.

Rogach, A. Quantum dots still shining strong 30 years on. ACS Nano. 8(2014): 6511–6512.

Rossetti, R., Nakahara, S. and Brus, L.E. Quantum size effects in the redox potentials, resonance Raman spectra, and electronic spectra of cadmium sulfide crystallites in aqueous solution. J. Chem. Phys. 79(1983): 1086–1088.

Rusch, D.B., Halpern, A.L., Sutton, G., Heidelberg, K.B., Williamson, S., Yooseph, S. et al. The sorcerer II global ocean sampling expedition: northwest Atlantic through eastern tropical Pacific. PLoS Biol. 5(2007): 398–431.

Sahel, J.A. and Roska, B. Gene therapy for blindness. Annu. Rev. Neurosci. 36(2013): 467–488.

Sasaki, J., Brown, L.S., Chon, Y.S., Kandori, H., Maeda, A., Needleman, R. et al. Conversion of bacteriorhodopsin into a chloride ion pump. Science 269(1995): 73–75.

Shimizu, Y., Inoue, A., Tomari, Y., Suzuki, T., Yokogawa, T., Nishikawa, K. et al. Cell-free translation reconstituted with purified components. Nat. Biotechnol. 19(2001): 751–755.

Sineshchekov, O.A., Govorunova, E.G., Li, H. and Spudich, J.L. Bacteriorhodopsin-like channelrhodopsins: Alternative mechanism for control of cation conductance. Proc. Natl. Acad. Sci. USA 114(2017): E9512–E9519.

Stoeckenius, W. and Rowen, R. A morphological study of *Halobacterium halobium* and its lysis in media of low salt concentration. J. Cell Biol. 34(1967): 365–393.

Stuart, J.A., Marcy, D.L., Wise, K.J. and Birge, R.R. Volumetric optical memory based on bacteriorhodopsin. Synth. Met. 127(2002): 3–15.

Tittor, J. and Oesterhelt, D. The quantum yield of bacteriorhodopsin. FEBS Lett. 263(1990): 269–273.

van Stokkum, I.H.M. and Lozier, R.H. Target analysis of the bacteriorhodopsin photocycle using a spectrotemporal model. J. Phys. Chem. B 106(2002): 3477–3485.

Wang, P., Chang, A.Y., Novosad, V., Chupin, V.V., Schaller, R.D. and Rozhkova, E.A. Cell-free synthetic biology chassis for nanocatalytic photon-to-hydrogen conversion. ACS Nano. 11(2017): 6739–6745.

Wax, T.J., Greco, J.A., Chen, S., Wagner, N.L., Zhao, J. and Birge, R.R. Tunable photocycle kinetics of a hybrid bacteriorhodopsin/quantum dot system. Nano Res. 12(2019): 365–373.

Wax, T. and Zhao, J. Optical features of hybrid molecular/biological-quantum dot systems governed by energy transfer processes. J. Mater. Chem. C. 7(2019): 6512–6526.

Weinert, T., Skopintsev, P., James, D., Dworkowski, F., Panepucci, E., Kekilli, D. et al. Proton uptake mechanism in bacteriorhodopsin captured by serial synchrotron crystallography. Science 365(2019): 61–65.

Wickstrand, C., Dods, R., Royant, A. and Neutze, R. Bacteriorhodopsin: Would the real intermediates please stand up? Biochim. Biophys. Acta 1850(2015): 536–553.

Wickstrand, C., Nogly, P., Nango, E., Iwata, S., Standfuss, J. and Neutze, R. Bacteriorhodopsin: Structural insights revealed using X-ray lasers and synchrotron radiation. Annu. Rev. Biochem. 88(2019): 59–83.

Wise, K.J., Gillespie, N.B., Stuart, J.A., Krebs, M.P. and Birge, R.R. Optimization of bacteriorhodopsin for bioelectronic devices. Trends Biotechnol. 20(2002): 387–394.

Ye, R., Xiang, C., Lin, J., Peng, Z., Huang, K., Yan, Z. et al. Coal as an abundant source of graphene quantum dots. Nat. Commun. 4(2013): 2943. doi: 10.1038/ncomms3943.

Yen, C.-W., Hayden, S.C., Dreaden, E.C., Szymanski, P. and El-Sayed, M.A. Tailoring plasmonic and electrostatic field effects to maximize solar energy conversion by bacteriorhodopsin, the other natural photosynthetic system. Nano Lett. 11(2011): 3821–3826.

Yutin, N. and Koonin, E.V. Proteorhodopsin genes in giant viruses. Biol. Direct. 7(2012): 34. doi: 10.1186/1745-6150-7-34.

CHAPTER 13

The Protonmotive Force in Geochemistry and the Origin Question

Is the Origin of Bioenergetics Connected with the Origin of Life?

13.1 Is bioenergetics more ancient than genetic code?

Except for viruses and phages, no life exists in the absence of the protonmotive force. Prokaryotic and eukaryotic cells have a common ability to create and maintain the protonmotive force. It is an equally universal and ancient feature of the last universal common ancestor (LUCA) as the genetic code. In this book's spirit, we shall assume that the dissipation of proton gradients is even more ancient than the genetic code. As with all other hypotheses about life origin, the involvement of the protonmotive force (pmf) also opens the chicken-and-egg problem. The protonmotive force is needed for the ATP synthesis and the synthesis of all cellular macromolecules, but the pmf cannot be established without membrane protein pumps. Is there any abiotic origin for the protonmotive force that can explain the synthesis of organic molecules?

The well-known origin of life hypothesis by Oparin (1957, 1961) is no longer the dominant hypothesis in the field. Still, it is worthwhile to mention the "primeval broth" or "primordial soup" idea of **coacervate** protein droplets that presumably accumulated at the surface of primitive oceans under the influence of UV radiation, day-night temperature cycling, evaporation-tides cycling, strong winds, and lightning in the Archean era. Some amino acids and even purine bases, ribose, and deoxyribose can be obtained in a laboratory using UV radiation or electric sparks acting on simpler ingredients in water solution, thus, without using enzymes (Oparin 1965). For instance, Stanley Miller performed the enzyme-less synthesis of 10 out of 20 proteinogenic amino acids (Miller 1955). These ten amino acids were: Gly, Ala, Ser, Asp, Glu, Val, Leu, Ile, Pro, and Thr. These are mostly small, hydrophobic, and polar amino acids. However, racemic mixtures were obtained, consisting of 50% D-isomers and 50% L-isomers (Oro et al. 1977). This result is expected according to the Second Law of Thermodynamics because the entropy is maximal for the racemic mixture containing 50% of both stereoisomers. Still, it opens a big problem about the origin of life, already recognized by Louis Pasteur in 1848, when he discovered the **optical activity** of natural organic compounds. He regarded it as the best borderline between the chemistry of inanimate and living matter (Pasteur 1848). Life needed free-energy investments from the very beginning to create nonequilibrium low probability homochiral structures. Asymmetry is pervasive today in the whole biosphere, and it is not restricted to proteins with their L-amino acids. Nucleic acids contain only D-ribose and D-deoxyribose, while phospholipids are also optically active. The chirality of biological molecules is the signature of life, just as Pasteur stated 170 years ago. However, the homochirality emergence at the Hadean or Archean eon remains an unsolved puzzle despite numerous speculations and experiments (Blackmond 2019).

Homochirality ensures that all present-day biological macromolecules are nonequilibrium structures; thus, subject to spontaneous racemization. This tendency can be literally seen by persons in advanced old age with weakened eyesight due to racemization in the long-lived lens proteins such as alpha-crystallin. By age 70, almost 5% of D-Ser and more than 9% of D-Asp and D-Asn (denoted as D-Asx) accumulate in normal lenses, enough to induce significant denaturation of lens proteins and also an indication that racemization is involved in the etiology of cataract (Hooi and Truscott 2011). There is a strikingly linear relationship between age and racemization percentage in proteins from normal and cataract lens with about 4% higher value of the 100D/(D+L) ratio for D-Asx isomers irrespective of age in the case of cataract lens.

Modern scenarios for a terrestrial origin of protocells propose shallow ponds of anoxic geothermal fields as "hatcheries" of the first cells rather than sites of "primordial soup" (Mulkidjanian et al. 2003, 2012, Deamer 2017, Joshi et al. 2017). Instead of protein droplets, the lipid vesicles are assumed as the earliest "chemical bags" containing concentrated metabolic precursors of biological macromolecules, and various energizing sources are proposed for their abiotic synthesis. The impacts of asteroids and comets during Hadean eon are often mentioned as possible sources of lipid molecules. Vesicles emerge by self-assembly of amphipathic and hydrophobic lipid monomers into the vesicular membrane. High enough lipid and low enough salt concentration is needed for this to happen. These conditions may have been satisfied in the environment of surface geothermal pools. Some authors support the Lipid World hypothesis for the terrestrial origin of life (Segré et al. 2001, Tessera 2017, Deamer 2017). The low salt condition was not satisfied in Hadean oceans. Primordial oceans had similar high salt concentrations of ~ 600 mM NaCl as present-day oceans. High salt concentration decreases the hydrophobic effect essential for the formation of membranes, vesicles, and liposomes, especially in a likely situation of primitive lipid monomers being simpler than present-day phospholipids of sphingolipids.

Nevertheless, the world of lipids is considerably richer than the world of proteinogenic amino acids. Therefore, it is possible that some lipid types can self-assemble into vesicles resistant to high salt concentration. Jordan et al. (2019) have shown that a mixture of amphipathic single-chain lipids with more complex head groups (possible candidate lipid molecules for forming proto-cells that are simpler than phospholipids) do form vesicles at modern ocean salinity within a wide pH range. Present-day living cells produce more than 100 thousand different lipid monomeric structures (Schmitt-Kopplin et al. 2019). The most costly and complex monomers to form are amphipathic phospholipid-like molecules. Without some simpler analogs of phospholipids, the stable membranes could not form, and geochemical proton-gradient means for primordial bioenergetics could not be transferred to bioenergetic membranes of protocells with the ability to spread beyond their geochemical hatcheries. Thus, abiotic synthesis of phospholipid-like molecules remains the problem for all "lipid-world first" origin of life theories.

However, the principal arguments against all "chemical bag" proposals for the terrestrial origin of life are, firstly, the high probability of frequent huge impacts during Late Heavy Bombardment in the Hadean Earth (Abramov and Mojzsis 2009). Such impacts vaporized ocean surface and all surface lakes and ponds. Depending on released energy, more or less efficient sterilization follows for any nescient protocells in surface environments. Secondly, all three life domains, archaeal cells, bacteria, and eukaryotic cells, are utterly dependent on chemiosmotic mechanisms for maintaining and reproducing their complex far-from-equilibrium structures. "Chemical bags" are devoid of membrane-based bioenergetics.

A natural approach to the origin of life question is to look for congruence among the top-down biochemical and bottom-up geochemical processes. Are there any exergonic (energy-releasing) reactions that could have served as meeting points? The acetyl CoA pathway is the only carbon fixation pathway common to archaea (methanogens) and bacteria (acetogens). Similar acetyl CoA pathways for CO_2 fixation are shared among all life domains, indicating their deeply ancient roots (Russell and Martin 2004). The H_2 and CO_2 are the only small input molecules for all acetyl CoA pathways leading to carbon fixation. These molecules were plentiful in diverse geochemical environments of the Hadean era. As they serve today, these molecules could have served at the emergence of life period to drive proto-metabolism and carbon fixation in organic molecules (Herschy et al. 2014, and references within; Hudson et al. 2020). Everett Shock has described this exergonic pathway of carbon fixation as "a free lunch you're paid to eat" (Shock et al. 1998). However, the present-day acetyl CoA pathway is dependent on the continuous operation of a chemiosmotic mechanism. One must ask then if intermittent and insecure free-energy inputs such as lightning and daylight UV radiation were ever enough to drive this or any other of multiple pathways proposed as to jump-start the development of life (Peretó 2012, 2019)?

One possible answer is that a persistent protonmotive force existed as a natural abiotic feature of hydrothermal vents located at a considerable depth under oceanic surfaces (Martin and Russell 2003, Sleep et al. 2004, Lane 2010, Lane et al. 2010). Such geologically active underwater vents were present during the Hadean and Archaean eons too. During these eons, oceans had high acidity. The acidity was due to a completely different atmosphere: the absence of oxygen and high carbon dioxide concentration. At the same time, more intensive volcanic activity during these eons likely produced a high number of geothermal vents similar to the present-day Lost City vent (Proskurowski et al. 2008).

The alkaline hydrothermal vents similar to the Lost City mounds at ocean bottom were well protected in Hadean eon from all asteroid and comet impacts capable of evaporating oceanic surfaces up to several hundred meters depth. Even more importantly, the Lost City structures contain very tight internal porous spaces for venting out a rich mixture of solvated minerals at a high pH. Inorganic membranes enclosing these surfaces are often fragile and thin. In the Hadean and Archaean era, these membranes separated the thousand times higher proton concentrations in acidic oceans than in vent fluids. Thus, the proton gradient of four to six pH units was maintained continuously. These proton gradient values are similar to or higher than the present-day protonmotive force of archaic cells and bacteria (Lane and Martin 2012). Namely, one pH unit corresponds to 59 mV (see Chapter 6), and four pH units are then equivalent to 236 mV potential.

Michael Russell postulated (Martin and Russell 2003, Russel 2006, Martin and Russell 2007, Russell et al. 2010, Russel 2018) that alkaline submarine hydrothermal vents served as "hatcheries" of the first proto-cells. He forcefully argued with Branscomb (Branscomb and Russell 2018a,b) that any living system is extremely far from equilibrium. This asymmetry is essential for present-day life. Hence, these authors concluded that some abiotic driving forces must have been in operation to maintain extreme disequilibrium situations during life emergence on Earth too. The authors stressed that this idea is the opposite of all ideas on how life originated as the "chemistry-in-a-bag" or "primordial soup," however energized. Martin and Russell famously remarked in 2003 that "once autoclaved a bowl of chicken soup at any temperature will never bring forth life". Even when supported with the RNA world idea, the prebiotic broth needed to contain a minimal set of dozen RNA molecules with about two thousand nucleotides to perform a primitive translation process. The probability for this to happen is calculated to be one over 1018th power of 10. The number 10^{1018} is an incomparably greater number than all the time, space, and the number of elementary particles in the universe. It means that the probability of accidentally getting this start-up set of RNA molecules was equal to zero during the first billion years of Earth's history (Wächtershäuser 2014).

Russell's proposal is specific concerning the possible chemical composition of abiotic membranes and abiotic reactions driven by highly nonequilibrium pH gradients and redox gradients. Alkaline (pH ~ 11) hydrothermal fluid heated from the mantle plumes at 100 to 120°C and rich in H_2, methane, and $HCOO^-$ bubbled in a colder acidic (pH ~ 5–6) ocean rich in CO_2, NO_3^-, and $H_2PO_4^-$. Inorganic membranes separating geothermal fluids from ocean water were composed of ferrous hydroxides, iron sulfides, iron-nickel, iron-nickel sulfides, olivine, pyrite, orthopyroxene, and serpentine minerals similar to ones found in the present-day Lost City mounds (Proskurowski et al. 2008). Elements Si, Mg, Ca, Zn, Mo, Co, Mn, and W were also present in thin mineral membranes contributing to their catalytic properties. Thin mineral membranes (several micrometers on average) are, however, thick (Lane 2017) compared to biological double-layer membranes (several nanometers in thickness). Mineral membranes form a catalytic labyrinth of interconnected pores. Also, mineral membranes are leaky and inefficient barriers lacking sufficient strength. They survived better when strengthened by polyalanines and some other simple peptides like AILSST capable of forming highly resistant beta-sheet surfaces and even self-replication like amyloid peptides (Greenwald and Riek 2012). Peptides have a stabilizing effect on RNA strands, too (Carny and Gazit 2011). These authors also found that the self-assembly process of homochiral-diphenylalanine is much more rapid than their hetherochiral diastereoisomers. Christian de Duve (2003) presciently stressed the importance of peptide-RNA interactions for the protometabolism leading to the FUCA (the *first* universal common ancestor).

13.2 Closing the gap between geochemistry and biochemistry

The emergence of life was undoubtedly a complex affair. All we mentioned up to now is not convincing enough to single out the alkaline hydrothermal vents as the best candidate sites for primitive bioenergetic processes at the time of life origin. How did they perform the enzyme-less synthesis of amino acids, sugars, nucleotides, and lipids, and why these small organic molecules were not diluted in the Hadean ocean before getting a chance for building up polypeptides, nucleic acids, and membranes? The chemiosmotic role of subsurface alkaline vents in life's origin is still hotly disputed (Jackson 2016, Lane 2017). Indeed, attempts to close the gap between geochemistry and biochemistry should take into account that biochemistry is mostly vectorial, while conventional synthetic chemistry is not (Herschy et al. 2014). Wächtershäuser proposed that the vectorial character of the pyrite-water interface at hydrothermal vents contributed to the development of vectorial biochemistry and the origin of life (Wächtershäuser 1992). The long life of hydrothermal vents for up to 100 thousand years ensured a continuous and constant free energy source for carbon fixation (Martin et al. 2014, Herschy et al. 2014). Interestingly, the inorganic cofactors present in present-day (Fe,Ni)S proteins that catalyze carbon fixation and pyruvate synthesis are similar to (Fe, Ni)S minerals greigite and mackinawite found in alkaline hydrothermal systems (Russell and Martin 2004, Harel et al. 2014, Herschy et al. 2014). Thus, there is some similarity between present-day metabolism with a central role for pyruvate and possible abiotic synthesis of formate and pyruvate under conditions prevailing in Hadean and Archaean geochemical reactors (Hudson et al. 2020).

An initially small yield of organic molecules formed inside vents and their dilution in ocean water is not the issue for concern. The synthesis can occur deep within the pore-containing labyrinth of the vent (Lane 2017). The yield quickly increases once a set of autocatalytic reactions is established (Huber and Wächtershäuser 2006). The pH gradient across a mineral surface is acidic on the ocean side (in pores connected to ocean water) and alkaline on the vent-fluid side (in pores connected to hot magmatic plumes). This pmf is enough to drive vectorial chemical synthesis in the sponge-like structure of hydrothermal mounds.

It is possible the virions emerged in deep vent environments before membrane enveloped protocells. Virions are inactive viruses or phages incapable of replication outside bacterial, archaeal, or eukaryotic cells (Berliner et al. 2018). There are about 10^{31} virions in present-day oceans alone (Suttle 2005). Thus, virions are the most abundant biological entities on Earth. Strangely, nobody knows how or where they fit in the tree of life. The most diverse morphology and genomics are found for archaeal viruses (Berliner et al. 2018). The symbiosis between some giant archaeal virus and infected archaeal cell may have been responsible for the emergence of the cell nucleus in first eukaryotic cells (see Chapter 11, section 11.7, second paragraph). But back to the life-origin question, it is interesting that the physiology and metabolism of archaea methanogens and hyperthermophiles closely mimic the geochemistry of alkaline hydrothermal vents (Lane 2017). Ribosomes in all cells also have nucleic acids encapsulated within proteins, resembling virions. Ribosome-like particles may have already been present in LUCA protocell ancestors.

There is no need to assume that the same geochemical environment was required for abiotic lipid synthesis and the synthesis of peptides, RNA, or proto-virions. A wide variety of microenvironments and conditions (pH, temperature, pressure, and oxidation states) existed on early Earth, and liquid water was present already about 4.3 billion years ago (Cleaves II et al. 2012). Proto-virions "infection" as a "communication" among different geothermal environments cannot be excluded during the Hadean era. Huge tides at that time (about 20 times higher than present-day tides due to considerably closer Moon) may have helped in mixing organic self-aggregates from different environments. Tsunami were likely to be stronger and more frequent than in the present-day period due to internal or external (asteroids impacts) causes. They could have also facilitated the interaction among peptide aggregates, nucleic acid aggregates, and vesicular lipid structures, no matter what were the sites of their geochemical origin.

Some of the present-day viruses are "nude," that is, without any external lipid envelope. Others "steal" the membrane from infected cells to gain the lipid cover over their protein-RNA or protein-DNA complex. Metabolic "nudity" of virions due to lack of any metabolic processes may be relevant for understanding whether proto-virions existed in the Hadean era and how they emerged. It is not likely that we shall find the fossil evidence of proto-virions or any other rudimentary life forms. As we have seen in this section, different geochemical processes, locations, and drives have been proposed for the abiogenic synthesis of organic molecules, which drove this production of unusual complexity (proto-virions included) before Darwinian evolution became even possible.

When we study the evolution possibilities at the molecular level, the far-from-equilibrium condition is even more important in the prebiotic era than for current living systems (Martin and Horvath 2013). The nonlinearity is another requirement for driving the system to an ordered conformation. Together, these two criteria can induce the replication or autocatalysis process (Pascal et al. 2013). The far-from-equilibrium state can persist only by continuous free-energy dissipation. Dissipative structures are not maintained and cannot perform any dissipation at the thermodynamic equilibrium. Dynamically maintained kinetic barriers should have existed to prevent the dissolution of self-organization, and the whole system decay into equilibrium. If such "Dynamic Kinetic Stabilization" existed during abiotic evolution, it was utterly dependent on the constant far-from-equilibrium environment to maintain hypercycles of complex catalytic networks (Oehlenschläger and Eigen 1997).

Kinetic gating and cycle-based thermodynamics may have been essential for the emergence and operation of molecular machines (Astumian 2018). However, overlooking dissipation is likely to be misguided despite the author's strong contrary statements: "The focus on dissipation is misguided. Dissipation is waste and drives nothing other than heating. The idea that systems adapt and evolve to dissipate as much energy as possible is simply wrong and must be discarded onto the scrap heap of incorrect theories" (Astumian 2018). Disregarding the dissipation as the major part of energy transduction in all far-from-equilibrium situations can be named the "Titanic delusion." It is the delusion that what cannot be seen below the surface can be safely ignored as a useless thing of no consequence.

It is true that some proteins acting as molecular machines exhibit amazingly high efficiency. Still, we have shown in Chapter 11 that an optimal efficiency of free-energy transduction is not only associated with a high dissipation but can be derived as the consequence of maximal partial entropy production theorem. Other authors stressed that the entropy production is an essential concept in the stochastic thermodynamics pertaining to the operation of molecular machines (Seifert 2012). Regarding life-origin proposals, it is not credible that considerably simpler molecular machines for harnessing free-energy gradients in the origin-of-life period were more efficient than their present-day relatives. Much more likely situation then was that the ratio of wasted to used energy for any organic synthesis was considerably higher in that period.

Thus, we must return to the same question why dissipation was crucial for the emergence of life as it is essential in the present-day bioenergetics? Is it possible that entropy production had some role in selecting the organic structures capable of increasing entropy production? The assumption underlying this question about the capability of far-from-equilibrium systems to facilitate entropy production increase in the universe opens the issue that is clearly in the realm of physics: why should the universe care about any local increase in the dissipation level? The dissipation rate cannot be the consequence of the Second Law of Thermodynamics. Unfortunately, the Second Law does not say anything about how fast the system approaches the thermodynamic equilibrium.

The evolution of all physical systems is coupled to the decrease of their free energy in the least time (Wang 2006, Lucia 2014). Thermodynamics is the only branch of classical physics concerned with evolution. It does not distinguish between animate and inanimate systems. The physical principle of maximum energy dispersal is equivalent to the maximal rate of entropy production with no demarcation lines to its application between animate and inanimate, incipient life, or developed living systems (Kaila and Annila 2008, Annila and Salthe 2010, Annila and Kolehmainen 2015). The origin of life must involve the dissipation of energy (Pascal et al. 2013) even to the extent that living systems should be regarded as manifestations of physical principles about dissipation intensity rather than ends in themselves (Annila and Baverstock 2014). Thus, the consideration of the physical tenets relevant for the origin of life gives the advantage to "metabolism-first" (Morowitz 2004, Anet 2004, Russel 2018) rather than the "replication-first" (Joyce 2002, Orgel 2004, Müller 2006, Wächtershäuser 2014) hypothesis. Concerning the choice of relevant physical principles, it is wise, in my opinion, to take the advice expressed by Annila and Kuismanen in 2009: "The objective is not to reduce biology to deterministic physics but to amend the formalism of physics with the holistic characteristic of biology."

13.3 The "dissipation-first" hypothesis for the origin of life

The "metabolism-first" hypothesis for the progression from non-living to living states of matter (Morowitz 2004) attracted the criticism focused at two points: a) a lack of biological evolvability (Orgel 2008), and b) a lack of specific nonequilibrium inorganic environment and energy input characteristics capable of driving the proposed enzyme-less reaction network (Adam et al. 2017). The environment-centric approach and entropy-centric-approach regarding the thermodynamic depth of abiotic energy inputs (Lloyd and Pagels 1988) are nicely connecting the emergence of chemical complexity with increased dissipation. The invisible sign of a thermodynamically deeper organic molecule is higher entropy generation during its prebiotic production (Adam et al. 2017). The thermodynamic depth is just one among more than 40 definitions of complexity (Lloyd 2001). It is particularly apt for considering how a given form of energy input from the abiotic environment can give rise to complex networks of mutually interacting molecules. The jungle of possible abiotic reactions before life emerged reflects a high molecular diversity produced by geochemistry (Cleaves II et al. 2012, Schmitt-Kopplin et al. 2019). Thus, some selection principle must have been in operation during prebiotic evolution. The excess energy from thermal or radioactive sources has been channeled to a relatively small number of organic chemical species enabling dissipative structuring and emergent behavior. More than ten different radioactive isotopes of heavy elements were present on early Earth in higher concentrations (Bassez 2017). They emitted high energy α-particles capable of producing crystal distortions (in minerals containing them) and water ionization. Radiolyzed water dissipated excess energy by inducing the O_2 creation and chemical synthesis of some organic compounds in the presence of transition metals. The formation of chemical bonds is the by-product of abiogenic energy transduction (Adam et al. 2017). Produced organic molecules regulated dissipation by retaining a small fraction of input free-energy and postponing the return to the ground state.

Instead of focusing on the efficiency of building up organic molecules, we can focus on the efficiency of consuming available thermodynamic gradients. Integration within a hierarchical organization is an efficient mechanism for funneling flows through the steepest descents in the free energy landscape for both inanimate and animate systems (Annila and Kuismanen 2009). The abiotic synthesis of larger organic compounds by far-from-equilibrium submarine hydrothermal fluids is linked to increased energy release (Shock and Canovas 2010). Life is just one of the myriad of entropy producing systems (Schneider and Sagan 2005), but one of the most efficient in free-energy dissipation and entropy export. At present, life is unique among all other entropy producing systems with its superb capability to regulate dissipation according to the availability of free-energy inputs. Thus, it is not a far-fetched inference to assume that the origin of life scenario also involved adaptating the first proto-metabolic mechanisms to fluctuating environmental conditions. This observation is just one step far from the conclusion that linkage between the higher dissipation potential and the potential for the synthesis of ever more complex organic compounds was already present at the origin of life.

In their phylogenetic approach, Weiss et al. (2016) identified genes common to archaea and bacteria that illuminate the physiology, metabolism, biochemistry, and habitat of our postulated LUCA ancestor. LUCA was able to harness geochemically produced ion gradients in hydrothermal environments. Methanogens and clostridial acetogens were probably growing already in the Hadean era just as they do today, that is, by using the exergonic reactions of methane, acetate, and formate synthesis. The geochemical synthesis of methane from H_2 and CO_2 was taking place before LUCA progenitors appeared. In this evolutionary scenario, the progenitors of methanogens hijacked the natural methane synthesis by accelerating it, with concomitant entropy production increase due to faster relaxation of high proton and Na^+ gradients. The natural abundance of subsurface hydrothermal flows rich in transition metals, sulfur, reactive C1 carbon species, H_2, and CO_2 may exist on other geologically active planetary bodies, thus raising our hope that life is universal too. It is a rather simple postulate that rock-water interactions, far from thermodynamic equilibrium, are conductive to carbon and energy transformations leading to the emergence of primitive life forms elsewhere in the universe.

In conclusion, the origin of bioenergetics (Lane 2017, Deamer 2017, Torday and Miller 2017), the origin of regulation capability (Cornish-Bowden and Cárdenas 2019), and the origin of order-promoting dissipation must have all been tightly coupled with the origin of life. The prebiotic evolution during roughly the first billion years of Earth's history formed the bridge between geochemistry and biochemistry with likely participation of the energy input from the surrounding Hadean atmosphere, Sun's UV irradiation, and the Hadean Earth's exposure to the Late Heavy Bombardment (Abramov and Mojzsis 2009) from the nearby universe. Dissipative self-organization is autocatalytic—more dissipative structures produce an increased dissipation and vice versa, but there is a delicate balance between internal growth and external forcing. Even in the very beginnings, life was a thin, fragile skin that simultaneously separated and linked together interior and exterior networks of geochemical and nuclear reactions.

Definitions and explanations from Chapter 13:

The **coacervate concept** refers to the colloidal particles in an aqueous solution. The dense phase (particles) is in thermodynamic equilibrium with the dilute (aqueous) phase. Chemical constituents of a dense phase can be lipids, proteins, nucleic acids, or some of their mixtures. Self-organization of coacervates due to phase separation of dense and dilute phases is often considered relevant for life's origin.

The **Optical activity** of a linearly polarized light is manifested as the rotation of the plane of polarization after light passes through chiral materials. Such materials contain chiral molecules lacking microscopic mirror symmetry. For chiral molecules in solution, the rotation angle is proportional to the path length and concentration. Circular dichroism spectroscopy involves circularly polarized light. It is widely used for investigating the secondary structure of peptides and proteins.

References

Abramov, O. and Mojzsis, S.J. Microbial habitability of the Hadean earth during the late heavy bombardment. Nature 459(2009): 419–422.

Adam, Z.R., Zubarev, D., Aono, M. and Cleaves, H.J. Subsumed complexity: abiogenesis as a by-product of complex energy transduction. Phil. Trans. R. Soc. A 375(2017): 20160348. http://dx.doi.org/10.1098/rsta.2016.0348.

Anet, F.A. The place of metabolism in the origin of life. Curr. Opin. Chem. Biol. 8(2004): 654–659.

Annila, A. and Kuismanen, E. Natural hierarchy emerges from energy dispersal. BioSystems 95(2009): 227–233.

Annila, A. and Salthe, S. Physical foundations of evolutionary theory. J. Non-equilib. Thermodyn. 35(2010): 301–321.

Annila, A. and Baverstock, K. Genes without prominence: a reappraisal of the foundations of biology. J. R. Soc. Interface 11(2014): 20131017. http://dx.doi.org/10.1098/rsif.2013.1017.

Annila, A. and Kolehmainen, E. On the divide between animate and inanimate. J. Syst. Chem. 6(2015): 2. doi: 10.1186/s13322-015-0008-8.

Astumian, R.D. Trajectory and cycle-based thermodynamics and kinetics of molecular machines: The importance of microscopic reversibility. Acc. Chem. Res. 51(2018): 2653–2661.

Bassez, M.-P. Anoxic and oxic oxidation of rocks containing Fe(II)Mg-silicates and Fe(II)-monosulfides as source of Fe(III)-minerals and hydrogen. Geobiotropy. Orig. Life Evol. Biosph. 47(2017): 453–480.

Berliner, A.J., Mochizuki, T. and Stedman K.M. Astrovirology: viruses at large in the universe. Astrobiology 18(2018): 207–223.

Blackmond, D.G. The origin of biological homochirality. Cold Spring Harb. Perspect. Biol. 11(2019): a032540. doi: 10.1101/cshperspect.a032540.

Branscomb, E. and Russell, M.J. Frankenstein or a submarine alkaline vent: who is responsible for abiogenesis? Part 1: What is life—that it might create itself? BioEssays 40(2018a): 1700179. doi: 10.1002bies.201700179.

Branscomb, E. and Russell, M.J. Frankenstein or a submarine alkaline vent: who is responsible for abiogenesis? Part 2: As life is now, so it must have been in the beginning. BioEssays 40 (2018b): 1700182. doi: 10.1002bies.201700182.

Carny, O. and Gazit, E. Creating prebiotic sanctuary: Self-assembling supramolecular peptide structures bind and stabilize RNA. Orig. Life Evol. Biosph. 41(2011): 121–132.

Cleaves II, H.J., Scott, A.M., Hill, F.C., Leszczynski, J., Sahai N. and Hazen, R. Mineral–organic interfacial processes: potential roles in the origins of life. Chem. Soc. Rev. 41(2012): 5502–5525.

Cornish-Bowden, A. and Cárdenas, M.L. Contrasting theories of life: Historical context, current theories. In search of an ideal theory. Biosystems (2019): 104063. doi: 10.1016/j.biosystems.2019.104063.

de Duve, C. A research proposal on the origin of life. Orig. Life Evol. Biosph. 33(2003): 559–574.

Deamer, D. The role of lipid membranes in life's origin. Life 7(2017): 5. doi: 10.3390/life7010005.

Greenwald, J. and Riek, R. On the possible amyloid origin of protein folds. J. Mol. Biol. 421(2012): 417–426.

Harel, A., Bromberg, Y., Falkowski, P.G. and Bhattacharya, D. Evolutionary history of redox metal-binding domains across the tree of life. Proc. Nat. Acad. Sci. USA 111(2014): 7042–7047.

Herschy, B., Whicher, A., Camprubi, E., Watson, C., Dartnell, L., Ward, J. et al. An origin-of-life reactor to simulate alkaline hydrothermal vents. J. Mol. Evol. 79(2014): 213–227.

Hooi, M.Y.S. and Truscott, R.J.W. Racemisation and human cataract. D-Ser, D-Asp/Asn and D-Thr are higher in the lifelong proteins of cataract lenses than in age-matched normal lenses. Age 33(2011): 131–141.

Huber, C. and Wächtershäuser, G. α-Hydroxy and α-amino acids under possible Hadean, volcanic origin-of-life conditions. Science 314(2006): 630–632.

Hudson, R., de Graaf, R., Rodin, M.S., Ohno, A., Lane, N., McGlynn, S.E. et al. CO_2 reduction driven by a pH gradient. Proc. Natl. Acad. Sci. USA 117(2020): 22873–22879.

Jackson, J.B. Natural pH gradients in hydrothermal alkali vents were unlikely to have played a role in the origin of life. J. Mol. Evol. 83(2016): 1–11.

Joshi, M.P., Samanta, A., Tripathy, G.R. and Rajamani, S. Formation and stability of prebiotically relevant vesicular systems in terrestrial geothermal environments. Life 7(2017): 51. doi: 10.3390/life7040051.

Jordan, S.F., Rammu, H., Zheludev, I.N., Hartley, A.M., Maréchal, A. and Lane, N. Promotion of protocell self-assembly from mixed amphiphiles at the origin of life. Nat. Ecol. Evol. 3(2019): 1705–1714.

Joyce, G.F. The antiquity of RNA-based evolution. Nature 418(2002): 214–221.

Kaila, V.R.I. and Annila, A. Natural selection for least action. Proc. R. Soc. A. 464(2008): 3055–3070.

Lane, N., Allen, J.F. and Martin, W. How did LUCA make a living? Chemiosmosis in the origin of life. BioEssays 32(2010): 271–280.

Lane, N. Why are cells powered by proton gradients? Nature Education 3(2010): 18. doi: 10.1038/46903.

Lane, N. and Martin, W.F. The origin of membrane bioenergetics. Cell 151(2012): 1406–1416.

Lane, N. Proton gradients at the origin of life. Bioessays 39(2017). doi: 10.1002/bies.201600217.

Lloyd, S. and Pagels, H. Complexity as thermodynamic depth. Ann. Phys. 188(1988): 186–213.

Lloyd, S. Measures of complexity: a nonexhaustive list. IEEE Control Syst. Mag. 21(2001): 7–8.
Lucia, U. The Gouy-Stodola theorem in bioenergetic analysis of living systems (irreversibility in bioenergetics of living systems). Energies 7(2014): 5717–5739.
Martin, W. and Russell, M.J. On the origins of cells: a hypothesis for the evolutionary transitions from abiotic geochemistry to chemoautotrophic prokaryotes, and from prokaryotes to nucleated cells. Phil. Trans. R. Soc. Lond. B 358(2003): 59–85.
Martin, W. and Russell, M. On the origin of biochemistry at an alkaline hydrothermal vent. Phil. Trans. R. Soc. Lond. B 362(2007): 1887–1925.
Martin, O. and Horvath, J.E. Biological evolution of replicator systems: Towards a quantitative approach. Orig. Life Evol. Biosph. 43(2013): 151–160.
Martin, W.F., Sousa, F.L. and Lane, N. Energy at life's origin. Science 344(2014): 1092–1093.
Miller, S.L. Production of some organic compounds under possible primitive Earth conditions. J. Amer. Chem. Soc. 77(1955): 2351–2361.
Morowitz, H.J. 2004. Beginnings of Cellular Life: Metabolism Recapitulates Biogenesis. Yale University Press, New Haven, CT,USA.
Mulkidjanian, A.Y., Cherepanov, D.A. and Galperin, M.Y. Survival of the fittest before the beginning of life: selection of the first oligonucleotide-like polymers by UV light. BMC Evol. Biol. 3(2003): 12. doi: 10.1186/1471-2148-3-12.
Mulkidjanian, A.Y., Bychkov, A.Yu., Dibrova, D.V., Galperin, M.Y. and Koonin, E.V. Origin of first cells at terrestrial, anoxic geothermal fields. Proc. Natl. Acad. Sci. USA 109(2012): E821–E830.
Müller, U.F. Re-creating an RNA world. Cell Mol. Life Sci. 63(2006): 1278–1293.
Oehlenschläger, F. and Eigen, M. 30 years later—a new approach to Sol Spiegelman's and Leslie Orgel's *in vitro* evolutionary studies. Dedicated to Leslie Orgel on the occasion of his 70-th birthday. Orig. Life Evol. Biosph. 27(1997): 437–57.
Oparin, A.I. 1957. The Origin of Life on the Earth, Oliver and Boyd, London, England.
Oparin, A.I. 1961. Life, its Nature, Origin, and Development. Academic Press, New York, USA. Translated from the Russian of Oparin (1924).
Oparin, A.I. The origin of life and the origin of enzymes. Adv. Enzymol. Relat. Areas Mol. Biol. 27(1965): 347–380.
Orgel, L.E. Prebiotic chemistry and the origin of the RNA world. Cri. Rev. Biochem. Mol. Biol. 39(2004): 99–123.
Orgel, L.E. The implausibility of metabolic cycles on the prebiotic Earth. PLoS Biol. 6(2008): e18. doi: 10.1371/journal.pbio.0060018.
Oro, J., Miller, S.J. and Urey, H.C. 1977. Energy conversion in the context of the origin of life. pp. 7–19. *In*: Buvet, R. (ed.). Living Systems as Energy Converters. North Holland, New York, NY, USA.
Pascal, R., Pross, A. and Sutherland, J.D. Towards an evolutionary theory of the origin of life based on kinetics and thermodynamics. Open Biol. 3(2013): 130156. doi.org/10.1098/rsob.130156.
Pasteur, L. Relation qui peut exister entre la forme crystalline et la compostion chimique, et sur las cause de la polarization rotatoire. Ann. Chim. Phys. 26(1848): 535–538.
Peretó, J. Out of fuzzy chemistry: from prebiotic chemistry to metabolic networks. Chem. Soc. Rev. 41(2012): 5394–5403.
Peretó, J. 2019. Prebiotic chemistry that led to life. pp. 219–233. *In*: Kolb, V.M. (ed.). Handbook of Astrobiology. CRC Press, Boca Raton, USA.
Proskurowski, G., Lilley, M.D., Seewald, J.S., Früh-Green, G.L., Olson, E.J., Lupton, J.E. et al. Abiogenic hydrocarbon production at Lost City hydrothermal field. Science 319(2008): 604–607.
Russell, M.J. First life. American Scientist 94(2006): 32–39.
Russell, M.J., Hall, A.J. and Martin, W. Serpentinization as a source of energy at the origin of life. Geobiology 8(2010): 355–371.
Russell, M.J. and Martin, W. The rocky roots of the acetyl-CoA pathway. Trends Biochem. Sci. 29(2004): 358–363.
Russell, M.J. Green rust: The simple organizing 'seed' of all life? Life 8(2018): 35. doi: 10.3390/life8030035.
Schmitt-Kopplin, P., Hemmler, D., Moritz, F., Gougeon, R.G., Lucio, M., Meringer, M. et al. Systems chemical analytics: introduction to the challenges of chemical complexity analysis. Faraday Discuss. 218(2019): 9–28.
Schneider, E.D. and Sagan, D. 2005. Into the Cool: Energy Flow, Thermodynamics and Life. University of Chicago Press,Chicago, IL, USA.
Segré, D., Ben-Eli, D., Deamer, D.W. and Lancet, D. The lipid world. Orig. Life Evol. Biosph. 31(2001): 119–145.
Seifert, U. Stochastic thermodynamics, fluctuation theorems and molecular machines. Rep. Prog. Phys. 75(2012): 126001. doi: 10.1088/0034-4885/75/12/126001.
Shock, E.L., McCollom, T.M. and Schulte, M.D. 1998. The emergence of metabolism from within hydrothermal systems. pp. 59–76. *In*: Wiegel, J. and Adams, M.W.W. (eds.). Thermophiles: the Keys to Molecular Evolution and the Origin of Life. Taylor & Francis, London, UK.

Shock, E. and Canovas, P. The potential for abiotic organic synthesis and biosynthesis at seafloor hydrothermal systems. Geofluids 10(2010): 161–192.

Sleep, N.H., Meibom, A., Fridriksson, Th., Coleman, R.G. and Bird, D.K. H_2-rich fluids from serpentinization: Geochemical and biotic implications. Proc. Natl. Acad. Sci. USA 101(2004): 12818–12823.

Suttle, C.A. Viruses in the sea. Nature 437(2005): 356–361.

Tessera, M. Research program for a search of the origin of Darwinian evolution: Research program for a vesicle-based model of the origin of Darwinian evolution on prebiotic early Earth. Orig. Life Evol. Biosph. 47(2017): 57–68.

Torday, J.S. and Miller Jr, W.B. The resolution of ambiguity as the basis for life: A cellular bridge between Western reductionism and Eastern holism. Prog. Biophys. Mol. Biol. 131(2017): 288–297.

Wang, Q.A. Maximum entropy change and least action principle for nonequilibrium systems. Astrophys. Space Sci. 305(2006): 273–281.

Wächtershäuser, G. Groundworks for an evolutionary biochemistry: The iron-sulphur world. Prog. Biophys. Mol. Biol. 58(1992): 85–201.

Wächtershäuser, G. The place of RNA in the origin and early evolution of the genetic machinery. Life 4(2014): 1050–1091.

Weiss, M.C., Sousa, F.L., Mrnjavac, N., Neukirchen, S., Roettger, M., Nelson-Sathi, S. et al. The physiology and habitat of the last universal common ancestor. Nat. Microbiol. 1(2016): 16116. doi: 10.1038/nmicrobiol.2016.116.

Chapter 14

Integrating Glycolysis with Oxidative Phosphorylation by Hexokinases

14.1 Hexokinases are ancient housekeeping enzymes

We shall examine in this chapter how ATP hydrolysis can drive the "forbidden" reaction. The often mentioned example of coupled reactions (see Chapter 3) in biochemical textbooks is glucose activation by hexokinase enzyme, which adds the phosphate group. It is the initial and obligatory reaction of glycolysis that is usually presented as a "very simple example." The reaction integrates glycolysis with oxidative phosphorylation. Indeed, it looks simple enough when written as a chemical equation:

$$\text{Glucose} + \text{MgATP}^{2-} \longrightarrow \text{Glucose-6-PO}_4^{2-} + \text{MgADP}^{-} + \text{H}^{+}$$

Standard free-energy changes are compared to study this and other coupled reactions with classical methods of equilibrium thermodynamics. Then, the positive free-energy change for the uphill reaction of glucose phosphorylation is more than compensated with almost three times higher negative free-energy change associated with ATP hydrolysis. Under physiological conditions, the net free-energy change for hexokinase reaction is even more negative. This exergonic or "energy-releasing" reaction would still be too slow in a cell (almost one-year-long reaction half-life) if not speeded up by hexokinase. How much is it speeded up? Stockbridge and Wolfenden (2009) found that yeast hexokinase produces rate enhancement (k_{cat}/k_{non}) of 4.7 x 10^{13} in comparison with nonenzymatic reaction rate constant k_{non}. That was the confirmation of an early insight: enzymes are the specific enhancers of the reaction rate (Koshland 1958).

Hexokinases are ancient housekeeping enzymes present in all cells for billions of years due to their essential function of trapping the glucose and other simple carbohydrates (Roy et al. 2019). Phosphorylation of sugars inhibits their passive diffusion across the cell membrane. Sugars with added two negative charges remain safely trapped in the cytoplasm for subsequent glycolysis. Human hexokinases I and II (HK-I and HK-II) reversibly associate with mitochondria at a convenient location to trap just released ATP molecules produced by that organelle. HK-I and HK-II have different metabolic roles. Their expression and intracellular distribution also differ from tissue to tissue. HK-I is mainly associated with the outer mitochondrial membrane and is committed to glycolysis. HK-II dissociates from mitochondria for high enough G6P concentration and translocates between mitochondria and cytoplasmic compartments for better regulation of glycogen formation (John et al. 2011). It is ubiquitous in the brain and red blood cells due to the importance of catabolic reactions for neuronal cells and erythrocytes. The HK-II is also highly expressed in insulin-sensitive tissues of skeletal and cardiac muscle with a dual role of promoting glycolysis when mitochondrially-bound and promoting glycogen formation in other locations. Hexokinases

I, II, and III have about 250 times higher affinity for glucose than HK-IV (the glucokinase). The Michaelis constant K_M is in the millimolar range for glucokinase and in the micromolar range for other hexokinases. Evolutionary ranking of hexokinases was also performed with the result that the glucokinase is the simplest of four isoforms and probably the closest to a putative common ancestor (Cárdenas et al. 1998). As for some other enzymes (see Chapter 10), biological evolution improved the catalytic efficiency and increased substrate specificity of hexokinases (Cárdenas et al. 1998). The same authors have very convincingly shown in that paper that none of the four hexokinase isoforms exhibits the straightforward Michaelis-Menten kinetics, although it is a good enough approximation for HK-I and HK-II.

Hexokinase product, the glucose 6-phosphate, is a potent natural inhibitor of cytoplasmic HK-II. The best artificial inhibitors of hexokinases are also the best inhibitors of glycolysis. Glucose-deprivation effects of one such molecule, the 2-deoxy-D-glucose, have been studied for a long time, even as a chemotherapeutic drug (Aft et al. 2002). HK-II is overexpressed in cancer cells, which metabolize glucose at an enhanced rate (Mathupala et al. 2006). One ingenious application of biochemistry and medical physics in a clinical setting is the tumors imaging *in vivo* via **positron emission tomography** to detect the binding of a radioactive analog of 2-deoxyglucose, the 1-[^{18}F]fluoro-2-deoxy-D-glucose, to HK-II (Young et al. 1999). Another application of the same technique predicts age-associated cognitive decline (de Leon et al. 2001). The 3-bromopyruvate inhibitor of HK-II is the chemotherapeutic drug blocking glycolysis in tumor cells, thus halting or slowing tumor growth (Chen et al. 2009).

The telltale sign that enzyme is regulated is when the hyperbolic curve for the reaction rate dependence on substrate concentration (v(S)) is no longer seen. For instance, Monod-Wyman-Changeux (MWC) kinetics for cooperative allosteric regulation of oligomeric enzymes leads to the sigmoidal v(S) curve. The idea that ligand can cause enzyme conformational change also implied dynamical changes in multimeric enzymes. Consequently, the original MWC model was generalized to dynamical nonequilibrium situations (Marzen et al. 2013). The surprise for biochemists was that such allosteric changes were also possible in the monomeric enzyme hexokinase (Niemeyer et al. 1975, Cárdenas 2015). The authors proposed that the sigmoidal kinetics of glucokinase increases enzyme's efficiency in the range of rapidly changing physiological substrate concentrations. Sigmoidal kinetics and its possible relationship with metabolic regulation, entropy production, and growth under limited supply of nutrients were discussed in Chapter 9, section 9.5, and in Chapter 11.

Using Terrel Hill's method (Hill 1977), it would be easy to calculate how entropy production contributions are distributed in all catalytic steps once when all or majority of rate constants are estimated in a corresponding catalytic scheme. Several questions have to be answered beforehand. One of them is about the possibility of wasteful (slip) transitions producing only ATP hydrolysis. Steitz proposed that the induced fit mechanism ensured almost total elimination of wasteful ATP hydrolysis (Bennett and Steitz 1978, Steitz et al. 1981). A small ATPase activity was detected in the absence of sugar, but this slippage transition is not included in common schemes for hexokinase catalytic mechanism.

In any case, coupled ATP hydrolysis and glucose phosphorylation reactions, catalyzed by this single extraordinary enzyme—the hexokinase, fully deserve all attention devoted to it by biochemists, biophysicists, and medical researchers through many past decades. We mentioned above that there are four isoforms of hexokinase in mammals. One of them, the glucokinase, acts as a glucose sensor besides mediating the initial step of glycolysis. We present here one possible kinetic scheme for the glucokinase mechanism of action in the form of Hill's diagram (Figure 14.1). There is no clear cut separation between the driving cycle and the driven cycle, as in Figure 3.1 from Chapter 3. Driving (ATP hydrolysis) and driven reaction (glucose phosphorylation) are interleaved (Branscomb et al. 2017). Also, the last product released can be glucose-6-phosphate instead of ADP (Gregoriou et al. 1981). Our scheme (Figure 14.1) favors the Branscomb et al. (2017) "escapement mechanism" hypothesis about a common nature of enzymatic free-energy conversions, namely that

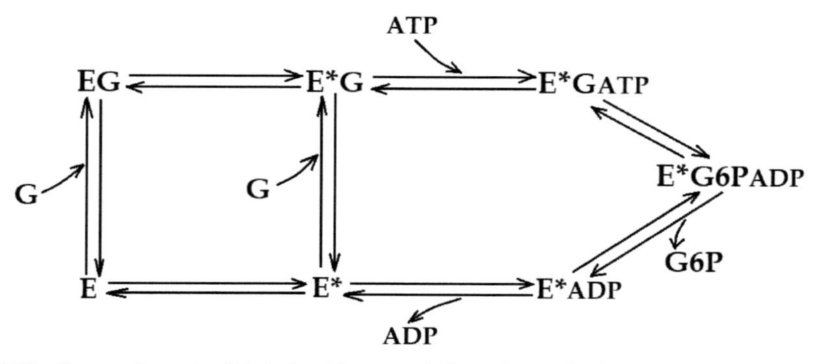

Figure 14.1: Hill's diagram for a simplified glucokinase catalytic cycle. Productive cycling is assumed in a clockwise direction with $MgADP^-$ as the last product released (Monasterio and Cárdenas 2003). Low-affinity state E and high-affinity state E* are two different open conformations induced by substrates G (glucose) and ATP to undergo a big conformational change toward catalytically competent closed conformations. A great number of water molecules are then removed from the enzyme active site to facilitate product formation: glucose-6-phosphate (G6P).

the full completion of the driven process triggers the completion of the driving process. It is the **conditional completion** mechanism.

14.2 Conformational changes associated with water ejection and entropy production

Thomas Steitz reported with Bennett (1978) that a large conformational change occurs upon glucose binding to hexokinase. A smaller subunit of the enzyme in an open conformation has remarkable freedom to rotate, twist and finally clamp tightly together with a larger subunit after glucose is caught up in a middle grove (Kamata et al. 2004). The enzyme has found local energy minimum in its open conformation, but it is not the final folded state, which is much more compact and globular-like (Kamata et al. 2004, Cárdenas 2015). Reaching the final closed state is gated by substrates (glucose and ATP), which triggers a dissociation avalanche of H_2O molecules from the protein. Upwards of 100 water molecules are released from the enzyme in binding substrates and taken up again from solution during conformational changes in a single catalytic cycle (Rand 1992). The expulsed mass of water is almost 10 times greater than the mass of expulsed products. It indicates that hexokinase is, like the ribosome, a predominantly "physical" catalyst (Stockbridge and Wolfenden 2009) whose mechanism of action is to bring together substrates in a water-poor environment. We can think of it as water ejecting and water-attracting nanomachine with about 300 water pumping turns per second in the case of yeast hexokinase (Todd and Gomez 2001).

The calorimetric technique these authors used to measure reaction rates is an excellent experimental verification for the proportionality between the reaction rate and generated thermal power. The thermal flux multiplied by overall free-energy change (the driving force) gives the entropy production for biological pumps (Kjelstrup et al. 2005). Active transport pumps are usually located in topologically closed membranes and are essential for bioenergetic conversions. However, in previous chapters, we have seen that cytoplasmic enzymes' catalytic rates are also proportional to entropy production. The initial release of many water molecules upon substrate binding is the first drive of the hexokinases' catalytic cycle as a significant hydrophobic force of essentially entropic nature. Directional water ejection is the pumping activity. Thus, hexokinases are biological pumps despite being cytoplasmic enzymes. The preferred direction (outside of grove tunnel) for water ejection opens the possibility of increased diffusion for active hexokinases. Enhanced diffusion was found for some other enzymes which can move toward regions with higher substrate concentrations (Sengupta et al. 2013). **Chemotaxis** for hexokinase was indeed observed recently (Mohajerani et al. 2018). These authors proposed that the thermodynamic driving force for hexokinase chemotaxis arises from entropically favored expulsion of water outside the enzyme pocket upon glucose binding. Water release from enzyme surface results in a significant entropy increase because bulk water molecules have considerably greater freedom of movement than surface-bound water. Hence, it is not surprising that entropy changes have an equally important role as enthalpy changes in the study of thermodynamics and dynamics of hexokinase catalytic mechanism using the reaction rate theory (Laidler and King 1983, Stockbridge and Wolfenden 2009).

To be more specific, the activation energy E_a from the Arrhenius equation for the reaction rates $k = A \cdot \exp(-E_a/RT)$ contains the change in activation entropy term, which determines the rate-limiting transition state barrier for the product release step. It affects local unfolding into more open conformation concomitant with product release. Entropically-favored local unfolding connected to product release has been confirmed for adenylate kinase (Saavedra et al. 2018). These authors named it the dynamic-based allosteric tuning mechanism when they examined the regulatory role of amino acid substitutions at surface sites distant from the substrate-binding site. The mechanism applies to hexokinases too.

14.3 Is hexokinase-2 the gatekeeper for life and death?

Integrating glycolysis with oxidative phosphorylation is another vital function of HK-II. Its association with mitochondria is with the ATP synthasome. The ATP synthasome complex consists of at least four proteins that are all together connecting inner and outer mitochondrial membrane: inner membrane ATP-synthase and adenine nucleotide translocator (ANT), mitochondrial creatine kinase (MtCK) in the intermembrane mitochondrial space, and outer membrane VDAC porin (VDAC is the acronym for "voltage-dependent anion channel") to which HK-II reversibly binds (Mathupala et al. 2009, Guzun et al. 2012, Koit et al. 2017). That is how HK-II gets preferential access to mitochondrially synthesized ATP molecules. It may be enough to explain how hexokinase can discriminate between mitochondrially generated ATP and extramitochondrial ATP molecules (Wilson 2003). HK-II contributes to respiratory control together with the mitochondrial creatine kinase (MtCK), which also has preferential access to ATP generated by mitochondrion and can also regulate the ADP concentration in the immediate vicinity to adenine nucleotide translocase (ANT) (Figure 14.2). In the presence of creatine, the ADP flux through ANT enhances the rate of oxidative phosphorylation (Timohhina et al. 2009) and corresponding entropy production.

Hexokinase III is the least known of four hexokinase isoforms. It can be found in the cytoplasm and the nucleus. Specifically, it is associated with the perinuclear compartment (PNC) in the periphery of the nucleolus (Wilson 2003). PNC, with its HK-III, is highly enriched in malignant cells with metastatic capacity (Pollock and Huang 2010). As HK-II, the HK-I and HK-III enzymes also contain an allosteric binding site for the inhibition by their product glucose-6-phosphate. The HK-III is also inhibited with its substrate, the glucose, when glucose concentration approaches the millimolar range (Cárdenas et al. 1998).

Glucokinase (HK-IV) activators are potential anti-diabetic drugs for patients with type II diabetes but with some unwanted side effects, such as high incidence of hypoglycemia (Nakamura and Terauchi 2015). No natural glucokinase activator was ever found. Dramatic effects for human health are probably due to the absence of product feedback inhibition for HK-IV (Cárdenas 2015). We mentioned in section 14.1 that HK-IV acts as a molecular sensor. It connects glucose metabolism to insulin release. However, it is very sensitive to oxidation by reactive oxygen species (Cárdenas 2015). Together with α-synuclein, HK-IV belongs among the most sensitive enzymes to irreversible carbonylation, the process with a major adverse influence at normal and accelerated aging (Krisko and Radman 2019). More than 100 natural amino acid substitutions causing type 1 diabetes (before 25 years of age) or permanent neonatal diabetes are distributed all over human glucokinase sequence containing 465 amino acid residues.

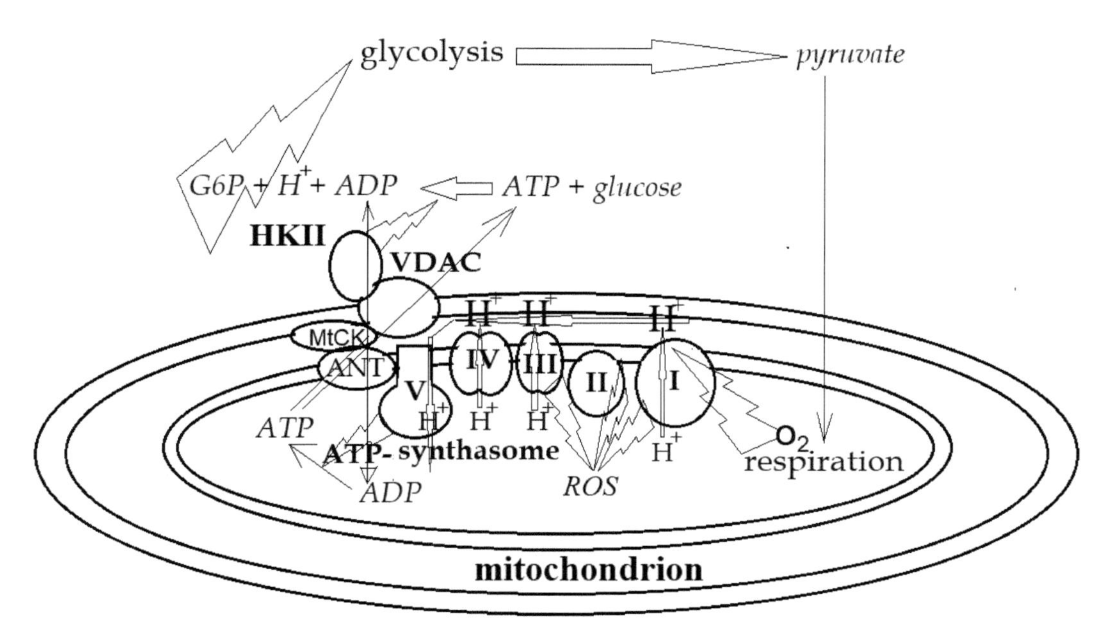

Figure 14.2: Simplified scheme for the mitochondrial interactosome. This illustration aims to stress the role of hexokinase II (HK-II) reversible association with voltage-dependent anion channel (VDAC) in regulating proton, ADP, and ATP fluxes driven by respiratory and ATP-synthasome supercomplexes. HK-II connects glycolysis with respiration via respiratory chain complexes I to IV and ATP-synthasome consisting of H^+-ATP-synthase (complex V), adenine nucleotide translocase (ANT), mitochondrial creatine kinase (MtCK), and VDAC. Metabolites are glucose-6-phosphate (G6P), ATP, ADP, inorganic phosphate P_i (not shown), and pyruvate. ROS is a common abbreviation for all reactive oxygen species that are produced mainly by complex I and III (Efremov and Sazanov 2011, Quinlan et al. 2013). The mitochondrion is just a cartoon representing idealized outer membrane bilayer (outer ellipses) and inner membrane bilayer (inner ellipses) connected by VDAC porin.

Remarkably, HK-II becomes insensitive to product G6P inhibition when the enzyme is bound to mitochondria. The malignancy potential is associated with increased expression of HK-II. Cancer cells are using the opportunity to get more fuel for their propensity to metabolize glucose to lactic acid at a high rate in the presence of oxygen ("Warburg effect") and in the absence of feedback that controls the system (Seyfried 2015). German chemist Otto Warburg (see Chapter 15) concluded that "the cancer cells can obtain approximately the same amount of energy from fermentation (i.e. glycolysis to lactic acid) as from respiration" (Warburg 1956). A more recent estimate is that mitochondria supply from 40 to 75% of ATP requirements for cancer cells (Mathupala et al. 2010) in comparison to about 90% of ATP requirements for normal cells.

Peter Pedersen and his students determined how the prevention of product inhibition of HK-II leads to cancer's phenotype of elevated glycolysis in the presence of oxygen (Pedersen 2007). G6P inhibits HK-II in the cytoplasm but not when the enzyme is bound to the VDAC porin. Tumor cells can even contain an increased number of mitochondria than corresponding healthy cells (Koit et al. 2017). However, cancer cells' mitochondria have frequent respiratory complex I deficiencies (Ramsay et al. 2011, Lemarie and Grimm 2011). The problem with the mitochondria from cancer cells is that the HK-II-VDAC association can suppress the mitochondrial-mediated apoptosis of these cells. HK-II helps to keep mitochondrial permeability transition pore (MPT pore) in closed conformation when bound to the ATP-synthasome complex (Bernardi et al. 2015). The identity of MPT pore is still hotly debated (Baines and Gutiérrez-Aguilar 2018, Carroll et al. 2019). Recent observations support the idea that monomers or dimmers of H^+ -ATP-synthase can be prompted to undergo conformational changes conducive to the formation of the MPT megachannel formation in the inner mitochondrial membrane (Urbani et al. 2019, Carraro et al. 2019).

MPT pores have never been reported in prokaryotes. These pores and their dynamics are the unique and critically important invention of eukaryotic cells enabling the development of all multicellular organisms because every individual organism's development is always tightly connected to programmed cell death. Mitochondrial permeability transition, when transition pores are continuously open, can spell the death of mitochondria and the whole cell that contains mitochondria. In that sense, HK-II enzymes are gatekeepers for life and death with a vital function to integrate death signals and commit cells to the point of no return with a dissipation peak characteristic for cell death (Bernardi et al. 2015).

In addition to several natural death signals, the artificial death signals have also been found. The peptide TAT-HK prevents HK-II association with outer mitochondrial membrane VDAC porin (Chiara et al. 2008). It is composed of VDAC-binding domain of HK-II, that is, the first 15 amino acid residues from HK-II, which are linked to the cell-penetrating TAT peptide (TAT CPP) to allow its easy entry in the cell:

MIASHLLAYFFTELN–RKKRRQRRR
Mitochondrial-binding peptide–HIV-1 TAT CPP peptide

Some other peptides targeting mitochondria in malignant cells are potential therapeutic peptides (Constance and Lim 2012). Cancer cells need a plentiful supply of free energy for their unchecked proliferation. Despite the Warburg effect, respiratory metabolism of mitochondria is crucial for cancer survival. Thus, malignant mitochondria and their protein-protein interactions contributing to cancer phenotype are a key target for chemotherapy. Small and cheap chemotherapeutic molecules remain the preferred choice for pharmaceutical companies. However, small molecules are often not suitable for disrupting specific cancer-driving protein-protein interactions without some harmful side-effects. That is why "smart" anticancer peptides, able to cause mitochondrial apoptosis, such as TAT-HP, are getting ever more attention in the scientific community interested in advancing precision medicine (Deslouches and Di 2017, Zhang et al. 2019).

A hallmark of mitochondrial apoptosis is the release of cytochrome c into the cytoplasm. That is seen after the TAT-HK peptide application (Chiara et al. 2008). Organs and cells in our body with the highest metabolic rate are the brain and neurons. Neurons are always poised at the brink of catastrophic autodestruction due to their high activity. It is not accidental that HK-II is often named the brain hexokinase. Its high presence in brain cells regulates both the active life and death of neurons.

Two additional abilities of HK-II (compared to high-affinity hexokinases I and III) are important for regulating bioenergetics of normal and cancer cells. As we mentioned earlier, all high-affinity hexokinases derive their evolutionary origin from simpler glucokinase type enzymes. This evolutionary event arose as an ancestral gene duplicated at the time of invertebrates' origin leading to hexokinases with two domains and doubling of molecular weight for HK I, II, and III (Cárdenas et al. 1998). The initial result was that HK I, II, and III had two active sites for glucose phosphorylation and the capacity to double the G6P product formation rate. Only the isoform II retained the catalytic capacity in both domains. In contrast, isoforms I and III lost the N-terminal domain's catalytic capacity but gained the additional capacity for feedback regulation by ADP, inorganic phosphate, and G6P (Mathupala et al. 2010). Doubling the rate of product formation is especially useful for fetal tissues genetically programmed for fast growth at the expense of plentiful nutrients provided by the mother. Embryonic cells preferentially use aerobic glycolysis. That is why most cancers revert to fetal type isoforms of crucial metabolic enzymes. This biochemical shift enables tumors to bypass metabolic constraints even under adverse physiological conditions such as hypoxia (low oxygen concentration). Vander Heiden et al. proposed in 2009 that all proliferating cells use aerobic glycolysis with highly expressed HK-II.

To sum up this chapter and connect it with previous chapters, let us see what we learned about nonequilibrium thermodynamics applications to coupled biochemical reactions. Textbooks' picture about one uphill and another downhill half-reaction is somewhat misleading. None of these reactions would produce anything in any reasonable time. The enzyme knots them together tightly in space, time, and function so that entropy production is increased up to billion-billion-fold. The biological evolution is not so blind, after all. It probes innumerable pathways for increasing entropy production until the most probable one is found. That is usually the pathway of more complex organization and tighter coupling enabling faster entropy increase of the universe. The biological coupling agent is the enzyme whose catalytic efficiency, complexity, and regulatory capability increased during evolution. It increased because maintaining and spreading small complexity islands of living cells in a vast universe is one way to obey the thermodynamic quirk of far-from-equilibrium situations: the most probable and fastest approach to thermodynamic equilibrium always involves some degree of transient structural ordering intrinsically coupled to overall entropy increase. Thus, biology and physics meet together when scientists from both fields recognize the entropy production increase as the single invisible coupling agent hidden behind myriad different marvelous but transitory living structures.

The entropy production increase is not always a positive development for us. Some cancer phenotypes are connected to entropy production increases when the absence of feedback inhibition or its inactivation is used by cancer cells to proliferate (Montero et al. 2016). However, both normal and cancer cells have an in-built ability to rapidly adopt the entropy production level to environmental conditions changes. Hexokinases are just one class of highly regulated enzymes with a pivotal role in connecting entropy production with bioenergetic free-energy transduction. These enzymes work according to our proposal that entropy production regulation is the hallmark of life in the light of other attempts to define life (see the previous chapter). In previous chapters, we have explored many other enzyme-catalyzed bioenergetic circuits as concrete examples of how life on Earth complied with the entropy-increasing universal evolution facilitated by structure-forming aspects. While stars cannot control their gradual or explosive entropy export to the universe, cells' bioenergetics regulate their entropy production at a nearly constant high level for prolonged periods. In the next chapter, we shall examine how an even simpler enzyme than hexokinase can connect the bioenergetics of cancer, aging, and brain.

Definitions and explanations from Chapter 14:

A driven process's conditional completion is a more complex feedback mechanism than the reaction-rate control by slow rate-limiting step, the last one in a biochemical reaction cycle. The last step enslaves the rate of all previous steps, including much faster reaction initiation needed for driving all subsequent steps. The driven process can also enslave the driving process so that the full completion of the driven process triggers the completion of the driving process (the "escapement mechanism" hypothesis). The respiratory control in mitochondrial bioenergetics is the example of a conditional completion feedback regulation (see Chapter 7, Definitions and explanations).

Chemotaxis is a cell's or organism's ability to move in response to a chemical stimulus. Bacteria need to swim toward higher concentrations of nutrients. Chemotaxis is critical during all phases of development for multicellular organisms. Some peptides and proteins are chemoattractants for leukocytes. Enzymatic chemotaxis is an enhanced diffusion of enzymes towards regions of higher substrate concentration.

Positron emission tomography (PET) is an imaging method used in nuclear medicine. It requires the injection of a radioactive isotope (the tracer) in a patient's body and the detection of gamma rays produced by the annihilation of isotope-released positrons with electrons in their vicinity. A dedicated computer program performs the 3D reconstruction of tracer concentration within the body. It produces tomography imaging (PET) from multiple projectional radiographs. Radiolabeled molecular probes are also used to combine specific radioactive tracers with selective binding to certain tumor types or brain receptors.

References

Aft, R.L., Zhang, F.W. and Gius, D. Evaluation of 2-deoxy-D-glucose as a chemotherapeutic agent: mechanism of cell death. Br. J. Cancer 87(2002): 805–812.

Baines, C.P. and Gutiérrez-Aguilar, M. The still uncertain identity of the channel-forming unit(s) of the mitochondrial permeability transition pore. Cell Calcium 73(2018): 121–130.

Bennett, W.S. and Steitz, T.A. Glucose-induced conformational change in yeast hexokinase. Proc. Nati. Acad. Sci. USA 75(1978): 4848–4852.

Bernardi, P., Rasola, A., Forte, M. and Lippe, G. The mitochondrial permeability transition pore: channel formation by F-ATP synthase, integration in signal transduction, and role in pathophysiology. Physiol. Rev. 95(2015): 1111–1155.

Branscomb, E., Biancalani, T., Goldenfeld, N. and Russell, M. Escapement mechanisms and the conversion of disequilibria; the engines of creation. Phys. Rep. 677(2017): 1–60.

Carraro, M., Checchetto, V., Szabó, I. and Bernardi, P. F-ATP synthase and the permeability transition pore: fewer doubts, more certainties. FEBS Lett. 593(2019): 1542–1553.

Carroll, J., He, J., Ding, S., Fearnley, I.M. and Walker, J.E. Persistence of the permeability transition pore in human mitochondria devoid of an assembled ATP synthase. Proc. Natl. Acad. Sci. USA 116(2019): 12816–12821.

Cárdenas, M.L., Cornish-Bowden, A. and Ureta, T. Evolution and regulatory role of the hexokinases. Biochim. Biophys. Acta 1401(1998): 242–264.

Cárdenas, M. Understanding mechanisms of enzyme co-operativity: The importance of not being at equilibrium. Perspect. Sci. 4(2015): 10–16.

Chen, Z., Zhang, H., Lu, W. and Huang, P. Role of mitochondria-associated hexokinase II in cancer cell death induced by 3-bromopyruvate. Biochim. Biophys. Acta 1787(2009): 553–560.

Chiara, F., Castellaro, D., Marin, O., Petronilli, V., Brusilow, W.S., Juhaszova, M. et al. Hexokinase II detachment from mitochondria triggers apoptosis through the permeability transition pore independent of voltage-dependent anion channels. PLoS ONE 3(2008): e1852. doi: 10.1371/journal.pone.0001852.

Constance, J.E. and Lim, C.L. Targeting malignant mitochondria with therapeutic peptides. Ther. Deliv. 3(2012): 961–979.

de Leon, M.J., Convit, A., Wolf, O.T., Tarshish, C.Y., DeSanti, S., Rusinek, H. et al. Prediction of cognitive decline in normal elderly subjects with 2-[^{18}F]fluoro-2-deoxy-D-glucose/positron-emission tomography (FDG/PET). Proc. Natl. Acad. Sci. USA 98(2001): 10966–10971.

Deslouches, B. and Di, Y.P. Antimicrobial peptides with selective antitumor mechanisms: prospect for anticancer applications. Oncotarget 8(2017): 46635–46651.

Efremov, R.G. and Sazanov, L.A. Respiratory complex I: 'steam engine' of the cell? Curr. Opin. Struct. Biol. 21(2011): 532–540.

Gregoriou, M., Trayer, I.P. and Cornish-Bowden, A. Isotope-exchange evidence for an ordered mechanism for rat-liver glucokinase, a monomeric cooperative enzyme. Biochemistry 20(1981): 499–506.

Guzun, R., Gonzalez-Granillo, M., Karu-Varikmaa, M., Grichine, A., Usson, Y., Kaambre, T. et al. Regulation of respiration in muscle cells *in vivo* by VDAC through interaction with the cytoskeleton and MtCK within mitochondrial interactosome. Biochim. Biophys. Acta 1818(2012): 1545–1554.

Hill, T.L. 1977. Free Energy Transduction in Biology: The Steady State Kinetic and Thermodynamic Formalism. Academic Press, New York, NY, USA.

John, S., Weiss, J.N. and Ribalet, B. Subcellular localization of hexokinases I and II directs the metabolic fate of glucose. PLoS ONE 6(2011): e17674. doi: 10.1371/journal.pone.0017674.

Kamata, K., Mitsuya, M., Nishimura, T., Eiki, J.-I. and Nagata, Y. Structural basis for allosteric regulation of the monomeric allosteric enzyme human glucokinase. Structure 12(2004): 429–438.

Kjelstrup, S., Rubi, J.M. and Bedeaux, D. Energy dissipation in slipping biological pumps. Phys. Chem. Chem. Phys. 7(2005): 4009–4018.

Koit, A., Shevchuk, I., Ounpuu, L., Klepinin, A., Chekulayev, V., Timohhina, N. et al. Mitochondrial respiration in human colorectal and breast cancer clinical material is regulated differently. Oxid. Med. Cell Longev. (2017): 1372640. doi: 10.1155/2017/1372640.

Koshland Jr., D.E. Application of a theory of enzyme specificity to protein synthesis. Proc. Natl. Acad. Sci. USA 44(1958): 98–104.

Krisko, A. and Radman, M. Protein damage, ageing and age-related diseases. Open Biol. 9(2019): 180249. doi.org/10.1098/rsob.180249.

Laidler, K.L. and King, M.C. Development of transition state theory. J. Phys. Chem. 87(1983): 2657–2664.

Lemarie, A. and Grimm, S. Mitochondrial respiratory chain complexes: apoptosis sensors mutated in cancer. Oncogene 30(2011): 3985–4003.

Marzen, S., Garcia, H.G. and Phillips, R. Statistical mechanics of Monod–Wyman–Changeux (MWC) models. J. Mol. Biol. 425(2013): 1433–1460.

Mathupala, S.P., Ko, Y.H. and Pedersen, P.L. Hexokinase II: Cancer's double-edged sword acting as both facilitator and gatekeeper of malignancy when bound to mitochondria. Oncogene 25(2006): 4777–4786.

Mathupala, S.P., Ko, Y.H. and Pedersen, P.L. Hexokinase-2 bound to mitochondria: Cancer's stygian link to the "Warburg effect" and a pivotal target for effective therapy. Semin. Cancer Biol. 19(2009): 17–24.

Mathupala, S.P., Ko, Y.H. and Pedersen, P.L. The pivotal roles of mitochondria in cancer: Warburg and beyond and encouraging prospects for effective therapies. Biochim. Biophys. Acta. 1797(2010): 1225–1230.

Mohajerani, F., Zhao, X., Somasundar, A., Velegol, D. and Sen, A. A theory of enzyme chemotaxis: From experiments to modeling. Biochemistry 57(2018): 6256–6263.

Monasterio, O. and Cárdenas, M. Kinetic studies of rat liver hexokinase D ('glucokinase') in non-co-operative conditions show an ordered mechanism with MgADP as the last product to be released. Biochem. J. 371(2003): 29–38.

Montero, S., Martin, R.R., Guerra, A., Casanella, O., Cocho, G. and Nieto-Villar, J.M. Cancer glycolysis I: Entropy production and sensitivity analysis in stationary state. J. Adenocarcinoma 1(2016): 1.

Nakamura, A. and Terauchi, Y. Present status of clinical deployment of glucokinase activators. J. Diabetes Invest. 6(2015): 124–132.

Niemeyer, H., Cárdenas, M.L., Rabajille, E., Ureta, T., Clark-Turri, L. and Peñaranda, J. Sigmoidal kinetics of glucokinase. Enzyme 20(1975): 321–333.

Pedersen, P.L. Warburg, me and hexokinase 2: Multiple discoveries of key molecular events underlying one of cancers' most common phenotypes, the "Warburg Effect", i.e., elevated glycolysis in the presence of oxygen. J. Bioenerg. Biomembr. 39(2007): 211–222.

Pollock, C. and Huang, S. The perinucleolar compartment. Cold Spring Harb. Perspect. Biol. (2010); 2: a000679. doi: 10.1101/cshperspect.a000679.

Quinlan, C.L., Perevoshchikova, I.V., Hey-Mogensen, M., Orr, A.L. and Brand, M.D. Sites of reactive oxygen species generation by mitochondria oxidizing different substrates. Redox Biology 1(2013): 304–312.

Ramsay, E.E., Hogg, P.J. and Dilda, P.J. Mitochondrial metabolism inhibitors for cancer therapy. Pharm. Res. 28(2011): 2731–2744.

Rand, R.P. Raising water to new heights. Science 256(1992): 618.

Roy, S., Vega, M.V. and Harmer, N.J. Carbohydrate kinases: A conserved mechanism across differing folds. Catalysts 9(2019): 29. doi: 10.3390/catal9010029.

Saavedra, H.G., Wrabl, J.O., Anderson, J.A., Li, J. and Hilser, V.J. Dynamic allostery can drive cold adaptation in enzymes. Nature 558(2018): 324–328.

Sengupta, S., Dey, K.K., Muddana, H.S., Tabouillot, T., Ibele, M.E., Butler, P.J. et al. Enzyme molecules as nanomotors. J. Am. Chem. Soc. 135(2013): 1406–1414.

Seyfried, T.N. Cancer as a mitochondrial metabolic disease. Front. Cell Dev. Biol. 3(2015): 43. doi: 10.3389/fcell.2015.00043.

Steitz, T.A., Shoham, M. and Bennett, W.S. Structural dynamics of yeast hexokinase during catalysis. Phil. Trans. R. Soc. Lond. B 293(1981): 43–52.

Stockbridge, R.B. and Wolfenden, R. The intrinsic reactivity of ATP and the catalytic proficiencies of kinases acting on glucose, *N*-acetylgalactosamine, and homoserine. J. Biol. Chem. 284(2009): 22747–22757.

Timohhina, N., Guzun, R., Kersti Tepp, K., Monge, C., Varikmaa, M., Vija, H. et al. Direct measurement of energy fluxes from mitochondria into cytoplasm in permeabilized cardiac cells *in situ*: some evidence for mitochondrial interactosome. J. Bioenerg. Biomembr. 41(2009): 259–275.

Todd, M.J. and Gomez, J. Enzyme kinetics determined using calorimetry: A general assay for enzyme activity? Anal. Biochem. 296(2001): 179–187.

Urbani, A., Giorgio, V., Carrer, A., Franchin, C., Arrigoni, G., Jiko, C. et al. Purified F-ATP synthase forms a Ca^{2+}-dependent high-conductance channel matching the mitochondrial permeability transition pore. Nat. Commun. 10(2019): 4341. doi.org/10.1038/s41467-019-12331-1.

Vander Heiden, M.G., Cantley, L.C. and Thompson, C.B. Understanding the Warburg Effect: The metabolic requirements of cell proliferation. Science 324(2009): 1029–1033.

Warburg, O. On the origin of cancer cells. Science 123(1956): 309–314.

Wilson, J.E. Isozymes of mammalian hexokinase: structure, subcellular localization and metabolic function. J. Exp. Biol. 206(2003): 2049–2057.

Wirth, C., Brandt, U., Hunte, C. and Zickermann, V. Structure and function of mitochondrial complex I. Biochim. Biophys. Acta 1857(2016): 902–914.

Young, H., Baum, R., Cremerius, U., Herholz, K., Hoekstra, O., Lammertsma, A.A. et al. Measurement of clinical and subclinical tumour response using [18F]-fluorodeoxyglucose and positron emission tomography: review and 1999 EORTC recommendations. European Organization for Research and Treatment of Cancer (EORTC) PET Study Group. Eur. J. Cancer 35(1999): 1773–1782.

Zhang, C., Yang, M. and Ericsson, A.C. Antimicrobial peptides: Potential application in liver cancer. Front. Microbiol. (2019): 1257. doi: 10.3389/fmicb.2019.01257.

Chapter 15

Bioenergetics of the Brain, Aging, and Cancer Cells as Bridged by α-synuclein

When famous physicists, like Erwin Schrödinger and Sir Roger Penrose (Hameroff and Penrose 2014), entered into biology, they were not modest in choosing problems to tackle. Arguably, the questions of what is life and how cognitive functions arose are the most challenging problems in biology. Francis Crick struggled up to the very last day of his life with the question of where consciousness resides in a brain. Out of thousands of functionally important proteins in our brain, we shall examine just one in this chapter. However complex, the biochemical context cannot be avoided while discussing the mystery of memories (Eichenbaum 2017) in health and disease or the mystery how an extremely rich functionality emerges within a fixed anatomical structure (Park and Friston 2013). More than ten thousand papers have been published about synucleins (Surguchov 2019) after they were discovered in 1988 (Maroteaux et al. 1988). We shall see that α-synuclein is involved in almost everything we can think of and literally in our thinking. It is surprising and disconcerting that its normal physiological function or functions are still poorly understood. In the case of human α-synuclein expressed by the SNCA gene, the UniProtKB database (SYUA_HUMAN) has annotations for 25 molecular functions and 80 biological processes. We shall mostly focus on bioenergetic aspects. My interest in this small protein awoke thanks to conversations with Miro Radman and other colleagues from the Mediterranian Institute for Life Sciences, so this chapter will have some personal touch too. Among multiple other proteins, synucleins are connected with neurotoxicity, aging, and malignancy. An acceleration of bioenergetics for dopaminergic neurons can lead to neurodegenerative disorders, a decrease in bioenergetics occurs with the aging process, and a switch in bioenergetics is associated with cancer.

15.1 Hallmarks of aging and cancer

Very few people can see something special in you, some exceptional talent or virtue which you are not even aware of having. They are precious gifts in your life. One such person in my life is Alec Keith. He noticed my love for bioenergetics before I realized having it in myself. He encouraged the theoretical physicist DJ to publish his first biological paper using some leftover dusty Warburg apparatus for measuring respiration (Juretić 1976). At 85 years of age, he recently claimed about his 19-years old self, "There was nothing special about me." Through all these years, he saved the letter of gratitude given to him during the Korean war by Park Bo Sun in joyful anticipation of a new school year. This young teacher hoped that Alec and other UN soldiers would soon help establish a peaceful Korea.

Alec paid a very high price. The next year he was, for the rest of his life, in a wheelchair. I first met him in 1972 as my professor and future mentor at Penn State University. At that time, my goal was to get an education in biophysics in order to learn why life developed in the universe. With my physics background, I was well aware that physicists consider life's complex structure as having an extremely low probability. I was also a polio victim like Alec but could walk with difficulties, and I also thought about myself in the same terms: "There is nothing special about me." He tried to convince me that there was something special about me and provided crucial support for the PhD thesis research.

The second time I met him, about 15 years afterward, Alec was mostly done with publishing about 100 original interdisciplinary papers, and he went on to hold about 200 patents and develop pharmaceutical companies with a market capitalization in billions of dollars. At least one of his recent patents (US10252743) is an oral supplement nutraceutical containing chosen exotic mushroom species, known in traditional oriental medicine for its antitumor effects. Alec observed its synergistic effect in boosting immune function, which may help inhibit or reduce tumor cells' growth and possibly curtail some other diseases related to premature aging. I mentioned this as a parallel development in our lives. Thanks to him, I was also, for a good part of my life, the professor of biophysics keenly aware of how potent are natural chemical compounds developed by long biological evolution, especially when acting in synergy.

Alec is known today as a generous philanthropist. For instance, in 2004, he donated 1.6 million dollars to Arizona State University Cancer Research Institute. Together with many other honest and sometimes heroic stories about truly dedicated scientist G. Robert (Bob) Pettit, this story can be found in Dr. Robert S. Byars's book "Waging War on Cancer." I lost regular contact with Alec during the partial occupation of disarmed Croatia in 1991 by heavily armed Milošević's communists from Serbia, similar to war initiated by North Korean communists in 1950s. It makes me sad because I can only repeat Dr. Pettit's words: "Alec is a brilliant guy and a great friend."

The rise in cancer incidence has become one of the leading causes for decreasing the quality of life in modern society. Cancer treatments are, as a rule, expensive and with an uncertain outcome. It is mainly due to cancer cells' ability to develop resistance and the inability of healthy tissues and organs to tolerate long-lasting chemotherapeutic drug treatments, which are not selective enough for killing only cancer cells. Cancer causes a shortening of lifespan and is intimately connected with aging. We would expect that the hallmarks of aging are very similar to the hallmarks of cancer. In a seminal review paper, Hanahan and Weinberg (2011) listed ten hallmark capabilities of cancer: deregulating cellular bioenergetics, avoiding immune destruction, genome instability and mutation, tumor-promoting inflammation, enabling replicative immortality, avoiding growth suppressors, resisting cell death, inducing **angiogenesis**, activating invasion and metastasis, and sustaining proliferative signaling. Aging in mammals also includes numerous hallmarks. The main bioenergetic hallmarks are mitochondrial dysfunction and exponential accumulation of somatic mutations (Milholland et al. 2015), in particular, with the high mutation incidence in the mitochondrial DNA, which is less well protected from reactive oxygen species and less well repaired than nuclear DNA (Kennedy et al. 2012). Respiration is a double-edged sword acting as both a vital and potentially damaging influence, which accumulates with years. Mitochondrial respiratory chain diverts up to 2% of consumed O_2 into the superoxide and hydrogen peroxide production. Other hallmarks of aging are **epigenetic** alterations, loss of proteostasis, altered nutrient-sensing, stem cell exhaustion, cellular senescence, **telomere** attrition, loss of cell-cell communication (López-Otin et al. 2013), misfolded or carbonylated proteins, protein aggregation (Ludtmann et al. 2018), and the modulation of the misfolded protein level (Proctor et al. 2016, Perić et al. 2017).

One common hallmark is the high somatic mutation burden. Some tumors can accumulate several hundred mutations (Fox and Loeb 2010), and high mutation frequency is a dominant feature for many tissues in old age. The gain of pathological (toxic) function due to somatic mutations is common mechanism in aging and neoplasms' development. It is only partially avoided in some vertebrates. Simultaneously delayed senescence and absence of cancers are probably not accidental among rare long-lived small mammals such as naked mole rats (Biliński et al. 2015).

Another common hallmark is the change of bioenergetics and the change in regulating bioenergetics. Although bioenergetics is not restricted to mitochondria, we shall consider first the changes in mitochondrial bioenergetics because this research field, initiated almost 100 years ago, is still in vigorous development. Cancer cells and tissues exhibit a pronounced change in cellular bioenergetics connected to mitochondrial dysfunction. Healthy cells have very effective free-energy transduction. For each glucose molecule catabolized to pyruvate, healthy cells with mitochondria can produce up to 36 ATP molecules. Tumor glycolysis is different. Cancer cells can perform aerobic fermentation, first described by Otto Heinrich Warburg (1927) as a shift to primitive yeast-like metabolism (Koppenol et al. 2011, Blum and Kloog 2014). Warburg was awarded the Nobel Prize in Physiology or Medicine in 1931 for his discovery of cytochrome c oxidase. My observation, mentioned in the first section of Chapter 8, was that the mold *Neurospora crassa* also shifts its metabolism toward fermentation and accumulation of fatty acid deposits when respiration is blocked by cyanide or inhibited by lack of phosphatidylcholine, an essential phospholipid for the proper activity of cytochrome c oxidase (Juretić 1976).

Warburg did not describe tumor tissue metabolism as primitive. His judgment from the research spanning three decades was that "tumor cell is more versatile than the normal cell" in choosing between glycolysis and respiration (Warburg 1956). Since glycolysis produces only 2 ATP molecules per consumed glucose molecule, cancer cells had to discover a compensation mechanism in hypoxic conditions characteristic of tumors *in vivo*. It is the upregulation of membrane-embedded glucose transporters (Phan et al. 2014) and increased glucose uptake for up to 20 times. Hexokinase II is also overexpressed, reinforcing the glycolytic phenotype (see Chapter 14, section 14.1). Tumor cells have a considerably higher efficiency of glycolysis than healthy cells (Luo et al. 2006). This high free-energy transduction efficiency can be maintained regardless of oxygen supply, enabling cancer cells with a flexible response capability to the changing microenvironment and a metabolic advantage over healthy cells.

15.2 Alpha-synuclein connection to neurodegenerative diseases

Connecting age-related decrease in bioenergetics to conformational changes in certain proteins is still an ambitious challenge. Detailed analysis in several cases led to the discovery of conformational changes enabling protein aggregation, protein misfolding, or protein transmembrane insertion connected to oligomerization, membrane pore formation, and toxic response (Tsigelny et al. 2012, Lim et al. 2015, Proctor et al. 2016, Fusco et al. 2017, Ludtmann et al. 2018). A fascinating example is the amyloid-β and α-synuclein role in mitochondrial dysfunction and neurodegenerative diseases (Ryan et al. 2015, Kawamata and Manfredi 2017). The $A\beta_{1-42}$ is the most toxic species of all amyloid-β peptides, which arise from the natural processing of the amyloid precursor protein (APP). The α-synuclein is a small protein with only 140 amino acid residues, so that it can be considered a longer peptide as well. Accumulation in mitochondria of $A\beta_{1-42}$ or α-synuclein are causes for defective mitochondrial bioenergetics and impaired neuronal activity in Alzheimer's disease (AD), Parkinson's disease (PD), amyotrophic lateral sclerosis (ALS), frontotemporal dementia (FTD), and Huntington's disease (HD). Alpha-synuclein has been implicated not only in the PD, but also in other α-synucleinopathies: dementia with Lewy bodies (DLB), progressive supranuclear palsy (PSP), and multiple system atrophy (MSA), all with movement disorder phenotype (Cheng et al. 2011, Kovach 2016). The most frequent age-related diseases, PD are AD, are affecting around 2% of the population older than 60 years (de Lau and Breteler 2006).

These two peptides, $A\beta_{1-42}$ and α-synuclein, can exhibit profound changes in their secondary and tertiary structures. For instance, α-synuclein shifts from around 3% to more than 70% of α-helix content upon lipid binding. It is a useful ability to perform numerous regulatory functions in a brain or for postulated host defense physiological function for these and some other amyloid peptides (Kagan et al. 2012), but potentially toxic when oligomers and aggregates of these peptides produce uncontrolled permeability increase of host cell membranes.

In precision medicine endeavors, numerous antibodies have been produced against pathological modifications of α-synuclein (Valera et al. 2016). The rationale for the development of active, passive, and T-cell based immunotherapy is the detection of toxic α-syn oligomers in the plasma membrane, extracellular space, and blood sera of Parkinson's disease patients. These efforts to use immunotherapy and vaccination had only limited success in targeting deleterious molecular conformers or as early biomarkers for developing neurodegenerative disorders. Novel immunotherapeutic approaches are hindered by cross-reactivity with the physiological forms of monomeric α-synuclein (Kovach 2016).

An important α-synuclein function is the regulation of synaptic transmission among neurons by acting as a safety brake on neurotransmission (Benskey et al. 2016). Among total cytosolic proteins in the central nervous system, it comprises up to 1%, which is just another confirmation about the dominant role of inhibitory pathways in developed brains. While mitochondria localization is detected for α-synuclein and amyloid-β (Nakamura 2013), their normal physiological function in mitochondria is still an active research topic. They may be involved in the quality control process to clear damaged mitochondria. They may also be involved in innate host defense against bacteria leading to the collateral damage of host mitochondria, organelles which originated as bacterial endosymbionts. The potency of amyloid-β is similar or higher than well known LL-37 human antimicrobial peptide in inhibiting the growth of the bacteria *Streptococcus pneumonia*, which can cause pneumococcal meningitis, and in inhibiting the growth of the fungi *Candida albicans*, which can cause candidal meningitis (Soscia et al. 2010).

15.3 Low concentration of regulation signal ions and molecules

As a rule, ions, small molecules, peptides, and proteins are present in low concentration when used by cells as regulation signals. Ergo, dynamic changes in their localization are hard to detect. For instance, cells have a near-vacuum state for calcium around 0.1 μM, while external fluids, such as blood, have calcium concentrations in the millimolar range. Mild activation of Ca^{2+} influx in a neuronal cell supports neuronal survival and signaling. Also, a limited generation of reactive oxygen and nitrogen species mediates normal neuronal function (Nakamura and Lipton 2011). However, allowing too high calcium concentration in the cytoplasm and excessive generation of free radicals is linked to death-signals for a cell, causing regulated cell destruction (apoptosis).

A good example is also α-synuclein. When present in micromolar or overexpressed level (Ryan et al. 2015), it promotes mitochondrial-dependent apoptosis leading to neuronal degradation, but nanomolar levels of α-synuclein protect neurons from oxidative stress, nitrosative (nitric oxide) stress, and glutamate toxicity. The threshold α-syn concentration for the formation of potentially toxic oligomers is only about 90 nM (Plotegher et al. 2014). Conflicting reports about α-synuclein inhibiting (Shaltouki et al. 2018) or stimulating mitophagy (mitochondrial turnover) (Grassi et al. 2018) are probably due to the underappreciated dynamics of post-translational peptide phosphorylation. A significant fraction of α-synuclein interacts with mitochondria. Thus, it is likely that synuclein, its post-translational modifications, and numerous interactions with mitochondrial membrane-associated proteins (Kawamata and Manfredi 2017) are involved in maintaining the balance in neuronal bioenergetics (Nakamura 2013). That particular balance can be named mitochondrial homeostasis because it is a physiological regulation among fusion, fission, mitophagy, and transport of mitochondria.

15.4 Alpha-synuclein oligomers can dissipate the mitochondrial membrane potential

Depolarization of bacteria and mitochondria by cationic antimicrobial peptides (AMPs) leads to bioenergetics' collapse (Westerhoff et al. 1989a). The membrane poration mechanism is probably different for amyloid peptides because the amyloid-β has a net negative charge under physiological conditions, while most AMPs are cationic. A rare exception is human AMP dermcidin. With 17% or lesser identity to synucleins, it is an anionic AMP unrelated to synucleins in primary structure, but possible similarities in the mechanism of membrane poration were not examined. Oligomerization of anionic AMPs may activate their antimicrobial activity and defeat the microbial resistance mechanisms (Soscia et al. 2010). Similar to the mechanism of action for some peptide antibiotics (Juretić et al. 1989, Juretić 1990), α-synuclein oligomers can cooperate in causing mitochondria depolarization and dysfunction (Ghio et al. 2016, Kawamata and Manfredi 2017). Either slowly or quickly, most antibiotics and toxins trigger a cascade of events that culminates in the formation of membrane nonselective pores, an irreversible event, which quickly leads to the demise of cellular bioenergetics and cell death.

There were some interesting coincidences in discoveries of synucleins and peptide antibiotics, initially considered entirely different research fields. Alpha-synuclein was discovered in 1988 by Maroteaux et al. (1988), while peptide antibiotics magainins were found in 1987 (Zasloff 1987, Soravia et al. 1988). I discovered the cooperative and synergetic activity of magainins in membrane-potential dissipation (Juretić et al. 1989, Juretić 1990, Westerhoff et al. 1995). This observation was confirmed in numerous studies afterward, first by Williams et al. (1990), who observed the magainins-caused release of carboxyfluorescein from liposomes. Three different methods to detect synergy were used by Zerweck et al. (2018), under the assumption that the synergy is due to peptide-peptide physical interactions. These authors found that the critical sequence motif GxxxG must be present in the PGLa antimicrobial peptide to cause synergy with the magainin 2 peptide. Motifs with small amino acids G, A, S of the type [G,A,S]xxx[G,A,S] are known to be important for the dimerization of peptide helical segments in the membrane environment (Walters and DeGrado 2006). Other authors concluded that anionic membranes with a suitable negative intrinsic curvature catalyze the self-organizing behavior of some antimicrobial peptides (Leber et al. 2018). Alpha-synuclein has four small motifs of the type [G,A]xxx[G,A] in imperfect repeats R1, R2, and R4 of its N-terminal 60 amino acids (Figure 15.1). Synergistic behavior has been observed for this and other amyloidogenic peptides and proteins (Biza et al. 2017). For instance, amyloid-β can seed the polymerization process of α-synuclein (Tsigelny et al. 2008) and tau protein (Guo et al. 2006).

Cancer, aging, and neurodegenerative diseases have in common that some peptides and proteins can gain toxic function, usually after somatic mutations. However, even wild-type peptides and proteins can gain toxic function when overexpressed. We mentioned the example of hexokinase-2 from cancer cells in the previous chapter. It is the case for α-synuclein overexpression in older age neurodegenerative diseases, while γ-synuclein overexpression is associated with malignancy (Ahmad et al. 2007). These peptides are similar in their N-terminal region and different in their C-terminal part. The C-terminal part of α-synuclein has a role in preventing peptide's aggregation, and it appears to have the chaperone activity (Ahmad et al. 2007).

15.5 Structure-function dissection of α-synuclein domains

Human α-synuclein is an intrinsically disordered protein (Theillet et al. 2016). The N-terminal half of α-synuclein has a hydrophobic core that can change its conformation from α-helical to π-helical and to β-strand when interacting with a membrane (Tsigelny et al. 2012). It can also form oligomers and aggregates in cytoplasm and upon contact with membranes starting from mostly disordered conformation in the solution. Dynamic changes in conformation, depending on the environment,

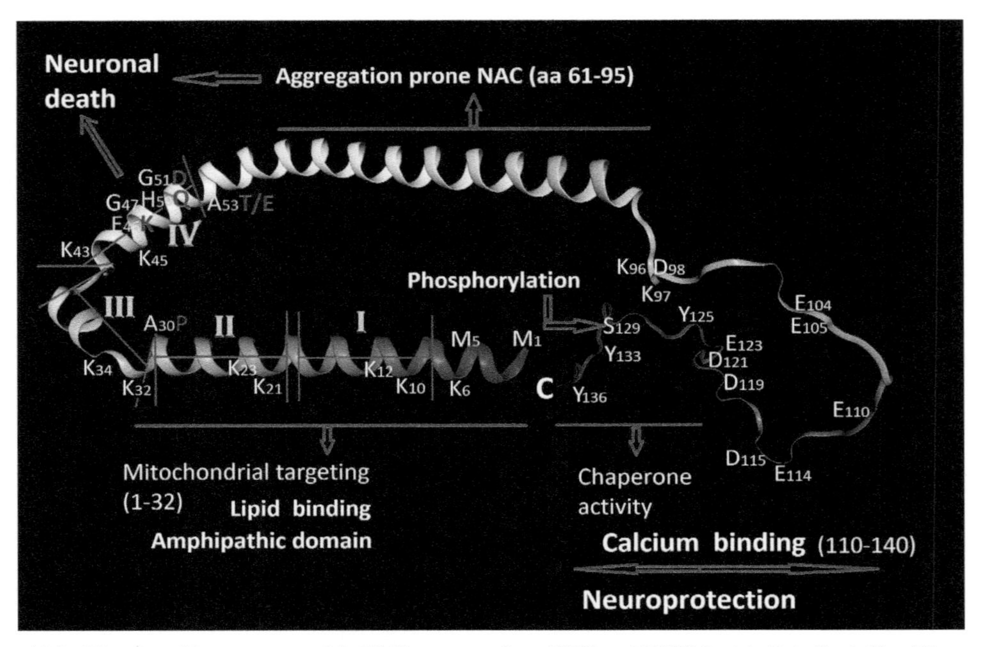

Figure 15.1: We selected human α-synuclein NMR structure from 2KK8 and 1XQ8 Protein Data Bank files (Ulmer et al. 2005, Rao et al. 2010) and visualized it by the NGL viewer (Rose and Hildebrand 2015, Rose et al. 2018). The single letter code is used to label amino acids (M = Methionine, G = Glycine, A = Alanine, Y = tyrosine, S = Serine, H = Histidine, positively charged K = Lysine, negatively charged E = Glutamate, and D = Aspartate). Four amphipathic repeats of 11 residues at the N-terminal (Roman numerals I to IV) are found in the 9–52 α-synuclein segment of bony vertebrates. The fourth repeat contains the GxxxG motif, which is known to promote dimerization and oligomerization in the presence of a membrane environment. The region with cationic repeats encompasses the mitochondrial targeting signal 1–32 and extends to alanine 53. It contains the most frequent amino acid substitutions causing Parkinson's disease and neuronal death (red letters). The peptide is a Janus-faced molecule associated with toxic and beneficial effects. It also has the chameleon's ability for conformational changes from disordered monomers into helix-turn-helix V-shape oligomerization upon contact with highly curved membranes or self-aggregation in beta-sheets seeded by residues 61–95 and eventually encompassing residues 31–109. Residues 96–140 are prone to calcium-binding, carbonylation (lysine 96), phosphorylation (serine 129), and nitration (tyrosines 125, 133, 136) associated with aging, but they also have a chaperone and neuroprotective ability.

is the characteristic feature of the α-synuclein and other polypeptides belonging to the unique class of natively unfolded and disordered proteins which have a limited degree of folding in solution, but are capable of chameleon-like transformation upon contact with a membrane (Pfefferkorn et al. 2012). Physical features and conformations of a peptide and a membrane are crucial for a successful peptide-membrane interaction. Alpha-synuclein preferentially binds to highly curved membranes with anionic phospholipids (Bendor et al. 2013). These are stressed membranes, which synuclein can stabilize against fusing (Ghio et al. 2016) or induce the negative curvature as preparation for endocytosis (Vargas et al. 2014). It then changes its N-terminal conformation into a broken helix shape suitable for the accommodation to a curved surface of synaptic vesicles, which belong to the smallest cargo-containing vesicles in our bodies (Bendor et al. 2013).

The first 32 amino acids at the N-terminal of α-synuclein contain mitochondrial targeting signal peptide (Devi et al. 2008). Amino acid substitutions associated with early-onset of Parkinson's disease are the most frequent in the segment 43–53. Residues 61–95 from the NAC segment can seed aggregates (fibrils) with misfolded beta-sheet structures (Figure 15.1). Such aggregates are also associated with neurodegenerative diseases named synucleinopathies. Peptide N-terminal half contains seven imperfect repeats of 11 residues. The first four repeats are depicted in Figure 15.1 with Roman numerals as important weakly alpha-amphipathic and weakly lipid-binding segments with a consensus KTKEGV motif. These repeats are uniquely found in the synucleins of jawed vertebrates (fishes, amphibians, reptiles, birds, and mammals) when Prosite search is done with the query "[GES]-**K**-[TA]-**K**-[EQ]-**G**-**V**-[VAL]-[AEYH]-[AVG]-[AVG]" (terms in square brackets mean that only those single-letter amino acid codes are accepted that are found inside brackets). In the fourth repeat from K43 to A53, the "small motif" GXXXG is always present and may promote the oligomerization of synuclein monomers in a membrane environment (Walters and DeGrado 2006). It has confounded the investigators that the oligomerization of helical segments in a membrane environment is important for the normal physiological function of synuclein and also for its toxicity associated with synucleinopathies. One can conclude that α-synuclein is the Janus-faced molecule with toxic and beneficial effects.

Studies about the structure-function relationship for the synuclein C-terminal half have revealed an equally intricate pattern. The C-terminal sequence segment of 32 residues (109–140) contains acidic tandem repeats of 16-residues (109–124 and 125–140) that confer Ca^{2+} binding activity (Nielsen et al. 2001). The C-terminal residues 125–140 have a chaperone (Kim et al. 2002) and neuroprotective ability (Leng and Chuang 2006). Nitration (tyrosines 125, 133, 136) and carbonylation (lysine 96) (Jang et al. 2010) changes are not only age-related features (McCormack et al. 2012), but are possibly also responsible for pathologically active forms of the molecule. The calcium-binding segment (residues 109–140) and phosphorylation sites (serine 87, serine 129, tyrosine 125) have an important role in many physiological functions of α-synuclein (Okochi et al. 2000, Nielsen et al. 2001, Bendor et al. 2013). However, the serine 129 phosphorylated molecule has been connected with the progression of Lewy-type histopathology characteristic for Lewy body diseases (DLB and PD) (Walker et al. 2013).

In general, an enemy acting from inside to destroy cellular bioenergetics is more dangerous than an external enemy. For instance, cancer and other inherent age-related diseases are more dangerous for us and more difficult to handle by our immune system than external harmful microbes for which mammals have developed multiple defenses through eons. Interdisciplinary research and a combination of experimental and theoretical approaches seem to be necessary to unravel the mechanism for α-synuclein regular and toxic activity. Modern-day molecular dynamics simulations in experts' hands are an excellent new microscopy tool for examining the dynamics of peptide-membrane interaction (Juretić et al. 2018, Ulmschneider and Ulmschneider 2018) when combined with NMR or X-ray spectroscopy, biochemical and ultrastructural analysis.

In particular, wild-type α-synuclein and its mutant A53T molecule were recently examined by MD simulations for their membrane-perforation ability (Tsigelny et al. 2012). The N-terminal part of α-synuclein (residues 1–60) has a net positive charge and the propensity to form amphipathic α-helices. It is attracted by negatively charged phospholipid polar heads of phosphatidylinositol, phosphatidylglycerol, phosphatidylserine, and cardiolipin from the inner mitochondrial membrane (Ghio et al. 2016). The electrostatic energy of phospholipids-α-synuclein interaction increases one order of magnitude when peptide leaves water upon entering into membrane bilayer. It happens due to a sharp decrease in the dielectric constant at the water-membrane interface and in the membrane interior. Hydrophobic interactions are also important for the deeper entrance of α-synuclein hydrophobic core (residues 61–95) in a membrane, which has a chameleon propensity to form either α-helix or β-sheet secondary structure. Forming membrane pores from α-synuclein oligomers is quicker and almost over during only about 10 ns for the mutant A53T (Tsigelny et al. 2012). The familial (inherited) form of Parkinson's disease is associated with the mutation causing A53T substitution (Ryan et al. 2015). Accelerated self-assembly of α-synuclein with this mutation can result in membrane-bound oligomers with antiparallel β-sheets forming their core structures or into elongated fibrils containing parallel β-sheets (Ghio et al. 2016). It is still debated what is more toxic in the case of Parkinson's disease. Toxic structures can be the pore-forming α-synuclein oligomers depolarizing inner mitochondrial membrane or cytoplasmic deposits (termed Lewis bodies) of elongated amyloid fibrils. Bendor et al. (2013) suggested that α-synuclein oligomers are responsible for toxicity rather than fibrils.

15.6 Misfolding, mobility, promiscuous interactions, and prion diseases

Normal physiological functions of α-synuclein are still being investigated, in part due to an extraordinarily long list of its interacting partners (Benskey et al. 2016, Burré et al. 2018). In addition to auto-association, α-synuclein interacts with other synucleins. One beneficial function of β-synuclein is to prevent α-synuclein from aggregation (Cheng et al. 2011). It appears that α-synuclein is the central hub involved in so many interactions and pathways that the old dogma from the 1960s about linear causal chain "one gene, one protein, one function" is proven to be utterly wrong in this case. Its moniker "synuclein" (SYNapse + NUCLEus) means that this small protein was first discovered in purified synaptic vesicles and at the nuclear envelope of nerve cells' nucleus (Maroteaux et al. 1988), indicating to authors that it may be involved in coordinating nuclear and synaptic events. Nuclear localization was difficult to reproduce in subsequent studies, and α-synuclein has been detected in denucleated cells, such as red blood cells, in blood plasma, platelets, lymphocytes, cerebrospinal fluid, and many other organs in addition to the brain (Burré et al. 2018). The extraordinary mobility of α-synuclein was confirmed in neuronal cells. It was found to be widely distributed among cytoplasmic space and almost all membrane-containing compartments of some neuronal cell classes. Dynamic attachment and disengagement from membrane surfaces and mobility among different vesicles and organelles also imply remarkable structural flexibility for synuclein (Mor et al. 2016).

Different synuclein species are not equally distributed within cells and extracellular space, within different cell types and separate compartments of the same cell type. Export from a cell is one mechanism developed by evolution to eliminate mutated or toxic synucleins with deleterious amino acid substitutions or post-translational modifications. Another is to store it in insoluble inert aggregates inside a cell. The third mechanism of attaching the ubiquitin peptide to synucleins is shared with the majority of proteins destined for the degradation via the proteosome after being misfolded or modified (Burré et al. 2018). Preferred targets for ubiquitination of α-synuclein are lysines from repeats I, II, III, and IV (Figure 15.1). As mentioned before, these sequence regions contain mitochondrial targeting signal 1–32 and the most frequent amino acid substitutions causing Parkinson's disease and neuronal death.

Observations of the α-synuclein presence in the extracellular space and neuron to neuron propagation evoked the possibility of prion-like spread for some self-associated synuclein species (Burré et al. 2018). The prion-like spread among neurons has also been observed for amyloid-β and microtubule-associated protein tau (Frost et al. 2009). Prions, synuclein, tau, and Aβ proteins are all connected with neurodegenerative diseases, such as Creutzfeldt–Jakob disease (CJD) and Bovine spongiform encephalopathy (BSE), causing the motor and cognitive dysfunction due to cell death of long-living neuronal cells. The BSE is named the "mad-cow" disease since it first spread among cows in England when farmers invented the cannibalistic practice of feeding these herbivores with recycled bovine carcasses (Prusiner 1997). These devastating disorders are classified according to proteins causing them (prion diseases, synucleinopathies, tauopathies), or according to early sites of pathology (ALS, FTD, BSE), or according to authors who first identified them (AD, PD, HD, CJD). Their common feature of protein misfolding, leading to positive feedback seeds for polymerization (self-seeding), prompted Stanley B. Prusiner (the Nobel Prize in Physiology and Medicine 1997 winner) and his colleagues to extend the term "prion diseases" to all of these disorders (Prusiner 2017).

Do 100 trillion bacteria in our intestines have anything to say about functional or dysfunctional interaction of prion-like proteins with our brain cells? Apparently, they do (Houser and Tansey 2017). Neurodegenerative diseases are multi-system disorders tightly connected with neuroinflammation and higher levels of, for instance, phosphorylated and aggregated α-synuclein in intestines. The persistent intestinal inflammation is not only connected to the PD disorder, but it can start many years before the onset of PD symptoms. Leaking guts, leaking blood-brain barrier, and accelerated aging with higher cancer incidence have all been associated with inflammatory conditions.

The interconnection of amyloid-β, prions, and α-synuclein oligomers has a deeper molecular basis as all of them have been proposed to mediate the level of metabotropic glutamate receptors mGluR5 (Price et al. 2010, Teich et al. 2015, Scheckel and Aguzzi 2018). In the next section, we shall see why these interactions are important for normal neuronal signaling and also for neurodegeneration.

15.7 Abnormal protein-protein interactions can lead to a bioenergetic collapse

Abnormal protein-protein interactions attracted the interest of scientists and research-funding resources. In the case of neurodegenerative diseases, a fundamental insight is still very limited concerning the normal physiological role of amyloidogenic proteins and mechanisms on how it can go amiss. About 40 human proteins can form abnormal aggregates (fibrils) *in vivo* and *in vitro* (Biza et al. 2017). It is a well documented but strange observation. The prevention of inappropriate protein-protein interactions has been one of the major goals for biological evolution and selection, likely to be of equal importance in preventing wrong cell-cell interactions in multi-cellular organisms. Failure of the former or latter leads to bioenergetic collapse, cell death, organ inactivation, and organism death. One may conclude that amyloidogenic peptides and proteins provide normal physiological functions of such importance, that their inherent instability is mostly controlled to a satisfactory degree during reproductive age. The amyloid deposits are probably not as toxic as some oligomer intermediates, acting as seeds for membrane permeabilization and the extension of amyloid fibrils. There is a very fine dynamic balance between monomers' beneficial influence and higher-order oligomers' harmful activity. Stronger interest and investment in revealing the regular physiological role of amyloidogenic peptides and proteins should help understand and prevent triggers that derail this role toward neurodegenerative diseases (AD, PD, FTD, ALS, HD, CJD, and other diseases), accelerated aging, and cancer.

There are several additional features, which are shared with protein-indicators of prion diseases. One common feature is the earlier mentioned possibility that some of the prion-like proteins may have host defense antimicrobial function in addition to their other central nervous system roles (Soscia et al. 2010, Kagan et al. 2012, Park et al. 2016). Another is many interaction partners and multiple phenotypic traits that complicate understanding the functional importance of prions (Peggion et al. 2018) and prion-like proteins. Intensive investigation of neurodegenerative diseases revealed some physiological functions for these proteins. For instance, more than 20 tauopathies (Zhou et al. 2017) and associated presynaptic dysfunctions implicated derailing of tau protein from its primary role of regulating axonal microtubule dynamics (Weingarten et al. 1975, Frost et al. 2009). Cellular cytoskeleton homeostasis and regulation are of particular importance for neuronal cells with numerous extensions (dendritic branching) and axon length comparable to animal body length for some motor neurons. Neuronal protein tau contributes to stabilizing microtubule assembly, but hyperphosphorylated tau (P-tau) inhibited microtubule-promoting tau activity and even destroyed already present microtubules (Alonso et al. 2018). Tau and its post-translational modifications, together with other prion-like amyloid proteins, probably play an important role in axonal transport, which is essential for the survival of a neuron. Alpha-synuclein also interacts with tubulin and motor proteins (kinesin, dynein, septin-7) using microtubules as intracellular transport railways or for other purposes (Burré et al. 2018). It shifts the balance of tau species toward microtubule-inhibitory P-tau and promotes tau aggregation, while tau can seed α-synuclein aggregation. Alpha-synuclein ability to interact with tubulin from neuronal dendrites and cell body, as tau protein does, opens an intriguing possibility that α-synuclein is involved in postulated tubulin-associated memory encoding and even in moments of conscious experience (Hameroff and Penrose 1996, 2014).

During action-potential stimulation, α-synuclein quickly disperses from synaptic boutons (Fortin et al. 2005). The synuclein contributes to synaptic transmission regulation by facilitating the recycling of synaptic vesicles (Benskey et al. 2016). Thus, it is implicated in another costly free-energy transduction process besides mitochondria recycling, which we mentioned earlier (Grassi et al. 2018). Why is so much free-energy invested in synuclein-mediated neuronal dynamics? One possibility is its involvement in neuronal plasticity with the ultimate conversion of short-term to long term memory (Cheng et al. 2011). Support for this hypothesis is the change of α-synuclein level in brain regions of songbirds involved in learning songs (George et al. 1995). Peripheral blood samples of professional musicians had an upregulated SNCA gene for α-synuclein after their performance with string instruments that lasted about two hours (Kanduri et al. 2015). An overlap was found between singing-regulated genes of songbirds and upregulated genes after music performance by musicians. SNCA and several other genes are linked to dopamine metabolism. Dopaminergic neurons control neuronal pathways essential for motor behavior by releasing the dopamine neurotransmitter. In turn, repeated dopamine release leads to intensely pleasurable emotions and enhanced synaptic plasticity of musicians (Rosenkranz et al. 2007, Järvelä 2018). The possible molecular mechanism is a potent modulation of alpha-synuclein monomers assembly by dopamine binding only to extended α-synuclein conformation (Illes-Toth et al. 2013). It was suggested that α-synuclein is necessary for long-term memory and good cognitive function (Saleh et al. 2015). The suggestion seemingly contradicts observations that α-synuclein is primarily associated with cognitive decline during various neurodegenerative disorders. The synuclein location may be important. Extracellular α-synuclein oligomers impair memory and long-term potentiation (Martin et al. 2012).

With its low molecular weight, the α-synuclein can easily pass through the nuclear pores of the nuclear envelope and reach chromosomes if direct interaction with them is needed to establish the long-term memory. Oxidized synuclein binds to DNA and can cause DNA strand breaks (Surguchov 2019). Alpha-synuclein can even bind to its own mRNA, thus modulating the SNCA gene expression (Surguchev and Surguchov 2017). Less direct, but equally pivotal are the alterations in the expression of the membrane-bound glutamate receptors mGluR5 (Price et al. 2010). The accumulation of α-synuclein leads to an increased number of mGluR5 receptors at the cell surface and over-activation of the downstream signaling pathway. Activated by the excitatory neurotransmitter glutamate, this pathway triggers the cascade of events to release the intracellular messenger calcium from intracellular stores. In turn, calcium activates a large number of calcium-sensitive proteins. One of them phosphorylates the cyclic-AMP responsive element-binding protein (CREB). It is the main protein hub serving as a key regulator of local and global synaptic plasticity (Lisman et al. 2018). The phosphorylation of CREB's serine-133 increases CREB-DNA binding activity and enhances excitability (Grimes and Jope 2001). The mitochondrial dysfunction and glutamate excitotoxicity occur when some threshold is breached in the neuronal activity (Cassano et al. 2016). In other words, neurons can die when over-excited through overly active glutamate receptors.

Still, we must ask what is so special about human dopaminergic neurons to make them essential for the brain's activity despite being vulnerable to the excitatory pathway activators? Each such mature neuron has up to 2.5 million synapses. Thus, its hyper-branched architecture makes millions of connections with other brain cells. An excessive branching with a huge number of mitochondria and synapses cannot be robust with respect to oxidative stress and mitochondrial dysfunction during periods of voracious free-energy demands (Uchihara and Giasson 2016). Evolution has used hyperbranching of phylogenetically ancient networks to achieve widespread innervation of the cerebral cortex, basal ganglia, and hippocampus, thus exposing some frailty of these archaic networks (Diederich and Parent 2012). One possibility is that "reptilian brain" and "primate brain" in each of us still have some communication issues that are exposed under stress. Present-day research favors comparative studies assuming that the simpler structure of the reptilian brain can suggest solutions to unsolved neuroscience questions (Naumann et al. 2015).

Alcohol-craving patients do expose their liver and brain to additional stress. They showed an increased expression of the SNCA gene in their blood. Several studies identified the SNCA gene as a genetic link to alcoholism (Rotermund et al. 2017). As mentioned earlier, CREB transcriptional activator was induced in transgenic mice expressing human synuclein. For the development of different addictions, CREB is regarded as a key modulator. Patients affected by depression and post-traumatic stress also exhibited alterations in the CREB signaling pathway. CREB expression decreased both in animals and humans with post-traumatic stress disorder (PTSD) (Martini et al. 2013, Fenster et al. 2018). Stress can cause the translocation of α-synuclein to the nucleus (Surguchov 2015), where oxidized synuclein can bind to DNA and possibly interfere with the CREB-DNA binding. We described above the alternative indirect pathway for the α-synuclein-CREB interaction via mGluR5 receptors. These receptors' dysregulation has been recently connected to suicidal ideation (Davis et al. 2019). The present availability of quick methods to sample and analyze blood from the PTSD patients to detect CREB and SNCA expression can be used in suicide risk management. For instance, Croatia lost more than three thousand relatively young veterans – brave defenders of our country during the 1990–1995 aggression, who committed suicide after surviving all odds stacked against them, including long imprisonments (for many of them) with daily tortures in Serbian concentration camps. After almost 30 years, suicide risk is still high among the remaining population of Croatian veterans exposed to marginalization, slanders, and unjust trials for invented war crimes.

15.8 The paradoxical nature of brain cell bioenergetics

A single large neuron is nursed and maintained in the far-from-equilibrium state by thousands, if not millions of other cell types. The peculiar nature of brain cell bioenergetics is seen as a precarious balance between neuron's long and happy life at the cliff's edge of frequent depolarizations through action potentials. Depolarizations are obligatory for the communication function, and at the same time, can be a dangerous trigger for chain-reactions, which can lead to neuron's death. Prion-like proteins are involved in both aspects of a neuron's life. Like α-synuclein, some act as tethers to connect a train of presynaptic vesicles (Lautenschläger et al. 2018) or regulate the connection between mitochondria and endoplasmic reticulum (Kawamata and Manfredi 2017). Synuclein preference for highly curved membranes makes it likely that α-synuclein plays a role in intracellular trafficking, docking, fusion, and fission of organelles (Bridi and Hirth 2018). Apparently, α-synuclein can serve as a generalized workhorse key for locking in or enabling many different connections between such membranes.

Intercellular trafficking with the involvement of synucleins may also exist even to the extent of organelles' exchange between different cells. It is an active research area about intercellular communication between animal cells mediated by tunneling nanotubes (Rustom et al. 2004, Gerdes and Carvalho 2008, Rostami et al. 2017). This phenomenon, named "cellular parabiosis", can be either beneficial or harmful for the organism depending on what healthy cells are getting rid of and what cancer cells are gaining as a high-quality material from healthy cells (Portela et al. 2019, Radman 2019). All of the mentioned roles in trafficking require enhanced free-energy usage and dissipation. They can place some neuron cell types at high risk in case of free-energy failure.

Metal ions above some optimal concentrations serve as triggers for neuron dysfunction. Increased levels of iron (Fe), copper (Cu), zinc (Zn), and manganese (Mn) are associated with neuronal death in AD, PD, and other neurodegenerative disorders (Cahill et al. 2009, Gentile et al. 2018). Copper binds to the methionine motif $\mathbf{M_1}$DVF$\mathbf{M_5}$ at the N-terminal first five amino acid residues and to two other anchoring-sites containing histidine (H_{50}) or methionine (M_{116} and M_{127}) of α-synuclein (Gentile et al. 2018) (Figure 15.1). It is not an accidental or isolated case. The regulation of metal ion homeostasis is one of the many physiological functions of α-synuclein, prion, amyloid-β, and other prion-like proteins, which are all metal-binding proteins (Sadakane and Kawahara 2018). Binding of metal ions can get out of hand in neurodegenerative diseases, even in the case of essential metals mentioned above, known to play a crucial role in cellular free-energy transduction and the brain's bioenergetics. All essential metal-ions are tightly regulated to have optimal concentrations in different brain functional parts to maintain their involvement in highly intensive nerve cells' bioenergetics and avoid toxicity at a higher exposure level. For instance, α-synuclein is the ferrireductase, which catalyzes Fe (III) conversion to Fe (II). Making Fe (II) available for membrane enzymes involved in bioenergetics promotes the production of reactive oxygen species (ROS) (Davies et al. 2011). ROS participate in the intracellular signal transduction (Finkel 2011), but excess ROS causes oxidative stress. The β-synuclein may be able to reduce cellular sensitivity to free radicals produced by Fe (II) and to strengthen the protective effect of iron storage protein ferritin (Cahil et al. 2009, Angelova et al. 2018, Sadakane and Kawahara 2018).

Overexposure to Mn does not trigger progressive neurodegeneration, although the resulting manganism disorder has similar symptoms as Parkinson's disease. Fe^{3+}, Cu^{2+} and Mn^{2+} all bind to α-synuclein and prion protein (Sadakane and Kawahara 2018). Toxic metals ions of aluminum (Al) can also bind to amyloid prion-like proteins (Sadakane and Kawahara 2018), while other neurotoxic metals, such as mercury (Hg), lead (Pb), and arsenic (As), have other molecular targets. Comorbidities or mixed pathologies are then often seen when brain tissue is examined because aggregation of α-synuclein, tau protein, amyloid-β can lead to a variety of distinct synucleopathies, tauopathies, and Alzheimer disease-related pathology profiles.

15.9 Regulatory hotspots control the entropy production rate

Coming back to mitochondrial involvement in death or life signals, health, cancer, or neurological diseases, let us recall the Chapter 14 (section 14.3) discussion about enigmatic permeability transition pores (MPT). Dysfunctional mitochondria accumulate in cancer cell lines and produce more reactive oxygen species than healthy mitochondria. Paradoxically, their regulatory mechanisms change in such a way to prolong the lifetime of cancer cells (Bonora and Pinton 2014). Suppressing the opening of MPT is one evolutionary invention for avoiding aging. It is the facility of pre-invasive cancer cells that can be potentially useful for anti-aging research if we can inhibit tumorigenesis and metastasis at the same time. Mitochondrial permeability transition (Hunter and Haworth 1979) leads to the rapid dissipation of organelle's membrane potential with a corresponding peak in entropy production. We observed the same effect in 1989 after adding antimicrobial peptides to isolated rat liver mitochondria (Westerhoff et al. 1989b). Two well-known activators of MPT are increased mitochondrial calcium concentration and oxidative stress. It appears that MPTs and their environment of inner mitochondrial membrane protein complexes are regulatory hotspots for organelle's survival and the survival of the cell containing them. To underscore the importance of the inner mitochondrial membrane, Peter Rich estimated that a single human body has, on average, 14 thousand square meters of it (Rich 2003).

In conclusion, the observed epidemic of neurological illnesses during the last 30 years (Schofield 2017) is, in part, connected to amyloid prion-like protein misfolding. It is often due to the synergistic action of neurotoxicant molecular species, and it is always associated with bioenergetic collapse as a trigger for neuronal cell death. When, for whatever reasons, excess heat cannot be released in the environment, then neurons die, the organism dies, the ecosystem dies, and even a biosphere can die off. Tight regulation of entropy production by cells, organs, and organisms is an absolute requirement for the maintenance of far-from-equilibrium homeostasis with optimized efficiency values for all life-maintaining processes. All entropy produced within topologically closed compartments, i.e. within a limited volume, must be exported by them too. It does not lead to the minimum entropy production, but it does lead to the need that maximum entropy production values are subject to all constraints preventing its explosive increase and subsequent collapse. This principle is faithfully observed by all living beings, except human civilization. We may have already triggered a cataclysmic and unstoppable global heating due to fossil fuel burning and broken constraints of Earth's ability to release excess heat into the coldness of outer space (Schellnhuber 2008). Unfortunately, money-making is the primary driver of the climatic apocalypse, mainly because it does not recognize biological and thermodynamic constraints in-built into natural mechanisms to removal excess heat. It is not much different from an unhealthy addiction to drugs, which can stimulate our brain cells to produce excess activity triggering their death, except that nobody is willing to stop driving cars and stop producing garbage without complete recycling.

Even if we cease all industrial activities, the runaway global heating will only speed up. Too many positive feedback loops are already acting in synergy. Fewer aerosols in the atmosphere will remove one of the few remaining negative feedbacks (negative feedbacks have a positive influence in containing explosion-driving positive feedbacks). It will be helpful to learn from nature in ever more detail on how biological evolution has solved the same problem of exporting produced internal heat. This problem increased in scope with the development of endothermy, homeothermy, and larger brains. The epidemic of neurological diseases and other civilization-associated diseases (obesity, diabetes, hypertension, autoimmune diseases, osteoporosis, certain cancer types, novel viruses, and multidrug-resistant bacteria) could not have come at a more inconvenient time. The burdens for our health of toxified air, the water we drink, the food we eat, and extreme climatic events are intimately connected with a civilization and globalization process grounded in fossil fuel exploitation. From a geological perspective, the burning of fossil fuels is just one of the many explosions initiated by the industrial revolution through the last several centuries. It is a very short time compared with hundreds of millions of years used by nature to ensure the safe underground storage of fossil fuels. Unfortunately, scientific and industrial revolution are tightly connected to a profit pathology scenario. Limited natural resources are destroyed and converted into money and power accumulation for some, while forcing us all to pay for last-ditch efforts to avoid Earth's toxification consequences. Incidentally, if we are lucky, our descendants will still be able to live in a far less-supporting hothouse world. It is the situation that Earth experienced about 55 million years ago due to catastrophic carbon release, which, however, happened at a ten times slower rate with a moderate species extinction (Bowen et al. 2015).

Definitions and explanations from Chapter 15:

Angiogenesis is a normal and vital physiological process during growth, tissue formation, and wound healing. Tumor cells usually promote angiogenesis because the process forms new blood vessels bringing essential nutrients and oxygen to fast-growing malignant cancers.

Epigenetic changes are functionally relevant alterations in the genome that do not alter the DNA sequence. The control mechanisms producing such changes are DNA methylation and all ways of activating and silencing different genes. Barbara McClintock was the first scientist to suggest that it is possible to inherit genome activity changes that are not caused by alterations in DNA (see the explanation about telomeres).

Telomere protects the end of the chromosome. The telomere sequence of nucleotides TTAGGG repeats itself several thousand times in humans, but decreases in length with age. That telomere attrition process happens during each cell replication with concomitant shortening. Three lady scientists, Barbara McClintock, Elizabeth Blackburn, and Carolyn Widney Greider, were awarded the Nobel Prize in Physiology or Medicine (McClintock in 1983, the other two in 2009). McClintock discovered the telomere protective role. Blackburn and Greider discovered telomerase, the enzyme that protects telomeres. The present consensus is that telomeres are nucleoprotein structures whose shortening is related to human aging and age-related diseases (Turner et al. 2019).

References

Ahmad, M., Attoub, S., Singh, M.N., Martin, F.L. and El-Agnaf, O.M.A. γ-Synuclein and the progression of cancer. FASEB J. 21(2007): 3419–3430.

Alonso, A.D., Cohen, L.S., Corbo, C., Morozova, V., Elldrissi, A., Phillips, G. et al. Hyperphosphorylation of tau associates with changes in its function beyond microtubule stability. Front. Cell. Neurosci. 12(2018): 338. doi: 10.3389/fncel.2018.00338.

Angelova, D., Jones, H. and Brown, D. Levels of alpha- and beta-synuclein regulate cellular susceptibility to toxicity from alpha-synuclein oligomers. FASEB J. 32(2018): 995–1006.

Bendor, J., Logan, T. and Edwards, R.H. The function of α-synuclein. Neuron. 79(2013): doi: 10.1016/j.neuron.2013.09.004.

Benskey, M.J., Perez, R.G. and Manfredsson, F.P. The contribution of alpha synuclein to neuronal survival and function—Implications for Parkinson's disease. J. Neurochem. 137(2016): 331–359.

Biliński, T., Paszkiewicz, T. and Zadrag-Tecza, R. Energy excess is the main cause of accelerated aging of mammals. Oncotarget 6(2015): 12909–12919.

Biza, K.V., Nastou, K.C., Tsiolaki, P.L., Mastrokalou, C.V., Hamodrakas, S.J. and Iconomidou, V.A. The amyloid interactome: Exploring protein aggregation. PLoS ONE 12(2017): doi: 10.1371/journal.pone.0173163.

Blum, R. and Kloog, Y. Metabolism addiction in pancreatic cancer. Cell Death Dis. 5(2014): e1065. doi: 10.1038/cddis.2014.38.

Bonora, M. and Pinton, P. The mitochondrial permeability transition pore and cancer: molecular mechanisms involved in cell death. Front. Oncol. 4(2014): 302. doi: 10.3389/fonc.2014.00302.

Bowen, G.J., Maibauer, B.J., Kraus, M.J., Röhl, U., Westerhold, T., Steimke, A. et al. Two massive, rapid releases of carbon during the onset of the Palaeocene–Eocene thermal maximum. Nat. Geosci. 8(2015): 44–47.

Bridi, J.C. and Hirth, F. Mechanisms of α-synuclein induced synaptopathy in Parkinson's disease. Front. Neurosci. 12(2018): 80. doi: 10.3389/fnins.2018.00080.

Burré, J., Sharma, M. and Südhof, T.C. Cell biology and pathophysiology of α-synuclein. Cold Spring Harb. Perspect. Med. 8(2018): doi: 10.1101/cshperspect.a024091.

Byars, R.S. 2015. Waging War on Cancer: Dr. Pettit's Lifelong Quest to Find Cures. Friesen Press, Victoria, BC, Canada,

Cahill, C.M., Lahiri, D.K., Huang, X. and Rogers, J.T. Amyloid precursor protein and alpha synuclein translation, implications for iron and inflammation in neurodegenerative diseases. Biochim. Biophys. Acta. 1790(2009): 615–628.

Cassano, T., Pace, L., Bedse, G., Lavecchia, A.M., De Marco, F., Gaetani, S. et al. Glutamate and mitochondria: Two prominent players in the oxidative stress-induced neurodegeneration. Curr. Alzheimer Res. 13(2016): 185–197.

Cheng, F., Vivacqua, G. and Yu, S. The role of alpha-synuclein in neurotransmission and synaptic plasticity. J. Chem. Neuroanat. 42(2011): 242–248.

Davies, P., Moualla, D. and Brown, D.R. Alpha-synuclein is a cellular ferrireductase. PLoS ONE 6(2011): e15814. doi: 10.1371/journal.pone.0015814.

Davis, M.T., Hillmer, A., Holmes, S.E., Pietrzak, R.H., DellaGioia, N., Nabulsi, N. et al. *In vivo* evidence for dysregulation of mGluR5 as a biomarker of suicidal ideation. Proc. Natl. Acad. Sci. USA 116(2019): 11490–11495.

Devi, L., Raghavendran, V., Prabhu, B.M., Avadhani, N.G. and Anandatheerthavarada, H.K. Mitochondrial import and accumulation of α-synuclein impair complex I in human dopaminergic neuronal cultures and Parkinson disease drain. J. Biol. Chem. 283(2008): 9089–9100.

de Lau, L.M.L. and Breteler, M.M.B. Epidemiology of Parkinson's disease. Lancet Neurol. 5(2006): 525–535.

Diederich, N.J. and Parent, A. Parkinson's disease: acquired frailty of archaic neural networks? J. Neurol. Sci. 314(2012): 143–151.

Eichenbaum, H. Prefrontal cortex: A mystery of belated memories. Curr. Biol. 27(2017): R418–R420.

Fenster, R.J., Lebois, L.A.M., Ressler, K.J. and Suh, J. Brain circuit dysfunction in post-traumatic stress disorder: from mouse to man. Nat. Rev. Neurosci. 19(2018): 535–551.

Finkel, T. Signal transduction by reactive oxygen species. J. Cell Biol. 194(2011): 7–15.

Fortin, D.L., Nemani, V.M., Voglmaier, S.M., Anthony, M.D., Ryan, T.A. and Edwards, R.H. Neural activity controls the synaptic accumulation of α-synuclein. J. Neurosci. 25(2005): 10913–10921.

Fox, E.J. and Loeb, L.A. Lethal mutagenesis: Targeting the mutator phenotype in cancer. Semin. Cancer Biol. 20(2010): 353–359.

Frost, B., Jacks, R.L. and Diamond, M.I. Propagation of tau misfolding from the outside to the inside of a cell. J. Biol. Chem. 284(2009): 12845–12852.

Fusco, G., Chen, S.W., Williamson, P.T.F., Cascella, R., Perni, M., Jarvis, J.A. et al. Structural basis of membrane disruption and cellular toxicity by α-synuclein oligomers. Science 358(2017): 1440–1443.

Gentile, I., Garro, H.A., Ocanña, S.D., Gonzalez, N., Strohäker, T., Schibich, D. et al. Interaction of Cu(I) with the Met-X3-Met motif of alpha-synuclein: binding ligands, affinity and structural features. Metallomics 10(2018): 1383–1389.

George, J.M., Jin, H., Woods, W.S. and Clayton, D.F. Characterization of a novel protein regulated during the critical period for song learning in the zebra finch. Neuron. 15(1995): 361–372.

Gerdes, H.-H. and Carvalho, R.N. Intercellular transfer mediated by tunneling nanotubes. Curr. Opin. Cell Biol. 20(2008): 470–475.

Ghio, S., Kamp, F., Cauchi, R., Giese, A. and Vassallo, N. Interaction of α-synuclein with biomembranes in Parkinson's disease—role of cardiolipin. Prog. Lipid Res. 61(2016): 73–82.

Grassi, D., Howard, S., Zhou, M., Diaz-Perez, N., Urban, N.T., Guerrero-Given, D. et al. Identification of a highly neurotoxic α-synuclein species inducing mitochondrial damage and mitophagy in Parkinson's disease. Proc. Natl. Acad. Sci. USA 115(2018): E2634–E2643.

Grimes, C.A. and Jope, R.S. CREB DNA binding activity is inhibited by glycogen synthase kinase-3β and facilitated by lithium. J. Neurochem. 78(2001): 1219–1232.

Guo, J.-P., Arai, T., Miklossy, J. and McGeer, P.L. Aβ and tau form soluble complexes that may promote self aggregation of both into the insoluble forms observed in Alzheimer's disease. Proc. Natl. Acad. Sci. USA 103(2006): 1953–1958.

Hameroff, S. and Penrose, R. 1996. Orchestrated objective reduction of quantum coherence in brain microtubules: The "Orch OR" model for consciousness. pp. 507–540. *In*: Hameroff, S.R., Kaszniak, A.W. and Scott, A.C. (eds.). Toward a Science of Consciousness—The First Tucson Discussions and Debates. MIT Press, Cambridge, MA, USA.

Hameroff, S. and Penrose, R. Consciousness in the universe. A review of the 'OrchOR' theory. Phys. Life Rev. 11(2014): 39–78.

Hanahan, D. and Weinberg, R.A. Hallmarks of cancer: The next generation. Cell 144(2011): 646–674.

Houser, M.C. and Tansey, M.G. The gut-brain axis: is intenstinal inflammation a silent driver of Parkinson's disease pathogenesis? NPJ Parkinsons Dis. 2017 Jan 11. doi: 10.1038/s41531-016-0002-0.

Hunter, D.R. and Haworth, R.A. The Ca^{2+}-induced membrane transition in mitochondria. I. The protective mechanisms. Arch. Biochem. Biophys. 195(1979): 453–459.

Illes-Toth, E., Dalton, C.F. and Smith, D.P. Binding of dopamine to alpha-synuclein is mediated by specific conformational states. J. Am. Soc. Mass Spectrom. 24(2013): 1346–1354.

Jang, A., Lee, H.J., Suk, J.E., Jung, J.W., Kim, K.P. and Lee, S.J. Non-classical exocytosis of alpha-synuclein is sensitive to folding states and promoted under stress conditions. J. Neurochem. 113(2010): 1263–1274.

Järvelä, I. Genomics studies on musical aptitude, music perception, and practice. Ann. N.Y. Acad. Sci. 1423(23 March 2018). doi: 10.1111/nyas.13620.

Juretić, D. Cyanide-resistant respiration of a *Neurospora crassa* membrane mutant. J. Bacteriol. 126(1976): 542–543.

Juretić, D., Hendler, R.W., Zasloff, M. and Westerhoff, H.V. Cooperative action of magainins in disrupting membrane-linked free-energy transduction. Biophys. J. 55(1989): 572a.

Juretić, D. Antimicrobial peptides of the magainin family: membrane depolarization studies on *E. coli* and cytochrome oxidase liposomes. Studia Biophysica 138(1990): 79–86.

Juretić, D., Sonavane, Y., Ilić, N., Gajski, G., Goić-Barišić, I., Tonkić, M. et al. Designed peptide with a flexible central motif from ranatuerins adapts its conformation to bacterial membranes. Biochim. Biophys. Acta Biomembr. 1860(2018): 2655–2668.

Kagan, B.L., Jang, H., Capone, R., Arce, F.T., Ramachandran, S., Lal, R. et al. Antimicrobial properties of amyloid peptides. Mol. Pharm. 9(2012): 708–717.

Kanduri, C., Kuusi, T., Ahvenainen, M., Philips, A.K., Lähdesmäki, H. and Järvelä, I. The effect of music performance on the transcriptome of professional musicians. Sci. Rep. 5(2015): 9506. doi: 10.1038/srep09506.

Kawamata, H. and Manfredi, G. Proteinopathies and OXP HOS dysfunction in neurodegenerative diseases. J. Cell Biol. 216(2017): 3917–3929.

Kennedy, S.R., Loeb, L.A. and Herr, A.J. Somatic mutations in aging, cancer and neurodegeneration. Mech. Ageing Dev. 133(2012): 118–126.

Kim, T.D., Paik, S.R. and Yang, C.H. Structural and functional implications of C-terminal regions of α-synuclein. Biochemistry 41(2002): 13782–13790.

Koppenol, W.H., Bounds, P. L. and Dang, C.V. Otto Warburg's contributions to current concepts of cancer metabolism. Nat. Rev. Cancer 11(2011): 325–327.

Kovach, G.G. Molecular pathological classification of neurodegenerative diseases: Turning towards precision medicine. Int. J. Mol. Sci. 17(2016): 189. doi: 10.3390/ijms17020189.

Lautenschläger, J., Stephens, A.D., Fusco, G., Ströhl, F., Curry, N., Zacharopoulou, M. et al. C-terminal calcium binding of α-synuclein modulates synaptic vesicle interaction. Nat. Commun. 9(2018): 712. doi: 10.1038/s41467-018-03111-4.

Leber, R., Pachler, M., Kabelka, I., Svoboda, I., Enkoller, D., Vácha, R. et al. Synergism of antimicrobial frog peptides couples to membrane intrinsic curvature strain. Biophys. J. 114(2018): 1945–1954.

Leng, Y. and Chuang, D.-M. Endogenous α-synuclein is induced by valproic acid through histone deacetylase inhibition and participates in neuroprotection against glutamate-induced excitotoxicity. J. Neurosci. 26(2006): 7502–7512.

Lim, L., Lee, X. and Song, J. Mechanism for transforming cytosolic SOD1 into integral membrane proteins of organelles by ALS-causing mutations. Biochim. Biophys. Acta 1848(2015): 1–7: doi.org/10.1016/j.bbamem.2014.10.002.

Lisman, J., Cooper, K., Sehgal, M. and Silva, A.J. Memory formation depends on both synapse-specific modifications of synaptic strength and cell-specific increases in excitability. Nat. Neurosci. 21(2018): 309–314.

López-Otín, C., Blasco, M.A., Partridge, L., Serrano, M. and Kroemer, G. The hallmarks of aging. Cell 153(2013): 1194–1217.

Ludtmann, M.H.R., Angelova, P.R., Horrocks, M.H., Choi, M.L., Rodrigues, M., Baev, A.Y. et al. Alpha-synuclein oligomers interact with ATP synthase and open the permeability transition pore in Parkinson's disease. Nat. Commun. 9(2018): 2293. doi: 10.1038/s41467-018-04422-2.

Luo, L., Molnar, J., Ding, H., Lv, X. and Spengler, G. Physicochemical attack against solid tumors based on the reversal of direction of entropy flow: an attempt to introduce thermodynamics in anticancer therapy. Diagn. Pathol. 1(2006): 43. doi: 10.1186/1746-1596-1-43.

Maroteaux, L., Campanelli, J.T. and Scheller, R.H. Synuclein: A neuron-specific protein localized to the nucleus and presynaptic terminal. J. Neurosci. 8(1988): 2804–2815.

Martin, Z.S., Neugebauer, V., Dineley, K.T., Kayed, R., Zhang, W., Reese, L.C. et al. α-Synuclein oligomers oppose long-term potentiation and impair memory through a calcineurin-dependent mechanism: relevance to human synucleopathic diseases. J. Neurochem. 120(2012): 440–452.

Martini, C., Da Pozzo, E., Carmassi, C., Cuboni, S., Trincavelli, M.L., Massimetti, G. et al. Cyclic adenosine monophosphate responsive element binding protein in post-traumatic stress disorder. World J. Biol. Psychiatry. 14(2013): 396–402.

McCormack, A.L., Mak, S.K. and Di Monte, D.A. Increased α-synuclein phosphorylation and nitration in the aging primate substantia nigra. Cell Death Dis. 3(2012): e315. doi: 10.1038/cddis.2012.50.

Milholland, B., Auton, A., Suh, Y. and Vijg, J. Age-related somatic mutations in the cancer genome. Oncotarget 6(2015): 24627–24635.

Mor, D.E., Ugras, S.E., Daniels, M.J. and Ischiropoulos, H. Dynamic structural flexibility of α-synuclein. Neurobiol. Dis. 88(2016): 66–74.

Nakamura, T. and Lipton, S.A. Redox modulation by S-nitrosylation contributes to protein misfolding, mitochondrial dynamics, and neuronal synaptic damage in neurodegenerative diseases. Cell Death Differ. 18(2011): 1478–1486.

Nakamura, K. Alpha-synuclein and mitochondria: Partners in crime? Neurotherapeutics 10(2013): 391–399.

Naumann, R.K., Ondracek, J.M., Reiter, S., Shein-Idelson, M., Tosches, M.A., Yamawaki, T.M. et al. The reptilian brain. Curr. Biol. 25(2015): R317–R321.

Nielsen, M.S., Vorum, H., Lindersson, E. and Jensen, P.H. Ca^{2+} Binding to α-synuclein regulates ligand binding and oligomerization. J. Biol. Chem. 276(2001): 22680–22684.

Okochi, M., Walter, J., Koyama, A., Nakajo, S., Baba, M., Iwatsubo, T. et al. Constitutive phosphorylation of the Parkinson's disease associated alpha-synuclein. J. Biol. Chem. 275(2000): 390–397.

Park, H.-J. and Friston, K. Structural and functional brain networks: From connections to cognition. Science 342(2013): 1238411. doi: 10.1126/science.1238411.

Park, S.C., Moon, J.C., Shin, S.Y., Son, H., Jung, Y.J., Kim, N.H. et al. Functional characterization of alpha-synuclein protein with antimicrobial activity. Biochem. Biophys. Res. Commun. 478(2016): 924–928.

Peggion, C., Stella, R., Chemello, F., Massimino, M.L., Arrigoni, G., Cagnin, S. et al. The prion protein regulates synaptic transmission by controlling the expression of proteins key to synaptic vesicle recycling and exocytosis. Mol. Neurobiol. (20 August 2018). doi.org/10.1007/s12035-018-1293-4.

Perić, M., Lovrić, A., Šarić, A., Musa, M., Dib, P.B., Rudan, M. et al. TORC1-mediated sensing of chaperone activity alters glucose metabolism and extends lifespan. Aging Cell 16(2017): 994–1005.

Pfefferkorn, C.M., Jiang, Z. and Lee, J.C. Biophysics of α-synuclein membrane interactions. Biochim. Biophys. Acta 1818(2012): 162–171.

Phan, L.M., Yeung, S.-C.J. and Lee, M.-H. Cancer metabolic reprogramming: importance, main features, and potentials for precise targeted anti-cancer therapies. Cancer Biol. Med. 11(2014): 1–19. doi: 10.7497/j.issn.2095-3941.2014.01.001.

Plotegher, N., Gratton, E. and Bubacco, L. Number and brightness analysis of alpha-synuclein oligomerization and the associated mitochondrial morphology alterations in live cells. Biochim. Biophys. Acta 1840(2014): 2014–2024.

Portela, M., Venkataramani, V., Fahey-Lozano, N., Seco, E., Losada-Perez, M., Winkler. F. et al. Glioblastoma cells vampirize WNT from neurons and trigger a JNK/MMP signaling loop that enhances glioblastoma progression and neurodegeneration. PLoS Biol. 17(2019): doi.org/10.1371/journal.pbio.3000545.

Price, D.L., Rockenstein, E., Ubhi, K., Phung, V., MacLean-Lewis, N., Askay, D. et al. Alterations in mGluR5 expression and signaling in Lewy Body Disease and in transgenic models of alpha-synucleinopathy—Implications for excitotoxicity. PLoS ONE 5(2010): e14020. doi: 10.1371/journal.pone.0014020.

Proctor, E.A., Fee, L., Tao, Y., Redler, R.L., Fay, J.M., Zhang, Y. et al. Nonnative SOD1 trimer is toxic to motor neurons in a model of amyotrophic lateral sclerosis. Proc. Natl. Acad. Sci. USA 113(2016): 614–619.

Prusiner, S.B. Prion diseases and the BSE crisis. Science 278(1997): 245–251.

Prusiner, S.B. 2017. Prion Diseases. Cold Spring Harbour Laboratory Press. New York, NY, USA.

Radman, M. Cellular parabiosis and the latency of age-related diseases. Open Biol. 9(2019): 180250. doi.org/10.1098/rsob.180250.

Rao, J.N., Jao, C.C., Hegde, B.G., Langen, R. and Ulmer, T.S. A combinatorial NMR and EPR approach for evaluating the structural ensemble of partially folded proteins. J. Am. Chem. Soc. 132(2010): 8657–8668.

Rich, P. The cost of living. Nature 421(2003): 583–583.

Rose, A.S. and Hildebrand, P.W. NGL Viewer: a web application for molecular visualization. Nucleic Acids Res. 43(2015): W576–W579.

Rose, A.S., Bradley, A.R., Valasatava, Y., Duarte, J.M., Prlić, A. and Rose, P.W. NGL viewer: web-based molecular graphics for large complexes. Bioinformatics 34(2018): 3755–3758.

Rosenkranz, K., Williamon, A. and Rothwell, J.C. Motorcortical excitability and synaptic plasticity is enhanced in professional musicians. J. Neurosci. 27(2007): 5200–5206.

Rostami, J., Holmqvist, S., Lindström, V., Sigvardson, J., Westermark, G.T., Ingelsson, M. et al. Human astrocytes transfer aggregated alpha-synuclein via tunneling nanotubes. J. Neurosci. 37(2017): 11835–11853.

Rotermund, C., Reolon, G.K., Leixner, S., Boden, C., Bilbao, A. and Kahle, P.J. Enhanced motivation to alcohol in transgenic mice expressing human α-synuclein. J. Neurochem. 143(2017): 294–305.

Rustom, A., Saffrich, R., Markovic, I., Walther, P. and Gerdes, H.-H. Nanotubular highways for intercellular organelle transport. Science 303(2004): 1007–1010.

Ryan, B.J., Hoek, S., Fon, E.A. and Wade-Martins, R. Mitochondrial dysfunction and mitophagy in Parkinson's: from familial to sporadic disease. Trends Biochem. Sci. 40(2015): 200–210.

Sadakane, Y. and Kawahara, M. Implications of metal binding and asparagine deamidation for amyloid formation. Int. J. Mol. Sci. 19(2018): 2449. doi: 10.3390/ijms19082449.

Saleh, H., Saleh, A., Yao, H., Cui, J., Shen, Y. and Li, R. Mini review: linkage between α-Synuclein protein and cognition. Transl. Neurodegener. 4(2015): 5. doi: 10.1186/s40035-015-0026-0.

Scheckel, C. and Aguzzi, A. Prions, prionoids and protein misfolding disorders. Nat. Rev. Genet. 19(2018): 405–418.

Schellnhuber, H.J. Global warming: Stop worrying, start panicking? Proc. Natl. Acad. Sci. USA 105(2008): 14239–14240.

Schofield, K. The metal neurotoxins: An important role in current human neural epidemics? Int. J. Environ. Res. Public Health 14(2017): 1511. doi: 10.3390/ijerph14121511.

Shaltouki, A., Hsieh, C.H., Kim, M.J. and Wang, X. Alpha-synuclein delays mitophagy and targeting Miro rescues neuron loss in Parkinson's models. Acta Neuropathol. 136(2018): 607–620.

Soravia, E., Martini, G. and Zasloff, M. Antimicrobial properties of peptides from *Xenopus* granular gland secretions. FEBS Lett. 228(1988): 337–340.

Soscia, S.J., Kirby, J.E., Washicosky, K.J., Tucker, S.M., Ingelsson, M., Hyman, B. et al. The Alzheimer's disease-associated amyloid β-protein is an antimicrobial peptide. PLoS ONE 5(2010): e9505. doi: 10.1371/journal.pone.0009505.

Surguchev, A.A. and Surguchov, A. Synucleins and gene expression: Ramblers in a crowd or cops regulating traffic? Front. Mol. Neurosci. 10(2017): 224. doi: 10.3389/fnmol.2017.00224.

Surguchov, A. Intracellular dynamics of synucleins: "here, there and everywhere". Int. Rev. Cell Mol. Biol. 320(2015): 103–169.

Surguchov, A. Protein-DNA interaction: One step closer to understanding the mechanism of neurodegeneration. J. Neurosci. Res. 97(2019): 391–392.

Teich, A.F., Nicholls, R.E., Puzzo, D., Fiorito, J., Purgatorio, R., Fa', M. et al. Synaptic therapy in Alzheimer's disease: A CREB-centric approach. Neurotherapeutics 12(2015): 29–41.

Theillet, F.-X., Binolfi, A., Bekei, B., Martorana, A., Rose, H.M., Stuiver, M. et al. Structural disorder of monomeric α-synuclein persists in mammalian cells. Nature 530(2016): 45–50.

Tsigelny, I.F., Crews, L., Desplats, P., Shaked, G.M., Sharikov, Y., Mizuno, H. et al. Mechanisms of hybrid oligomer formation in the pathogenesis of combined Alzheimer's and Parkinson's diseases. PLoS ONE 3(2008): e3135. doi: 10.1371/journal.pone.0003135.

Tsigelny, I.F., Sharikov, Y., Wrasidlo, W., Gonzalez, T., Desplats, P.A., Crews, L. et al. Role of α-synuclein penetration into the membrane in the mechanisms of oligomer pore formation. FEBS J. 279(2012): 1000–1013.

Turner, K.J., Vasu, V. and Griffin, D.K. Telomere biology and human phenotype. Cells 8(2019): 73. doi: 10.3390/cells8010073.

Uchihara, T. and Giasson, B.I. Propagation of alpha-synuclein pathology: hypotheses, discoveries, and yet unresolved questions from experimental and human brain studies. Acta Neuropathol. 131(2016): 49–73.

Ulmer, T.S., Bax, A., Cole, N.B. and Nussbaum, R.L. Structure and dynamics of micelle-bound human alpha-synuclein. J. Biol. Chem. 280(2005): 9595–9603.

Ulmschneider, J.P. and Ulmschneider, M.B. Molecular dynamics simulations are redefining our view of peptides interacting with biological membranes. Acc. Chem. Res. 51(2018): 1106–1116.

Valera, E., Spencer, B. and Eliezer Masliah, E. Immunotherapeutic approaches targeting amyloid-β, α-synuclein, and tau for the treatment of neurodegenerative disorders. Neurotherapeutics 13(2016): 179–189.

Vargas, K.J., Makani, S., Davis, T., Westphal, C.H., Castillo, P.E. and Chandra, S.S. Synucleins regulate the kinetics of synaptic vesicle endocytosis. J. Neurosci. 34(2014): 9364–9376.

Walker, D.G., Lue, L.F., Adler, C.H., Shill, H.A., Caviness, J.N., Sabbagh, M.N. et al. Changes in properties of serine 129 phosphorylated α-synuclein with progression of Lewy-type histopathology in human brains. Exp. Neurol. 240(2013): 190–204.

Walters, R.F.S. and DeGrado, W.F. Helix-packing motifs in membrane proteins. Proc. Natl. Acad. Sci. USA 103(2006): 13658–13663.

Warburg, O., Wind, F. and Negelein, E. The metabolism of tumors in the body. J. Gen. Physiol. 8(1927): 519–530.

Warburg, O. On the origin of cancer cells. Science 123(1956): 309–314.

Weingarten, M.D., Lockwood, A.H., Hwo, S.-Y. and Kirschner, M.W. A protein factor essential for microtubule assembly. Proc. Nat. Acad. Sci. USA 72(1975): 1858–1862.

Westerhoff, H.V., Juretić, D., Hendler, R.W. and Zasloff, M. Magainins and the distruption of membrane-linked free-energy transduction. Proc. Natl. Acad. Sci. USA 86(1989a): 6597–6601.

Westerhoff, H.V., Hendler, R.W., Zasloff, M. and Juretić, D. Interactions between a new class of eukaryotic antimicrobial agents and isolated rat liver mitochondria. Biochim. Biophys. Acta 975(1989b): 361–369.

Westerhoff, H.V., Zasloff, M., Rosner, J.L., Hendler, R.W., De Wall, A., Vaz Gomez, A. et al. Functional synergism of the magainins PGLa and magainin-2 in *Escherichia coli*, tumor cells and liposomes. Eur. J. Biochem. 228(1995): 257–264.

Williams, R.W., Starman, R., Taylor, K.M., Gable, K., Beeler, T., Zasloff, M. et al. Raman spectroscopy of synthetic antimicrobial frog peptides magainin 2a and PGLa. Biochemistry 29(1990): 4490–4496.

Zasloff, M. Magainins, a class of antimicrobial peptides from *Xenopus* skin: Isolation, characterization of two active forms, and partial cDNA sequence of a precursor. Proc. Natl. Acad. Sci. USA 84(1987): 5449–5453.

Zerweck, J., Strandberg, E., Kukharenko, O., Reichert, J., Bürck, J., Wadhwani, P. et al. Molecular mechanism of synergy between the antimicrobial peptides PGLa and magainin 2. Sci. Rep. 7(2018): 13153. doi: 10.1038/s41598-017-12599-7

Zhou, L., McInnes, J., Wierda, K., Holt, M., Herrmann, A.G., Jackson, R.J. et al. Tau association with synaptic vesicles causes presynaptic dysfunction. Nat. Commun. 8(2017): 15295. doi: 10.1038/ncomms15295.

CHAPTER 16

Retrospections, Contrasting Viewpoints, Incentives, Challenges, Prospects, and Conclusions

16.1 Life is the evolution and multiplication of dissipation-steering systems

An open nonequilibrium system needs continuous power input to survive universal entropy-increasing tendencies. Its output to the surrounding environment depends on system properties. The Second Law requires that the entropy generation is the major part of the system's output irrespective of its structural features. Irreversible processes generate entropy in the system's interior. Hence, exported entropy can always be expressed as entropy production. High overall entropy production value may indicate that a highly active system has more or less complex ways of transforming input power into generated output entropy. Entropy rise manifests itself mostly as heat. It is indeed a wasted work—a form of useless energy. It is not useless as the evidence that something interesting is going on in the system—dynamics that should be explored to discover what internal fluxes, forces, and structures contributed to free-energy input, entropy production, and entropy export.

Furthermore, although wasted work and wasted power are contained within the total entropy production, the signatures of potentially useful work and power are also contained. There is no wasted work without absorbed work, and there is no self-organization via dissipation in the absence of absorbed work (England 2015). England and coworkers (Kachman et al. 2017) adopted the name "dissipative adaptation" for the phenomenon in which the system's structure-dynamics patterns bear the signature of the highly work-absorbing states that the system had to traverse during its history. The experimental realization of dissipative adaptation was recently achieved, such that increased energy dissipation (drive) leads to increasing self-organization (Te Brinke et al. 2018). The human design of dynamic nanosystems based on dissipative adaptation is only in its infancy due to our inability to achieve anything similar to living-cell control over far-from-equilibrium processes. For instance, the respiratory control in mitochondrial bioenergetics implies tight coupling and fine regulation for substrate oxidation (oxidation affinity), ATP turnover (phosphorylation affinity), and the electrochemical potential difference of ions (usually the protonmotive force) (Nath 2019a). It is defined as stimulation of oxygen consumption by ADP (**state 3**, active respiration) and abrupt decrease when mitochondria have converted enough ADP into ATP (**state 4**, resting respiration).

Kachman et al. (2017) and Horowitz and England (2017) used simulated chemical reactions to conclude that kinetically stable behaviors are finely tuned to the sufficiently strong external thermodynamic drive. Enough complex chemical reaction networks can spontaneously find rare far-from-equilibrium steady states by harvesting the maximum energy possible from the environment. These authors proposed that the self-emergence of adaptive resonance results in energy-seeking lifelike behavior patterns when matched to the drive frequency of absorbed work. When coupled, toy autocatalytic cycles lead to self-replication and proliferation (Sarkar and England 2019). Hopefully, a future investigation will clarify the requirements for the dissipative self-organization in realistic far-from-equilibrium bioenergetic conversions.

The unstated assumption in the games with toy reaction networks is that they should follow the lead of biological evolution, which maximizes energy extraction efficiency from life's environment. The assumption sounds reasonable to the physicists interested in the physics of living systems. The snag is nature's tendency to be more sophisticated in its behavior patterns than expected. There is no clear cut general optimization strategy for the maximal efficiency of energy extraction or conversion in bioenergetic structures and processes produced by biological evolution. On the contrary, if we let our eyes rise above computer games and look through the window, lush greenery can be seen—a living example that green leaves are always rejecting to use the significant part of incoming light energy from the middle part of the light spectrum (Arp et al. 2020, Duffy 2020). Plants ignore the most energy-rich part of sunlight because maximal free-energy transduction efficiency was never the evolutionary goal for them or photosynthetic microorganisms. The stability against external noise and safety protection features were major leads during the evolution of photosynthesis—leads, which resulted in optimal efficiency values considerably lower from maximal possible efficiency. Life has developed multiple strategies for steering the dissipation of absorbed work toward its maintenance and replication. None of the persistance mechanisms are perfect—most channel only the small percentage of input free-energy into synthetic pathways or the work output. The rest is dissipated, thus increasing the entropy of life's surroundings.

A simpler example of potential power production due to dissipative drive is mixing up of freshwater and seawater. It is the textbook case of entropy production. However, as Ye and coworkers have recently shown (Ye et al. 2019), a part of it can be diverted into power production by a mixing entropy battery that taps into salt gradients. Without complete optimization, they obtained satisfactory net energy recovery efficiency (the power output percentage) that approaches 10%. The prospect is to transform the laboratory invented technique into affordable and durable technology that could harness the renewable blue energy from oceans for electricity production by the coastal wastewater treatment plants.

16.2 Love and hate for irreversible entropy increase

One can wonder why entropy production and energy dissipation are frequently dismissed or overlooked in physics, biology, and bioenergetics. Scientists are emotional beings, even more than laypersons. However, hating entropy increases, entropy generation, entropy production, and dissipation is just like hating all products of evolution from atoms to life, and the whole universe. It is like hating yourself or hating God, just an unreasonable subjective emotion. Entropy increase is an invisible hand that should not influence us or anything else, but it does. It is the reason for the evolution of everything in the universe, including life. Due to its invisibility, entropy changes are easy to disregard. But can we safely ignore everything that seems invisible and useless to us? Besides bringing life to live, entropy production also brings death, another non-scientific reason for discounting it. Deep inside, we would all like to live forever. Thus, it is convenient to close the eyes to the reality that nothing lasts forever and that we are mortal beings, to ignore our own evolution and uncertain nondeterministic future lurking ahead of us. The best way to deal with that problem of personal mortality is to bring the support of the most intelligent persons that ever lived, like Newton, Einstein, and the fathers of quantum mechanics, who incorporated time invariance and the absence of evolution in the very core of their famous time-symmetric equations. They all dreamed about the static universe—the universe devoid of evolution, while life was a mystery to them as eloquently written by Albert Einstein:

"The most beautiful thing we can experience is the mysterious. It is the source of all true art and science. He to whom the emotion is a stranger, who can no longer pause to wonder and stand wrapped in awe, is as good as dead—his eyes are closed. The insight into the mystery of life, coupled though it be with fear, has also given rise to religion. To know what is impenetrable to us really exists, manifesting itself as the highest wisdom and the most radiant beauty, which our dull faculties can comprehend only in their most primitive forms—this knowledge, this feeling is at the center of true religiousness" (Einstein 1931).

Entropy definition and entropy changes kept appearing in various disguises during the 19th and 20th centuries—an amazing resurrection ability for the physical quantity causing headaches to countless students. A firm connection of entropy changes to the mystery of life and informatics was finally established around the middle of the previous century. The 21st century has seen a gradual approach to an ever better connection between entropy production, the evolution of life, and the evolution of the universe.

16.3 The predictive power of causal entropic principle

In fundamental physics, Raphael Bousso learned the art of connecting thermodynamics to general relativity and quantum theory directly from his mentor Stephen Hawking, the doyen of physics that produced the insight about black hole entropy and radiation via Hawking radiation. About ten years ago, professor Bousso published results about the history of entropy production in the universe with the remarkable claim that the cosmological constant can be predicted from the causal entropic principle (Bousso et al. 2007, Bousso and Harnik 2010). The cosmological constant is the infinitesimal amount of gravitationally repulsive energy infusing the vacuum of empty space. It is interpreted as dark energy. In terms of Planck units, the cosmological constant $\Lambda = 3.7 \times 10^{-122}$ is indeed very close to zero, but the evidence for dark energy reserves for it a whopping 68% of the present-day universe's energy. For a long time, it was a major embarrassment for quantum physicists. Its predictions of vacuum energy were initially some 120–122 orders of magnitude higher from observationally determined value. That was considered "the worst theoretical prediction in the history of physics" (Hobson et al. 2006). Involved calculations, providing pleasure to theoretical physicists reading the review about that mystery (Martin 2012), resulted in the mismatch between theoretical expectations and astronomical observations of "only" 55 orders of magnitude. The conclusion was that zero-point vacuum fluctuations exist in Nature despite a huge mismatch with their astronomical consequences. That mystery for physics was known quite some time ago. In desperation, Steven Weinberg, the Nobel laureate in Physics 1979, stated: "Perhaps Λ must be small enough to allow the Universe to evolve to its present, nearly empty and flat state because otherwise there would be no scientists to worry about it (Weinberg 1987)." That anthropic condition selects only those universes in which galaxies and life can exist. It also places the sharp upper bound on the effective cosmological constant. As Weinberg noticed, the anthropic upper bound for Λ is still several orders of magnitude higher from the observed value.

Neither Weinberg (1987) nor Martin (2012) mentioned another mystery in physics—the entropy problem (low initial entropy of the universe and large present-day entropy). Some authors argued that huge entropy increase during the formation of black holes makes the entropy production relevant for the evolution of the universe (Lineweaver 2014, Patel and Lineweaver 2020). That viewpoint is in accord with accumulated evidence that black holes govern galaxies' evolution (Tombesi et al. 2015). Bousso, however, observed that the black hole's entropy does not belong to the causal space-time diamond—the largest causally connected region of the universe (Bousso et al. 2007). At the same time, he did not want to divorce thermodynamics from brave attempts to unify quantum physics and general relativity. The prediction of the cosmological constant in accord with observations would be an essential step in that direction. He noticed that entropy production is a much more basic and physically well-grounded quantity that can replace the ill-defined numbering of intelligent observers required by anthropic condition.

The entropy production proxy for observers reverses the negative feelings about a life-dissipation connection (that life prefers maximal efficiency and minimal dissipation). More to the point, the entropy production history in the causal diamond of the observable universe indicated that "physical parameters are most likely to be found in the range of values for which the total entropy production within a causally connected region is maximized" (Bousso et al. 2007). Bousso named that statement the causal entropic principle and apologized to other authors who used the "entropic principle" terminology in an unrelated context. He could have named it the "maximum entropy production principle," but as we well know from this book, the MaxEP principle was also appropriated by other authors for a multitude of different applications. The insightful papers by Bousso and coauthors (Bousso et al. 2007, Bousso and Harnik 2010) used the causal entropic principle to predict the cosmological constant within one standard deviation from the observed value. In that picture, free energy is converted into "entropic quanta." Entropy production is the signature for such conversions, the highest due to dust heated by starlight. We are that dust, as nicely explained in the Bible, but not so insignificant dust. Our present fossil fuel burning activity produces more entropy during one minute than the merger of huge black holes (each containing about 10 million Sun masses) in the same time frame—the merger producing detectable space oscillations in the whole universe (Bousso and Harnik 2010, Weiss 2018, Barish 2018, Thorne 2018). Thus, the authors of that estimate clearly understood that "intelligent observers" can be much more efficient in entropy production than stars or mind-boggling astronomical events producing the cosmos quakes. They also understood the advantages of using nonequilibrium thermodynamics for estimating when causal entropy production (a proxy for the formation of complex structures) peaked in our universe. Their entropic principle (a version of the MaxEP principle) predicted that most observers live near the time of maximum entropy production.

16.4 Earth's-specific geosphere-biosphere connection does not exist anywhere else in the universe

In biology, enzymes are macromolecules acting as workhorses (some would say matchmakers) to perform all activities cells need to survive and reproduce. That is the reason why we focused on enzymes in this book. Enzymes perform substrate to product bioenergetic transformations. They also perform free-energy transduction from one form into another when embedded into topologically closed membranes. They are responsible for bridging life to the universe. Our biosphere is so intimately connected to the geosphere that any alien intelligence can confirm the existence of life on Earth by detecting its imprint on our planet's atmosphere (Sagan et al. 1993). The presence of life profoundly influenced the surface mineral composition, with the consequence that the probability P of duplicating Earth's mineralogy anywhere else in the cosmos is estimated as $P \approx 10^{-322}$ (Hystad et al. 2015, Hazen et al. 2015). The conclusion about the life-mineralogy connection (Earth's biosignature) is the following: the specifics of our planet's life and mineral diversity are unique in the universe despite the high probability of life-supporting habitability and mineralogy existing elsewhere in our and other galaxies (Hazen 2017, Chan et al. 2019). Bioenergetics may be universal in the universe, but our biosphere and geosphere are not. There is only one Gaia (Lovelock 2009), and our duty is to nurse and cherish it. Lovelock's early prophetic warnings about Gaia's vanishing face have been followed, confirmed, and given an increasingly urgent note by a great number of concerned scientists, including myself (https://scientistswarning.forestry.oregonstate.edu/).

16.5 The "birth canal" of biology

Let us revisit the putative proto-ribosome role in life's origin (Chapter 13, section 13.2, and Chapter 14, section 14.2). As present-day ribosomes, their first ancestors must have also contained a combination of RNA and polypeptides. Still, it is far from clear how that idea can be used to close the gap between prebiotic chemistry and biochemistry. Hud and his collaborators suggested that the common core of proto-ribosome emerged through chemistry alone (Bowman et al. 2015). They stressed that the translation is universally conserved among all branches of life. Translation-performing structure—the ribosome is also unrivaled in conservation. Slight changes in the core ribosome size through eons are the proxy for complexity (Bowman et al. 2015). Besides, ribosomes are predominantly "physical catalysts" acting as entropy traps (Sievers et al. 2004, Schroeder and Wolfenden 2007), and Brownian machines that couple conformational changes, driven by thermal energy, to directional movements (Zaccai et al. 2016).

If the chemical coevolution of RNA and polypeptides was enough for proto-ribosome emergence, it challenges the traditional RNA-world hypothesis for life's origin. We mentioned earlier (Chapter 13, section 13.1) the stabilizing effect the peptides have on RNA strands and the preferential assembly of homochiral polypeptides (Carny and Gazit 2011). The oldest living fossils are genetic code and the common core of all ribosomes—supporting the Nick Hud vision of what chemical evolution had to produce before the initiation of Darwinian evolution. No fancy biochemical tricks of present-day enzymes are needed for the ribosomal peptidyl transferase center, which works like a dehydrator. Removing a water molecule to form the peptide bond is an entropy-driven process because it requires desolvation of surfaces forming the functional center with concomitant entropy increase of liberated water molecules. Additional entropy increase occurs during nascent polypeptide folding at the end of its transfer through the ribosome tunnel (Jennaro et al. 2014). The ribosomal exit tunnel co-evolved with protein folding, thus justifying its poetic description "the birth canal of biology (Bowman et al. 2020)."

All of these considerations implied for Hud and his collaborators the preference for the heating-cooling, day-night, or wet-dry cycles in the primordial soup of organic compounds, possibly near or at the Earth's surface. Another proto-life branch with energy-harvesting ability was also required before the Darwinian world could emerge (Chapter 13). That branch had a primitive chemiosmotic ability, but it had to become free from its geochemical nesting place, possibly at the subsurface alkaline hydrothermal vents. The chemiosmosis and topologically closed membrane were needed for the emergence of highly nonequilibrium GTP-GDP and ATP-ADP inside concentrations. In the same chapter, we also mentioned the possibility of proto-virions "infection" as a "communication" among different geothermal environments during the Hadean era. The similarity of virions to ribosomes brings to mind how the fruitful marriage between two proto-life branches could have occurred. Proto-ribosomes and membrane vesicles able to maintain an excess of "high-energy" molecules like ATP and GTP were ideally suited to jump-start replication, transcription, and translation when joined together in the FUCA or LUCA ancestor for all life (see Chapter 13). For the present-day synthesis of a small protein with about 200 amino acid residues, almost 1000 ATP and GTP molecules must be hydrolyzed (Berg et al. 2002). The GTP hydrolysis, in particular, accelerates the tRNA translocation 40-fold, but there is a loose coupling between the energy-conversion in the rearrangements of the elongation factor (where GTP is bound) and the conformational changes on the ribosome during the translocation process (Rodnina et al. 2020). It is another confirmation of the ribosome's nanomachine as a Brownian machine performing the atypical mechano-catalysis with the exceptionally high entropic contribution.

Considering what is common to using available free energy gradients toward more efficient synthesis and folding of first polypeptides, the answer is again the "entropic principle." As for the whole universe, far-from-equilibrium entropy increase during chemical and biological evolution at the young Earth's surface did not leave other visible traces except for the conserved processes (genetic code, translation) and selected homochiral polymers (nucleic acids, polypeptides).

16.6 Dissecting entropy production is not an exercise in futility

When kinetic modeling of an enzyme catalytic activity can be performed, the calculation of entropy production is not just an exercise in futility. This book's focus on entropy production caused by bioenergetics suggests that we should explore what life is doing to its environment at least as much as to life itself. Entropy production does manifest itself as produced heat exported to the environment and in the biosignatures of increased cycling of atoms and small molecules crucial for life. It is also the signature of essential processes inside the observed system when all contributions to overall dissipation are taken into account. All living systems are useless if judged only according to their heat production. Cars are worse than useless when final driving results are expressed as the production of toxic molecules, greenhouse gases, and vibrant ecological systems' conversion into dead infrastructures such as roads, parking places, and garages. While cars are easy to see for everyone, it is not so with nanocurrents and chemical affinities bridging bioenergetics to the rest of the universe. The dissection of overall entropy production for enzymes and bioenergetic systems provides answers to the question of what are the rate-limiting steps in the free-energy transduction responsible for the final—seemingly useless result of increased entropy production. It identifies those nanocurrents and dynamical dissipative patterns that are essential for maintaining living system homeostasis. Thus, we have a positive answer to the question: Can their extent of entropy production identify the evolved macromolecular transitions and dissipative pathways critical for regulating the biomolecular function? Together with other authors (Weber et al. 2015), we have found that dominant entropy-producing transitions are inseparably coupled to nonequilibrium conformational switches, rate-limiting steps, and signaling processes enabling the intelligent-like behavior of living cells.

We can recognize the precariously balanced (marginaly stable) dissipative self-organization, dissipative self-assembly, self-regulation, and dissipative adaptation as a causal connection to what is communicated to the world at large—an increased dissipation (England 2015, Horowitz and England 2017, Van Rossum et al. 2017). In general, out-of-equilibrium dissipative self-assembly enables the spatial, temporal, and kinetic control over the formation of intrinsically unstable structures. It is just what life needs, not only for homeostasis, but also for self-healing, development, replication, and evolution. The human-made systems for dissipative self-assembly are still scarce, although the profits can be immense (De and Klajn 2018).

The part of input power can be channeled into the output power. In this book, we have described several examples of how a driving force-flux couple is transformed into a driven force-flux couple. Free-energy transduction is responsible for both increasing and decreasing the overall entropy production. The increase would not happen in the absence of the enzyme performing energy conversion. The decrease is the price paid for generating driven force-flux pairs and slowing down the energy dispersion. Increases and decreases are interconnected so that much higher entropy production increase takes place in an enzyme's presence during longer periods. Driving-driven roles are interchanged at each additional step in the hierarchy for free-energy conversion: driven force-flux couple becomes the driving one, and a new driven force-flux couple appears. The best example in bioenergetics is how driving photon or electron flux is first converted into the protonmotive force and proton flux. Next, the proton flux becomes the driving one leading to the flux of synthesized ATP molecules mediated by ATP-synthase.

16.7 Is the output power maximized?

ATP-synthases are biological nanomachines. Recently, the variational principle was introduced with the claim that the evolution of biological nanomachines was guided by the maximization of the power output (Wagoner and Dill 2019a,b). Authors defined the power output as:

$$P\,(w, \Delta\mu, \tau\,) = N \cdot w/T = \Delta G_{tot} \cdot w/(\Delta\mu \cdot T) \sim w/(\Delta\mu \cdot T) \sim w/(\Delta\mu \cdot \tau) = (w/\Delta\mu) \cdot J$$

where ΔG_{tot} is the consumed energy of the muscle fiber, for instance, that consumes free energy $\Delta\mu$ by N **myosin II** motors, w is the work output of each individual motor, T is the temperature, and τ is the cycle time of a single motor. The cycle time is the inverse value of the cycle flux J.

The proportionality is assumed between cycle time and temperature. Furthermore, the authors assumed that input free energy $\Delta G_{tot} = N \cdot \Delta\mu$ is fixed. If the number of nanomotors N is also fixed, there is no need to retain the consumed energy value $\Delta\mu$ in the denominator. When $\Delta\mu$ is retained, the last right-hand term contains the thermodynamic efficiency $w/\Delta\mu$ of the nanomotor operation. If it is not retained, the power output is proportional to the product of work (force) w and the cycle flux

J. The cycle flux has a nonlinear dependence on input and output force. The output force multiplied with the flux is the productive (useful) part of the entropy production in the nanomachine's operating cycle.

Wagoner and Dill (2019a) variational principle requires a numerical variation of the output force (work) until the output power is maximized. Schmiedl and Seifert explained in their 2008 paper why maximal power output must exist for the optimal output force F. As the output force (work) is varied, the power vanishes for F = 0 and for the stall force when flux (speed) vanishes. There is no other choice but maximal output power somewhere between the level flow (F = 0) and static head state (v = 0). We conclude that maximizing individual nanomachines' power output is mathematically equivalent to maximizing those contributions to overall entropy production (positive or negative) that are channeled into the power output. The mathematical equivalence translates into two faces of the same physical reality—power production and entropy production. In the limiting case of perfect free-energy conversion, entropy production is power production. Produced currents may be useful for chemical synthesis or for producing work, but ultimately their permanent effect is the irreversible entropy increase of the universe. When energy conversion is not perfect, as it is not in all realistic cases, increased entropy production is connected to increased power production. In terms of operational fluxes and forces (Chapter 3), the power gain in cells and organs is nothing else but a smaller negative part of total dissipation associated with these living structures. Only positive contributions to entropy production are seen in the environment (the universe) if we can neglect the work produced by living systems on the ambient. Also, there would be no power gain nor temporary power storage in far-from-equilibrium gradients of organic molecules inside cells without much larger dissipation. Maximal entropy production in the crucial steps for producing output power ensures the increase in the output power. The reverse logic also applies. If the output power is maximized in the steady-state, the dissipation is increased. It does not have to be the maximal total dissipation because dissipation is a more general concept than power. Some dissipative channels disconnected from power production (slip transitions) can always exist.

Qian et al. (2008) considered the total input power for biological nanomotors. That quantity is the amount of chemical energy consumed per unit time or the rate of free energy dissipation—the quantity Td_iS/dt we used throughout this book. The power output can be found as the useful work done to the environment in unit time, but there is also the wasted energy output or true entropy production. Thus, these authors divided the total input power in terms of dissipation and power output. Since the dissipation was also the power input, as in many other examples elucidated in this book, we cannot dismiss it out of hand by naming it the wasted power or wasted work. One can even ask if a perfect dissipator exists in nature. Perhaps, it can be an open system far enough from the thermodynamic equilibrium. Some authors have found perfect dissipators in black holes (Candelas and Sciama 1977, Davies 1978), others in the self-replicating living systems (England 2013). Adrian Bejan included the whole "Earth engine" among perfect dissipators (Bejan and Lorente 2010) that are purely dissipative systems (Bejan 2017).

Wagoner and Dill (2019a) took a different approach to maximize only the useful work output. They examined a dozen biological nanomachines while searching for the optimal way to achieve high speed without sacrificing efficiency. These were: *E. coli* F_0F_1-ATPase, animal F_0F_1-ATPase, chloroplast F_0F_1-ATPase, Na-K ATPase, sarco/endoplasmic reticulum Ca^{2+}-ATPase pump (SERCA), proton PP_i pump, Na^+/Ca^{2+}-K^+ exchanger (NCKX), plasma membrane Ca^{2+}-ATPase pump (PMCA), V ATPase, plasma membrane H^+-ATPase (PM H^+ pump), Na^+-Ca^{2+} exchanger of cardiac sarcolemma (NCX), and myosin II. They found that optimal performance can be achieved by splitting a large free-energy step into more numerous, smaller substeps. Another kinetic advantage was the direct coupling between downhill (driving) and uphill (driven) processes through conformational changes and transition-state location. Interestingly, we found the same performance advantages by using the transitional entropy production theorem for the chloroplast ATPase (Dewar et al. 2006) and some highly efficient enzymes not usually included among biological nanomachines (Chapter 10). Correlating catalytic efficiency with the entropy production increase and the evolutionary distance increase from a putative common ancestor (Juretić et al. 2019a) looks like a promising approach for firmer connection between biologically observed parameters of evolution and the constraints imposed by thermodynamics and kinetics.

The reason the Wagoner and Dill procedure (2019a) works well is the tradeoff between thermodynamic efficiency (or secondary force output) and produced flux. With increasing efficiency and stronger force output, the cycle time becomes unphysiologically long, and the flux declines precipitously. With the flux value being a proxy for speed, it is the tradeoff between nanomachines' speed and efficiency. Without such a tradeoff, it would not be possible to maximize entropy production (Kleidon 2009) or power production. The authors cited the known procedures for finding the maximal power output of molecular and thermal machines (Schmiedl and Seifert 2008) but did not mention the connection to the maximal entropy production principle. They admitted that *free-energy dissipation* is needed to maintain the nonequilibrium steady state of biological nanomachines. Their optimization approach has considerably greater generality. It can be applied to other energy conversion processes far from thermodynamic equilibrium, not only to molecular and thermal machines. Our example of equivalence between power production and entropy production in the simplified (devoid of unproductive steps) scheme for the first irreversible photosynthetic steps (Chapter 8) is one of the more important applications. We also found that the thermodynamic efficiency (output to input force ratio) must be strictly limited to some optimal value to get the reasonable optimal values for electron/proton flux. Our optimization procedure focused at finding maximum entropy production only for the charge separation steps leading to the chemiosmotic power production, an approach similar to the Wagoner and Dill (2019a) method in seeking the best compromise (the tradeoff) between optimal speed and optimal ratio of output to the input force.

We emphasized in this book the natural tendency of physical and biological processes to increase the entropy production. A large number of authors noticed how the emergence of dissipation driven structures helps in increasing the rate of entropy production. Quasi-stable structural patterns and even **fractal** geometries are often seen. The driven fluxes and forces are contributing to increased dissipation together with emerged nonequilibrium structures promoting such secondary fluxes and forces. Thus, it is not only the engineers' love for their favorite machines' maximal power production. Biological evolution also led to optimized energy-converting structures exhibiting close to maximal possible power production. We propose that the maximization of power production is coupled to an increase in absorbed work and total entropy production. Thus, the input free-energy cannot be regarded as the constant parameter except for more or less artificial steady-state conditions. For 11 biological nanomachines selected by Wagoner and Dill (2019a), the input free-energy is the ion electrochemical gradient or nonequilibrium concentrations of ATP and ADP. The homeostasis of active cells can maintain the stability of chosen nonequilibrium steady states with selected constant input free energy gradients only for limited periods. Development, biological evolution, and dissipative adaptation would not be possible if criticality is never reached to break down extant and establish new steady states. The dissipative adaptation (England 2015) works during evolution as an extension of the Second Law of Thermodynamics, resulting in increasing self-organization that accelerates coupled evolutions: universal (thermodynamic) and specific (biological).

One contrary example is enough in physics to falsify any hypothesis, but biology and biological evolution are more complex. Light-activated proton pumps like bacteriorhodopsin were not optimized for maximal power production (Chapter 3, section 3.3, and Chapter 12). However, some of the numerous biological nanomachines may operate in the far-from-equilibrium regime that is indeed close to maximal power production and maximal free-energy transduction efficiency. Even if natural evolution did not optimize all biological ion pumps and nanomotors in that direction, Wagoner and Dill's (2019a,b) research might give a route to optimize them for biotechnological applications (see Chapter 12 for some attempts at biotechnological applications of bacteriorhodopsin). For this book's goal, it matters that power production is closely related to the partial entropy production responsible for the increase in the secondary (driven) force. Yet, maximizing the power production is not likely to be the strategy nature used during biological evolution. Nature's preference was for increased dissipation even if it meant a significant decrease in the power production.

The conundrum of how it can be that input power appears as entropy production is expressed in the bilinear expression X_kJ_k (sum implied over all conjugated force-flux pairs) for dissipation (see Chapter 3, section 3.1). It implies system-environment interaction through external forcing (X_k) and internal responses (J_k). Furthermore, it also implies that the major part of input power is converted into internal and exported dissipation, while a minor part can be transformed into internal patterns (dissipative structures) and exported power. The power transforming role of open far-from-equilibrium systems cannot be seen from the verbal definition of their high entropy production: entropy increase due to irreversible processes inside the system. All living systems we encounter on Earth are self-amplifying power transformers. An obligatory coupling of power transformation to dissipation is the reason why we can study chosen entropy production contributions to discover crucial power input and power production output pathways for biological macromolecules, living systems, and ecosystems.

16.8 Is the entropy generation minimized?

The minimization of entropy generation is the favorite end design practice for many engineers. It started during the industrial revolution as the design of more efficient machines with higher power output. Gradually, the entropy concept was developed together with the Second Law formulation by well-known fathers of classical thermodynamics Carnot, Clausius, Kelvin, and others, some of them engineers. It stands to reason that lesser entropy generation in the ambient saves money because the engine's performance is improved, be it a simple steam turbine or a complex one in a nuclear power plant. The energy crisis from the 1970s revived the interest in the minimization methods of entropy generation. The methods are also known as **exergy** (available work) analysis.

The best known and most prolific author of research results, textbooks, and popular books about that topic is the distinguished professor Adrian Bejan, the engineer with all degrees from the prestigious MIT University, now at Duke University (Lage et al. 2008). According to his own words, he experienced a moment of deep insight in 1996 (Bejan 1997). It was a dramatic epiphany requiring the invention of a new word to describe it, just as we mentioned in Chapter 11 (sections 11.7 and 11.8) for novel concepts introduced by Paul Boyer. Adrian Bejan constructed a new word "constructal" as an antonym for the word fractal (Bejan 1997). The significance of the constructal idea morphed and developed through years toward theory, fourth law, the law, and finally, the "first principle of physics" (Bejan 2016a). But identical verbal formulation survived: "For a finite-size flow system to persist in time (to live), it must evolve freely such that it provides greater access to its currents (Bejan 1997, 2017)." The words "to live" meant that thermodynamics and geometric optimization methods deserved to be extended to describe living systems (Bejan 1997). As Eric Karsenti observed (2008), the engineers who enter the field of biology tend to speak of design, but examples of biological self-organization are not the evidence for a thought out design. In his recent 2017 paper (Bejan 2017), Bejan stressed that "the constructal law is universally valid precisely because it is not a statement of optimality and end design." The constructal law's mathematical formulation focused on minimal overall flow resistance or minimal entropy generation rate (Bejan and Lorente 2004). The verbal formulation is easier to generalize because it describes the evolution of configuration in freely morphing thermodynamic systems when flows are free to find easier pathways offering less resistance.

Surprisingly so, Adrian Bejan classified many of the proposed variational principles from irreversible thermodynamics, engineering, physiology, and biology into the "end design" or "destiny" concepts common to religious fundamentalism (Bejan 2016b). Regarding biological evolution, he claimed in the same paper that maximization of entropy generation rate is blatantly wrong because "the evolution of the system flow organization is toward a smaller entropy generation rate, not a greater one" (Bejan 2016b). He allowed that geophysics is an exception by citing the Paltridge contribution to the formulation of the maximum entropy production principle and its applications to global atmospheric circulation (Paltridge 1975). Somehow, that example of the whole Earth as a perfect dissipater is incorporated in the constructal law (Bejan and Lorente 2010). It allowed for the claim in that paper that the constructal law is far more general than the 'maximum entropy production' principle.

Interestingly, some of Bejan's ardent followers (Lucia and Maino 2013, Lage et al. 2018) connected the maximization (not the minimization) of entropy generation with the least action principle and the evolution of metabolic networks toward states of maximum entropy production (Unrean and Srienc 2011, and references within). In his publications, Lucia pointed out that entropy generation is not the thermodynamic property, nor a thermodynamic potential of a system, because it depends on the state of both the system and its reference environment. The same holds for the exergy concept. Still, the entropy generation is useful to describe the irreversibility of the system's interaction with its surroundings (Lucia and Sciubba 2013). It provides a global description of irreversible processes by linking the entropy increase to the dissipated work in the surroundings at some reference temperature T_{ref} (Lucia 2012). To make peace between contrasting viewpoints, Lucia suggested that minimum entropy generation is related to the system, while maximum entropy generation is related to the interaction between the system and the environment (Lucia 2012). Two logical problems are connected with that interpretation. The first one is the absence of any entropy change in a steady-state system—meaning that the entropy generation rate in the system must be identical to entropy production and to the time differential of the entropy generated by the system in the environment. Minimal entropy production of the system must be exported to preserve the steady-state, but minimal entropy transfer cannot create maximal entropy generation in the environment. The second one is that entropy generation is calculated as the imprint on the environment because the system is regarded as a black box when this and other older concepts from thermodynamics (exergy) are used. Thus, the concept of entropy generation related only to the system does not make much sense, while entropy production due to irreversible processes in the system does.

What makes sense is the possibility to perform the accounting for all the entropy generating streams of the system. A map can be drawn then on how the destruction of exergy is distributed over the engineering system of interest (Bejan 2002). In this book, we strived to drive home the idea that the same approach is profitable for dissecting a chosen enzyme's entropy production contributions to discover what internal nanofluxes have major importance or are rate-limiting for the enzyme performance. I did not use the end design assumption of engineers about the desirability of minimal dissipation and maximal efficiency.

Umberto Lucia extended Bejan's ideas into widely separated fields, from energy footprints of quantum thermodynamics to cancer bioenergetics (Lucia and Grisolia 2019 and references within) and entropy generation by all living systems resulting in the footprint on the environment, due to irreversibility (Lucia 2014). His excursion into the bioenergetics of cancer cells (Lucia and Deisboeck 2018, and references within) resulted in offering conceptual anticancer therapy (Lucia et al. 2014). Some shortcomings of Lucia et al. (2014) biochemical models are the reversed direction of proton pumping for plasma-membrane located V-ATPase (H^+ ions are pumped outside the cell, not inside the cell) and the misidentification of principal ion pumps responsible for establishing the membrane potential (**Na^+, K^+-ATPase** is fundamental for the maintenance of plasma-membrane potential in animal cells, not the **vacuolar V-ATPase**). Thus, I do not expect that the proposed conceptual anticancer therapy can work in practice, despite the constructed edifice of equations based on a constructal theory approach. However, the constructal law served to argue that an external magnetic field can be used to "modify the rotation of the diseased cell towards the normal one" (Lucia et al. 2014).

Bejan's excursion into the fundamental question of what is life (Bejan 2016a) has a strange déjà vu quality. He did not like Schrödinger's approach to that question, but not because Schrödinger lacked a more profound knowledge of chemistry and biology. He bravely stated: "Unlike Schrodinger, I will place this question firmly in the realm of physics—the science of everything."

According to Bejan, his constructal law is the answer to the question: what is life. He brings all questions and answers about the hierarchical or self-organization under a single scientific umbrella: the constructal law. The answers boil down to the urge for more and easier movement of everybody and everything. Thus, he offers a new definition of life: life is a movement that evolves freely. It must have the ability and freedom to change flow configuration, morph, evolve, spread, and retreat. The life phenomenon and the fingerprint of constructal law are seen everywhere, including rivers, raindrops, lighting, snowflakes, air turbulence, animal design, and technology, concluding that life is older than the biosphere because inanimate flows populated the Earth before the emergence of the first living cells. Consequently, the "Physics of Life" book explores the evolution of technology as common to the evolution of life and other life-like natural phenomenons.

To first-year students of classical mechanics, the teachers (including myself) liked to repeat "pay attention to forces." I have no direct knowledge, but it seems reasonable to assume that Adrian Bejan instructed his students to pay attention to fluxes in the system and how they can be modified for easier flows and lesser dissipation. It is the point of partial agreement and significant divergence between his approach and the take-home message from this book. Easier flow through the hierarchy of free-energy transducers depends on details of how fluxes and forces are organized and regulated. Stronger flows are associated with greater, not lesser free energy dissipation, and free energy transduction efficiency can increase together with the entropy production increase. The regulatory strategy of living systems evolves and improves their ability to access the incoming flows, distribute the internal flows according to variable priorities, and control the output flows toward the surrounding environment. The overall strategy maintains the dynamic stability of a life-cycle. All of this requires a higher, not lesser, potential for a fast increase of internal entropy production and exported entropy generation in favorable circumstances.

16.9 Fine regulation of brain's bioenergetics and heat production

Our brain is an excellent example of extremely intensive bioenergetics coupled with equally strong dissipation. It may always be close to self-organized criticality (Bak 1996, Ma et al. 2019). That organ is highly dependent on the dynamic stabilization of unstable far-from-equilibrium states. The price to be paid is a quick deletion of sensory information. Monteforte and Wolf (2010) found a strikingly large entropy production due to the complex connectivity of the excitatory-inhibitory network of neurons in the cerebral cortex. Researchers were able to calculate how quickly an activity pattern is lost through tiny changes. Forgetting is very quick (in seconds) due to chaotic dynamics and high associated dissipation. It is likely to have an important role in the short-term stabilization of the essential sensory input and longer-term sleep memory consolidation (Hopfield et al. 1983). Brain cells produce more excess entropy than any other cell type. Mitochondria-rich neurons produce high amounts of entropy and reactive oxygen species (ROS) simultaneously. During evolution, animal brains developed sophisticated local and global blood-flow regulation to eliminate the excess heat and toxic ROS compounds. The brain's importance for survival ensured that exported entropy waves coming from the brain are instead directed to other organs. That may be the missing lead why the whole body and brain oscillate in synchrony between wakeful and sleeping state. Sleep deprivation kills faster than food deprivation—a trait common to us and fruit flies. A recently put hypothesis is that some organs (gut) cannot completely neutralize toxic ROS compounds during wakeful periods (Vaccaro et al. 2020).

The hypothesis does not solve the mystery of molecular requirements for changes from consciousness to sleep, or to reversible coma after general anesthesia (Franks 2008). Pavel et al. discovered this year (2020) that several general anesthetics disrupt the specific class of lipid rafts and activate the TREK-1 potassium channel. Thus, the integrity of fragile and almost untraceable lipid rafts (see Chapter 7, section 7.7) may be responsible for existence, reversible loss (during sleep), or more serious loss of consciousness (coma, anesthesia). It is a sobering thought about how fragile our consciousness is while driving a car or a politician's consciousness while deciding the fate of millions. The TREK family of potassium channels control cell excitability (Djillani et al. 2019). The TREK-1 is unique by having at least four control sites and an unusual selectivity filter sensitive to conformational changes induced by pressure, temperature, acidity, lipids, hormones, and so on—the list increases every year with new discoveries (Lolicato et al. 2017, Pope et al. 2020). It acts as a central switch in the brain—the membrane located interaction partner with numerous proteins involved in many physiological and pathological processes.

We are very sensitive even to slight changes in our heat production. Too little, we are freezing to death, too much, and we are sick with a high fever. Fine regulation of heat production is vital for most organisms. Different maximal entropy production principles have been applied in this book to examine their usefulness in living cells' bioenergetics. Extremes are not always the best solution in biology. Various safety valves and regulatory mechanisms take care of the upper-limit entropy production in the unicellular organisms, healthy cells, and cancer cells (Niebel et al. 2019). Limits and safety valves are not imposed to increase the free-energy transduction efficiency, but ensure that cells can safely export all of the internally produced entropy. Metabolic networks may well tend to maximize the entropy production rate (Dewar 2010, Vallino 2010, Bordel and Nielsen 2010, Unrean and Srienc 2011, Franklin et al. 2012). However, our simulations suggested that theoretical upper limits for the entropy production are approached only for essential pathways and transitions between functional states. Different safety valves take care that overall entropy production limits are far below the theoretical maximum. Nevertheless, minimal entropy production is far from being a desirable goal for any cell type with an active metabolism. An analogy of the static-head steady-state with close to minimal entropy production cannot be approached at all. A membrane dielectric breakdown and cell death occur if the protonmotive force is high enough to substantially decrease the entropy production.

16.10 The lower limit for bioenergetics of dormant cells

Dormant cells with extremely slow metabolism and correspondingly low entropy production have been found during the past few decades. Some dormant cells using only the maintenance metabolism have been estimated to survive for about 10^8 years (Price and Sowers 2004). Microbial communities indeed persist in sub-seafloor sediment in the presence of very low energy fluxes. Some have been recently revived after about 100 million years of survival in stasis (Morono et al. 2020). Bacterial and archaeal microbes extracted from deeper locations beneath the seafloor have basal metabolic activity close to the theoretical minimum (Hoehler and Jørgensen 2013, LaRowe and Amend 2015, Bradley et al. 2020). For the least active methanogens, the minimum power to stay alive hovers just above a zeptowatt, or 10^{-21} watt. By comparison, a human body at rest uses about 100 watts. There are many examples of exotic metabolic pathways developed by evolution to ensure the minimalistic maintenance mode metabolism in harsh conditions (Rowe et al. 2015, Lam et al. 2019, Ray et al. 2020).

Are these examples enough to associate the minimum entropy production with the microbes using the minimal power to remain viable? In a constant energy-poor environment, the basal metabolic activity is needed to repair defects arising from the metastability of all organic molecules used by life. Power demand and required work to perform repairs do not increase with time, but free-energy transduction efficiency is low, and corresponding entropy increases in the environment accumulate with time. We must take a more general viewpoint—the whole biomass functions as the enzyme. In an energy-poor environment, it does not change itself. It only dissipates available electrochemical gradients over geological periods. Dormant biomass still gives rise to circulation patterns maintained by entropy export to its surroundings. Low efficiency and low yield of organic compounds do not preclude that total entropy production over eons of stasis is high enough to leave the geological imprint in the global habitats. For instance, biotic changes in carbon and sulfate cycling from deeply buried basalt are the telltale signs of chemosynthetic, entropy-releasing metabolic activity by sulfate reducers and methanogens (Lever et al. 2013).

Some authors proposed that an optimum amount of biomass exists to maximize the overall energy dissipation rate with the proviso that both quantities are constrained with available free-energy sources (Vallino 2010, Vallino and Algar 2016). The other caveat Vallino mentioned (2010) is the capability of biological systems to predict the future. Temporal strategies until conditions improve include spore formation by microbes. At the other end of the spectrum are certain animals and plants' strategies to store resources, expecting worsening winter conditions. Physical systems cannot predict the future, biological can. When averaged over time, a biological strategy leads to greater entropy production.

16.11 The upper limit for the bioenergetics' entropy production

We developed the theoretical method to calculate maximum entropy production for each transition between functional states of biomacromolecules. The method produces the optimal values of kinetic constants, catalytic constant, and catalytic efficiency—the values that are often close to but higher from observed data. Extended calculations of the same type can predict the upper limit for the bioenergetics' entropy production (Niebel et al. 2019). Besides scholarly value, there may be a practical benefit to identify the critical pathways and transitions that can produce the greatest increase in catalytic efficiency when subjected to directed evolution experiments. Directed evolution is not so much restricted by natural upper rate limits at which cells can free themselves from internally produced entropy. Also, biotechnological methods can bypass biological safety valves.

Paradoxically perhaps, the rate-limiting transitions are the most important contributors to overall entropy production. Faster kinetic steps are enslaved and regulated by the slowest rate-limiting steps. Regulation is helped in the presence of imperfections, such as the slippage mechanism mentioned in Chapter 3. Nelson et al. (2000) used the V-ATPase investigation to suggest that utilizing the imperfect (the proton slip) as a virtue is a fundamental property of living systems. We can consider a high entropy production by living cells as an imperfection if we disregard how much the physiological complexity increase depended on ingenious ways life found during evolution to increase the dissipative self-organization.

For that to happen, life had to develop the ability to control energy harvesting from the ambient. Life's environment has a bipolar aspect. A strong-enough free-energy gradient must exist in the surroundings to drive biochemical reactions. The other face of the ambient is its malleability to the self-serving activity of enough-plentiful living systems (Gaia hypothesis, Introduction). In Chapter 9, section 9.6, we presented the "side reaction" viewpoint about life. For chemists, the side reaction term means less important, driven by the more important reaction. How can any side reaction control a much more energetic and faster main reaction? Is it natural for a driven process to take control and enslave the driving process? When living systems are subsumed as all "side reactions" with the ability to assert internal and external control, the answer to that question is positive.

16.12 Allocating dissipation

From many examples of driven reactions extending their control over driving reactions, we mentioned the hexokinase case (Chapter 14, Branscomb et al. 2017) and the respiratory control transitions between mitochondrial states 3 and 4 (Chapter 7). Is there any way to detect a common footprint for all driven processes controlling the driving processes? Driving reactions, as ATP hydrolysis, are the major contributors to the entropy production. Even the simplest cells have the virtuoso's ability to control in time and space the consumption of the ATP-like molecules to allocate dissipation for driving prioritized uphill processes. Dissipation allocation is used by cells to activate signaling proteins and pathways (Weber et al. 2015), improve the speed of molecular machines (Brown and Sivak 2017), and increase the catalytic efficiency of enzymes during their evolution (Juretić et al. 2019a,b). Dissipative handles left by evolution (Weber et al. 2015) are the best fingerprint for tracing cellular control switches crucial for survival.

Extremes are easiest to detect. The control is so good that an astonishingly wide range of metabolic activities exists. Hibernation and **estivation** are examples of life's ability to maintain an extremely low metabolic activity during prolonged cold or hot and dry periods. At the opposite end of the spectrum is the vigorous growth of microorganisms when the maximal amount of free energy can be acquired to be channeled into the uphill synthetic pathways (lesser amount) and exported as entropy generation (higher amount). Thus, life can either slow down or speed up the driving process after both internal needs and external conditions are assessed to arrive at the best strategy for a considered living system. Heat production is a crude but effective way of measuring the footprint of all metabolic controls expressed by cells or an organism. Accounting for the most important internal fluxes and forces enables the entropy production calculations and identifies essential switch points controlling dissipation rise or decrease.

Molecular switches can either activate or inhibit signaling systems leading to entropy production amplification or decrease. The dissipation decrease occurs during stressful conditions, such as starvation. The conditional completion of the driven process is one evolutionary innovation that controls the entropy production decrease (see Chapter 14). When regulatory networks maintain a high (maximal) level of entropy production for a longer time scale, it is seemingly contrary to Leonid Martyushev's conclusion about the MaxEP principle being valid only for short time scales and small volumes (Martyushev 2013). Two solutions are possible: a) biological regulation of dissipation is not about MaxEP but about increased or decreased total entropy production, and (b) empirically shown MaxEP during longer time intervals (Vallino 2010, Vallino and Algar 2016) should not be judged by the restrictions for local MaxEP principles.

Local analysis is not sufficient to answer how global stability and homeostasis is maintained by living systems (Fang and Wang 2020). During eons of biological evolution, organisms become better adapted to their environment. The energy cost of biochemical feedback control mechanisms is expressed as entropy production because the adaptation is necessarily a nonequilibrium process that breaks the detailed balance and the time-reversal symmetry (Lan et al. 2012). These authors found the exponential relationship between the adaptation error and the energy dissipation rate for *Escherichia coli* chemotaxis. Hence, the adaptation accuracy is not very sensitive to the pool size of free-energy carriers that are consumed to push the system "uphill" in a noisy environment. A modest 20% thermodynamic efficiency is enough for *E. coli* cells to have the maximum adaptation accuracy of ~ 99% in chemotaxis. During starvation, the driving process of consuming free-energy rich molecules is considerably weakened with the concomitant decrease in entropy production, but a high adaptation accuracy survives almost unchanged (Lan et al. 2012). That is an interesting confirmation for our thesis that maximizing efficiency was never the goal of biological evolution, even in this case, when dissipation is decreased, not increased. In general, the total entropy production of all regulatory networks is precisely the thermodynamic cost for maintaining the stability, robustness, and vital functions of the nonequilibrium system (Fang and Wang 2020). Well known "selfish gene" concept (Dawkins 1976) may have to be replaced with the system biology concept (Kitano 2002) of optimized networks for the dissipation allocation. Multiple layers of regulations are evolutionary better conserved than any DNA or protein structures, suggesting that they are essential and traceable to optimal dissipation allocation (Abudukelimu et al. 2017).

16.13 Quantum thermodynamics on the horizon?

Linear-response theory breaks down for quantum systems. In semiconductors, the combination of thermodynamics and quantum mechanics provides optimal solutions even for macroscopic thermoelectric devices (Whitney 2015). For instance, the maximal efficiency of a quantum thermoelectric is found for given power output, and high efficiency is found at the quantum bound on maximal power output. When nonlinear (quantum) effects are strong, as they are for nonlinear phonon and photon effects, maximum efficiency coincides with maximum power (Whitney 2015).

Thermodynamics of light was an essential insight that led Planck and Einstein to the quantum unification of wave and particle theories (Anglin 2010). Similarly, the relationship between black holes and thermodynamics (Hawking 1976) provided insight into the unification possibility for gravitation and quantum physics. It turned out that the entropy of black hole is truly enormous, despite black hole characterization with only the mass, charge, and spin, as if a huge star became an elementary particle after collapsing into a black hole. If all elementary particles are converted into something more fundamental at the black hole surface, that would explain a huge entropy increase. However, black hole entropy opens the questions about the information-loss paradox (Maldacena 2020) and the existence of the entropy quantum (Kirwan 2004).

Interest in stochastic and quantum thermodynamics revived during the past two decades (Gemmer et al. 2009), sometimes leading to sweeping generalizations. Beretta (2020) proposed the 4th law of thermodynamics. His "Steepest Entropy Ascent" principle aspires to bridge the gap between classical and quantum thermodynamics, microworld, and macroworld. In the macroworld, Beretta's principle is equivalent to the maximum entropy production principle. Participants in recent developments in stochastic thermodynamics regard the processes with negative entropy production as a strong indication that Beretta's principle is not valid. Beretta and others (Esposito et al. 2010) concluded, on the other hand, that the concept of negative entropy production is not valid irrespective of the small number of particles in the system (Beretta 2020).

It is not a trivial endeavor to extend classical thermodynamics and statistical mechanics methods to active biological systems exhibiting dissipation and fluctuations. Mandal et al. (2017) published one possible pathway to a consistent definition of far-from-equilibrium entropy production within the framework of stochastic thermodynamics. Their goal was to develop the groundwork for obtaining verifiable predictions of emergent collective properties such as dissipative self-organization and pattern formation. Joining thermodynamic descriptions from different research fields will take more time but is certainly relevant for a broad scope of evolutionary topics, from microscopical and astronomical to biological.

16.14 Bioenergetics of photosynthesis and respiration from the entropy production viewpoint

The very first immediate steps after the photon absorption by the chromophore are of quantum nature. Excited quasiparticles (excitons) are then channeled toward the charge separation pathway. Bioenergetics of photosynthesis profited from the application of our maximal transitional entropy production theorem, proving that the photosynthetic cycle's first irreversible (charge-separating) steps already produce a massive amount of entropy (Juretić et al. 2019b). In the face of confusion regarding the entropy production in photosynthesis (Chapters 5 and 8), that result helped clarify why photosynthetic energy transformations are at least a million times more intensive than the intensity of energy transformations taking place in an average star's equivalent mean volume.

In the bioenergetics of respiration, the most important reaction is oxidative ATP synthesis by F_0F_1-ATP synthase. We used the MaxENT principle and the possibility offered by the MTEP theorem (Juretić et al. 2019b) to find the maximal entropy production in the chosen transition between enzyme functional states. The selected transition was the most important one for connecting driving proton flux to driven flux of synthetized ATP molecules. Experimental data and simulations suggested that the reaction is compatible with the increased overall entropy production and the maximum entropy production in the ATP synthesis step (Dewar et al. 2006).

When oxidative phosphorylation by rat liver mitochondria is considered, Sunil Nath used linear relationships between fluxes and forces (Nath 1998). His first conclusion was that evolution tunes the system for minimum dissipation. He described that insight as a delightful, beautiful, and powerful variational principle. The system stays as close to equilibrium as possible to minimize dissipation at the stationary state of zero net proton flux. Not surprisingly so, because it is the static head steady state. When he also considered the nonohmic (nonlinear) proton leak, the dissipation still exhibited a minimum for the output force variation. Among other experimental results supporting his findings, Nath mentioned the Jörg Stucki paper from 1980. The optimization of oxidative phosphorylation remained the focus of Nath's interests 20 years later when he published three papers about that topic (Nath 2019a,b,c). In all of them, he used linear nonequilibrium thermodynamics. The reviewers of his manuscript submitted to Biophysical Chemistry asked for clarification about Stucki's contribution (Stucki 1980). The clarification cited two of my papers (Dewar et al. 2006, Dobovišek et al. 2011) and two of Nath's recent papers (Nath 2018, 2019b). Citations were included in the claim: "In contrast to Stucki's finding of minimum dissipation (Φ) at the optimal steady state, a maximum Φ has been obtained for the operating steady state in oxidative phosphorylation, given the constraints" (Nath 2019a).

What changed during the past two decades that the same author now proposes the opposite thermodynamic interpretation for similar oxidative phosphorylation data? Nath advanced the biothermokinetic approach that included some molecular and kinetic details of energy transduction (Nath 2019c, 2021). It permitted the simultaneous observation of maximum output power, maximum free energy dissipation, and optimal free-energy transduction efficiency under the imposed constraints at a particular (optimal) value of the force ratio (Nath 2019c). Free-energy transduction efficiency and dissipation both increase when mitochondria is stimulated by ADP addition to leave the resting state 4 and enter the metabolically active state 3. Sunil Nath interpreted his experimental results in terms of the least action principle formulated as an integral over the dissipation function multiplied with a relevant time for oxidative phosphorylation (Nath 2019b).

16.15 What should be maximized, partial, or total entropy production to get an insight into the self-organized establishment of order?

Throughout this book, we took pains to emphasize that we never looked for maximal <u>total</u> entropy production. All MaxEP principles claim to deal with the total entropy production of the system. Stijn Bruers carefully examined different formulations of MaxEP principles in his PhD thesis (Bruers 2007a). He recognized that our approach of seeking maximal <u>partial</u> entropy production is shared by some other formulations of the MaxEP principle despite claims to the contrary. Does a firm foundation for different MaxEP principles exist or not is still a matter of controversy. Some experts in the field of irreversible thermodynamics regard the application successes of MaxEP as an unexplained curiosity. Researchers with a good overview of the field admitted that applications have been in an ad hoc manner (Dewar et al. 2014, Ban and Shigeta 2019). The same can be claimed for the definition and applications of the partial entropy production concept. For instance, Polettini et al. (2016) applied the phrase "partial entropy production" to the least amount of entropy that a system sustaining current J can produce in a given steady state. It is the minimum possible entropy production, not the chosen contribution to total EP.

Shiraishi and Sagawa (2015) and Otsubo et al. (2020) opined that we cannot observe and are not interested in all of the microscopic transitions. Thus, the total amount of entropy production is not within reach and may not be relevant for calculations in many nonequilibrium systems. Therefore, within the theoretical framework of stochastic thermodynamics, they considered a subset of all possible transitions and associated partial entropy production of subsystems. The name they adopted was partially masked nonequilibrium dynamics. Their goal was to examine if it is still possible to obtain fundamental results such as the fluctuation theorem (Crooks 1999, Jarzynski 2000) and thermodynamic uncertainty relations (Barato and Seifert 2015, Horowitz and Gingrich 2020). Indeed, the total entropy production can be decomposed into the subsets of transitions (additivity property) so that partial entropy production satisfies the fluctuation theorem (Shiraishi and Sagawa 2015). Moreover, generalized thermodynamic uncertainty relations were also satisfied (Otsubo et al. 2020).

Our goal in this book was different. We first asked which one, out of all possible steady-state currents in the system, is producing some useful or interesting output. Our second question was which transitions in the system contribute most to total entropy production increase after selected current is optimized by maximum EP requirement for these transitions. Both questions converged toward the same answer—the transitions involved in producing the power output are those that cause the most significant increase in overall entropy production. Without requiring the MaxEP in the total entropy production, we found the computational method to simultaneously optimize the rate-limiting steps in kinetic schemes and increase overall EP.

Different goals, methods, and model systems used in Sagawa and collaborators' contributions do not hide the importance of examining the conditions for the existence of the extremal entropy production value for a transition between chosen states of a system. Such a research program within the theoretical framework of stochastic thermodynamics did not take off yet. Our contributions to that topic described in this book (MTEP theorem and applications) are limited to the mean values of various quantities for the steady state kinetics (neglecting fluctuations).

Ban and Shigeta (2019) initiated their research about various flow patterns by asking whether partial or total entropy production determines the evolution of a system driven from equilibrium by temperature difference. They ascertained which component of entropy production should be maximized to predict the system behavior. Hence, they used a modified MaxEP principle dealing with the partial entropy production, an approach similar in spirit to the MTEP theorem's applications presented in this book. They ignored smaller entropy production due to viscosity and focused on thermal convection producing flow patterns called hydrothermal waves. Hydrothermal waves are perpendicular to the driving force (the temperature gradient). A theoretical justification was postponed for later work. In any case, it is of interest that an entropy production selection rule can be used to determine the source of the partial entropy production essential for the system's evolution and its thermodynamic state.

There are several intriguing examples (besides MTEP theorem applications) when dynamic stability can be characterized by maximal partial entropy production. Paltridge climate model is one well known older example (Paltridge 1979). Bruers (2007b) noted that not the total EP is maximized in Paltridge's formulation of the MaxEP variational principle. Namely, only the atmospheric heat transport processes from the equator to the pole are considered while looking for an optimal transport coefficient similar to the observed one. The emergence of the turbulent transfer of energy in a driven climate system is associated with a dazzling pattern formation and greatly increased dissipation until dynamic stability is reached (Paltridge 2001). Thus, the choice of processes constrained to maximize partial entropy production is clear. These are processes that create dissipative structures, that is, the self-organized establishment of order. Kleidon et al. (2014) asked the related question: "With the breadth of competing processes shaping the Earth's system, how do we know which entropy production should be maximized?" Their answer was to look for the maximum rate by which heat can be converted into mechanical work because there should be an associated (partial) entropy production.

Far from equilibrium phase transitions can be the source of the self-organized establishment of order. Martyushev and Konovalov (2011) proposed that a necessary condition for a nonequilibrium phase transition is larger entropy production in the final nonequilibrium phase. The hypothesis is presumably based on the MaxEP principle. Two caveats should be taken into account. The first one is assumed linear or cubic relationships between fluxes and forces by these authors. The second one is that larger total entropy production can result from the maximization of partial entropy production, which is crucial for the system's evolution. For biological systems, nonlinear feedback with exponential flux-force relationships is necessary for phase transitions from bistability to excitability and oscillations (Ehrmann et al. 2019).

Historical development toward quantitative studies of first and second-order phase transitions in chemical systems started with the Schlögl model reactions (Schlögl 1972). The first model consists of reactions $X + A \leftrightarrow 2X$, $X \leftrightarrow B$. More complex toy reactions of the second Schlögl model are $2X + A \leftrightarrow 3X$ and $X \leftrightarrow B$. The models contain only one variable concentration X. In both cases, the system is assumed to be in contact with reservoirs of particles A and B. Concentrations of A and B are somehow maintained at the constant level by appropriate feeding of the reactor. All chemical species are assumed to exist as rarefied gases. The second trimolecular reaction scheme is autocatalytic. In a review paper about ten years ago, Vellela and Qian (2009) used the words "the canonical example of a chemical reaction system" to describe Schlögl's model. Other authors (Endres 2015) used even stronger words "the well characterized biochemical Schlögl model" while admitting the absence of enzymatically driven reactions. For almost 50 years, the motivation to study Schlögl's model was always the same wishful self-assurance about its relevance for far-from-equilibrium biological phenomena. Examples of assumed relevance are the bistable behaviors such as heart models and visual perception (Vellela and Qian 2009), self-activating genes and phosphorylation-dephosphorylation cycle (Endres 2015), and the biochemical switching and oscillations (Erhmann et al. 2019, Nguyen and Seifert 2020).

The motivation to study the nonequilibrim phase transitions by a minimal toy model containing only one variable remained unchanged from Schlögl's contribution (1972) to the present days. How can entropy production be calculated for Schlögl's models, and how does it change during phase transitions? Is that model complex enough to serve as a starting point to describe the coupling of driving and driven biochemical reactions? The first question was examined by Tomé and de Oliveira (2018). Entropy production and reaction flux displayed a jump to much higher values during phase transition. Concerning the second question, Ban and Shigeta (2019) turned to general observation (Ziegler 1977) that nonequilibrium processes can be divided into two types: compound and complex. The former are uncoupled processes such that each elementary process depends only on the corresponding flux, not on all of the system's fluxes. Martyushev and Seleznev (2014) claimed that the maximum entropy production principle is valid for complex but not for uncoupled compound processes.

Noa et al. (2019) noticed that a specific entropy production signature is a reliable tool for categorizing continuous and discontinuous phase transitions far from thermodynamic equilibrium. They obtained the bistable behavior in the entropy production for the class of models described by the logistic equation. As solutions of logistic differential equations, logistic curves are a particular case of sigmoidal functions we mentioned before (sections 9.5, 11.11, and 14.1). In each instance, the S-shaped curve was the sign of self-ordering, cooperative kinetics, and metabolic regulation. Logistic curves describe the growth and decline of "species" competing for limited resources available to an open system interacting with an environment. There is initial slow growth, subsequent mild or steep jump, and final saturation in each logistic curve. In complex networks, jumps in some order parameters are also viewed as jumps in the entropy production (Noa et al. 2019). Thus, the discontinuities in the order parameter are also presented as the discontinuities in the entropy production. Complex networks considered by Noa et al. (2019) have questionable relevance to complex biological networks representing active matter, except for one common feature: the misalignment parameter introduced by these authors when discussing possible applications. Remarkably, the entropy production increase occurs until a sharp maximum is reached for the optimal value of the misalignment parameter with a subsequent decrease after a further increase of that parameter.

We have now come to the crux of the living systems' magic incisively described in Max Delbrück words: "The closer one looks at these performances of matter in living organisms, the more impressive the show becomes. The meanest living cell becomes a magic puzzle box full of elaborate and changing molecules ..." (Delbrück 1949, 1970). The present-day insight speaks about molecular machines and nanomotors performing purposeful tasks while immersed into and exposed to violent thermal chaos (Hoffmann 2016). It does not look like order-from-order molecular mechanism proposed by Erwin Schrödinger (see Chapters 4 and 5), not disorder from order evolution outlook of conventional thermodynamics (Shimizu 1979). Andrew Huxley suggested in his 1957 paper that muscle transduces random thermal motion into directed mechanical motion (Hoffman 2016). The conversion of heat flux into mechanical work happens both in a natural setting (Kleidon et al. 2014) and in human-made machines driven by temperature differences. Still, the random thermal motion is not the heat flux. Huxley's suggestion looks like a horrible violation of the Second Law until the statistical formulation of that law is considered. The Second Law does not prohibit microscopic and occasional direct conversion of heat into mechanical work. It prohibits the continuous repetition of that mechanism. An input of externally supplied chemical energy is needed to complete the cycle and reset the molecular machine's mechanism to its starting position. For motor proteins myosin and actin, ATP hydrolysis can initiate detachment to dissociate the motor (myosin) from actin filaments. Myosin is then described as a machine that rectifies random thermal motion by converting it into directed motion. The cycle time of the motor is such that thermal gradients cannot be sustained. The motor operates isothermally and must be kept out of thermodynamic equilibrium by providing the chemical energy for a crucial reset step. The ADP release step is the rate-limiting reset step in the actomyosin's mechanochemical ATPase cycle (Jacobs et al. 2011). It allows for the displacement of actin filaments and walking of myosin along actin. It is a strain sensitive step proposed already by Huxley (1957) to explain the nonlinear force-velocity relationship for muscle contraction (Hill 1938). The similarity to rate-limiting steps we often mentioned in this book (previously in this chapter and Chapters 8, 10, 11, 12, and 14) comes immediately to mind together with the question if associated (partial) entropy production is also a significant contribution to total EP of the mechanochemical cycle.

There are multitudes of various mechanochemical cycles that are operating in each of our cells. Before examining energetics and dissipation, a few words about the structural organization are in order. Crystal-like organization of actin, myosin, and accessory giant proteins titin and nebulin is best ordered in filments from cross-striated muscles (Dasbiswas et al. 2018). Titin is the largest of all elongated human proteins, with its length of more than 1.5 μm and a total of more than 34 thousand amino acid residues. However, contractility can arise without the classical sarcomeric actomyosin organization in striated muscles (Lenz et al. 2012). Purified solutions of microtubules and kinesin can form extensile networks that become contractile in the presence of dynein motor protein (Forster et al. 2015). Microtubules extend to separate chromosomes during the **anaphase stage of the cell cycle** division. Animal cells precisely regulate in time and space their morphology and self-movement by highly cooperative and localized actomyosin contraction or extension (Linsmeier et al. 2016, Lenz 2020). Lesser rigidity and buckling propensity of actin filaments make them more suitable than tubulin filaments to store and release bending elasticity (Belmonte et al. 2017). The observed intrinsic disorder in actin filaments (Bremer et al. 1991), several myosin loops (Geeves 2016), and some accessory proteins (Tolkatchev et al. 2019) can help in elucidating the structural basis for their suitability in generating and sensing mechanical forces. We already mentioned intrinsic disorder in another example of order-promoting disorder (section 7.8.2).

Myosin II filaments are ordered to a different degree in stritated muscles, smooth muscles, and non-muscle cells. Regulation details and all inducers and inhibitors are still not entirely known for myosin II assembly into filaments and ordering of actin networks (Dasbiswas et al. 2018). But understanding the production and dissipation of contractile stress is essential for getting an insight into how actomyosin networks mediate tunable control of long-range cellular responses (McFadden et al. 2017). That topic is under active investigation, notwithstanding the well-researched biochemistry of motor proteins and accessory molecules such as alpha-actinin cross-linking proteins (Murrell et al. 2015, Popov et al. 2016, Bidone et al. 2017). Henkes et al. (2020) published one example of how the mechanical stress conversion into dissipation can result in dynamical pattern formation by epithelial cells. They assumed that cell-surface friction is the main dissipation source for migrating confluent cells on a solid substrate growth medium. Correlations in cell movements are then simulated as overdamped dry dynamics. Theoretical predictions are in quantitative agreement with observed velocity correlation distances of ten or more cell sizes (around 100 μm) at constant temperature conditions (37°C). Thus, partial entropy production selected by Henkes et al. (2020) is the most important source for the evolution of long-range correlation patterns. These authors did not attempt to calculate the partial entropy production associated with a cell crawling speed. However, they cited Garcia et al.'s (2015) observation of a pronounced maximum in the velocity correlation length for the optimal mean cellular velocity. It would be interesting to see if the maximal partial entropy production requirement predicts the observed optimal velocity value for the best dynamic self-organization of crawling cells. One would expect that the optimal foraging strategy emerges from an optimal compromise between cell-cell and cell-substrate interactions by the best dissipative adaptation to the environment.

Étienne et al. (2015) presented a captivating picture of whole non-muscle cells behaving like viscoelastic liquid motors. The actomyosin cortex of non-muscle cells also shows a Hill-like nonlinear force-velocity response (Hill 1938) to changes in the environment's mechanical properties. The major part of myosin power input (ATP hydrolysis) is internally dissipated. The minor part of at most 5% is being transmitted to the cell environment as mechanical power. The authors estimated that only about one part per thousand of the total power involved in cell metabolism is invested in the myosin activity. Hence, they speculated that the evolution of actomyosin networks in non-muscle cells never minimized the dissipation costs. The evolutionary pressure on this cost was low, but it was high for the cell's quick adaptability to mechanical challenges. Remarkably, the non-muscle actomyosin network is regulated to fluidize and flow whenever cells need it. Invested chemical power is seen as internal dissipation and exported heat. Internal dissipation pathways include viscous-like dissipation, internal creep, transverse modes of F-actin fluctuations, longitudinal modes of F-actin fluctuations, polymerization-depolymerization dynamics for actin filaments, and other partial contributions to total entropy production.

Intriguingly so, the entropy production rate is maximized in the non-contractile, stable state of actomyosin network architecture (Seara et al. 2018). Thus, the dissipation is responsible for maintaining the dynamic stability of actomyosin networks. That *tour de force* combination of the novel experimental setup and theoretical simulations challenges the prevailing model of the stable state by molecular motors oriented parallel to F-actin. The stable state, Seara et al. (2018) observed, is characterized by the misalignment between myosin and actin filaments. When the corresponding order parameter is calculated, it assumes a low value after everything is stabilized. Hence, the misalignment parameter must have an optimal high value for maximum entropy production due to transverse deformations of actin filaments (fluctuations, bending, and plucking perpendicular to some average F-actin axis). There are enhanced actin fluctuations in a non-contractile state associated with the enhanced entropy production rate. These connections of order, alignment, or misalignment parameter changes with the entropy production changes are reminiscent of what Noa et al. (2019) found for complex networks. Discontinuities in order or alignment parameters are also seen in the entropy production rate. A possible reason in biological systems is that criticality must be reached when the system is most easily pushed in the direction of order-to-disorder or disorder-to-order transitions with desired long-range effects. Dissipation is the ultimate destination of all free energy flows in actomyosin network systems, from the chemical potential of ATP-ADP couple to intermediate free-energy storage in the form of mechanical stress and concentration gradients of myosin filaments and cross-linkers (Floyd et al. 2019).

Transitorily increased mechanical energy of the actomyosin network is a useful "secondary force" paid for with the partial dissipation of chemical energy as heat. In that sense, high entropy production is an investment, acting as a catalyst for an enhanced turnover (Bier et al. 2016). In the actomyosin active network, partial entropy production is channeled into the filament bending work that is first increased in the non-contractile state and then released into the contractile state (Seara et al. 2018). The apparent contradiction between EP increase or decrease in the final ordered state has its origin in partial EP calculations. If the focus is on transverse fluctuations of actin filaments, the EP decreases in the contractile state. When the focus is on longitudinal sliding movements in the contractile state, the corresponding partial EP is increased due to directed cooperative flow.

At the end of this section, we can ask, is there any equation of life, and if it is, what breathes life into the equation? The question is a homage to Stephen Hawking's question about the unified theory of the universe: "What is it that breathes fire into the equations and makes a universe for them to describe?" (Hawking 1988). We can propose the conjecture that partial entropy productions causing nonequilibrium pattern formation are responsible for marginal dynamic stability that, on the one hand, maximizes the corresponding dissipation and, on the other hand, breathes the life in equations describing disorder-to-order transitions. The Second Law does not forbid the dissipative emergence of orderly self-organized patterns. The dynamics of actomyosin networks is one well-researched example of how it can happen in the macroscopic world of biologically relevant contractile fluid (Étienne et al. 2015).

16.16 The turbulence and active matter challenge

"It's a place where no rules apply, or at least they haven't figured' em out yet"—this citation from the Jeannette Walls book (The Glass Castle) is as good an introduction as any to the physics of turbulence in complex systems. Remarkably persistent turbulent flows are seen everywhere, from microscopic biological systems (Doostmohammadi et al. 2017) to contrails after an aircraft already passed high above our head. Also, in centuries-old Jupiter's Great Red Spot. However, the transition from laminar to turbulent flow remains one of the major unsolved problems in classical physics and technological applications, which still defies a thorough understanding (Frisch 2004, Eames and Flor 2011). Given the lack of comprehensive mathematical characterization of turbulence in conventional fluids, it is hardly surprising that active nonequilibrium fluids are even less understood (Wensink et al. 2012, Bratanov et al. 2015). Long-range space-time interactions and correlations in such complex systems result in multifractal geometry patterns, weak chaos, and long memory.

Chavanis et al. (1996) considered a "general maximum entropy production principle" to explain the hydrodynamical vortices' robustness observed in a plethora of geophysical or astrophysical phenomena. They invoked Robert and Sommeria's paper from 1992, who proposed the variational principle for the diffusion fluxes in two-dimensional fluids—for given diffusion energy, the system tends to maximize its entropy production by a pattern-forming distribution of its fluxes. The system relaxes toward stationary states with maximum entropy in the spirit of Jaynes's ideas (Jaynes 1980). Robert and Sommeria have chosen physicists' time-honored practice in modeling complex systems and processes—selecting as many constraints and simplification as needed (but not more) to reproduce the essential aspects of an observed phenomenon. For turbulent flow, the reproduction of observed self-organization of vorticity (rotation tendency) required a careful choice of conserved quantities (constants of motion) as the constraints for setting up and solving the variational problem. The irreversibility is taken into account by maximizing the mixing entropy. Relevant partial entropy production for the system's evolution is taken into account by minimizing the diffusion energy and conserving the energy of vorticity fluxes. A similar approach was used by Chavanis et al. (1996), who extended it to the study of galaxy dynamics.

The long-range self-organization in turbulent flow has another name that we widely used in this book—the dissipative structures (see Introduction, Chapters 1, 5, 9, 13, and section 16.15 from this chapter). Kondepudi et al. (2020) pointed out that an increased entropy production rate coincides with such structures' perseverance. All dissipative structures are autocatalytic because their growth proceeds by the process-products catalyzing their own production. Any nonequilibrium growth implies the time evolution of forces and flows. This fact alone has a greater significance than the possibility of forming the products of flows and forces and interpreting it as a dissipation or entropy production. Complex internal responses to changing external fields are masked if we pay attention only to the total entropy increase. It would only lead to the usual interpretations from thermostatics and mechanics that entropy increase is nothing else but increased disorder, decreased useful energy, and reduced efficiency.

Dissipative structures are better described as cooperative and self-adjustable driven processes, adapting to changes in the environment, recovering after perturbations (self-healing), and achieving resonant amplification after a certain threshold forcing degree has been reached. These are all active matter characteristics, a special class of self-propelled bio-analog or biological dissipative structures. Self-sustained active turbulence in dense microbial suspensions is dominated by short-range interactions and shares some qualitative characteristics with classical turbulence on small scales, forming the vortex patterns of turbulent motion (Wensink et al. 2012). When density fluctuations are large, large-scale dynamic coherence such as swarming or flocking is observed differing from classical turbulence (Dombrowski et al. 2004). This kind of non-linear self-organization requires modeling flocking behavior and dissipation to define a new turbulence class (Bratanov et al. 2015).

Autonomous and directional motility is a highly sought-after property for future applications (Saper and Hess 2020). Doostmohammadi et al. (2017) demonstrated how turbulent channel flows could transition from an ordered vortex-lattice flow state to active puffs giving birth to ever-new gusts. Above some critical activity, puffs span the entire channel system. The transition to active turbulence corresponds to the emergence of the orientational order. Namely, strong enough activity leads to spontaneous symmetry breaking and unidirectional flow in a confined environment. These authors used the vorticity field as the most meaningful quantity to study turbulent flows and a surrogate for corresponding partial dissipation without ever mentioning or calculating entropy-related quantities. When intricate puff patterns arise, a corresponding entropy should decrease despite an overall entropy production increase. In their next paper, Doostmohammadi et al. (2018) review focused on those hydrodynamic theories that describe the microtubule-kinesin mixture's properties. Active turbulence was also characterized by high vorticity and vorticity correlations.

By the way, the "Active nematic" title of that review requires a brief explanation. We did not mention liquid crystals before in this book nor their nematic phase when molecules in the crystal align in loose parallel lines. Monomers in the crystal are oriented in a crystal-like way, although the nematic system flows like a liquid. Thread-like topological defects are observed in the nematic phase. Biological membranes are one special class of liquid crystals consisting of amphipathic molecules (mainly phospholipids). When devoid of motor proteins, artificial vesicles and liposomes are equilibrium membrane structures. Active nematics are much more interesting life-like structures. They are active matter liquid crystals maintained out of thermodynamic equilibrium by extracting free energy from surroundings. That is how active nematics perform mechanical work.

Active matter physicists assume that active nematics are relevant for a better understanding of fundamental biological processes. For instance, Doostmohammadi et al. (2018) concluded their review by citing the Saw et al.'s (2017) finding of how stress mechanics can govern epithelial cells' life and death. The extrusion process prevents the accumulation of superfluous epithelial cells, but it was not clear how it is controlled. Saw et al. (2017) discovered that spontaneously formed topological defects generate mechanical stress, which is enough to trigger apoptosis.

The biophysicist Andreas Bausch and his collaborators created a minimalistic cell model with biomechanical function (Keber et al. 2014). It was nothing else but microtubules and kinesin molecules encapsulated within lipid vesicles. The ATP was also supplied in the experimental set-up to fuel the motor protein kinesin. The authors were able to produce a myriad of non-equilibrium dynamical states by controlling activity and vesicle deformability. One of the authors, professor Zvonimir Dogic and his group have previously shown that active networks of microtubules and kinesin exhibit internally generated fluid flows, autonomous motility, and enhanced transport as long as chemical fuel (ATP) is present (Sanchez et al. 2012).

Surface streaming flows of such active nematics resemble cytoplasmic streaming observed in large cells and germ cells. This year Dogic's and other research groups extended the study of active microtubules-kinesin networks to three-dimensional model systems (Dulcos et al. 2020). They found even richer collective dynamics of flow patterns. Dominant novel topological structures emerged as expanding and self-annihilating loops. A still richer dynamics is possible in the presence of contractile actomyosin networks (see section 16.15). Tjhung et al. (2012) studied how contractile stresses create motility in 2D and 3D models of actomyosin cell extracts. The simulation results for 3D contractile droplets are deemed relevant for the large-scale tissue flow during wound healing and embryonic development in animal cells. A droplet containing a disordered actomyosin network can exhibit directional movement resembling the bulk motility mechanism and spreading of cancer cells

(Le Goff et al. 2020). The emerging field of active matter therapeutics shows promise in treating cancer and heart diseases (Ghosh et al. 2020).

Why active matter research is a worthy research challenge for the crosstalk of biology, physics, biophysics, and bioenergetics? From the biology viewpoint, the most interesting active matter research is performed with living cells. We can mention self-synchronized oscillations of dense bacterial swarms at large scales (www.quantamagazine.org/swirling-bacteria-linked-to-the-physics-of-phase-transitions-20170504/; Chen et al. 2017), large-scale velocity correlations of confluent epithelial cells (Henkes et al. 2020), apoptotic cell extrusion induced by mechanical stress (Saw et al. 2017), dynamic feedback between mechanical stresses and cell migration in wound healing and cancer (La Porta and Zapperi 2019), and morphogenesis in regenerating *Hydra* (Maroudas-Sacks et al. 2020). Converting chemical into mechanical energy is just one of the many pathways cells employ in energy conversion and dissipation cascades. It is not the end in itself, but the means for performing the multitude of biological functions.

From the physics viewpoint, life and movements concepts are interconnected in the self-propulsion property of active nematics leading to the emergence of visually attractive and exotic patterns of motion. Physicists and biophysicists take for granted that, for instance, perfectly clean preparations of just two proteins, such as tubulin and kinesin, can be isolated by biochemists from a bovine brain. After attracting active matter research funding, they play with these motor protein networks by using complex experimental setup and sophisticated theoretical tools. To their delight, they can create the perfect brainstorm in one tiny droplet of active matter. The active matter research is the focal point for the most difficult unsolved problem from classical physics and the state of the art biology research. As we stated before, the emergence of turbulence with associated dynamic patterns is a challenging physics problem. When turbulence is expressed as the dissipative ordering of macromolecular patterns, it becomes one of the most intriguing biological questions, which is how long-range intracellular and intercellular organization can emerge from short-range protein-protein interactions.

16.17 What is the appropriate statistical entropy definition for complex systems?

Far-from-equilibrium thermodynamics of active matter is the motivation to develop appropriate theoretical tools beyond the preoccupation with flow models of classical hydrodynamics. However, some physicists are reluctant to use Boltzmann's, Gibbs, or Shannon's entropy expressions in the field of turbulence research. Boltzmann-Gibbs's theory of statistical mechanics is excellent in describing the thermal equilibrium of simple systems. Its foundation is additive Boltzmann-Gibbs entropy (see Chapter 3 for entropy definitions):

$$S_{BG} = -k\sum_{i=1}^{W} p_i ln p_i$$

where p_i is the probability that the system is in the state *i*, W is the total number of (microscopic) states, and k is a positive constant equal to the Boltzmann constant k_B in physics (when state i corresponds to energy E_i), while in informatics k = 1 leads to the Shannon information entropy S_I (when possibilities and their probabilities are counted to find missing information). The normalization condition requires: $\Sigma_i\, p_i = 1$. It is easy to verify that entropic additivity $S_{BG}(A+B) = S_{BG}(A) + S_{BG}(B)$ is satisfied for the system composed of two independent subsystems A and B (Tsallis 2019).

However, the additivity relationship fails in complex systems with strongly correlated elements or subsystems. The BG theory is then inadequate, and S_{BG} entropy expression is questionable for such systems. For instance, both classical and quantum turbulence involve long-range correlations. In quantum systems, there is a strong quantum entanglement of its subsystems. Such complex many-body interactions suggested the need for different entropy definition and a different approach for calculating entropy changes. Constantino Tsallis introduced in 1988 the generalization of Boltzmann-Gibbs statistics (Tsallis 1988) by postulating the entropy expression S_q:

$$S_q = \frac{k}{q-1}\{1 - \sum_{i=1}^{W} p_i^q\}$$

where q is the entropic index related to the degree of non-extensivity. Tsallis entropy is, in general, non-additive. The limit $q \rightarrow 1$ recovers the regular S_{BG} entropy. Extended thermodynamics, with its S_q entropy, is described as a solid foundation for the turbulence study (Egolf and Hutter 2018). When system elements experience strong or global correlations, the application of nonextensive statistical mechanics with $q \neq 1$ has a better chance to provide agreement with observations than standard Boltzmann-Gibs statistics. The indices q are determined by fitting when precise dynamics are unknown (Tsallis 2012).

Generalized statistical distributions for nonequilibrium dynamics have also been proposed by Kaniadakis (2001) and Beck and Cohen (2003). Corresponding partition functions and velocity distributions are then obtained in a stationary state by standard optimization (variational) techniques from statistical physics. From the knowledge of the **partition function**, one can predict all mean macroscopic quantities. Predicted distributions have good agreement with observed distributions. For instance, the Tsallis statistics has been successfully applied to many difficult problems encountered in nature and in artificial systems. Readers can find an excellent overview at the http://tsallis.cat.cbpf.br/biblio.htm site. The main advantage of Tsallis over Boltzmann statistics is the former's surprising ability to describe an out of equilibrium quasi-steady situation amenable to evolutionary development. Possible shortcomings of Tsallis and other superstatistics are changes in generalized entropy definitions that are adjusted through some entropic index to fit each special situation or process of interest. However, the Tsallis' q parameter has been extracted from the first principles in several cases (Tsallis 2012). We collected several application examples in Table 16.1 to show the wide range of q-parameter values used for modeling typical complex systems.

Table 16.1: A wide range of q-parameter values obtained from Tsallis' statistics.

System	Research topic	q-value	Reference
2-D turbulence	Subdiffusion Normal diffusion Superdiffusion	< 5/3 5/3 > 5/3	Egolf and Hutter 2018
Radio frequency ion trap	$^{136}Ba^+$ spatial distributions cooled by various buffer gases	1.03–1.87	DeVoe 2009
Drift turbulence in electron plasma	Velocity profile for pure-electron plasma in strong magnetic field	0.5	Boghosian 1996
Multifractality of the sizes of eddies	Fully developed classical turbulence	0.237	Arimitsu and Arimitsu 2000
Dislocations in defect turbulence	Undulation chaos in heated fluid layer	1.5	Daniels et al. 2004
Protein folding	Self-organized criticality and fractal dimensions of protein chains	1.74	Moret 2011
Migration of canine kidney cells	Collective motions of cells	1.2	Lin et al. 2020
Cell migration (*Hydra* cells)	Correlated type anomalous diffusion	1.5	Upadhyaya et al. 2001
Stellar clusters	Non-Gaussian velocity distribution	1–3	Carvalho et al. 2010
Galaxy clusters	Peculiar velocity of galaxy clusters	0.23	Lavagno et al. 1998
Whole universe	Universe evolution—dark energy	1.8–2.2	Sharma and Shrivastava 2021

The protein folding application mentioned in Table 16.1 deals with fractal dimensions (Moret 2011). That is not surprising because Tsallis generalization of Boltzmann-Gibbs statistics was inspired by multifractal concepts from the field of complex dynamics (Tsallis 1988). In his 1995 paper, Tsallis demonstrated that the $S_q = 0$ case is equivalent to $q = d_f/d$, where d is the space dimension, and d_f is the fractal dimension in that space. He also noticed that long-range interactions responsible for complex dynamics are likely to modify the role of extensive and intensive thermodynamic variables (see Chapter 3). These remarks help to understand the surprising success of Tsallis' thermostatistics in reproducing the dynamical pattern distributions in non-equilibrium physical situations, when thermostatics should not be applicable (Nauenberg 2003). Predicted and observed velocity distributions deviate from Maxwell-Boltzmann distribution, known as a Gaussian probability distribution for molecules' kinetic energy (Tsalis 2012, 2019).

Regarding turbulence studies, for highly developed turbulence, the distribution of velocities and accelerations along particle trajectories is non-Gaussian in classical (La Porta et al. 2001) and quantum turbulence (La Mantia et al. 2013). Fully developed classical turbulence with high Reynolds numbers (a convenient parameter for predicting flow patterns) corresponds to turbulent kinetic energy and high local dissipation close to the onset of chaos. La Porta et al. (2001) explained what it means for a mosquito caught in a moderate wind of no more than 5 ms^{-1}. Every 15 seconds, it will experience accelerations 15 times the acceleration of gravity (150 ms^{-2}). The smart mosquito would prefer to cling to any surface, be it a glade of grass or our nourishing skin.

Is Tsallis statistics preferable to Terrel Hill's nanothermodynamics when dealing with small but complex systems relevant for bioenergetics? Oddly enough, these two approaches are claimed to be equivalent (García-Morales et al. 2005). Hill's nanothermodynamics has the priority since it was conceived in 1962 and refined in his famous book "Thermodynamics of Small Systems" (Hill 1964) to take into account the energetic contributions of surface, size, and edge effects important only for small systems (Hill 2001a,b, Carrete et al. 2008, Guisbiers 2019). The transition from macro to micro or nano-world studies revealed how physical and chemical changes are affected by strong correlations. In small correlated systems, the whole is greater than the sum of its parts. That non-extensivity property can be considered by redefining the internal energy (Hill's approach) or redefining the entropy (Tsallis' approach). Common to both approaches is the insight that small system's thermodynamic properties are usually different in different environments (Hill 2001b). In terms of nanothermodynamic variables, a small system has more degrees of freedom. In particular, there are additional contributions to entropy or internal energy. The entropy will be larger because of interaction with the environment and fluctuations in N (elements, molecules, nanoparticles) (Hill 2001a).

Hill's thermodynamics approach is different from Tsallis's approach to thermostatics when thermodynamic forces and entropy production are considered. Entropy production is absent from thermostatics. The correct form of thermodynamic forces is controversial in Tsallis thermodynamics (Nauenberb 2003, García-Morales et al. 2005). Entropy production is defined as the entropy to time ratio in the long-time limit (Tsallis et al. 2005a). That definition does not need thermodynamic forces and fluxes (Fuentes et al. 2011). The criticism also focuses on phenomenological parametrization (fitting) to obtain the q-probability distribution. The distribution is then not grounded in basic physical principles regarding turbulence and the emergence of other dissipative structures in driven systems. Furthermore, there is no physical justification for maximizing Tsallis q-entropy in far-from-equilibrium situations (Nauenberg 2003). Fractal dimensions of metastable dynamic patterns are often alluring to scientists, but it does not warrant using thermostatistics as a replacement for nonequilibrium thermodynamics. On the other hand, the central concept from nonequilibrium thermodynamics—the entropy production—does not require any change in its bilinear force times flux expression when different entropy definitions are used. Indeed, Terrel Hill used the classical Boltzmann-Gibbs entropy despite his early introduction of non-extensivity in nanothermodynamics.

As far as I can tell, connecting Hill's approach to nanothermodynamics with Tsallis' is still a promising research avenue. Entropy is different from other extensive functions in its context-dependence and non-extensivity for small and strongly correlated systems with long-range interactions. However, for some complex, scale-free structures, the S_q entropy with $q \neq 1$ is the one which is extensive, and not the S_{GB} entropy (Tsallis 2005b). Understanding irreversible entropy changes may profit from the marriage between Hill's and Tsallis' approach. For a given ensemble of small systems, a future research focus can be the entropy production associated with elementary internal transitions driven by external forces. As we mentioned before in this book, the bilinear flux-force product named entropy production is proportional to the power input. Some cyclic pathways for the power output can enhance the power input, contributing to metastable energy storage and increased overall entropy production. An optimal power level output can be discovered when selected internal transitions are maximized for their entropy production. We can expect that both Tsallis and Boltzmann-Gibs entropy will increase and possibly reach the metastable state's maximal value. Only future research can answer the question of which entropy expression will better describe the optimized state.

16.18 The relevance of the least action principle in bioenergetics

The least action principle—the most general principle of physics—should also be relevant in chemistry and biology. Arto Annila proposed that everything is ultimately composed of the quanta of actions (Annila 2016). The maximum energy dissipation principle is the special case of the least action principle (Moroz 2012). The entropy production of a microscopic path is the part of the action integral, which is odd under time reversal (Dewar 2003, Bruers 2007b, Seifert 2012). When friction can be neglected, the least action principle in mechanics is equivalent to time-invariant conservation laws and equations of motion. Symmetries discovered by physicists implied corresponding conservation laws (**Noether's theorem**). When dissipation cannot be neglected in chemical and biological processes, the action integral must be modified by adding the irreversible action, but the least action principle still holds. Roldán et al. (2015) noticed that entropy production is inversely proportional to the minimal time needed to decide whether the process runs forward or backward in time. The thermodynamic arrow of time of a nonequilibrium stochastic process follows from the steady-state EP and vice versa.

The essential result in nonequilibrium situations is the generalized fluctuation-dissipation theorem involving entropy production (Crooks 1999, Seifert 2012). The theorem tells us that the pathways with positive entropy production are exponentially more probable than the pathways associated with the time-reversed entropy production. Thus, a firm foundation for bioenergetics is some form of the maximum entropy production principle that takes into account all of the constraints and regulatory controls. In simple models of multistable states, the favored steady state is the one with the highest entropy production (Endres 2017). The physical foundation for prebiotic and biological evolution is an appropriately modified last action principle that governs both evolutions. It suggests the maximum entropy increase (Wang 2006). The evolution of chemical and biological complexity accelerated the entropy increase in the universe. The evolution of self-organized systems far from equilibrium has been observed as the evidence for the maximum entropy production principle (Belkin et al. 2015).

16.19 How bioenergetics bridges life to the universe?

Do we have cogent thoughts about energy transformations inherent to bioenergetics at the end of this book? There is no life without dissipation. Dissipation is associated with each biochemical pathway and each free-energy transformation in living cells. Since dissipation vanishes in thermostatic situations when thermodynamic equilibrium is established, we can certainly question the thermostatics' relevance for describing any bioenergetic process. The rarely mentioned hallmark of bioenergetics is broken time-reversal invariance and broken detailed balance (Battle et al. 2016). Dissipation breaks both of these conditions. At constant or nearly constant temperatures, dissipation can be calculated as entropy production. Entropy production must reach some minimal level compatible with the persistence of life. That is no problem in the presence of enduring force gradients. A natural tendency to dissipate all force gradients is obviously used by living cells. Flows are produced, and biological evolution organized them into biochemical cycles. Force times flow products are the input power connecting life to its ambient. That bilinear form (see Chapter 3) also describes the input power conversion into the output entropy production when life is regarded as a black box energy transformer. But life is not an immutable black box. It better resembles a controlled explosion of chain reactions to dissipate energy. Based on the results presented in this book, we can say that life is a self-controlled structure-building explosion of improving the harvesting of available power to convert it into the maximal possible dissipation. Life transforms potential into real entropy production by continually expanding and changing itself.

Such a general overview of bioenergetics is certainly disappointing to any engineer and any biologist. Engineers will complain that entropy generation is useless, and biologists will complain about the absence of insight into how biological structures could arise from so described energy transformations. Both criticisms are valid. Cells perform mechanical work, and internal structures are built by shifting some flows into the anabolic biosynthetic pathways. It may well be that mechanical and biosynthetic work are minor side reactions regarding overall energy flow redistribution through living systems. Still we would like to understand how life channels energy into productive pathways.

Throughout this book, we tried to illuminate through examples of how energy channeling occurs. It is time now to gather all threads together into a common picture. Sometimes life puts brakes on entropy production rates, and sometimes it does not. The dissection of all entropy production contributions helped to distinguish the former from the latter scenario. We focused on the contribution of the most elementary step to entropy production and examined which steps are selected by bioenergetic nanomachines to eliminate all brakes for maximum dissipation in these elementary processes. That is how we achieved the central result of this monograph. Selected transitions by biological evolution are those that accomplish two coupled goals at the same time. One is the primary outcome of increasing overall entropy production. Increased internal entropy is quickly and efficiently exported to the environment, thus contributing to the universe's entropy increase. That can be the consequence of the proposed Second Law generalization about the fastest possible entropy increase. The other goal is minor but vital. It consists of channeling input power into cellular structures and activities. These essential pathways lead to the cascade of free energy conversions with brakes, rate-limiting steps, and generally postponed dissipation until functional power output has run its biological course.

Enzyme evolution, coupled to the catalytic efficiency and energy conversion increases, is also connected to the increased entropy production (Juretić et al. 2019a,b). In that sense, too, the bioenergetics of present-day cells and organisms bridges the life-universe boundary. We have somewhat neglected ribonucleic acids during the cell cycle due to the traditional focus of bioenergetics on membrane enzymes performing active transport. Let us correct that omission. The hallmark of life we did not yet mention in this book is the dimension expansion-contraction cycle. The free energy of dimensionless quanta (photons) is partially transformed into the one-dimensional polypeptide and nucleic acid chains. A much larger part of photon free energy is converted by life into the spreading wave of low energy photons (heat radiation). One-dimensional chains of lipids, amino acids, and nucleotides are anything but random. They have an in-built tendency to form beautiful 2D and 3D macromolecular entities such as membranes and polypeptide sheets. These events' precise timing adds the time as the fourth dimension, and the regulation as the other of life's rarely mentioned hallmarks. Replication involves rejuvenation and contraction to the one-dimensional objects—the offspring DNA molecules, with subsequent repetition of multi-dimensional expansion. Hence, physics changes the focus to the invisible but major driving events for dimensional expansion: absorbed work, entropy production, and exported excess entropy, accompanied by the space-time expansion-contraction cycles. The cycles are far from the reversible repetition of everything. They occur in far-from-equilibrium situations and would not be possible without external drives fueling the translation-transcription dance of DNA, RNA, and protein molecules. The contraction part serves to prepare the expansion part of the cycle, spreading away in ever stronger waves through the space-time continuum. In the language of entropy changes, life accelerates the spread of entropy increase in those parts of the universe where it exists. Biological evolution is tightly coupled to such bioenergetic processes that speed up the universal thermodynamic evolution. Life firmly reigns over all of the available dissipative pathways. When needed, it can also restrict dissipation close to the theoretical lower limit compatible with its survival. We should humbly admit that we have insight only about a small part of the whole picture of how bioenergetics forms bridges across life, death, and the universe.

Definitions and explanations from Chapter 16:

The **Anaphase stage of the cell cycle** is the stage when replicated chromosomes are split and separated in preparation for cell division.

Estivation is a state of a lowered metabolic rate during summer time desiccation and damages caused by high temperatures. Although many different animals practice dormancy and inactivity during stressful dry and hot periods, biochemists are aware of high challenges that need to be overcome. Biochemical reactions are greatly speeded up with even a slight rise in temperature. Instead, estivating animals reduce the turnover of macromolecules and stabilize them.

Exergy is the maximum amount of work that can be extracted from the open system during the interaction with an ideal reference environment. It is not conserved as energy, despite being measured in the same units (Joules), and it is not the property of the system. Exergy transfers are evaluated at the system's boundary. Other names for exergy are available energy, usable energy, utilizable energy, work capability, or work potential. Common to the Gibbs free energy is that exergy is also lost due to irreversibility and dissipation. However, exergy is not the thermodynamic potential of the system, like free energy or enthalpy.

Fractal patterns or objects exhibit self-similarity—their structure is invariant under scaling.

Myosin II motor proteins self-assemble into mini filaments and interact with flexible actin filaments. In the ATP presence, myosin II motors move toward the "plus end" of active filaments. Cell division, contraction, and motility of eukaryotic cells would be hardly possible in the absence of myosin II motors because these proteins are active crosslinking agents using ATP hydrolysis to generate necessary force needed for these processes. Myosin motors operate far from thermodynamic equilibrium. That is the reason why they can generate forces on F-actin (filamentous actin composed of actin proteins monomers or dimmers). Actin filaments in non-muscle cells are constantly being polymerized from one end (plus end) and depolymerized from the other (minus end). In the absence of ATP, the rigor state is established when myosin remains attached to actin filaments.

Na^+, K^+-ATPase is an integral membrane enzyme pump for sodium and potassium ions. It transports three Na^+ ions out and two K^+ ions in the cell for each hydrolyzed ATP molecule. Na^+, K^+-ATPase is mainly responsible for creating the membrane potential in animal cells, depolarization, repolarisation, and the regulation of cytoplasmic ionic composition. It is the major free-energy consumer of the brain. Up to 70% of produced ATP in brain cells are used to drive the Na^+, K^+-ATPase activity. For the review of plasma-membrane ion pumps dependent on ATP hydrolysis, see Morth et al. (2011).

Noether's theorem was derived by Emmy Noether (1882–1935), the descendent of a distinguished family of German Jews and one of the best lady mathematicians that ever lived. In the simplified description, the theorem states: if a system has a continuous symmetry generated by local action, then there are corresponding quantities that are conserved with time. The theorem provided the common foundation for all of the known physical conservation laws. Frau Noether established with her theorem the research program that will last as long as science exists—finding the system's symmetries with corresponding conserved quantities for all systems evolving according to the principle of least action. Gender discrimination during her adult life in the early 20th century prevented her from getting a salary and professorship despite all of her groundbreaking contributions to mathematics and physics.

The **partition function** in statistical mechanics encodes the partitioning of probabilities among different microstates. Its numerical value depends on the degrees of freedom a system has. When partition function is known in classical or quantum physics, mean values of the system's total energy, free energy, entropy, pressure, and other thermodynamic variables can all be calculated.

State 3 respiration is the active metabolic state of rapidly respiring mitochondria. Fast ATP synthesis occurs in the presence of a substrate, ADP, inorganic phosphate, and oxygen.

State 4 respiration is the resting metabolic state of mitochondria with decreased respiration because enough ADP has been converted into ATP.

Vacuolar (H^+)-ATPase has the proton-pumping function driven by ATP hydrolysis. It is the integral membrane protein located mainly in cellular vacuoles (hence the V-ATPase notation), but also in the cell-surface membrane (plasma membrane). In lysosomes, the enzyme provides the acidic environment essential for protein degradation and recycling. Plasma-membrane located enzymes perform active proton transport from the cytoplasm to the extracellular space. The acidification of the extracellular space is, however, beneficial for tumor cell survival and invasiveness. More details about V-ATPases can be found in the recent McGuire et al. (2017) review.

References

Abudukelimu, A., Mondeel, T.D.G.A., Barberis, M. and Westerhoff, H.V. Learning to read and write in evolution: from static pseudoenzymes and pseudosignalers to dynamic gear shifters. Biochem. Soc. Trans. 45(2017): 635–652.

Anglin, J. Quantum optics: Particles of light. Nature 468(2010): 517–518.

Annila, A. Natural thermodynamics. Physica A 444(2016): 843–852.

Arimitsu, T. and Arimitsu, N. Analysis of fully developed turbulence in terms of Tsallis statistics Phys. Rev. E 61(2000): 3237–3240.

Arp, T.B., Kistner-Morris. J., Aji, V., Cogdell, R.J., van Grondelle, R. and Gabor, N.M. Quieting a noisy antenna reproduces photosynthetic light-harvesting spectra. Science 368(2020): 1490–1495.

Bak, P. 1996. How Nature Works: The Science of Self-Organized Criticality. Springer-Verlag, New York, NY, USA.

Ban, T. and Shigeta, K. Thermodynamic analysis of thermal convection based on entropy production. Sci. Rep. 9(2019): 10368. doi.org/10.1038/s41598-019-46921-2.

Barato, A.C. and Seifert, U. Thermodynamic uncertainty relation for biomolecular processes. Phys. Rev. Lett. 114(2015): 158101. doi: 10.1103/physrevlett.114.158101.

Barish, B.C. Nobel Lecture: LIGO and gravitational waves II. Rev. Mod. Phys. 90(2018): 040502. doi: 10.1103/RevModPhys.90.040502.

Battle, C., Broedersz, C.P., Fakhri, N., Geyer, V.F., Howard, J., Schmidt, C.F. and MacKintosh, F.C. Broken detailed balance at mesoscopic scales in active biological systems. Science 352(2016): 604–607.

Beck, C. and Cohen, E.G.D. Superstatistics. Physica A 322(2003): 267–275.

Bejan, A. Constructal-theory network of conducting paths for cooling a heat generating volume. Int. J. Heat Mass Transfer 40(1997): 799–816 (published on 1 Nov. 1996).

Bejan, A. Fundamentals of exergy analysis, entropy generation minimization, and the generation of flow architecture. Int. J. Energy Res. 26(2002): 545–565.

Bejan, A. and Lorente, S. The constructal law and the thermodynamics of flow systems with configuration. Int. J. Heat Mass Transfer 47(2004): 3203–3214.

Bejan, A. and Lorente, S. The constructal law of design and evolution in nature. Phil. Trans. R. Soc. B 365(2010): 1335–1347.

Bejan, A. 2016a. The Physics of Life: The Evolution of Everything. St. Martin's Press, New York, NY, USA.

Bejan, A. Constructal thermodynamics. Int. J. Heat Technol. 34(2016b), Special Issue 1: S1–S8. http://dx.doi.org/10.18280/ijht.34S101.

Bejan, A. Evolution in thermodynamics. Appl. Phys. Rev. 4(2017): 011305. doi: 10.1063/1.4978611.

Belkin, A., Hubler, A. and Bezryadin, A. Self-assembled wiggling nano-structures and the principle of maximum entropy production. Sci. Rep. 5(2015): 8323. doi: 10.1038/srep08323.

Belmonte, J.M., Leptin, M. and Nédélec, F. A theory that predicts behaviors of disordered cytoskeletal networks. Mol. Syst. Biol. 13(2017): 941. doi: 10.15252/msb.20177796.

Beretta, G.P. The fourth law of thermodynamics: steepest entropy ascent. Phil. Trans. R. Soc. A 378(2020): 20190168. doi.org/10.1098/rsta.2019.0168.

Berg, J.M., Tymoczko, J.L. and Stryer, L. 2002. Biochemistry, 5th Edition. Freeman, New York, NY, USA.

Bidone, T.C., Jung, W., Maruri, D., Borau, C., Kamm, R.D. and Kim, T. Morphological transformation and force generation of active cytoskeletal networks. PLoS Comput. Biol. 13(2017): e1005277. doi: 10.1371/journal.pcbi.1005277.

Bier, M., Lisowski, B. and Gudowska-Nowak, E. Phase transitions and entropies for synchronizing oscillators. Phys. Rev. E 93(2016): 012143. doi: 10.1103/physreve.93.012143.

Boghosian, B.M. Thermodynamic description of the relaxation of two-dimensional turbulence using Tsallis statistics. Phys. Rev. E 53(1996): 4754–4763.

Bordel, S. and Nielsen, J. Identification of flux control in metabolic networks using non-equilibrium thermodynamics. Metab. Eng. 12(2010): 369–377.

Bousso, R., Harnik, R., Kribs, G.D. and Perez, G. Predicting the cosmological constant from the causal entropic principle. Phys. Rev. D 76(2007): 043513. doi: 10.1103/physrevd.76.043513.

Bousso, R. and Harnik, R. Entropic landscape. Phys. Rev. D 82(2010): 123523. doi: 10.1103/physrevd.82.123523.

Bowman, J.C., Hud, N.V. and Williams, L.D. The ribosome challenge to the RNA world. J. Mol. Evol. 80(2015): 143–161.

Bowman, J.C., Petrov, A.S., Frenkel-Pinter, M., Penev, P.I. and Williams, L.D. Root of the tree: the significance, evolution, and origins of the ribosome. Chem. Rev. 120(2020): 4848–4878.

Bradley, J.A., Arndt, S., Amend, J.P., Burwicz, E., Dale, A.W., Egger, M. et al. Widespread energy limitation to life in global subseafloor sediments. Sci. Adv. 6(2020): eaba0697.

Branscomb, E., Biancalani, T., Goldenfeld, N. and Russell, M. Escapement mechanisms and the conversion of disequilibria; the engines of creation. Phys. Rep. 677(2017): 1–60.

Bratanov, V., Jenko, F. and Frey, E. New class of turbulence in active fluids. Proc. Natl. Acad. Sci. USA 112(2015): 15048–15053.

Bremer, A., Millonig, R.C., Sütterlin, R., Engel, A., Pollard, T.D. and Aebi, U. The structural basis for the intrinsic disorder of the actin filament: the "lateral slipping" model. J. Cell Biol. 115(1991): 689–703.

Brown, A.I. and Sivak, D.A. Allocating dissipation across a molecular machine cycle to maximize flux. Proc. Natl. Acad. Sci. USA 114(2017): 11057–11062.

Bruers, S. 2007a. Energy and Ecology. On entropy production and the analogy between fluid, climate and ecosystems (PhD Thesis). Katholieke Universiteit Leuven, Leuven, Belgium.

Bruers, S. A discussion on maximum entropy production and information theory. J. Phys. A: Math. Theor. 40(2007b): 7441–7450.

Candelas, P. and Sciama, D.W. Irreversible thermodynamics of black holes. Phys. Rev. Lett. 38(1977): 1372–1375.

Carny, O. and Gazit, E. Creating prebiotic sanctuary: Self-assembling supramolecular peptide structures bind and stabilize RNA. Orig. Life Evol. Biosph. 41(2011): 121–132.

Carvalho, J.C., Silva, R., do Nascimento jr., J.D., Soares, B.B. and De Medeiros, J.R. Observational measurement of open stellar clusters: A test of Kaniadakis and Tsallis statistics. EPL 91(2010): 69002. doi: 10.1209/0295-5075/91/69002.

Carrete, J., Varela, L.M. and Gallego, L.J. Nonequilibrium nanothermodynamics. Phys. Rev. E 77(2008): 022102. doi: 10.1103/PhysRevE.77.022102.

Chan, M.A., Hinman, N.W., Potter-McIntyre, S.L., Schubert, K.E., Gillams, R.J., Awramik, S.M. et al. Deciphering biosignatures in planetary contexts. Astrobiology 19(2019): 1075–1102.

Chavanis, P.H., Sommeria, J. and Robert, R. Statistical mechanics of two-dimensional vortices and collisionless stellar systems. Astrophys. J. 471(1996): 385–399.

Chen, C., Liu, S., Shi, X.-Q., Chaté, H. and Wu, Y. Weak synchronization and large-scale collective oscillation in dense bacterial suspensions. Nature 542(2017): 210–214.

Crooks, G.E. Entropy production fluctuation theorem and the nonequilibrium work relation for free energy differences. Phys. Rev. E 60(1999): 2721–2726.

Daniels, K.E., Beck, C. and Bodenschatz, E. Defect turbulence and generalized statistical mechanics. Physica D 193(2004): 208–217.

Dasbiswas, K., Hu, S., Schnorrer, F., Safran, S.A. and Bershadsky, A.D. Ordering of myosin II filaments driven by mechanical forces: experiments and theory. Phil. Trans. R. Soc. B 373(2018): 20170114. doi.org/10.1098/rstb.2017.0114.

Davies, P.C.W. 1978. Space—time singularities in cosmology and black hole evaporations. pp 74–93. *In*: Fraser, J.T., Lawrence, N. and Park, D.A. (eds.). The Study of Time III. Springer, New York, NY, USA. https://doi.org/10.1007/978-1-4612-6287-9_4.

Dawkins, R. 1976. The Selfish Gene. Oxford University Press, Oxford, UK.

De, S. and Klajn, R. Dissipative self-assembly driven by the consumption of chemical fuels. Adv. Mater. (2018): 1706750. doi: 10.1002/adma.201706750.

DeVoe, R.G. Power-law distributions for a trapped ion interacting with a classical buffer gas. Phys. Rev. Lett. 102(2009): 063001. doi: 10.1103/physrevlett.102.063001.

Delbrück, M. A physicist looks at biology. Trans. Conn. Acad. Arts Sci. 38(1949): 173–190. Reprinted In: Cairns, J., Stent, G.S. and Watson, J.D. (eds.). 1966. Phage and the Origins of Molecular Biology. Cold Spring Harbor Laboratory of Quantitative Biology. Cold Spring Harbor Laboratory Press, N.Y., USA.

Delbrück, M. A physicist's renewed look at biology: Twenty years later. Science 168(1970): 1312–1315.

Dewar, R.C. Information theory explanation of the fluctuation theorem, maximum entropy production and self-organized criticality in non-equilibrium stationary states. J. Phys. A: Math. Gen. 36(2003): 631–641.

Dewar, R.C. Maximum entropy production and plant optimization theories. Phil. Trans. R. Soc. B 365(2010): 1429–1435.

Dewar, R.C., Juretić, D. and Županović, P. The functional design of the rotary enzyme ATP synthase is consistent with maximum entropy production. Chem. Phys. Lett. 430(2006): 177–182.

Dewar, R.C., Lineweaver, C.H., Niven, R.K. and Regenauer-Lieb, K. (eds.). 2014. Beyond the Second Law. Springer-Verlag, Berlin, Heidelberg, Germany.

Djillani, A., Mazella, J., Heurteaux, C. and Borsotto, M. Role of TREK-1 in health and disease, focus on the central nervous system. Front. Pharmacol. 10(2019): 379. doi: 10.3389/fphar.2019.00379.

Dobovišek, A., Županović, P., Brumen, M., Bonačić Lošić, Ž., Kuić, D. and Juretić, D. Enzyme kinetics and the maximum entropy production principle. Biophys. Chem. 154(2011): 49–55.

Dombrowski, C., Cisneros, L., Chatkaew, S., Goldstein, R.E. and Kessler, J.O. Self-concentration and large-scale coherence in bacterial dynamics. Phys. Rev. Lett. 93(2004): 098103. doi: 10.1103/PhysRevLett.93.098103.

Doostmohammadi, A., Shendruk, T.N., Thijssen, K. and Yeomans, J.M. Onset of meso-scale turbulence in active nematics. Nat. Commun. 8(2017): 15326. doi: 10.1038/ncomms15326.

Doostmohammadi, A., Ignés-Mullol, J., Yeomans, J.M. and Sagués, F. Active nematics. Nat. Commun. 9(2018): 3246. doi: 10.1038/s41467-018-05666-8.

Duclos, G., Adkins, R., Banerjee, D., Peterson, M.S.E., Varghese, M., Kolvin, I. et al. Topological structure and dynamics of three-dimensional active nematics. Science 367(2020): 1120–1124.

Duffy, C.D.P. The simplicity of robust light harvesting. Science 368(2020): 1427–1428.

Eames, I. and Flor, J.B. New developments in understanding interfacial processes in turbulent flows. Phil. Trans. R. Soc. A 369(2011): 702–705.

Egolf, P.W. and Hutter, K. Tsallis extended thermodynamics applied to 2-d turbulence: Lévy statistics and q-fractional generalized Kraichnanian energy and enstrophy spectra. Entropy 20(2018): 109. doi: 10.3390/e20020109.

Ehrmann, A., Nguyen, B. and Seifert, U. Interlinked GTPase cascades provide a motif for both robust switches and oscillators. J. R. Soc. Interface 16(2019): 20190198. doi.org/10.1098/rsif.2019.0198.

Einstein, A. 1931. Living Philosophies. Simon and Schuster, New York, NY, USA.

Endres, R.G. Bistability: Requirements on cell-volume, protein diffusion, and thermodynamics. PLoS ONE 10(2015): e0121681. doi: 10.1371/journal.pone.0121681.

Endres, R.G. Entropy production selects nonequilibrium states in multistable systems. Sci. Rep. 7(2017): 14437. doi: 10.1038/s41598-017-14485-8.

England, J.L. Statistical physics of self-replication. J. Chem. Phys. 139(2013). doi: 10.1063/1.4818538.

England, J.L. Dissipative adaptation in driven self-assembly. Nat. Nanotechnol. 10(2015): 919–923.

Esposito, M., Lindenberg, K. and Van den Broeck, C. Entropy production as correlation between system and reservoir. New J. Phys. 12(2010): 013013. doi: 10.1088/1367-2630/12/1/013013.

Étienne, J., Fouchard, J., Mitrossilis, D., Bufi, N., Durand-Smet, P. and Asnacios, A. Cells as liquid motors: mechanosensitivity emerges from collective dynamics of actomyosin cortex. Proc. Natl. Acad. Sci. USA 112(2015): 2740–1245.

Fang, X. and Wang, J. Nonequilibrium thermodynamics in cell biology: Extending equilibrium formalism to cover living systems. Annu. Rev. Biophys. 49(2020): 227–246.

Floyd, C., Papoian, G.A. and Jarzynski, C. Quantifying dissipation in actomyosin networks. Interface Focus 9(2019): 20180078. doi.org/10.1098/rsfs.2018.0078.

Foster, P.J., Fürthauer, S., Shelley, M.J. and Needleman, D.J. Active contraction of microtubule networks. eLife 4(2015): e10837. doi: 10.7554/eLife.10837.

Franklin, O., Johansson, J., Dewar, R.C., Dieckmann, U., McMurtrie, R.E., Brännström, Å. et al. Modeling carbon allocation in trees: a search for principles. Tree Physiol. 32(2012): 648–666.

Franks, N.P. General anaesthesia: from molecular targets to neuronal pathways of sleep and arousal. Nat. Rev. Neurosci. 9(2008): 370–386.

Frisch, U. 2004. Turbulence. Cambridge University Press, Cambridge, UK.

Fuentes, M.A., Sato, Y. and Tsallis, C. Sensitivity to initial conditions, entropy production, and escape rate at the onset of chaos. Phys. Lett. A 375(2011): 2988–2991.

Garcia, S., Hannezo, E., Elgeti, J., Joanny, J.-F., Silberzan, P. and Gov, N.S. Physics of active jamming during collective cellular motion in a monolayer. Proc. Natl. Acad. Sci. USA. 112(2015): 15314–15319.

García-Morales, V., Cervera, J. and Pellicer, J. Correct thermodynamic forces in Tsallis thermodynamics: connection with Hill nanothermodynamics. Phys. Lett. A 336(2005): 82–88.

Geeves, M.A. Review: The ATPase mechanism of myosin and actomyosin. Biopolymers 105(2016): 483–491.

Gemmer, J., Michel, M. and Mahler, G. 2009. Quantum Thermodynamics. Emergence of Thermodynamic Behavior Within Composite Quantum Systems (Second Edition). Springer, Berlin, Germany.

Ghosh, A., Xu, W., Gupta, N. and Gracias, D.H. Active matter therapeutics. Nano Today (April 2020): 100836. doi: 10.1016/j.nantod.2019.100836.

Guisbiers, G. Advances in thermodynamic modelling of nanoparticles. Adv. Phys. X 4(2019): 1668299. doi.org/10.1080/23746149.2019.1668299.

Hawking, S.W. Black holes and thermodynamics. Phys. Rev. D 13(1976): 191–197.

Hawking, S.W. 1988. A Brief History of Time. Bantam Books, New York, N.Y., USA.

Hazen, R.M., Grew, E.S., Downs, R.T., Golden, J. and Hystad, G. Mineral ecology: Chance and necessity in the mineral diversity of terrestrial planets. Can. Mineral. 53(2015): 295–324.

Hazen, R.M. Chance, necessity and the origins of life: a physical sciences perspective. Phil. Trans. R. Soc. A 375(2017): 20160353. doi.org/10.1098/rsta.2016.0353.

Henkes, S., Kostanjevec, K., Collinson, J.M., Sknepnek, R. and Bertin, E. Dense active matter model of motion patterns in confluent cell monolayers. Nat. Commun. 11(2020): 1405. doi: 10.1038/s41467-020-15164-5.

Hill, A.V. The heat of shortening and the dynamic constants of muscle. Proc. R. Soc. Lond. B Biol. Sci. 126(1938): 136–195.

Hill, T.L. (1964). Thermodynamics of Small Systems. Dover Publications, New York, NY, USA.

Hill, T.L. Perspective: Nanothermodynamics. Nano Lett. 1(2001a): 111–112.

Hill, T.L. A different approach to nanothermodynamics. Nano Lett. 1(2001b): 273–275.

Hobson, M.P., Efstathiou, G.P. and Lasenby, A.N. 2006. General Relativity: An Introduction for Physicists, p. 187. Cambridge University Press, Cambridge, UK.

Hoehler, T.M. and Jørgensen, B.B. Microbial life under extreme energy limitation. Nat. Rev. Microbiol. 11(2013): 83–94.

Hoffmann, P.M. How molecular motors extract order from chaos. Rep. Prog. Phys. 79(2016): 032601. doi: 10.1088/0034-4885/79/3/032601.

Hopfield, J.J., Feinstein, D.I. and Palmer, R.G. "Unlearning" has a stabilizing effect in collective memories. Nature 304(1983): 158–159.

Horowitz, J.M. and England, J.L. Spontaneous fine-tuning to environment in many-species chemical reaction networks. Proc. Natl. Acad. Sci. USA 114(2017): 7565–7570.

Horowitz, J.M. and Gingrich, T.R. Thermodynamic uncertainty relations constrain non-equilibrium fluctuations. Nat. Phys. 16(2020): 15–20.

Huxley, A.F. Muscle structure and theories of contraction. Prog. Biophys. Biophysical Chem. 7(1957): 255–318.

Hystad, G., Downs, R.T., Grew, E.S. and Hazen, R.M. Statistical analysis of mineral diversity and distribution: Earth's mineralogy is unique. Earth Planet. Sci. Lett. 426(2015): 154–157.

Jacobs, D.J., Trivedi, D., David, C. and Yengo, C.M. Kinetics and thermodynamics of the rate-limiting conformational change in the actomyosin V mechanochemical cycle. J. Mol. Biol. 407(2011): 716–730.

Jarzynski, C. Hamiltonian derivation of a detailed fluctuation theorem. J. Stat. Phys. 98(2000): 77–102.

Jaynes, E.T. The minimum entropy production principle. Ann. Rev. Phys. Chem. 31(1980): 579–601.

Jennaro, T.S., Beaty, M.R., Kurt-Yilmaz, N., Luskin, B.L. and Cavagnero, S. Burial of nonpolar surface area and thermodynamic stabilization of globins as a function of chain elongation. Proteins 82(2014): 2318–2331.

Juretić, D., Bonačić Lošić, Ž., Kuić, D., Simunić, J. and Dobovišek, A. The maximum entropy production requirement for proton transfers enhances catalytic efficiency for β-lactamases. Biophys. Chem. 244(2019a): 11–21.

Juretić, D., Simunić, J. and Bonačić Lošić, Ž. Maximum entropy production theorem for transitions between enzyme functional states and its applications. Entropy 21(2019b): 743. doi: 10.3390/e21080743.

Kachman, T., Owen, J.A. and England, J.L. Self-organized resonance during search of a diverse chemical space. Phys. Rev. Lett. 119(2017): 038001. doi: 10.1103/PhysRevLett.119.038001.

Kaniadakis, G. Non-linear kinetics underlying generalized statistics. Physica A 296(2001): 405–425.

Karsenti, E. Self-organization in cell biology: a brief history. Nat. Rev. Mol. Cell Biol. 9(2008): 255–262.

Keber, F.C., Loiseau, E., Sanchez, T., DeCamp, S.J., Giomi, L., Bowick, M.J. et al. Topology and dynamics of active nematic vesicles. Science 345(2014): 1135–1139.

Kirwan Jr, A.D. Intrinsic photon entropy? The darkside of light. Int. J. Eng. Sci. 42(2004): 725–734.

Kitano, H. Systems biology: a brief overview. Science 295(2002): 1662–1664.

Kleidon, A. Nonequilibrium thermodynamics and maximum entropy production in the Earth system. Applications and implications. Naturwissenschaften 96(2009): 653–677.

Kleidon, A., Zehe, E., Ehret, U. and Scherer, U. 2014. Earth system dynamics beyond the Second Law: Maximum power limits, dissipative structures, and planetary interactions. pp. 163–182. *In*: Dewar, R.C., Lineweaver, C.H. and Regenauer-Lieb, K. (eds.). Beyond the Second Law. Entropy Production of Non-equilibrium Systems. Springer-Verlag, Berlin, Germany.

Kondepudi, D.K., De Bari, B. and Dixon, J.A. Dissipative structures, organisms and evolution. Entropy 22(2020): 1305. doi: 10.3390/e22111305.

La Mantia, M., Duda, D., Rotter, M. and Skrbek, L. Lagrangian accelerations of particles in superfluid turbulence. J. Fluid. Mech. 717(2013): R9. doi: 10.1017/jfm.2013.31.

La Porta, A., Voth, G.A., Crawford, A.M., Alexander, J. and Bodenschatz, E. Fluid particle accelerations in fully developed turbulence. Nature 409(2001): 1017–1019.

La Porta, C.A.M. and Zapperi, S. 2019. Cell Migrations: Causes and Functions. Springer Nature Switzerland AG, Cham, Switzerland.

Lage, J.L. et al. Professor Adrian Bejan on his 60th birthday. Int. J. Heat Mass Transfer 51(2008): 5759–5761.

Lage, J.L. et al. Celebration of Professor Adrian Bejan on his 70th birthday. Int. J. Heat Mass Transfer 126(2018): 1377–1378.

Lam, B.R., Barr, C.R., Rowe, A.R. and Nealson, K.H. Differences in applied redox potential on cathodes enrich for diverse electrochemically active microbial isolates from a marine sediment. Front. Microbiol. 10(2019): 1979. doi: 10.3389/fmicb.2019.01979.

Lan, G., Sartori, P., Neumann, S., Sourjik, V. and Tu, Y. The energy-speed-accuracy tradeoff in sensory adaptation. Nat. Phys. 8(2012): 422–428.

LaRowe, D.E. and Amend, J.P. Power limits for microbial life. Front. Microbiol. 6(2015): 718. doi: 10.3389/fmicb.2015.00718.

Lavagno, A., Kaniadakis, G., Rego-Monteiro, M., Quarati, P. and Tsallis, C. Non-extensive thermostatistical approach of the peculiar velocity function of galaxy clusters. Astro. Lett. and Communications 35(1998): 449–455.

Le Goff, T., Liebchen, B. and Marenduzzo, D. Actomyosin contraction induces in-bulk motility of cells and droplets. Biophys J. 119(2020): 1025–1032.

Lenz, M., Thoresen, T., Gardel, M.L. and Dinner, A.R. Contractile units in disordered actomyosin bundles arise from F-Actin buckling. Phys. Rev. Lett. 108(2012): 238107. doi: 10.1103/PhysRevLett.108.238107.

Lenz, M. Reversal of contractility as a signature of self-organization in cytoskeletal bundles. eLife 9(2020): e51751. doi.org/10.7554/eLife.51751.

Lever, M.A., Rouxel, O., Alt, J.C., Shimizu, N., Ono, S., Coggon, R.M. et al. Evidence for microbial carbon and sulfur cycling in deeply buried ridge flank basalt. Science 339(2013): 1305–1308.

Lin, S.-Z., Chen, P.-C., Guan, L.-Y., Shao, Y., Hao, Y.-K., Li, Q. et al. Universal statistical laws for the velocities of collective migrating cells. Adv. Biosys. 4(2020): 2000065. doi: 10.1002/adbi.202000065.

Lineweaver, C.H. 2014. The entropy of the universe and the maximum entropy production principle. Chapter 22, pp. 415–428. *In*: Dewar, R.C., Lineweaver, C.H., Niven, R.K. and Regenauer-Lieb, K. (eds.). Beyond the Second Law. Springer-Verlag, Berlin, Germany.

Linsmeier, I., Banerjee, S., Oakes, P.W., Jung, W., Kim, T. and Murrell, M.P. Disordered actomyosin networks are sufficient to produce cooperative and telescopic contractility. Nat. Commun. 7(2016): 12615. doi: 10.1038/ncomms12615.

Lolicato, M., Arrigoni, C., Mori, T., Sekioka, Y., Bryant, C., Clark, K.A. et al. Jr. K2P2.1 (TREK-1)–activator complexes reveal a cryptic selectivity filter binding site. Nature 547(2017): 364–368.

Lovelock, J. 2009. The Vanishing Face of Gaia: A Final Warning. Basic Books, New York, NY, USA.

Lucia, U. Maximum or minimum entropy generation for open systems? Physica A 391(2012): 3392–3398.

Lucia, U. and Maino, G. Entropy generation in biophysical systems. EPL 101(2013): 56002. doi: 10.1209/0295-5075/101/56002.

Lucia, U. and Sciubba, E. From Lotka to the entropy generation approach. Physica A 392(2013): 3634–3639.

Lucia, U. The Gouy-Stodola theorem in bioenergetic analysis of living systems (Irreversibility in bioenergetics of living systems). Energies 7(2014): 5717–5739.

Lucia, U., Ponzetto, A. and Deisboeck, T.S. A thermo-physical analysis of the proton pump vacuolar-ATPase: the constructal approach. Sci. Rep. 4(2014): 6763. doi: 10.1038/srep06763.

Lucia, U. and Deisboeck, T.S. The importance of ion fluxes for cancer proliferation and metastasis: A thermodynamic analysis. J. Theor. Biol. 445(2018): 1–8. doi: 10.1016/j.jtbi.2018.02.019.

Lucia, U. and Grisolia, G. Time: a constructal viewpoint & its consequences. Sci. Rep. 9(2019): 10454. doi.org/10.1038/s41598-019-46980-5.

Ma, Z., Turrigiano, G.G., Wessel, R. and Hengen, K.B. Cortical circuit dynamics are homeostatically tuned to criticality *in vivo*. Neuron 104(2019): doi: 10.1016/j.neuron.2019.08.031.

Maldacena, J. Black holes and quantum information. Nat. Rev. Phys. 2(2020): 123–125.

Mandal, D., Klymko, K. and DeWeese, M.R. Entropy production and fluctuation theorems for active matter. Phys. Rev. Lett. 119(2017): 258001. doi: 10.1103/PhysRevLett.119.258001.

Maroudas-Sacks, Y., Garion, L., Shani-Zerbib, L., Livshits, A., Braun, E. and Keren, K. Topological defects in the nematic order of actin fibres as organization centres of *Hydra* morphogenesis. Nat. Phys. (2020): doi.org/10.1038/s41567-020-01083-1.

Martin, J. Everything you always wanted to know about the cosmological constant problem (but were afraid to ask). C. R. Physique 13(2012): 566–665.

Martyushev, L.M. and Konovalov, M.S. Thermodynamic model of nonequilibrium phase transitions. Phys. Rev. E 84(2011): 011113. doi: 10.1103/PhysRevE.84.011113.

Martyushev, L.M. Entropy and entropy production: Old misconceptions and new breakthroughs. Entropy 15(2013): 1152–1170.

Martyushev, L.M. and Seleznev, V.D. The restrictions of the maximum entropy production principle. Physica A 410(2014): 17–21.

McFadden, W.M., McCall, P.M., Gardel, M.L. and Munro, E.M. Filament turnover tunes both force generation and dissipation to control long-range flows in a model actomyosin cortex. PloS Comput. Biol. 13(2017): e1005811. doi.org/10.1371/journal.pcbi.1005811.

McGuire, C., Stransky, L., Cotter, K. and Forgac, M. Regulation of V-ATPase activity. Front. Biosci. (Landmark Ed). 22(2017): 609–622.

Monteforte, M. and Wolf, F. Dynamical entropy production in spiking neuron networks in the balanced state. Phys. Rev. Lett. 105(2010): 268104. doi: 10.1103/PhysRevLett.105.268104.

Moret, M.A. Self-organized critical model for protein folding. Physica A 390(2011): 3055–3059.

Morono, Y., Ito, M., Hoshino, T., Terada, T., Hori, T., Ikehara, M. et al. Aerobic microbial life persists in oxic marine sediment as old as 101.5 million years. Nat. Commun. 11(2020): 3626. doi: 10.1038/s41467-020-17330-1.

Moroz, A. 2012. The Common Extremalities in Biology and Physics: Maximum Energy Dissipation Principle in Chemistry, Biology, Physics and Evolution. Elsevier, London, UK.

Morth, J.P., Pedersen, B.P., Buch-Pedersen, M.J., Andersen, J.P., Vilsen, B., Palmgren, M.G. et al. A structural overview of the plasma membrane Na^+, K^+-ATPase and H^+-ATPase ion pumps. Nat. Rev. Mol. Cell Biol. 12(2011): 60–70.

Murrell, M., Oakes, P.W., Lenz, M. and Gardel, M.L. Forcing cells into shape: the mechanics of actomyosin contractility. Nat. Rev. Mol. Cell Biol. 16(2015): 486–498.

Nath, S. A thermodynamic principle for the coupled bioenergetic processes of ATP synthesis. Pure & Appl. Chem. 70(1998): 639–644.

Nath, S. Optimality principle for the coupled chemical reactions of ATP synthesis and its molecular interpretation. Chem. Phys. Lett. 699 (2018): 212–217.

Nath, S. Integration of demand and supply sides in the ATP energy economics of cells. Biophys. Chem. 252(2019a): 106208. doi: 10.1016/j.bpc.2019.106208.

Nath, S. Coupling in ATP synthesis: Test of thermodynamic consistency and formulation in terms of the principle of least action. Chem. Phys. Lett. 723(2019b): 118–122.

Nath, S. Entropy production and its application to the coupled nonequilibrium processes of ATP Synthesis. Entropy 21(2019c): 746. doi: 10.3390/e21080746.

Nath, S. Molecular-level understanding of biological energy coupling and transduction: Response to "Chemiosmotic misunderstandings." Biophys. Chem. 268(2021): 106496. doi: 10.1016/j.bpc.2020.106496.

Nauenberg, M. Critique of q-entropy for thermal statistics. Phys. Rev. E 67(2003): 036114. doi: 10.1103/PhysRevE.67.036114.

Nelson, N., Perzov, N., Cohen, A., Hagai, K., Padler, V. and Nelson, H. The cellular biology of proton-motive force generation by V-ATPases. J. Exp. Biol. 2003(2000): 89–95.

Nguyen, B. and Seifert, U. Exponential volume dependence of entropy-current fluctuations at first-order phase transitions in chemical reaction networks. Phys. Rev. E 102(2020): 022101. doi: 10.1103/PhysRevE.102.022101.

Niebel, B., Leupold, S. and Heinemann, M. An upper limit on Gibbs energy dissipation governs cellular metabolism. Nat. Metab. 1(2019): 125–132.

Noa, C.E.F., Harunari, P.E., de Oliveira, M.J. and Fiore, C.E. Entropy production as a tool for characterizing nonequilibrium phase transitions. Phys. Rev. E 100(2019): 012104. doi: 10.1103/physreve.100.012104.

Otsubo, S., Ito, S., Dechant, A. and Sagawa, T. Estimating entropy production by machine learning of short-time fluctuating currents. Phys. Rev. E 101(2020): 062106. doi: 10.1103/PhysRevE.101.062106.

Paltridge, G.W. Global dynamics and climate. Q. J. R. Meteorol. Soc. 101(1975): 475–484.
Paltridge, G.W. Climate and thermodynamic systems of maximum dissipation. Nature 279(1979): 630–631.
Paltridge, G.W. A physical basis for a maximum of thermodynamic dissipation of the climate system. Q. J. Meteorol. Soc. 127(2001): 305–313.
Patel, V.M. and Lineweaver, C.H. Entropy production and the maximum entropy of the Universe. Proceedings 46(2020): 11. doi: 10.3390/ecea-5-06672.
Pavel, M.A., Petersen, E.N., Wang, H., Lerner, R.A. and Hansen, S.B. Studies on the mechanism of general anesthesia. Proc. Natl. Acad. Sci. USA 117(2020): 13757–13766.
Polettini, M., Lazarescu, A. and Esposito, M. Tightening the uncertainty principle for stochastic currents. Phys. Rev. E 94(2016): 052104. doi: 10.1103/PhysRevE.94.052104.
Pope, L., Lolicato, M., Daniel, L. and Minor, D.L. Jr. Polynuclear ruthenium amines inhibit K_{2P} channels via a "finger in the dam" mechanism. Cell Chem. Biol. 5(2020): 5011–524.
Popov, K., Komianos, J. and Papoian, G.A. MEDYAN: Mechanochemical simulations of contraction and polarity alignment in actomyosin networks. PLoS Comput. Biol. 12(2016): e1004877. doi: 10.1371/journal.pcbi.1004877.
Price, P.B. and Sowers, T. Temperature dependence of metabolic rates for microbial growth, maintenance, and survival. Proc. Natl. Acad. Sci. USA 101(2004): 4631–4636.
Qian, M., Zhang, X., Wilson, R.J. and Feng, J. Efficiency of Brownian motors in terms of entropy production rate. EPL 84(2008): 10014. doi: 10.1209/0295-5075/84/10014.
Ray, A.E., Zhang, E., Terauds, A., Ji, M., Kong, W. and Ferrari, B.C. Soil microbiomes with the genetic capacity for atmospheric chemosynthesis are widespread across the Poles and are associated with moisture, carbon, and nitrogen limitation. Front. Microbiol. 11(2020): 1936. doi: 10.3389/fmicb.2020.01936.
Robert, R. and Sommeria, J. Relaxation towards a statistical equilibrium state in two-dimensional perfect fluid dynamics. Phys. Rev. Lett. 69(1992): 2776–2779.
Rodnina, M.V., Peske, F., Peng, B.-Z., Belardinelli, R. and Wintermeyer, W. Converting GTP hydrolysis into motion: versatile translational elongation factor G. Biol. Chem. 401(2020): 131–142.
Roldán, É., Neri, I., Dörpinghaus, M., Meyr, H. and Jülicher, F. Decision making in the arrow of time. Phys. Rev. Lett. 115(2015): 250602. doi: 10.1103/PhysRevLett.115.250602.
Rowe, A.R., Chellamuthu, P., Lam, B., Okamoto, A. and Nealson, K.H. Marine sediments microbes capable of electrode oxidation as a surrogate for lithotrophic insoluble substrate metabolism. Front. Microbiol. 5(2015): 784. doi: 10.3389/fmicb.2014.00784.
Sagan, C., Thompson, W.R., Carlson, R., Gurnett, D. and Hord, C. A search for life on Earth from the Galileo spacecraft. Nature 365(1993): 715–721.
Sanchez, T., Chen, D.T.N., DeCamp, S.J., Heymann, M. and Dogic, Z. Spontaneous motion in hierarchically assembled active matter. Nature 491(2012): 431–434.
Saper, G. and Hess, H. Synthetic systems powered by biological molecular motors. Chem. Rev. 120(2020): 288–309.
Sarkar, S. and England, J.L. Design of conditions for self-replication. Phys. Rev. E 100(2019): 022414. doi: 10.1103/PhysRevE.100.022414.
Saw, T.B., Doostmohammadi, A., Nier, V., Kocgozlu, L., Thampi, S., Toyama, Y. et al. Topological defects in epithelia govern cell death and extrusion. Nature 544(2017): 212–216.
Schlögl, F. Chemical reaction models for non-equilibrium phase transitions. Z. Physik. 253(1972): 147–161. doi:10.1007/BF01379769.
Schmiedl, T. and Seifert, U. Efficiency of molecular motors at maximum power. Europhys. Lett. 83(2008): 30005.
Schroeder, G.K. and Wolfenden, R. The rate enhancement produced by the ribosome: an improved model. Biochemistry 46(2007): 4037–4044.
Seara, D.S., Yadav, V., Linsmeier, I., Tabatabai, A.P., Oakes, P.W., Tabei, S.M.A. et al. Entropy production rate is maximized in non-contractile actomyosin. Nat. Commun. 9(2018): 4948. doi: 10.1038/s41467-018-07413-5.
Seifert, U. Stochastic thermodynamics, fluctuation theorems and molecular machines. Rep. Prog. Phys. 75(2012): 126001. doi: 10.1088/0034-4885/75/12/126001.
Sharma, U.K. and Srivastava, V. Tsallis HDE with an IR cutoff as Ricci horizon in a flat FLRW universe. New Astron. 84(2021): 101519. doi.org/10.1016/j.newast.2020.101519.
Shimizu, H. Dynamic cooperativity of molecular processes in active streaming, muscle contraction, and subcellular dynamics: the molecular mechanism of self-organization at the subcellular level. Adv. Biophys. 13(1979): 195–278.
Shiraishi, N. and Sagawa, T. Fluctuation theorem for partially masked nonequilibrium dynamics. Phys. Rev. E 91(2015): 012130. doi: 10.1103/PhysRevE.91.012130.
Sievers, A., Beringer, M., Rodnina, M.V. and Wolfenden, R. The ribosome as an entropy trap. Proc. Natl. Acad. Sci. USA 101(2004): 7897–7901.

Stucki, J. The optimal efficiency and the economic degrees of coupling of oxidative phosphorylation. Eur. J. Biochem. 109(1980): 269–283.

Te Brinke, E., Groen, J., Herrmann, A., Heus, H.A., Rivas, G., Spruijt, E. et al. Dissipative adaptation in driven self-assembly leading to self-dividing fibrils. Nat. Nanotechnol. 13(2018): 849–855.

Thorne, K.S. Nobel Lecture: LIGO and gravitational waves III. Rev. Mod. Phys. 90(2018): 040503. doi.org/10.1103/RevModPhys.90.040503.

Tjhung, E., Marenduzzo, D. and Cates, M.E. Spontaneous symmetry breaking in active droplets provides a generic route to motility. Proc. Natl. Acad. Sci. USA 109(2012): 12381–12386.

Tolkatchev, D., Smith, Jr., G.E. and Kostyukova, A.S. Role of intrinsic disorder in muscle sarcomeres. Prog. Mol. Biol. Transl. Sci. 166(2019): 311–340.

Tombesi, F., Meléndez, M., Veilleux, S., Reeves, J.N., González-Alfonso, E. and Reynolds, C.S. Wind from the black-hole accretion disk driving a molecular outflow in an active galaxy. Nature 519(2015): 436–438.

Tomé, T. and de Oliveira, M.J. Stochastic thermodynamics and entropy production of chemical reaction systems. J. Chem. Phys. 148(2018): 224104. doi: 10.1063/1.5037045.

Tsallis, C. Possible generalization of Boltzmann-Gibbs statistics. J. Stat. Phys. 52(1988): 479–487.

Tsallis, C. Nonextensive thermostatistics and fractals. Fractals 3(1995): 541–547.

Tsallis, C., Gell-Mann, M. and Sato, Y. Extensivity and entropy production. Europhys. News 36(2005a): 186–189.

Tsallis, C., Gell-Mann, M. and Sato, Y. Asymptotically scale-invariant occupancy of phase space makes the entropy S_q extensive. Proc. Natl. Acad. Sci. USA 102(2005b): 15377–15382.

Tsallis, C. Nonadditive entropy S_q and nonextensive statistical mechanics: Applications in geophysics and elsewhere. Acta Geophysica 60(2012): 502–525.

Tsallis, C. Beyond Boltzmann–Gibbs–Shannon in physics and elsewhere. Entropy 21(2019): 696. doi: 10.3390/e21070696.

Unrean P. and Srienc, F. Metabolic networks evolve towards states of maximum entropy production. Metab. Eng. 13(2011): 666–673.

Upadhyaya, A., Rieu, J.-P., Glazier, J.A. and Sawada, Y. Anomalous diffusion and non-Gaussian velocity distribution of *Hydra* cells in cellular aggregates. Physica A 293(2001): 549–558.

Vaccaro, A., Dor, Y.K., Nambara, K., Pollina, E.A., Lin, C., Greenberg, M.E. et al. Sleep loss can cause death through accumulation of reactive oxygen species in the gut. Cell 181(2020): 1307–1328.

Vallino, J.J. Ecosystem biogeochemistry considered as a distributed metabolic network ordered by maximum entropy production. Philos. Trans. R. Soc. Lond. B. Biol. Sci. 365(2010): 1417–1427.

Vallino, J.J. and Algar, C.K. The thermodynamics of marine biogeochemical cycles: Lotka revisited. Ann. Rev. Mar. Sci. 8(2016): 333–356.

Van Rossum, S.A.P., Tena-Solsona, M., van Esch, J.H., Eelkema, R. and Boekhoven, J. Dissipative out-of-equilibrium assembly of man-made supramolecular materials. Chem. Soc. Rev. 46(2017): 5519–5535.

Vellela, M. and Qian, H. Stochastic dynamics and non-equilibrium thermodynamics of a bistable chemical system: the Schlögl model revisited. J. R. Soc. Interface 6(2009): 925–940.

Ye, M., Pasta, M., Xie, X., Dubrawski, K.L., Xu, J., Liu, C. et al. Charge-free mixing entropy battery enabled by low-cost electrode materials. ACS Omega 4(2019): 11785. doi: 10.1021/acsomega.9b00863.

Wagoner, J.A. and Dill, K.A. Opposing pressures of speed and efficiency guide the evolution of molecular machines. Mol. Biol. Evol. 36(2019a): 2813–2822.

Wagoner, J.A. and Dill, K.A. Mechanisms for achieving high speed and efficiency in biomolecular machines. Proc. Natl. Acad. Sci. USA 116(2019b): 5902–5907.

Wang, Q.A. Maximum entropy change and least action principle for nonequilibrium systems. Astrophys. Space Sci. 305(2006): 273–281.

Weber, J.K., Shukla, D. and Pande, V.S. Heat dissipation guides activation in signaling proteins. Proc. Natl. Acad. Sci. USA 112(2015): 10377–10382.

Weinberg, S. Anthropic bound on the cosmological constant. Phys. Rev. Lett. 59(1987): 2607–2610.

Weiss, R. Nobel Lecture: LIGO and the discovery of gravitational waves I. Rev. Mod. Phys. 90(2018): 040501. doi.org/10.1103/RevModPhys.90.040501.

Wensink, H.H., Dunkel, J., Heidenreich, S., Drescher, K., Goldstein, R.E., Löwen, H. et al. Meso-scale turbulence in living fluids. Proc. Natl. Acad. Sci. USA 109(2012): 14308–14313.

Whitney, R.S. Finding the quantum thermoelectric with maximal efficiency and minimal entropy production at given power output. Phys. Rev. B 91(2015): 115425. doi: 10.1103/physrevb.91.

Zaccai, G., Natali, F., Peters, J., Řihová, M., Zimmerman, E., Ollivier, J. et al. The fluctuating ribosome: thermal molecular dynamics characterized by neutron scattering. Sci. Rep. 6 (2016): 37138. doi: 10.1038/srep37138.

Ziegler, H. 1977. An Introduction to Thermomechanics. North Holland, Amsterdam, The Netherlands.

Index

B

C

D

F

G

H

M

N

O

P

T